T0178546

Lecture Notes in Computer Science 14490

The series Lecture Notes in Computer Science (LNCS), including its subseries Lecture Notes in Artificial Intelligence (LNAI) and Lecture Notes in Bioinformatics (LNBI), has established itself as a medium for the publication of new developments in computer science and information technology research, teaching, and education.

LNCS enjoys close cooperation with the computer science R & D community, the series counts many renowned academics among its volume editors and paper authors, and collaborates with prestigious societies. Its mission is to serve this international community by providing an invaluable service, mainly focused on the publication of conference and workshop proceedings and postproceedings. LNCS commenced publication in 1973.

Zahir Tari · Keqiu Li · Hongyi Wu

Editors

Algorithms and Architectures for Parallel Processing

23rd International Conference, ICA3PP 2023
Tianjin, China, October 20–22, 2023
Proceedings, Part IV

 Springer

Editors
Zahir Tari
Royal Melbourne Institute of Technology
Melbourne, VIC, Australia

Keqiu Li
Tianjin University
Tianjin, China

Hongyi Wu
University of Arizona
Tucson, AZ, USA

ISSN 0302-9743   ISSN 1611-3349 (electronic)
Lecture Notes in Computer Science
ISBN 978-981-97-0858-1   ISBN 978-981-97-0859-8 (eBook)
https://doi.org/10.1007/978-981-97-0859-8

This Springer imprint is published by the registered company Springer Nature Singapore Pte Ltd.
The registered company address is: 152 Beach Road, #21-01/04 Gateway East, Singapore 189721, Singapore

Paper in this product is recyclable.

Preface

On behalf of the Conference Committee, we welcome you to the proceedings of the 2023 International Conference on Algorithms and Architectures for Parallel Processing (ICA3PP 2023), which was held in Tianjin, China from October 20–22, 2023. ICA3PP2023 was the 23rd in this series of conferences (started in 1995) that are devoted to algorithms and architectures for parallel processing. ICA3PP is now recognized as the main regular international event that covers the many dimensions of parallel algorithms and architectures, encompassing fundamental theoretical approaches, practical experimental projects, and commercial components and systems. This conference provides a forum for academics and practitioners from countries around the world to exchange ideas for improving the efficiency, performance, reliability, security, and interoperability of computing systems and applications.

A successful conference would not be possible without the high-quality contributions made by the authors. This year, ICA3PP received a total of 503 submissions from authors in 21 countries and regions. Based on rigorous peer reviews by the Program Committee members and reviewers, 193 high-quality papers were accepted to be included in the conference proceedings and submitted for EI indexing. In addition to the contributed papers, six distinguished scholars, Lixin Gao, Baochun Li, Laurence T. Yang, Kun Tan, Ahmed Louri, and Hai Jin, were invited to give keynote lectures, providing us with the recent developments in diversified areas in algorithms and architectures for parallel processing and applications.

We would like to take this opportunity to express our sincere gratitude to the Program Committee members and 165 reviewers for their dedicated and professional service. We highly appreciate the twelve track chairs, Dezun Dong, Patrick P. C. Lee, Meng Shen, Ruidong Li, Li Chen, Wei Bao, Jun Li, Hang Qiu, Ang Li, Wei Yang, Yu Yang, and Zhibin Yu, for their hard work in promoting this conference and organizing the reviews for the papers submitted to their tracks. We are so grateful to the publication chairs, Heng Qi, Yulei Wu, Deze Zeng, and the publication assistants for their tedious work in editing the conference proceedings. We must also say "thank you" to all the volunteers who helped us at various stages of this conference. Moreover, we were so honored to have many renowned scholars be part of this conference. Finally, we would like to thank all speakers, authors, and participants for their great contribution to and support for the success of ICA3PP 2023!

October 2023

Jean-Luc Gaudiot
Hong Shen
Gudula Rünger
Zahir Tari
Keqiu Li
Hongyi Wu
Tian Wang

Organization

General Chairs

Jean-Luc Gaudiot   University of California, Irvine, USA
Hong Shen   University of Adelaide, Australia
Gudula Rünger   Chemnitz University of Technology, Germany

Program Chairs

Zahir Tari   Royal Melbourne Institute of Technology,
                                      Australia
Keqiu Li   Tianjin University, China
Hongyi Wu   University of Arizona, USA

Program Vice-chair

Wenxin Li   Tianjin University, China

Publicity Chairs

Hai Wang   Northwest University, China
Milos Stojmenovic   Singidunum University, Serbia
Chaofeng Zhang   Advanced Institute of Industrial Technology,
                                      Japan
Hao Wang   Louisiana State University, USA

Publication Chairs

Heng Qi   Dalian University of Technology, China
Yulei Wu   University of Exeter, UK
Deze Zeng   China University of Geosciences (Wuhan), China

Workshop Chairs

Laiping Zhao   Tianjin University, China
Pengfei Wang   Dalian University of Technology, China

Local Organization Chairs

Xiulong Liu   Tianjin University, China
Yitao Hu   Tianjin University, China

Web Chair

Chen Chen   Shanghai Jiao Tong University, China

Registration Chairs

Xinyu Tong   Tianjin University, China
Chaokun Zhang   Tianjin University, China

Steering Committee Chairs

Yang Xiang (Chair)   Swinburne University of Technology, Australia
Weijia Jia   Beijing Normal University and UIC, China
Yi Pan   Georgia State University, USA
Laurence T. Yang   St. Francis Xavier University, Canada
Wanlei Zhou   City University of Macau, China

Program Committee

Track 1: Parallel and Distributed Architectures

Dezun Dong (Chair)   National University of Defense Technology,
                                  China
Chao Wang   University of Science and Technology of China,
                                  China
Chentao Wu   Shanghai Jiao Tong University, China

Chi Lin	Dalian University of Technology, China
Deze Zeng	China University of Geosciences, China
En Shao	Institute of Computing Technology, Chinese Academy of Sciences, China
Fei Lei	National University of Defense Technology, China
Haikun Liu	Huazhong University of Science and Technology, China
Hailong Yang	Beihang University, China
Junlong Zhou	Nanjing University of Science and Technology, China
Kejiang Ye	Shenzhen Institute of Advanced Technology, Chinese Academy of Sciences, China
Lei Wang	National University of Defense Technology, China
Massimo Cafaro	University of Salento, Italy
Massimo Torquati	University of Pisa, Italy
Mengying Zhao	Shandong University, China
Roman Wyrzykowski	Czestochowa University of Technology, Poland
Rui Wang	Beihang University, China
Sheng Ma	National University of Defense Technology, China
Songwen Pei	University of Shanghai for Science and Technology, China
Susumu Matsumae	Saga University, Japan
Weihua Zhang	Fudan University, China
Weixing Ji	Beijing Institute of Technology, China
Xiaoli Gong	Nankai University, China
Youyou Lu	Tsinghua University, China
Yu Zhang	Huazhong University of Science and Technology, China
Zichen Xu	Nanchang University, China

Track 2: Software Systems and Programming Models

Patrick P. C. Lee (Chair)	Chinese University of Hong Kong, China
Erci Xu	Ohio State University, USA
Xiaolu Li	Huazhong University of Science and Technology, China
Shujie Han	Peking University, China
Mi Zhang	Institute of Computing Technology, Chinese Academy of Sciences, China

Jing Gong	KTH Royal Institute of Technology, Sweden
Radu Prodan	University of Klagenfurt, Austria
Wei Wang	Beijing Jiaotong University, China
Himansu Das	KIIT Deemed to be University, India
Rong Gu	Nanjing University, China
Yongkun Li	University of Science and Technology of China, China
Ladjel Bellatreche	National Engineering School for Mechanics and Aerotechnics, France

Track 3: Distributed and Network-Based Computing

Meng Shen (Chair)	Beijing Institute of Technology, China
Ruidong Li (Chair)	Kanazawa University, Japan
Bin Wu	Institute of Information Engineering, China
Chao Li	Beijing Jiaotong University, China
Chaokun Zhang	Tianjin University, China
Chuan Zhang	Beijing Institute of Technology, China
Chunpeng Ge	National University of Defense Technology, China
Fuliang Li	Northeastern University, China
Fuyuan Song	Nanjing University of Information Science and Technology, China
Gaopeng Gou	Institute of Information Engineering, China
Guangwu Hu	Shenzhen Institute of Information Technology, China
Guo Chen	Hunan University, China
Guozhu Meng	Chinese Academy of Sciences, China
Han Zhao	Shanghai Jiao Tong University, China
Hai Xue	University of Shanghai for Science and Technology, China
Haiping Huang	Nanjing University of Posts and Telecommunications, China
Hongwei Zhang	Tianjin University of Technology, China
Ioanna Kantzavelou	University of West Attica, Greece
Jiawen Kang	Guangdong University of Technology, China
Jie Li	Northeastern University, China
Jingwei Li	University of Electronic Science and Technology of China, China
Jinwen Xi	Beijing Zhongguancun Laboratory, China
Jun Liu	Tsinghua University, China

Kaiping Xue	University of Science and Technology of China, China
Laurent Lefevre	National Institute for Research in Digital Science and Technology, France
Lanju Kong	Shandong University, China
Lei Zhang	Henan University, China
Li Duan	Beijing Jiaotong University, China
Lin He	Tsinghua University, China
Lingling Wang	Qingdao University of Science and Technology, China
Lingjun Pu	Nankai University, China
Liu Yuling	Institute of Information Engineering, China
Meng Li	Hefei University of Technology, China
Minghui Xu	Shandong University, China
Minyu Feng	Southwest University, China
Ning Hu	Guangzhou University, China
Pengfei Liu	University of Electronic Science and Technology of China, China
Qi Li	Beijing University of Posts and Telecommunications, China
Qian Wang	Beijing University of Technology, China
Raymond Yep	University of Macau, China
Shaojing Fu	National University of Defense Technology, China
Shenglin Zhang	Nankai University, China
Shu Yang	Shenzhen University, China
Shuai Gao	Beijing Jiaotong University, China
Su Yao	Tsinghua University, China
Tao Yin	Beijing Zhongguancun Laboratory, China
Tingwen Liu	Institute of Information Engineering, China
Tong Wu	Beijing Institute of Technology, China
Wei Quan	Beijing Jiaotong University, China
Weihao Cui	Shanghai Jiao Tong University, China
Xiang Zhang	Nanjing University of Information Science and Technology, China
Xiangyu Kong	Dalian University of Technology, China
Xiangyun Tang	Minzu University of China, China
Xiaobo Ma	Xi'an Jiaotong University, China
Xiaofeng Hou	Shanghai Jiao Tong University, China
Xiaoyong Tang	Changsha University of Science and Technology, China
Xuezhou Ye	Dalian University of Technology, China
Yaoling Ding	Beijing Institute of Technology, China

Yi Zhao	Tsinghua University, China
Yifei Zhu	Shanghai Jiao Tong University, China
Yilei Xiao	Dalian University of Technology, China
Yiran Zhang	Beijing University of Posts and Telecommunications, China
Yizhi Zhou	Dalian University of Technology, China
Yongqian Sun	Nankai University, China
Yuchao Zhang	Beijing University of Posts and Telecommunications, China
Zhaoteng Yan	Institute of Information Engineering, China
Zhaoyan Shen	Shandong University, China
Zhen Ling	Southeast University, China
Zhiquan Liu	Jinan University, China
Zijun Li	Shanghai Jiao Tong University, China

Track 4: Big Data and Its Applications

Li Chen (Chair)	University of Louisiana at Lafayette, USA
Alfredo Cuzzocrea	University of Calabria, Italy
Heng Qi	Dalian University of Technology, China
Marc Frincu	Nottingham Trent University, UK
Mingwu Zhang	Hubei University of Technology, China
Qianhong Wu	Beihang University, China
Qiong Huang	South China Agricultural University, China
Rongxing Lu	University of New Brunswick, Canada
Shuo Yu	Dalian University of Technology, China
Weizhi Meng	Technical University of Denmark, Denmark
Wenbin Pei	Dalian University of Technology, China
Xiaoyi Tao	Dalian Maritime University, China
Xin Xie	Tianjin University, China
Yong Yu	Shaanxi Normal University, China
Yuan Cao	Ocean University of China, China
Zhiyang Li	Dalian Maritime University, China

Track 5: Parallel and Distributed Algorithms

Wei Bao (Chair)	University of Sydney, Australia
Jun Li (Chair)	City University of New York, USA
Dong Yuan	University of Sydney, Australia
Francesco Palmieri	University of Salerno, Italy

George Bosilca — University of Tennessee, USA
Humayun Kabir — Microsoft, USA
Jaya Prakash Champati — IMDEA Networks Institute, Spain
Peter Kropf — University of Neuchâtel, Switzerland
Pedro Soto — CUNY Graduate Center, USA
Wenjuan Li — Hong Kong Polytechnic University, China
Xiaojie Zhang — Hunan University of Technology and Business, China
Chuang Hu — Wuhan University, China

Track 6: Applications of Parallel and Distributed Computing

Hang Qiu (Chair) — Waymo, USA
Ang Li (Chair) — Qualcomm, USA
Daniel Andresen — Kansas State University, USA
Di Wu — University of Central Florida, USA
Fawad Ahmad — Rochester Institute of Technology, USA
Haonan Lu — University at Buffalo, USA
Silvio Barra — University of Naples Federico II, Italy
Weitian Tong — Georgia Southern University, USA
Xu Zhang — University of Exeter, UK
Yitao Hu — Tianjin University, China
Zhixin Zhao — Tianjin University, China

Track 7: Service Dependability and Security in Distributed and Parallel Systems

Wei Yang (Chair) — University of Texas at Dallas, USA
Dezhi Ran — Peking University, China
Hanlin Chen — Purdue University, USA
Jun Shao — Zhejiang Gongshang University, China
Jinguang Han — Southeast University, China
Mirazul Haque — University of Texas at Dallas, USA
Simin Chen — University of Texas at Dallas, USA
Wenyu Wang — University of Illinois at Urbana-Champaign, USA
Yitao Hu — Tianjin University, China
Yueming Wu — Nanyang Technological University, Singapore
Zhengkai Wu — University of Illinois at Urbana-Champaign, USA
Zhiqiang Li — University of Nebraska, USA
Zhixin Zhao — Tianjin University, China

Ze Zhang	University of Michigan/Cruise, USA
Ravishka Rathnasuriya	University of Texas at Dallas, USA

Track 8: Internet of Things and Cyber-Physical-Social Computing

Yu Yang (Chair)	Lehigh University, USA
Qun Song	Delft University of Technology, The Netherlands
Chenhan Xu	University at Buffalo, USA
Mahbubur Rahman	City University of New York, USA
Guang Wang	Florida State University, USA
Houcine Hassan	Universitat Politècnica de València, Spain
Hua Huang	UC Merced, USA
Junlong Zhou	Nanjing University of Science and Technology, China
Letian Zhang	Middle Tennessee State University, USA
Pengfei Wang	Dalian University of Technology, China
Philip Brown	University of Colorado Colorado Springs, USA
Roshan Ayyalasomayajula	University of California San Diego, USA
Shigeng Zhang	Central South University, China
Shuo Yu	Dalian University of Technology, China
Shuxin Zhong	Rutgers University, USA
Xiaoyang Xie	Meta, USA
Yi Ding	Massachusetts Institute of Technology, USA
Yin Zhang	University of Electronic Science and Technology of China, China
Yukun Yuan	University of Tennessee at Chattanooga, USA
Zhengxiong Li	University of Colorado Denver, USA
Zhihan Fang	Meta, USA
Zhou Qin	Rutgers University, USA
Zonghua Gu	Umeå University, Sweden
Geng Sun	Jilin University, China

Track 9: Performance Modeling and Evaluation

Zhibin Yu (Chair)	Shenzhen Institute of Advanced Technology, Chinese Academy of Sciences, China
Chao Li	Shanghai Jiao Tong University, China
Chuntao Jiang	Foshan University, China
Haozhe Wang	University of Exeter, UK
Laurence Muller	University of Greenwich, UK

Lei Liu	Beihang University, China
Lei Liu	Institute of Computing Technology, Chinese Academy of Sciences, China
Jingwen Leng	Shanghai Jiao Tong University, China
Jordan Samhi	University of Luxembourg, Luxembourg
Sa Wang	Institute of Computing Technology, Chinese Academy of Sciences, China
Shoaib Akram	Australian National University, Australia
Shuang Chen	Huawei, China
Tianyi Liu	Huawei, China
Vladimir Voevodin	Lomonosov Moscow State University, Russia
Xueqin Liang	Xidian University, China

Reviewers

Dezun Dong
Chao Wang
Chentao Wu
Chi Lin
Deze Zeng
En Shao
Fei Lei
Haikun Liu
Hailong Yang
Junlong Zhou
Kejiang Ye
Lei Wang
Massimo Cafaro
Massimo Torquati
Mengying Zhao
Roman Wyrzykowski
Rui Wang
Sheng Ma
Songwen Pei
Susumu Matsumae
Weihua Zhang
Weixing Ji
Xiaoli Gong
Youyou Lu
Yu Zhang
Zichen Xu
Patrick P. C. Lee
Erci Xu

Xiaolu Li
Shujie Han
Mi Zhang
Jing Gong
Radu Prodan
Wei Wang
Himansu Das
Rong Gu
Yongkun Li
Ladjel Bellatreche
Meng Shen
Ruidong Li
Bin Wu
Chao Li
Chaokun Zhang
Chuan Zhang
Chunpeng Ge
Fuliang Li
Fuyuan Song
Gaopeng Gou
Guangwu Hu
Guo Chen
Guozhu Meng
Han Zhao
Hai Xue
Haiping Huang
Hongwei Zhang
Ioanna Kantzavelou

Jiawen Kang
Jie Li
Jingwei Li
Jinwen Xi
Jun Liu
Kaiping Xue
Laurent Lefevre
Lanju Kong
Lei Zhang
Li Duan
Lin He
Lingling Wang
Lingjun Pu
Liu Yuling
Meng Li
Minghui Xu
Minyu Feng
Ning Hu
Pengfei Liu
Qi Li
Qian Wang
Raymond Yep
Shaojing Fu
Shenglin Zhang
Shu Yang
Shuai Gao
Su Yao
Tao Yin
Tingwen Liu
Tong Wu
Wei Quan
Weihao Cui
Xiang Zhang
Xiangyu Kong
Xiangyun Tang
Xiaobo Ma
Xiaofeng Hou
Xiaoyong Tang
Xuezhou Ye
Yaoling Ding
Yi Zhao
Yifei Zhu
Yilei Xiao
Yiran Zhang
Yizhi Zhou

Yongqian Sun
Yuchao Zhang
Zhaoteng Yan
Zhaoyan Shen
Zhen Ling
Zhiquan Liu
Zijun Li
Li Chen
Alfredo Cuzzocrea
Heng Qi
Marc Frincu
Mingwu Zhang
Qianhong Wu
Qiong Huang
Rongxing Lu
Shuo Yu
Weizhi Meng
Wenbin Pei
Xiaoyi Tao
Xin Xie
Yong Yu
Yuan Cao
Zhiyang Li
Wei Bao
Jun Li
Dong Yuan
Francesco Palmieri
George Bosilca
Humayun Kabir
Jaya Prakash Champati
Peter Kropf
Pedro Soto
Wenjuan Li
Xiaojie Zhang
Chuang Hu
Hang Qiu
Ang Li
Daniel Andresen
Di Wu
Fawad Ahmad
Haonan Lu
Silvio Barra
Weitian Tong
Xu Zhang
Yitao Hu

Zhixin Zhao
Wei Yang
Dezhi Ran
Hanlin Chen
Jun Shao
Jinguang Han
Mirazul Haque
Simin Chen
Wenyu Wang
Yitao Hu
Yueming Wu
Zhengkai Wu
Zhiqiang Li
Zhixin Zhao
Ze Zhang
Ravishka Rathnasuriya
Yu Yang
Qun Song
Chenhan Xu
Mahbubur Rahman
Guang Wang
Houcine Hassan
Hua Huang
Junlong Zhou
Letian Zhang
Pengfei Wang
Philip Brown
Roshan Ayyalasomayajula

Shigeng Zhang
Shuo Yu
Shuxin Zhong
Xiaoyang Xie
Yi Ding
Yin Zhang
Yukun Yuan
Zhengxiong Li
Zhihan Fang
Zhou Qin
Zonghua Gu
Geng Sun
Zhibin Yu
Chao Li
Chuntao Jiang
Haozhe Wang
Laurence Muller
Lei Liu
Lei Liu
Jingwen Leng
Jordan Samhi
Sa Wang
Shoaib Akram
Shuang Chen
Tianyi Liu
Vladimir Voevodin
Xueqin Liang

Contents – Part IV

CSDSE: Apply Cooperative Search to Solve the Exploration-Exploitation Dilemma of Design Space Exploration

Kaijie Feng[(✉)] ⓘ, Xiaoya Fan ⓘ, Jianfeng An ⓘ, Haoyang Wang ⓘ, and Chuxi Li ⓘ

School of Computer Science, Northwestern Polytechnical University, Xi'an, China
{fkj921009,haoyangwang,lichuxi}@mail.nwpu.edu.cn,
{fanxy,anjf}@nwpu.edu.cn

Abstract. The design and optimization of deep neural network accelerators should sufficiently consider numerous design parameters and physical constraints that render their design spaces massive in scale and complicated in distribution. When confronted with the massive and complicated design spaces, previous works on design space exploration suffer from the exploration-exploitation dilemma and are unable to simultaneously assure optimization efficiency and stability. In order to solve the exploration-exploitation dilemma, we present a novel design space exploration method named CSDSE. CSDSE implements heterogeneous agents separately responsible for exploration or exploitation to search the design space cooperatively and introduces a weighted compact buffer that encourages agents to search in diverse directions and bolsters their global exploration ability. CSDSE is implemented to enhance accelerator design. Compared to former methods, it achieves latency speedups of up to 6.1x and energy reductions of up to 1.3x in different constraint scenarios.

Keywords: Design Space Exploration · Deep Neural Network Accelerator · Design Automation · Reinforcement Learning

1 Introduction

Deep neural network (DNN) accelerators may achieve superior energy efficiency than general computing platforms because of their elaborately optimized datapath and memory hierarchy [4,13,20]. Sophisticated accelerator design requires joint optimization of hardware architecture and dataflow organization to sufficiently enhance the computing/memory resource utilization and prevent suboptimal performance [7,8,18]. However, the joint design space of hardware and dataflow is extremely large and grows exponentially with increasing parameters. As we discuss in Sect. 2, the design space for a typical DNN accelerator exceeds the order of 10^{100}, posing challenges for defining the global optimum. This has opened up extensive research on efficient design space exploration (DSE) methods for DNN accelerators.

ⓒ The Author(s), under exclusive license to Springer Nature Singapore Pte Ltd. 2024
Z. Tari et al. (Eds.): ICA3PP 2023, LNCS 14490, pp. 1–23, 2024.
https://doi.org/10.1007/978-981-97-0859-8_1

DSE methods for DNN accelerators fall into two categories: exhaust search and heuristic search. Exhaust search based methods evaluates all design points to assure global optimization. However, when confronted with the massive design spaces, the tremendous sampling and evolution costs make it non-trivial to exhaust the entire space [14,15]. Therefore, multiple research concentrates on heuristic search based methods, such as Genetic Algorithm, Bayesian Optimization and Reinforcement Learning, to efficiently acquire available design points within limited periods. Reinforcement Learning (RL) is especially appealing as it achieves superior optimization efficiency for massive design spaces compared to other heuristics search algorithms [7,8,14].

In RL-based DSE method, an agent continuously sample design points to train its search policy for discovering better design points with improved performance metrics like latency, energy, and throughput. Nevertheless, previous works neglect the rarity of valuable design points caused by various physical constraints, such as compute/memory resources, power, area, and bandwidth. Their solutions have to choose between exploiting limited valuable design points or exploring unknown design points, and thus face the challenge of balancing optimization efficiency and stability. DSE methods that are inclined to store and repeatedly exploit known valuable design points for guiding the search around these points can achieve superior efficiency [7,8]. Due to the absence of exploration, however, the exploitation strategy may inevitably lead the search to local optimizations and incur subpar optimization results, which renders the optimization unstable. DSE methods that are inclined to enhance exploration to find more unknown samples have more chances to get rid of local optimizations [1,9,14]. But the exploration strategy may squander much time on worthless or unavailable samples if physical constraints should be taken into account, thereby reducing the optimization efficiency. It is challenging for current DSE methods to simultaneously assure optimization efficiency and stability, which is defined as the exploration-exploitation dilemma of DSE methods in this work.

In order to solve the dilemma of exploration-exploitation, in this work, we provide a novel RL-based DSE method named CSDSE (Cooperative Search based Design Space Exploration method) for the optimization of neural network accelerators. In order to strike a compromise between efficiency and stability, CSDSE introduces a cooperative search mechanism that launches heterogeneous agents with different policies to task exploitation or exploration simultaneously; it arranges agents to cooperate with each other via sharing collected design points, which not only enables efficient communication among agents but also improves the sample efficiency via agent parallelism. Meanwhile, CSDSE notices the various correlations between design parameters and performance metrics, and emphasizes the weight of high-correlation parameters while organizing the shared compact buffer, thereby maintaining sample variety and forcing agents to search in diverse directions for further improving global exploratory ability. We also implement CSDSE to optimize the latency and energy of DNN accelerators. Compared to human-designed architectures, CSDSE achieves up to 3.5x latency speedup and 1.8x energy decrease with the same amount of resources. Compared to former DSE methods based on Genetic Algorithm, Bayesian Optimization and other Reinforcement Learning algorithms, CSDSE achieves latency speedups of up to 5x, 6.1x, 5.8x and energy reductions of up to 1.1x, 1.2x, 1.3x on Cloud, Large-Edge, Small-Edge constraints.

2 Background and Motivation

2.1 Design Space of Deep Neural Network Accelerator

The design space of DNN can be divided into two subspaces: hardware architecture subspace and dataflow subspace.

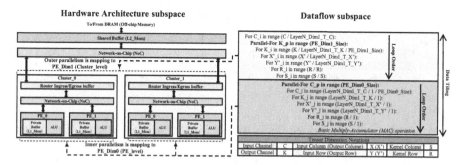

Fig. 1. The accelerator template and an instance of dataflow implementation. The detail parameter definitions are list in Table 1.

Hardware Architecture Subspace. The hardware architecture subspace consists of the design parameters that characterize computing component, memory hierarchy, network connection, and parallelism organization. Figure 1 illustrates an accelerator template: process elements (PEs) are the basic components that handle computing operations; PEs obtain data from the memory hierarchy, which consists of private buffers (L1 memory), global shared memory (L2 memory) and off-chip high-capacity DDR memory; PEs and memory hierarchy are connected by Network-on-Chip (NoC) with various topologies; PEs are organized into an array with multiple dimensions for supporting parallel computing.

Dataflow Subspace. The dataflow subspace consists of the design parameters that describe data tiling and scheduling. Figure 1 demonstrates an instance of the dataflow applied to the convolution loop procedure: data of input/output feature maps and kernel weights can be tiled into chunks that are available for storing on on-chip memories; the schedule order of data chunks can be rearranged to preserve specific data chunks from frequently transferring between on-chip and off-chip memory, which can efficiently enhance data reuse and reduce unnecessary time and energy consumption.

Massive and Complicated Design Spaces of DNN Accelerators. The optimization of DNN accelerators should jointly consider parameters of hardware architecture and dataflow of every single layer (defined as layer-wise optimization), which is the major reason that causes the scale expansion of design space [7,8,15,18]. Figure 2(a) concludes the design space scales of DNN accelerators for different DNN models, from which we can observe that the design space scale grows exponentially along with the layer depths, and the design space scale is even beyond the order of 10^{100}.

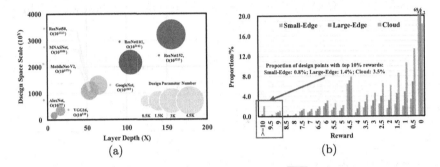

Fig. 2. Massive and complicated design spaces of DNN accelerators. (a) The design space scales of accelerators for different DNN models. The cycle areas denote the parameter numbers of the accelerators. (b) The design point distribution of DNN accelerator. We randomly sample 10000 design points from design space of the VGG16 accelerator and evaluate their values under various constraint scenarios. The definitions of design space and constraints are listed in Table 1 and Table 2. The value of design point is evaluated via reward: as formula 3 defines, design points with lower latency and satisfy all constraints are assigned higher rewards.

The optimization of DNN accelerators should also concern various physical constraints, including computing/memory resources, power, area, and bandwidth, which complicate the distribution of design space. Due to the strict constraints, valuable design points that contain superior performance and satisfy all constraints are relatively rarer than valueless design points that perform inferior performance or dissatisfy any constraints; as Fig. 2(b) reports, in the design space of the VGG16 accelerator, the design points with the top 10% rewards only occupy a proportion of 0.8% (under Small Edge constraint scenarios) to 3.5% (under Cloud constraint scenarios).

Due to the layer-wise optimization and strict physical constraints, the design spaces of DNN accelerators present the characteristics of massive scale and complicated distribution, which pose great challenges to the DSE methods. It is non-trivial for manual optimization or brute-force search to exhaust the entire space to find the optimal solution; thereby, it opens up a lot of research about efficient DSE methods based on heuristic search algorithms, which aim to find valuable design points in the shortest period possible.

2.2 Design Space Exploration Methods

Definition of Design Space Exploration. The design space exploration of DNN accelerators can be defined as a constrained optimization problem:

$$minimize \quad O(x)$$
$$s.t. \quad C_i(x) - Th_i < 0, \quad i \in [1, N_c] \tag{1}$$

where $x \in R^T$ represents the T-dimensional parameter vector of the design point, $O(x)$ denotes the objective function, $C_i(x)$ denotes the constraint function, Th_i is the constraint threshold, and N_c is the number of constraint functions.

According to the basic search algorithms, DSE methods of DNN accelerators can be classified into exhaust search based and heuristic search based methods.

Exhaust Search Based Methods. Exhaustive search methods sample and evaluate all design points, determining global optima. However, these methods may suffer from inferior search efficiency while being confronted with massive design spaces. Though previous works introduce compression mechanisms to prune the design space for enhancing efficiency [6,26], their solutions concentrate on the pruning of single layer but do not restrain the design space expansion caused by layer-wise optimization. When confronted with relatively limited design spaces, exhaust search based methods can be selected to guarantee global optimization. However, when confronted with the DNN accelerators that consider layer-wise optimization, exhaust search based methods may not be suitable for addressing the DSE problem.

Heuristic Search Based Methods. Heuristic search based methods use valid samples to guide subsequent searches; under certain objective distributions, heuristic search based methods can outperform exhaustive searches in efficiency. Extensive research is conducted for enhancing the efficiency of DSE based on various search algorithm, including Genetic Algorithm [15,18], Bayesian Optimization [22] and Reinforcement Learning [1,7–9,14]. Former works [7,8,14] further argue that, compared to other algorithms, Reinforcement Learning (RL) exhibits superior adaptation on long-horizon sequential decision problems and, consequently, may achieve greater efficiency on the DSE procedure of DNN accelerators which must sequentially consider enormous parameters and present similar characteristics to long-horizon sequential decision problems.

2.3 Challenge of Exploration-Exploitation Dilemma

RL presents better potential to handle the DSE of DNN accelerators. However, current RL-based DSE methods still face the challenge of simultaneously ensuring optimization efficiency and stability. Former works neglect the characteristic of the complicated distribution of accelerator design space and the phenomenon of the rarity of valuable design points, which may exacerbate the exploration-exploitation dilemma and thereby result in the efficiency-stability imbalance.

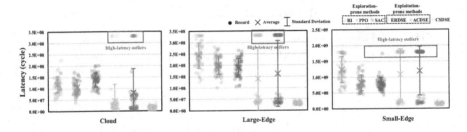

Fig. 3. Adaptability comparison of RL-based DSE methods. We implement different RL-based DSE methods to optimize the forward compute latency of accelerator on VGG16 model. Every selected method is evaluated 50 times and the result distributions are recorded. The definitions of design space parameters and constraint scenarios are listed in Table 1 and Table 2.

Former works ERDSE [8] and ACDSE [7], which use self-imitation learning [21] to focus their search around the neighborhood of current optimal design points, can be regarded as exploitation-prone DSE methods. While former works based on RL frameworks such as REINFORCE [1], PPO [9], and SAC [14], which do not apply prior-replay on collected design points and implement policy entropy [23] or maximum entropy model [11] to guarantee their continuous exploration capability, can be regarded as exploration-prone DSE methods. We conduct a preliminary experiment to compare the adaptability of different RL-based DSE methods on the latency optimization of a VGG16 accelerator. As Fig. 3 illustrates, exploitation-prone DSE methods ERDSE and ACDSE perform superior efficiency with finding better design points with lower latency in limited periods(compared to exploration-prone DSE methods, ERDSE and ACDSE achieve average 0.87x–2.75x and 0.8x–1.51x latency speedups). But their results also show larger deviations and contain more high-latency outliers (the average standard deviation of optimization results of ERDSE and ACDSE are 2.28x–3.16x and 2.95–3.27x larger than those of exploration-prone DSE methods), indicating their inferior stability. Exploration-prone DSE methods can reduce the deviation, but compared to ERDSE and ACDSE, they cannot achieve better optimization results and are relatively inefficient.

The massive and complicated design spaces of DNN accelerators intensify the exploration-exploitation dilemma of RL-based DSE methods, thereby limiting their efficiency and stability. In order to solve the dilemma, CSDSE is introduced under the following considerations: 1) former works with a single agent or homogeneous agents cannot be configured in different search strategies simultaneously, which motivates us to implement heterogeneous agents with separately tasking exploration and exploitation to cooperatively search the design space; 2) the sampling and evaluation costs of accelerator design points may be expensive [5,19], and thus it is essential for RL-based DSE methods to consider the reutilization of design points for enhancing optimization efficiency, which motivates us to implement a public buffer for enabling agents to share collected design points without incurring heavy communication burdens; 3) design points in the shared buffer should maintain variety so that agents can search in diverse directions to guarantee global optimization, which motivated us to cluster collected design points and highlight key design parameters for further enhancing the variety. In the following sections, we will introduce CSDSE and its methodology and implementation in detail.

3   CSDSE Method

3.1   Profile of CSDSE

The procedure of RL can be divided into two phases: 1) in sampling phase, agents, guided by policies $\pi(\theta)$, take actions a to interact with the environment to change states s, and receive rewards $r(s)$; 2) in learning phase, agents retrieve states and rewards as samples to train their policies in response for achieving higher rewards. Agents repeat the sampling-learning-sampling loop for continuously improve the reward value and thus achieve the purpose of optimization.

CSDSE is constructed based on RL, and its workflow still following the RL framework, as Fig. 4 illustrates:

Before exploration, DNN accelerator parameters, objective, constraints and metric evaluation model should be provided as the input of CSDSE. In the sampling phase, CSDSE simultaneously launches heterogeneous agents, under the guiding of diverse policies, to sequentially sample parameters (corresponding to actions in RL) to form the design points (corresponding to states in RL), and obtain performance metrics (corresponding to rewards in RL) from evaluation model (corresponding to environment in RL). Design points (states) and related metrics (rewards) are packed into the shared buffer as samples. In the learning phase, agents retrieves samples from the shared buffer following different strategies to train their policies respectively. Among the repeating of sampling phase and learning phase, agents continuously optimize their policies for finding better design points with superior performance metrics; until touching the upper bound of the period limit, CSDSE will terminate the exploration and output the final optimization result with the best metrics and related parameter settings.

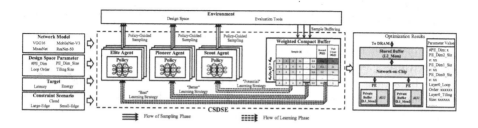

Fig. 4. Profile of CSDSE.

3.2 Definition of RL Framework

Before launching the exploration, CSDSE needs to define the necessary components of the RL framework, including actions, states, rewards, environment, and policies, according to the description of the DNN accelerator parameters, objective, constraints, and metric evaluation model.

Action and State. When RL is implemented for DSE, action a refers to the design parameter, and state s refers to the design point which is the combinations of design parameters with T dimensions, as formula 2 defines:

$$s = (a_1, a_2, a_3, ..., a_t, ..., a_T) \tag{2}$$

Reward. Reward $r(s)$ refer to the evaluation value of state s. Referring to former work [25], we define the reward function using formula 3:

$$r(s) = \frac{1}{O(s)} \cdot \prod \left(\frac{Th_i}{C_i(s)} \right)^{l_i}$$

$$where: \quad l_i \begin{cases} = 0, & C_i(s) \leq Th_i \\ > 0, & C_i(s) > Th_i \end{cases} \tag{3}$$

where $O(s)$ and $C_i(s)$ refer to the optimization function and constraint function about state s. $O(s)$ and $C_i(s)$ can be defined as any metric according to the requirements of users. For example, in this paper, we aim to optimize the forward computing latency $L(s)$ and energy consumption $E(s)$ of accelerators, and thus we define $O(s) = L(s)$ and $O(s) = E(s)$ (RL framework optimizes the objective by increasing the reward value, thereby we set $r(s) \propto \frac{1}{O(s)}$ to reduce the latency and energy). Meanwhile, PE number, memory capacity, and bandwidth are selected as constraints $C_i(s)$. Th_i refers to the constraint threshold defined by users, and l_i refers to the punishment factor that may reduce the rewards of design points that violate any constraints.

Policy. Policy $\pi(\theta)$ is defined as the probability distribution function of action, and θ is its weight parameter. In the sampling phase, agents feed the index of the action under sampling and current state into the policy function; then the policy function outputs the related probability distribution and predicted action value after the sampling under that distribution. In the learning phase, agents retrieve specific samples from the shared buffer to optimize the policy. Inspired by former works [7,8], CSDSE implements the self-imitation learning mechanism, in which agents can replay valuable samples collected from former policies to train the policy to improve convergence efficiency. Meanwhile, CSDSE combines policy entropy, motivated by former works [11,23], with self-imitation learning to force agents to reserve exploration ability and prevent agents from overfitting. Formula 4 defines the loss function for policy training:

$$L(\theta) = L_s(\theta) + L_e(\theta) \tag{4}$$

$$L_s(\theta) = - \sum_{t<T} [\ln \pi(a_t|s_t, \theta) \cdot (\sum_{t<T} \gamma^t \cdot r(s_t))] \tag{5}$$

$$L_e(\theta) = \sum_{t<T} [\sum_{a_t \in A_t} \pi(a_t|s_t, \theta) \cdot \ln \pi(a_t|s_t, \theta)] \tag{6}$$

$L_s(\theta)$ denotes the self-imitation learning component related to the selected sample s, which aims to guide the policy $\pi(\theta)$ to converge towards sample s ($\gamma \in (0,1)$ denotes the discount factor that guarantees the convergence of the long-horizon sample trajectory). $L_e(\theta)$ represents the policy entropy component, which aims to maintain the entropy of policy distribution and preserve the distribution from becoming deterministic. CSDSE also introduces Gaussian noise on policy $\pi(\theta)$ in the sampling phase to guarantee that agents can explore the neighbor area around sample s, thereby preserving their local exploration ability.

3.3 Cooperative Search and Heterogeneous Agents

CSDSE implements three kinds of agents to cooperatively search the design space: elite agent, which is in charge of exploiting the best samples and focuses on the neighbor area around these samples; pioneer agent, which is in charge of exploring regions around high reward samples; scout agent, which is in charge of exploring potential regions. In order to realize the diverse divisions, CSDSE schedules agents to retrieve samples from the shared buffer following different learning strategies:

Elite Agent. Elite agent follows the learning strategy called "Best". It acquires the sample with the highest reward from the shared buffer to train the policy network. Elite agent will check the shared buffer every epoch, and it will replace the holding sample when the best sample in the shared buffer is updated.

Pioneer Agent. Pioneer agent is implemented with the learning strategy called "Better". It selects samples with the top 10 rewards as candidates and retrieves one at random from them. Pioneer agent will hold the selected sample as the training input for a relatively long period (in this work, the period is set to 50 epochs) in order to sufficiently explore the neighbor region around the selected sample.

Scout Agent. Scout agent is scheduled to utilize the learning strategy called "Potential". It sorts samples according to the potential metric r_n defined in formula 7, and then acquires the sample with the highest rank. Metric $r_n(s)$ represents the potential of sample s; that is, the sample with higher reward $r(s)$ as well as lower visited count $n(s)$ (samples may be repeatedly sampled by agents, and the sampling number is defined as visited count) will be assigned higher potential metric r_n. Similar to pioneer agent, scout agent also periodically changes the holding sample.

$$r_n(s) = r(s)/\sqrt{n(s)} \tag{7}$$

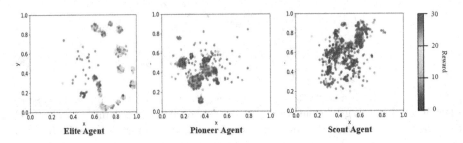

Fig. 5. The search regions of different agents. Elite agent, pioneer agent and scout agent cooperatively explore the design space of the Mobilenet-V2 accelerator. Every agent is assigned to sample 1000 design points. The t-SNE algorithms is implemented to embed the high-dimension design space into 2-dimension space for visualization (the x- and y-axes in the new coordinate system represent the relative positions).

Figure 5 illustrates an instance of the search regions of different agents. Elite agent is tasked with searching around samples with the highest reward; as a result, the samples collected by elite agent are more concentrated than those collected by other agents, and high reward samples occupy a relatively higher proportion. Conversely, scout agent is prone to explore potential unknown regions, resulting in a dispersed sample distribution. However, as compensation for exploration, the proportion of high reward samples collected by scout agent is rela-

---

Algorithm 1. Update of weighted compact buffer

Input: weighted compact buffer B, new sample s

1: Initialize neighboor cluster $L \leftarrow \{\}$
2: **if** $B = \{\}$ **then**
3:     append s in B
4:     $n(s) \leftarrow n(s) + 1$
5: **else**
6:     $update \leftarrow True$
7:     **for** s_i in B **do**
8:         **if** $d_w(s, s_i) \leq d_{th}$ **then**
9:             append s_i in L
10:             **if** $r(s) > r(s_i)$ **then**
11:                 $update \leftarrow update \land True$
12:             **else**
13:                 $update \leftarrow update \land False$
14:     **if** $update = True$ **then**
15:         delete L from B
16:         append s in B
17:         $n(s) \leftarrow n(s) + 1$
18:     **else**
19:         **for** s_i in L **do**
20:             $n(s_i) \leftarrow n(s_i) + 1$

---

tively lower than that of elite agent and pioneer agent. From a macro perspective, it is the mechanism of heterogeneous agents that ensures the global exploration of CSDSE; from a micro perspective, agents implement Gaussian noise to sustain the local exploration around selected samples. Therefore, CSDSE can adequately explore the entire design space.

3.4   Weighted Compact Buffer

As Fig. 3 demonstrates, former DSE methods ERDSE and ACDSE that tend to exploit precious samples indeed achieve better efficiency in massive and complicated design spaces. However, the lack of global exploration makes it easy for these methods to become mired in regions around myopic samples and fall into local optimization. Motivated by former works [2, 10], CSDSE introduces the compact buffer, as an effective complement to the implementation of heterogeneous agents, to increase the distance between samples and force agents to search in diverse regions in response to ensuring global exploration. Algorithm 1 illustrates the construction process of the compact buffer. CSDSE will check the distance between the collected samples and the new sample, and gather all samples within the threshold d_{th} into a cluster. Only if the reward of the new sample is greater than that of all other samples in the cluster will the buffer adopt the new sample and clean other samples.

In addition, CSDSE defines the weighted distance to highlight the value variations of high correlation parameters, enhancing the metric diversity of samples and preventing buffered samples from presenting similar metrics. CSDSE notices that different parameters present different correlations with metrics, as the example illustrated in Fig. 6. Parameters with high correlation coefficients may pose more significant influence on metrics than those of low correlation coefficients. However, in original compact buffer, low correlation parameters are assigned the same weight with high correlation parameters while computing the sample distance [2, 10]; that means, if the number of low correlation parameters are greater than high correlation parameters (this kind of situations commonly appear in the design spaces of DNN accelerators, because the dataflow parameters of loop schedule order and tiling size of each layer should be considered), samples with distance above the threshold will also present similar metrics, which may mislead agents to local optimization.

		Architecture parameters										Dataflow parameters																									
		#PE_Dim	PE_Dim0_Size	PE_Dim1_Size	PE_Dim2_Size	Para_O_C	Para_O_K	Para_O_X'	Para_O_Y'	Para_O_R	Para_O_R	Layer0_Dim0_O_C	Layer0_Dim2_O_Y'	Layer0_Dim2_O_R	Layer0_Dim2_O_S	Layer0_Dim0_O_R	Layer0_Dim0_O_Y'	Layer0_Dim0_O_K	Layer0_Dim1_O_X'	Layer0_Dim1_O_R	Layer0_Dim0_O_S	Layer0_Dim0_O_C	Layer0_Dim0_O_X'	Layer0_Dim0_O_Y'	Layer0_Dim0_O_R	Layer0_Dim0_O_S	Layer0_Dim2_O_C	Layer0_Dim2_O_K	Layer0_Dim1_O_Y'	Layer0_Dim1_O_K	Layer0_Dim1_O_C	Layer0_Dim1_O_Y'	Layer0_Dim1_O_X	Layer0_Dim1_O_K	Layer0_Dim1_O_X	Layer0_Dim1_O_Y'
Objectives	Latency	0.03	0.02	0.03	0.15	0.43	0.34	0.00	0.02	0.26	0.23	0.02	0.03	0.01	0.02	0.01	0.03	0.00	0.04	0.05	0.06	0.04	0.03	0.01	0.04	0.02	0.04	0.01	0.05	0.07	0.03	0.00	0.03	0.03	0.03	0.01	0.08
	Energy	0.02	0.02	0.02	0.04	0.10	0.18	0.05	0.00	0.06	0.08	0.00	0.02	0.02	0.03	0.01	0.03	0.03	0.02	0.01	0.02	0.01	0.02	0.01	0.11	0.05	0.02	0.02	0.04	0.01	0.00	0.01	0.00	0.01	0.03	0.02
Constraints	#PE	0.01	0.07	0.16	0.31	0.04	0.03	0.00	0.02	0.01	0.03	0.02	0.02	0.01	0.02	0.01	0.00	0.85	0.01	0.03	0.00	0.02	0.01	0.02	0.01	0.09	0.03	0.02	0.02	0.04	0.03	0.01	0.03	0.02	0.01	0.04	0.03
	L1_Mem	0.51	0.06	0.05	0.02	0.09	0.53	0.15	0.06	0.07	0.11	0.00	0.05	0.00	0.00	0.01	0.00	0.01	0.00	0.01	0.01	0.00	0.01	0.00	0.02	0.06	0.01	0.01	0.00	0.04	0.01	0.00	0.02	0.02	0.03	0.04	0.00
	L2_Mem	0.01	0.00	0.01	0.07	0.08	0.11	0.03	0.00	0.06	0.05	0.00	0.02	0.02	0.03	0.00	0.02	0.05	0.02	0.03	0.03	0.02	0.04	0.00	0.00	0.10	0.05	0.01	0.05	0.05	0.03	0.01	0.02	0.01	0.01	0.04	0.01

Fig. 6. An example of correlation difference between accelerator design parameters and metrics. The Spearman correlation coefficient is selected to evaluate the correlation difference. The DNN model is MobileNet-V2 and the parameter definitions are list in Table 1 (dataflow parameters of the first CNN layer are selected). 1000 samples are off-line collected via random search for computing the correlation coefficients. The judgment threshold between high and low correlation parameters are 0.1 empirically.

In order to get rid of this kind of situations, CSDSE increases the weight of high correlation parameters when computing the distance between samples. Formulas 8 and 9 defines the weighted distance d_w between sample s^i and s^j:

$$d_w(s^i, s^j) = \sqrt{\sum_{t<T} w_t \cdot (a_t^i - a_t^j)^2} \tag{8}$$

$$w_t = \begin{cases} \alpha \cdot [c(a_t)/ \sum_{a_p \in S_h} c(a_p)], & a_t \in S_h \\ (1-\alpha) \cdot (1/|S_l|), & a_t \in S_l \end{cases} \tag{9}$$

S_h denotes the set of high correlation parameters and S_l denotes the set of low correlation parameters. Prior to exploration, CSDSE will divide parameters into two sets according to the results of the correlation analysis between parameters and metrics on offline samples. Parameters with correlation coefficients above the threshold will be grounded in the high correlation set, and others will be grounded in the low correlation set. $c(a_t)$ denotes the average correlation coefficient between parameter a_t and metrics (metrics include objective and all constraints). α denotes the correction proportion of high correlation parameters. During the distance computing, parameters with higher correlation coefficients $c(a_t)$ will be assigned higher weights; their weights will not be diluted along with the increase of parameter numbers, as the weight accumulation of low correlation parameters will be limited by $1 - \alpha$. Therefore, the value variations of high correlation parameters can be reflected more clearly in distance variations, and the value variations of low correlation parameters will be largely disregarded.

4 Experiments and Analyses

4.1 Experiment Setup

We implement CSDSE to optimize the forward computing latency and energy of DNN accelerators (latency and energy are separately selected as the optimization objectives).

Table 1. Definitions of design space parameters

Type	Parameter	Range	Interval
Architecture	#PE_Dim	[1,3]	1
Architecture	PE_Dim2_Size	[2, 48]	2
Architecture	PE_Dim1_Size	[2, 48]	2
Architecture	PE_Dim0_Size	[2, 48]	2
Architecture	Para_O_{C, K, X', Y', R, S}[a,b]	[1, 6]	1
Dataflow	Layer0_Dim2_O_{C, K, X', Y', R, S}[c]	[1, 6]	1
Dataflow	Layer0_Dim1_O_{C, K, X', Y', R, S}	[1, 6]	1
Dataflow	Layer0_Dim0_O_{C, K, X', Y', R, S}	[1, 6]	1
Dataflow	Layer0_Dim2_T_{C, K, X', Y'}[d]	[1, Size(C, K, X',Y')][e]	\
Dataflow	Layer0_Dim1_T_{C, K, X', Y'}	[1, Size(C, K, X',Y')]	\
...	...	...	...
Dataflow	LayerN_Dim2_O_{C, K, X', Y', R, S}	[1, 6]	1
Dataflow	LayerN_Dim1_O_{C, K, X', Y', R, S}	[1, 6]	1
Dataflow	LayerN_Dim0_O_{C, K, X', Y', R, S}	[1, 6]	1
Dataflow	LayerN_Dim2_T_{C, K, X', Y'}	[1, Size(C, K, X',Y')]	\
Dataflow	LayerN_Dim1_T_{C, K, X', Y'}	[1, Size(C, K, X',Y')]	\

[a]**Tensor dimension denotes**: C, input channel; K, output channel; X', output column; Y', output row; R, kernel row; S, kernel column.

[b]**Parallelism order**: First #PE_Dim dimensions will be parallelized. For example, if #PE_Dim = 3 and Para_O_{C, K, X', Y', R, S} = {6, 6, 5, 4, 2, 1}, tensor dimension C, K and X' will be paralleled on PE dimension 2, 1 and 0.

[c]**Loop schedule order**: Tensor dimension with higher order will be scheduled at outer loop. For example, if Layer0_Dim0_O_{C, K', X', Y', R, S} = {5, 6, 3, 3, 2, 1}, the dimension loop schedule order of layer 0 on PE dimension 0 will be arranged as {K, C, X', Y', R, S}, in which K is the outermost dimension and S is the innermost one.

[d]**Tiling size**: Data is tiled into chunks for adapting limited memory capacity. The tiling sizes of tensor dimension R and S (that is, LayerN_Dimx_T_{R, S}) are fixed as the same as the dimension size Size(R) and Size(S) according to the evaluation tool requirement of MAESTRO [17]. Meanwhile, the tiling sizes of tensor dimensions on the innermost PE dimension (that is, LayerN_Dim0_T_{C, K, X', Y'}) are fixed on 1 for ensuring that there is only one MAC in a PE, which is referring to work [18]. Therefore, we do not include parameter LayerN_Dimx_T_{R, S} and LayerN_Dim0_T_{C, K, X', Y'} in the design space.

[e]**Tiling size range**: All divisors of Size(C, K, X',Y', R, S) will be chosen as action candidates. For example, once Size(C) = 32, then the range of LayerN_Dimx_T_C = {1,2,4,8,16,32}.

Deep Neural Network Models. VGG16 (VGG), ResNet-50 (RES), MobileNet-V2 (MBN), and MnasNet (MNAS) are selected as target DNN models for their various characteristics on parameter volume (dense models VGG, RES vs. sparse models MBN, MNAS), layer depth (shallow model VGG vs. deep models RES, MBN, MNAS) and network shape (regular models VGG, RES, MBN vs. irregular model MNAS), which are suitable to adequately evaluate CSDSE from different perspectives.

Design Space and Constraint Scenarios. Table 1 concludes the parametric definitions of DNN accelerators referring to former work [18]. The hardware architecture subspace consists of following parameters: PE dimension number (i.e. #PE_Dim), PE dimension size (i.e. PE_Dimx_Size) and parallelism order (i.e. Para_O_{C, K, X', Y', R, S},). The dataflow subspace of a single layer consists of following parameters: loop schedule order (i.e. LayerN_Dimx_O_{C, K, X', Y', R, S}) and tiling size (i.e. LayerN_Dimx_T_{C, K, X', Y'}) of tensor dimension.

Table 2 defines the constraints of PE/memory resources and bandwidth referring to human-designed accelerators. The constraint values of Cloud scenario is set referring to TPU [13], while the constraint values of Large-Edge and Small-Edge scenario are separately set referring to NVDLA-Large [20] and Eyeriss [4]. The bandwidth of accelerators are set according to human-designed accelerators. On the Cloud and Large-Edge scenario, the Network-on-Chip bandwidth (NoC_BW) is set to 256B/cycle and the Off-Chip bandwidth (Offchip_BW) is set to 64B/cycle. On the Small-Edge scenario, NoC_BW is set to 18B/cycle and Offchip_BW is set to 8B/cycle.

Table 2. Definitions of constraints

Scenarios	#PE	L1_Mem	L2_Mem
Cloud	65536	4 MB	24 MB
Large-Edge	1024	27 KB	512 KB
Small-Edge	168	512B	108 KB

Evaluation Model. MAESTRO [17] is an analytical model providing a line of metric evaluation tools that can be utilized in this work to evaluate accelerator designs. MAESTRO introduce a data-centric notation for represent accelerator dataflows with data mappings and reuses, which supports detailed dataflow analysis over arbitrary hierarchies of PEs. During the exploration, CSDSE samples design points and inputs them into MAESTRO, retrieving latency, energy, PE number and memory capacity to calculate rewards.

Baselines and Experiment Standards. We choose several algorithms, including Random Grid Search (RGS), Standard Genetic Algorithm (GA),

Bayesian Optimization (BO), REINFORCE (RI), Proximal Policy Optimization (PPO), Soft Actor-Critic (SAC), applied in related works as the comparison baselines. GA and BO are conventional optimization methods utilized in DSE domain, while RI, PPO and SAC are RL based algorithms used to handle long-horizon sequential decision problems. What's more, we choose two DSE domain specific methods, ERDSE and ACDSE, as the comparison references.

Table 3. Implementation details of baseline methods

Methods	Implementation details
GA	GA is implemented by Geatpy [12]. It is set with 50 populations and 20 generations. The crossover rate is set to 0.7 and the mutation rate is set to $1/T$, where T is the count of design parameters
BO	BO is implemented by HyperOpt [3]. Tree of Parzen Estimators (TPE) is selected as the model to construct the Bayesian Optimization for its good adaption for discrete variable. Other hyperparameters are set with suggested values
RL-based methods	In RL-based methods, including RI, PPO, SAC, ERDSE, ACDSE and CSDSE, A 4-layer Multi-Layer Perception (MLP) is selected for constructed the policy network. Referring to related works ERDSE [8] and ACDSE [7]: the hidden layer size of the network is [128, 64]; the learning rate is set to 0.001; the punishment factor l_i is set to 2; the discount factor γ is set to 0.99
CSDSE	CSDSE is implemented with 8 agents: 2 elite agents, 4 pioneer agents, 2 scout agents. For each agent, CSDSE requires a minimum of 2 to ensure that they can search in multiple directions. Here, more pioneer agents are used to enhance the exploration efficiency. The weighted distance threshold d_{th} is set to 0.33 for limiting the buffer size (which may not incur heavy burden on the buffer update) and the corrected proportion α is set to 0.8 for maintain the major contribution of high correlation parameters

During each evaluation, every method will sample 1000 design points (the experiment results show that most selected DSE methods may achieve relatively stable optimization results without density fluctuations before 1000 sample epochs) and the best results will be recorded. Every method will be evaluated 10 times separately on Cloud, Large-Edge, Small-Edge scenarios, and the average value of these 10 evaluation results will be collected as the optimization results. Table 3 lists the implementation details of experiments.

What's more, we will compare the latency and energy of auto-designed accelerator architectures of CSDSE with that of human-designed architectures published in former works, including TPU, NVDLA, Eyeriss, within the same resource constraints (for the sake of fairness, the dataflow description files are collected from the source code of MAESTRO and the performance metrics of these architectures are also evaluated via MAESTRO).

Experiment Device. Following are the details of our experiment devices. CPU: AMD Ryzen 5700G@3.8 Ghz with 8 cores and 16 threads; Memory: 16 GB DDR4@3200 MHz; Disk: 512 GB NVMe SSD.

4.2 Experiment Result and Analysis

Comparison of Human-Designed Accelerators. Figure 7 illustrates optimization results of the auto-designed accelerators searched by CSDSE within the same resource constraints as human-designed accelerators. For latency optimization, compared to TPU-like architecture, auto-designed accelerators achieve average 3.5x latency speedup; while compared to NVDLA-like and Eyeriss-like architectures, auto-designed accelerators achieve average 2.7x and 2.5x latency speedups separately. For energy optimization, auto-designed accelerators achieve similar results (average 0.9x and 1.0x) with TPU-like and NVDLA-like architecture; while compared to Eyeriss-like architectures, auto-designed accelerators still achieve average 1.8x energy reductions.

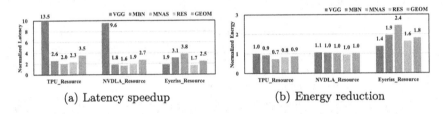

(a) Latency speedup   (b) Energy reduction

Fig. 7. Latency speedups and energy reductions of auto-designed accelerators of CSDSE. The latency and energy optimization results of CSDSE are selected as the normalization baselines. GEOM denotes the geometrical average results of different network models.

The auto-designed accelerators can select appropriate tensor dimensions for parallelization to improve PE utilization, as well as adaptively adjust the tiling size of every layer according to their various shapes, so as to sufficiently utilize on-chip memory to reduce the data transferring cost. Here we provide a case study for explaining the performance benefit of auto-designed accelerators. Table 4 and Table 5 presents an instance of auto-designed accelerator on VGG16 within the resource of TPU (the first and last CNN layers are presented). The auto-designed accelerator chooses to parallelize the tensor dimensions X', C, K; compared with TPU-like accelerator, it enhances the parallelism on dimension X' and thus it achieves 6.2x higher PE utilization on the first layer. TPU achieves higher PE utilization on the last layer, however, its bandwidth requirement (21930B/cycle) exceeds the limit of bandwidth setting (256B/cycle), which encumbers its performance. As for dataflow design, auto-designed accelerator chooses higher tiling size on dimension K, Y' on the first layer and adaptively improve the tiling size on dimension C on the late layer according the shape of network, which achieves 5.2x input reuse factor than TPU-like architecture.

Table 4. Accelerator design example on VGG16 within the resource of TPU

Accelerator	Acc from CSDSE	TPU-like Acc
#PE_Dim	3	2
PE_Dim2_Size	28	\
PE_Dim1_Size	48	256
PE_Dim0_Size	48	256
Parallelim_Order	{X', C, K}	{K, C}
Layer0[a]_Dim2_LoopOrder	{C, X', R, K, S, Y'}	\
Layer0_Dim1_LoopOrder	{R, K, S, X', Y', C}	{K, C, R, S, Y, X}
Layer0_Dim0_LoopOrder	{K, X', C, Y', R, S}	{K, C, Y, X, R, S}
Layer0_Dim2_TilingSize	{C = 3, K = 32, X'(P)[b] = 3, Y' = 74, R = 3, S = 3}	\
Layer0_Dim1_TilingSize	{C(P) = 1, K = 16, X' = 1, Y' = 2, R = 3, S = 3}	{C = 256, K(P) = 1, X = 3, Y = 3, R = 3, S = 3}
Layer0_Dim0_TilingSize	{C = 1, K(P) = 1, X' = 1, Y' = 1, R = 3, S = 3}	{C(P) = 1, K = 1, X = 3, Y = 3, R = 3, S = 3}
Layer12[c]_Dim2_LoopOrder	{C, R, X', Y', K, S}	\
Layer12_Dim1_LoopOrder	{Y', R, K, X', S, C}	{K, C, R, S, Y, X}
Layer12_Dim0_LoopOrder	{Y', R, C, X', S, K}	{K, C, Y, X, R, S}
Layer12_Dim2_TilingSize	{C = 32, K = 128, X'(P) = 4, Y' = 6, R = 3, S = 3}	\
Layer12_Dim1_TilingSize	{C(P) = 4, K = 32, X' = 1, Y' = 1, R = 3, S = 3}	{C = 256, K(P) = 1, X = 3, Y = 3, R = 3, S = 3}
Layer12_Dim0_TilingSize	{C = 1, K(P) = 1, X' = 1, Y' = 1, R = 3, S = 3}	{C(P) = 1, K = 1, X = 3, Y = 3, R = 3, S = 3}

[a]The tensor dimension size of layer0 is C = 3, K = 64, R = 3, S = 3, X = 224, Y = 224, X' = 222, Y' = 222.
[b]Tensor dimensions with notation 'P' will be paralleled on PE array.
[c]The tensor dimension size of layer12 is C = 512, K = 512, R = 3, S = 3, X = 14, Y = 14, X' = 12, Y' = 12.

Table 5. Comparison about key metrics[a] of auto-designed accelerator and TPU

Accelerator Layer	Acc from CSDSE Layer 0	TPU-like Layer 0	Acc from CSDSE Layer 12	TPU-like Layer 12
#PE	64512	65536	64512	65536
Avg utilized PEs	1184	192	768	65536
Input reuse factor[b]	206	190	3456	661
NoC_BW Req (B/cycle)	47	175	11	21930
Offchip_BW Req (B/cycle)	47	175	11	21930
Latency (cycle)	76089	1330668	471970	5329152
Energy (nJ)	15935	15865	559763	549667

[a]The key metric values are from the evaluation results of MAESTRO.
[b]Input reuse factor defines the number of input data local accesses per fetch.

Comparison of Different DSE Methods. Figure 8 presents the latency and energy optimization results of different DSE methods. For latency optimization, compared to other DSE methods, CSDSE achieves 1.9x (MNAS) to 5x (VGG) speedups in average on Cloud constraint, while the speedups on Large-Edge and Small-Edge are separately 1.4x (RES) to 6.1x (VGG) and 1.2x (MBN) to 5.8x (VGG). For energy optimization, CSDSE achieves up to 1.1x (MBN) speedup on Cloud constraint, and the speedups on Large-Edge and Small-Edge are separately 1.1x (VGG) to 1.2x (RES) and 1.1x (VGG) to 1.3x (MNAS). CSDSE achieves superior optimization results on all network models and constraint scenarios, demonstrating its ability to balance efficiency and stability.

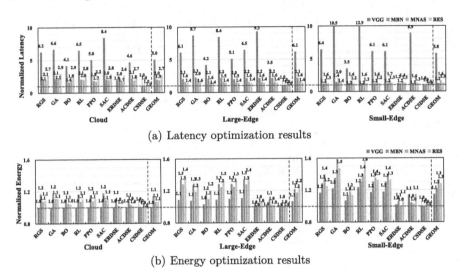

(a) Latency optimization results

(b) Energy optimization results

Fig. 8. Normalized latency and energy optimization results. The best result among all methods is chosen as the baseline, which is indicated by the red dotted line, and other results will be normalized according to the baseline. GEOM denotes the geometrical average of performance improvement acquired by CSDSE. (Color figure online)

Effectiveness of Cooperative Search. Table 6 compares the latency optimization results with and without cooperative search mechanism. Compare with single-agent methods ERDSE and ACDSE, multiple-agent methods, including CSDSE(E), CSDSE(P), CSDSE(S), CSDSE(NW) and CSDSE, achieve better average results due to the increase in sample count causes by the parallel sampling. Among homogeneous-agent methods, CSDSE(E) acquires better results than CSDSE(P) and CSDSE(S) on 67% of scenarios, demonstrating that elite agents with exploitation strategy perform better efficiency than exploration strategy. Heterogeneous-agent method CSDSE(NW) outperforms CSDSE(E) on scenarios with dense models and strict constraints (such as model MBN, MNAS, RES on Small-Edge constraints); CSDSE(E) may frequently stuck in local optimizations on massive complicated spaces, whereas heterogeneous agents can effectively avoid the local convergence. CSDSE improves the average results 1.12x, 1.15x, 1.16x than CSDSE(NW) due to the implementation of weighted compact buffer. The weight compact buffer highlights the variations of high correlation parameters, which guarantees the sample metric diversity and prevents the exploration from falling into the local optimization. From another point of view, the weighted compact buffer also encourages CSDSE to explore the subspace consisting of high correlation parameters (the scale of the subspace is far small than the original design space) and thus enhances the efficiency of CSDSE.

Time Costs Analysis of CSDSE. Figure 9(a) compares the time costs of CSDSE and other RL-based DSE methods. CSDSE is comparable to other methods in terms of total time cost (the total time cost of CSDSE is average 1.17x than other methods), and its cooperative search mechanism does not significantly increase the algorithm cost (the algorithm cost of CSDSE is average 0.81x than other methods). Agents in CSDSE cooperate with each other through the shared compact buffer; CSDSE only needs to handle the synchronization of buffer update without considering other extra synchronization consumption. Figure 9(b) analyzes the time cost constitution of CSDSE, demonstrating that the buffer update cost accounts for a relatively small proportion of total cost (0.18% to 0.28%). The time cost analysis also reveals that the evaluation of metrics contributes the most to the total time cost (85.78% to 91.14%), which is the reason for the time cost similarity of different methods. In this work, we utilize the analytical evaluation model MAESTRO that consumes only a few seconds for the evaluation of every epoch (2.6 s to 8.82 s); if future works employ cycle-accurate simulators or RTL-level implementations, of which the evaluation costs per epoch are typically in minutes or hours range [19, 24], the proportion of evaluation cost of CSDSE will increase and its algorithm cost can be ignored.

Table 6. Normalized latency optimization results[a] with and without cooperative search mechanism[b].

Constraint	Model	ERDSE	ACDSE	CSDSE(E)	CSDSE(P)	CSDSE(S)	CSDSE(NW)	CSDSE
Cloud	VGG	3.61	8.63	**1.00**	3.41	5.10	1.62	1.86
Cloud	MBN	2.64	2.76	**1.00**	1.71	2.01	1.26	1.29
Cloud	MNAS	1.66	1.83	1.30	2.03	1.79	1.36	**1.00**
Cloud	RES	2.60	2.66	1.68	1.90	2.44	1.36	**1.00**
Cloud	GEOM	2.53	3.28	**1.21**	2.18	2.58	1.39	1.24
Large-Edge	VGG	13.85	5.25	**1.00**	3.09	5.01	1.26	1.49
Large-Edge	MBN	2.17	2.17	1.59	1.89	1.90	1.15	**1.00**
Large-Edge	MNAS	1.60	1.60	1.19	1.59	1.60	1.46	**1.00**
Large-Edge	RES	1.42	1.42	1.22	1.40	1.38	1.23	**1.00**
Large-Edge	GEOM	2.87	2.26	1.23	1.90	2.14	1.27	**1.10**
Small-Edge	VGG	1.23	8.92	3.56	1.92	2.39	1.15	**1.00**
Small-Edge	MBN	1.17	1.06	1.17	1.08	1.14	1.17	**1.00**
Small-Edge	MNAS	1.45	1.63	1.43	1.50	1.52	1.25	**1.00**
Small-Edge	RES	1.25	1.25	1.25	1.25	1.25	1.10	**1.00**
Small-Edge	GEOM	1.27	2.10	1.65	1.41	1.51	1.16	**1.00**

[a]The best results are selected as baselines for normalization. We compute the geometrical average of the normalized latency of different models and record the result as GEOM.

[b]CSDSE(E), CSDSE(P), and CSDSE(S) are the variants of CSDSE that implemented with only elite, pioneer and scout agents. CSDSE(NW) denotes to CSDSE that implements only heterogeneous agents without weighted compact buffer.

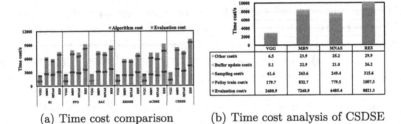

(a) Time cost comparison   (b) Time cost analysis of CSDSE

Fig. 9. Time cost comparison between RL-based methods. Here, for fairness, we fix the value of PE dimension number #PE_Dim = 3 bacause its variation may incur significant disturbance on evaluation cost.

5 Related Works

Design Space Exploration Methods. DSE methods based on exhaust search algorithms are commonly implemented in small-scale design spaces. Cong et al. [5] and Shao et al. [24] use exhaust search to ensure the global optimum. Dave et al. [6] introduce a holistic representation of the loop nests of dataflow accelerators which can drastically prune the design space for enabling exhaustive search. Compared with exhaust search algorithms, heuristic search algorithms may achieve superior performance for DSE in massive design spaces. Reagen et al. [26] provide an efficient DSE framework based on Bayesian Optimization for neural network accelerators. Kao et al. [15] and Lin et al. [18] implement Genetic Algorithm to handle the auto-design of neural accelerator architecture and dataflow optimization. However, heuristic optimization algorithms may suffer from the inefficiency when facing with the complicated design space because of the lack of valuable samples.

Design Space Exploration Methods Based on Reinforcement Learning. RL algorithms have gradually become popular in DSE domain due to their superior adaptability to constrained complicated design spaces with sparse rewards. Kao et al. [14] argue that RL, especially the policy-based algorithm REINFORCE, may achieve better efficiency than other heuristic search algorithms in DSE domain. Gao et al. [9] implement Actor-Critic algorithm PPO to tackle hardware resource allocation on heterogeneous systems. Feng et al. provide ERDSE and ACDSE [7,8] that refine REINFORCE algorithm to off-policy strategy for enhancing the exploitation ability, thereby achieving superior efficiency than other RL based methods. However, ERDSE and ACDSE cannot effectively balance the trade-off between exploitation and exploration, and thus they may frequently get stuck in local optimization.

Reinforcement Learning Methods for Balancing Exploration and Exploitation. Guo et al. [10] provide a method named DTSIL that encourages agent to learn diverse policies to prevent exploration from converging to local optimization. However, DTSIL is implemented with single agent that balances exploration and exploitation by stochastically modifying its policy, which inevitably wastes time on unavailable samples and affects its exploration efficiency. Meanwhile, DTSIL is designed for hard exploration games rather than the DSE of accelerators without considering the various correlations of design parameters, and thus its method may not adequately maintain the diversity between samples when implemented in the DSE domain. Badia et al. [2] introduce a RL agent named NGU that can simultaneously learn multiple directed exploration policies with different trade-offs between exploration and exploitation. But, NGU requires quite a lot of samples to train the value function for navigating the policy update, as argued in former works [8,14], of which the insufficient training of value function will affect the exploration efficiency. Krishnan et al. [16] apply multi-agent mechanism to handle the DSE problem of the microprocessor, which simultaneously implements multiple agents for exploring dependent subspaces. That multi-agent mechanism is introduced for reducing the design space scale rather than exploration in diverse directions, which may also suffer from the exploration-exploitation dilemma.

6 Conclusion

In CSDSE, heterogeneous agents cooperate with each other to solve the exploration-exploitation dilemma within massive and complicated design spaces. Elite agents ensure the exploitation of precious samples, while pioneer and scout agents enhance the exploration of potential regions. In addition, the weighted compact buffer is implemented to highlight the value variations of high correlation parameters, thereby improving sample diversity and exploration ability. Due to the cooperative search mechanism, CSDSE achieves up to 5x, 6.1x, 5.8x latency speedups and 1.1x, 1.2x, 1.3x energy reductions on Cloud, Large-Edge, and Small-Edge constraint scenarios compared to former DSE methods.

In current version, CSDSE should fix its agent division before being launched, thereby it cannot dynamically adjust agent count or type in runtime. This restricts the adaptive capacity and generalization of CSDSE while across DNN workloads, and it may potentially waste partial computility if the agent division is not reasonable. In future work, we are going to analyze the adaptive adjustment mechanism of agent division, in which the number or division of agents can be dynamically adjusted according to the exploration state, for further improving suitability and reducing the algorithm cost.

References

1. Abdelfattah, M.S., Dudziak, Ł., Chau, T., Lee, R., Kim, H., Lane, N.D.: Best of both worlds: AutoML codesign of a CNN and its hardware accelerator. In: 2020 57th ACM/IEEE Design Automation Conference (DAC), pp. 1–6. IEEE (2020)
2. Badia, A.P., et al.: Never give up: learning directed exploration strategies. arXiv preprint arXiv:2002.06038 (2020)

3. Bergstra, J., Yamins, D., Cox, D.: Making a science of model search: hyperparameter optimization in hundreds of dimensions for vision architectures. In: International Conference on Machine Learning, pp. 115–123. PMLR (2013)
4. Chen, Y.H., Krishna, T., Emer, J.S., Sze, V.: Eyeriss: an energy-efficient reconfigurable accelerator for deep convolutional neural networks. IEEE J. Solid-State Circuits 52(1), 127–138 (2016)
5. Cong, J., Wang, J.: PolySA: polyhedral-based systolic array auto-compilation. In: 2018 IEEE/ACM International Conference on Computer-Aided Design (ICCAD), pp. 1–8. IEEE (2018)
6. Dave, S., Kim, Y., Avancha, S., Lee, K., Shrivastava, A.: dMazeRunner: executing perfectly nested loops on dataflow accelerators. ACM Trans. Embed. Comput. Syst. 18(5s), 70 (2019). https://doi.org/10.1145/3358198
7. Feng, K., Fan, X., An, J., Li, C., Di, K., Li, J.: ACDSE: a design space exploration method for CNN accelerator based on adaptive compression mechanism. ACM Trans. Embed. Comput. Syst. (2022). https://doi.org/10.1145/3545177, Just Accepted
8. Feng, K., et al.: ERDSE: efficient reinforcement learning based design space exploration method for CNN accelerator on resource limited platform. Graph. Vis. Comput. 4, 200024 (2021)
9. Gao, Y., Chen, L., Li, B.: Spotlight: optimizing device placement for training deep neural networks. In: International Conference on Machine Learning, pp. 1676–1684. PMLR (2018)
10. Guo, Y., et al.: Memory based trajectory-conditioned policies for learning from sparse rewards. Adv. Neural. Inf. Process. Syst. 33, 4333–4345 (2020)
11. Haarnoja, T., Zhou, A., Abbeel, P., Levine, S.: Soft actor-critic: off-policy maximum entropy deep reinforcement learning with a stochastic actor. In: International Conference on Machine Learning, pp. 1861–1870. PMLR (2018)
12. Jazzbin, et al.: geatpy: the genetic and evolutionary algorithm toolbox with high performance in Python (2020)
13. Jouppi, N.P., et al.: In-datacenter performance analysis of a tensor processing unit. In: Proceedings of the 44th Annual International Symposium on Computer Architecture, pp. 1–12 (2017)
14. Kao, S.C., Jeong, G., Krishna, T.: ConfuciuX: autonomous hardware resource assignment for DNN accelerators using reinforcement learning. In: 2020 53rd Annual IEEE/ACM International Symposium on Microarchitecture (MICRO), pp. 622–636. IEEE (2020)
15. Kao, S.C., Krishna, T.: GAMMA: automating the HW mapping of DNN models on accelerators via genetic algorithm. In: 2020 IEEE/ACM International Conference on Computer Aided Design (ICCAD), pp. 1–9. IEEE (2020)
16. Krishnan, S., et al.: Multi-agent reinforcement learning for microprocessor design space exploration (2022)
17. Kwon, H., Chatarasi, P., Pellauer, M., Parashar, A., Sarkar, V., Krishna, T.: Understanding reuse, performance, and hardware cost of DNN dataflow: a data-centric approach. In: Proceedings of the 52nd Annual IEEE/ACM International Symposium on Microarchitecture, pp. 754–768 (2019)
18. Lin, Y., Yang, M., Han, S.: NAAS: neural accelerator architecture search. In: 2021 58th ACM/IEEE Design Automation Conference (DAC), pp. 1051–1056. IEEE (2021)

19. Muñoz-Martínez, F., Abellán, J.L., Acacio, M.E., Krishna, T.: STONNE: enabling cycle-level microarchitectural simulation for DNN inference accelerators. IEEE Comput. Archit. Lett. **20**(2), 122–125 (2021). https://doi.org/10.1109/LCA.2021.3097253
20. NVIDIA: Nvidia deep learning accelerator (2017). http://nvdla.org/
21. Oh, J., Guo, Y., Singh, S., Lee, H.: Self-imitation learning. In: International Conference on Machine Learning, pp. 3878–3887. PMLR (2018)
22. Reagen, B., et al.: A case for efficient accelerator design space exploration via Bayesian optimization. In: 2017 IEEE/ACM International Symposium on Low Power Electronics and Design (ISLPED), pp. 1–6. IEEE (2017)
23. Schulman, J., Wolski, F., Dhariwal, P., Radford, A., Klimov, O.: Proximal policy optimization algorithms. arXiv preprint arXiv:1707.06347 (2017)
24. Shao, Y.S., Xi, S.L., Srinivasan, V., Wei, G.Y., Brooks, D.: Co-designing accelerators and SoC interfaces using gem5-Aladdin. In: 2016 49th Annual IEEE/ACM International Symposium on Microarchitecture (MICRO), pp. 1–12. IEEE (2016)
25. Tan, M., et al.: MnasNet: platform-aware neural architecture search for mobile. In: Proceedings of the IEEE/CVF Conference on Computer Vision and Pattern Recognition, pp. 2820–2828 (2019)
26. Zheng, S., Liang, Y., Wang, S., Chen, R., Sheng, K.: FlexTensor: an automatic schedule exploration and optimization framework for tensor computation on heterogeneous system, ASPLOS 2020, pp. 859–873. Association for Computing Machinery, New York, NY, USA (2020)

Differential Privacy in Federated Dynamic Gradient Clipping Based on Gradient Norm

Yingchi Mao[1,2]([envelope]), Chenxin Li[2], Zibo Wang[2], Zijian Tu[2], and Ping Ping[1,2]

[1] Key Laboratory of Water Big Data Technology of Ministry of Water Resources, Hohai University, Nanjing, China
{yingchimao,pingping}@hhu.edu.cn
[2] School of Computer and Information, Hohai University, Nanjing, China
{221307040010,221307040022,201307040020}@hhu.edu.cn

Abstract. Federal learning achieves privacy preservation by adding noise to gradient. The noise needs to be clipped to prevent excessive noise from significantly affecting the accuracy of models. However, the imprecise clipped threshold affects the amount of gradient noise leading to degradation of model accuracy. In this paper, for reducing the impact of gradient noise on model accuracy, we propose a differential privacy in federated dynamic gradient clipping based on gradient norm method named DP-FedDGCN. DP-FedDGCN reduces the impact of the amount of gradient noise on the accuracy of the model by dynamically generating a clipped threshold to crop the gradients, achieving the trade-off between data protection and model accuracy. The experimental results show that the attacked accuracy remains consistent in the case of Dirichlet distribution parameters $\alpha = 1$, using MIA, ML-Leaks, and White-box inference attacks. Meanwhile, the average test accuracy outperforms the DP-FedAvg, DP-FedAGNC, and DP-FedDDC methods by about 2.46%, 1.07%, and 1.09%.

Keywords: Federated learning · Privacy budget · Differential privacy preservation · Gradient dynamic clipping

1 Introduction

Machine Learning (ML) requires large amounts of training data, yet it is difficult to share and use the data because they often contain private information such as faces and medical records. In order to train models using distributed private data, federated learning has emerged. Federated Learning (FL) is a machine learning approach that aims to address issues of data privacy and decentralization. In traditional centralized machine learning methods, data is commonly centralized on a central server for training purposes. However, this approach carries the risk of data breaches, especially when the data contains sensitive information, such as personal health records or financial transactions. Federated learning

Z. Tari et al. (Eds.): ICA3PP 2023, LNCS 14490, pp. 24–41, 2024.
https://doi.org/10.1007/978-981-97-0859-8_2

avoids the risk of data transfer and centralized storage by delegating model training to local devices (e.g., smartphones, sensors, or edge devices) for training. Figure 1 illustrates the privacy leakage problem in federation learning. During the model training process, the exchange of model parameters can still be the focus of privacy leakage attacks, and attackers can obtain the privacy of a data by deep mining gradient information. However, for small AI companies, privacy protection is not the main priority. How to improve model's performance with low privacy protection requirements is the problem to be solved.

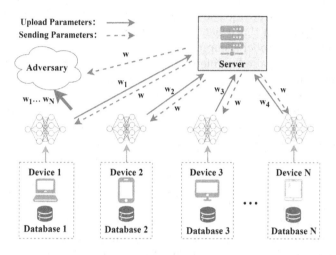

Fig. 1. Clients in federated learning may lead to privacy data leakage when uploading parameters.

The k-anonymity [1], l-diversity [2], and t-closeness [3] approaches make the quality and accuracy of data significantly degraded and vulnerable to other attacks, such as re-identification attacks and attribute inference attacks. Homomorphic encryption [4] does not degrade the accuracy of training [5], but it is inefficient to use homomorphic encryption in deep learning due to its computational complexity. Secure multi-party computation [5] is costly in terms of computational complexity and communication [6]. Differential privacy [7] performs one random substitution in the dataset, making it difficult to infer private information from the model gradients, which is competent in scenarios with low privacy protection needs.

In differential privacy, privacy budget ε which is a non-negative real number and noise level ζ are the parameters used to quantify the privacy protection level and noise intensity. Privacy budget ε measures the user's tolerance of privacy leakage when the information is released. The more the value of ε is, the lower the tolerance for privacy leakage is. Similarly, low ε means high tolerance to privacy leakage. Noise level ζ measures the amount of noise added. The more ζ is, the more the noise is. Specifically, given a privacy budget, the dataset is processed to a corresponding degree, which means adding the corresponding amount of noise. The added noise level ζ can be determined by the privacy budget ε with $\zeta = e^{-\varepsilon}$.

Meanwhile, the noise standard deviation σ determines the magnitude and distribution of the amount of noise. In differential privacy, noise usually uses Gaussian noise or Laplace noise, and the value of noise standard deviation σ is generally determined by the privacy budget ε and noise level ζ with $\sigma = \frac{\Delta_f}{\varepsilon}\sqrt{2ln(\frac{1.25}{\zeta})}$, where Δ_f is the global sensitivity. Traditional federated learning with differential privacy protection can effectively ensure privacy data security by adding noise to features [8], loss, and gradient levels [9]. However, they do not discuss the problem that when adding a large amount of noise perturbation to the gradients, the model accuracy is degraded.

Abadi [10] first proposed a differential privacy protection stochastic gradient descent algorithm (DP-FedAvg) based on differential privacy. Since there is no a priori bound on the gradient size for each client, the noise size varies widely on the gradient effect variation. DP-FedAvg can provide effective protection for sensitive privacy data by using a fixed gradient clipping method to crop limits on excessive gradients, which can effectively prevent the addition of larger noise [11–13]. Indiscriminate global fixed clipping can lead to distorted information at the cost of accuracy degradation. Since there are no a priori bounds, the effect of noise on gradients varies widely, leading to reduced accuracy. Setting an appropriate clipped threshold can be a good way to improve the training model accuracy. Low clipped threshold may lead to high bias and discard the information contained in the gradients. High clipped threshold will result in adding more noise. Dynamic adjustment of the clipped threshold is necessary.

There is no priori knowledge of the gradient clipped threshold and training data in the differential privacy method. DP-FedDGCN we proposed dynamically adjusts the clipped threshold based on the amount of change in the gradient norm of the current iteration. It speeds up the convergence of training and improves the accuracy of model training without reducing the privacy protection performance. The major contributions of this paper are as follows:

- We construct a dynamic clipped threshold function, considering the relationship between the actual gradient variation and the clipped threshold when the client trains the model.
- We fit clipped thresholds for asymptotic decay based on the overall trend of gradient norm changes to realize the dynamic clipped threshold adjustment.
- Experiments on MINST, CIFAR-10 and CIFAR-100 datasets demonstrate the good performance improvement of DP-FedDGCN under a high privacy budget.

The rest of this paper is organized as follows. Section 2 describes the work related to differential privacy preservation. Section 3 presents the system model of DP- FedDGCN. Section 4 describes the DP- FedDGCN approach in detail. Section 5 evaluates the performance of DP-FedDGCN based on a comparison with other methods. Finally, Sect. 6 presents the conclusions of this paper.

2   Related Work

Differential privacy protection stochastic gradient descent (DP-SGD) [10] method is the pioneer of integrating differential privacy into federation learning. DP-SGD adds differential privacy noise during the models training and data to federation learning. NbAFL [14] enables global data to be perturbed by certain Gaussian noise and adjusts the noise variance appropriately to meet the global differential privacy requirements. Guerraoui [15] argues that the SGD algorithm relies on the number of parameters of the ML model, making the training of large models infeasible. Yuan [16] proposes a parallel stochastic gradient descent algorithm to achieve privacy preservation in dynamic networks. DPFL [17] sets different privacy budgets based on imbalance data for different users, and then updates the training parameters for each user. However, there is no intrinsic control over the added noise, and the results are unsatisfactory. Therefore, incorporating only differential privacy techniques into the federated learning framework can add an inappropriate amount of noise during training, leading to a decrease in model usability and privacy preservation.

Trade-offs between model usability and privacy protection performance have been a long-standing effort of federated learning differential privacy methods for many years. Abadi [10] introduces "moment accounting" to constrain the loss of added privacy noise and investigated the hyperparameter tuning problem for privacy preservation [18]. Augenstein [19] uses "moment accounting" to propose a generative model of federated learning with differential privacy. Although the above methods show good usability with some specific parameter settings, they can not be extended to arbitrary models. In order to train arbitrary models with different privacy, PATE [20] trains the respective teacher models on independent privacy data and then extracts the teacher model predictions from some public samples to train a student model with privacy guarantees. However, PATE requires the public data to follow exactly the same distribution as the privacy data and requires a portion of the test data as public data, which results in poor model privacy. In addition, the performance is not acceptable for complex datasets.

DP-FedAvg [10] computes the gradient of each sample $g(x_i)$, and then derives the gradient clipping value by $g(x_i)/\max(1, \|g(x_i)\|_2 /C)$, where C is the clipped threshold. For some multilayer complex neural networks, the gradient norm can be large (that is proportional to the number of model parameters). The clipped threshold can significantly impact on the model results and even reduce the model usability in severe cases. DP-FedAGNC [21] uses the average $L2$ norm of the previous batch of gradients as the clipped threshold for the next batch. Nevertheless, the additional privacy budget for mean calculation may negatively affect the availability of the model. DP-FedDDC [22] uses a nearly linear decay function to set the clipped threshold and also adaptively adjusts the noise scaler to obtain better usability. However, it only performs well with smaller privacy budgets and does not work well with larger privacy budget.

3   Preliminary

3.1   Threat Model

Classic Member Inference Attack Method [23]: The target of the attack is to determine whether the records are part of the target model training dataset, and the attack is evaluated by randomly reconstructing the data from the target training and test datasets.

ML-Leaks (Adversary 1) Method [24]: It uses the shadow model to perform membership inference. The attacker divide the shaded data and set into sums, and then trains the shaded model based on the data. For each record in the model, the attacker selects three largest a posteriori values from the model output and marks them with "1" or "0" to train the attack model. Finally, the attacker inputs the three maximum posterior values into the attack model to obtain predictions for the members.

White-Box Inference Method [25]: The attacker trains the attack model on training and test datasets using various components to learn the differences in member inferences. In federated learning, the attack processes several inputs from the target model per round, and then captures the correlation between the parameters of each iteration.

3.2   Gradient Cropping

In a complete federated learning, the entire training network system consists of a central server and k local clients. Each of which has a local privacy dataset D_k. The global goal of differential privacy federated learning is to add Gaussian noise to the gradients and train a good global model M. Due to the addition of Gaussian noise in the gradients, the noise size has a significant impact on the model usability, where the sensitivity Δ_f is defined by

$$\Delta f = \max_{D,D'} \|M_D - M_{D'}\|_2 \tag{1}$$

where D and D' are adjacent datasets and M_D, $M_{D'}$ are the training models of the global model M on datasets D and D'.

Before adding noise, the gradients ∇ is clipped according to a clipped threshold C by

$$\nabla' = \frac{\nabla}{\max(1, \frac{\|\nabla\|_2}{C})} \tag{2}$$

where ∇ is the gradient after cropping. Since the gradients are cropped based on the ratio of the gradient norm and the clipped threshold. By finding the gradient norm law, the clipped threshold C is dynamically adjusted so that the clipped threshold can follow the gradient norm change dynamically.

For a better description of the problem we research in this paper, a list of the main symbolic parameters of this chapter is shown in Table 1.

Table 1. Summary of the main notation.

Symbol	Symbol Meaning
k	Number of local clients
D_k	Local datasets
B	Size of local batch
M	Central Server Global Model
M_k	Local model
T	Total iterations of training iterations
E	Number of local iterations
R	Total iterations of client and server communication
δ	Relaxation factor
η	Learning rate
Δ_f	Global Sensitivity
σ	Noise standard deviation
ε	Privacy Budget
$N(\mu, \sigma^2)$	Gaussian noise distribution
g_t	Gradient of the tth iteration
ρ_t	Adaptive decay rate for the tth iteration
L_{tLoss}	The tth iteration clipping loss
F_t	The decay coefficient of the tth iteration
C_t	The clipped threshold of tth iteration

4   Differential Privacy in Federated Dynamic Gradient Clipping Based on Gradient Norm

4.1   Overall Framework

Each client uses local private data for local model training. The gradient of the model are cropped according to the clipped threshold computed in the previous round after training. The noise obtained by the clipped threshold is added to the clipped gradient. The specific federated learning procedure is shown in Fig. 2.

The gradient dynamic clipping process can be decomposed into the following phases:

1) Construct the dynamic clipped threshold function.
2) Construct the gradient decay clipped threshold fitting curvature.
3) Compute the dynamic clipped threshold.

The above steps are executed sequentially in each round of model training until the final iteration completes the dynamic clipping process of model training. The detail process is shown in Fig. 2.

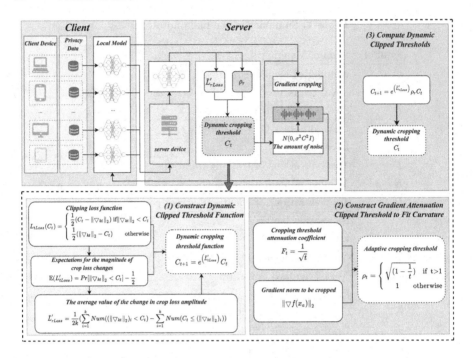

Fig. 2. The figure shows the framework of federated learning and the three phases of gradient clipping.

4.2 Relationship Between Gradient Variation and Clipped Threshold

During the model training procedure, the gradient of each client changes all the time. The dynamically adjusted clipped threshold reflects the gradient variation during the training procedure. It is assumed that the training procedure is carried out for T iterations with k clients involved. The gradient of the current iteration on each client is ∇_{kt}, where $t \in T$. For each clipped threshold C_t, the clipping loss function L_{tLoss} is obtained with

$$L_{tLoss}(C_t) = \begin{cases} \frac{1}{2}(C_t - \|\nabla_{kt}\|_2) & \text{if } \|\nabla_{kt}\|_2 < C_t \\ \frac{1}{2}(\|\nabla_{kt}\|_2 - C_t) & \text{otherwise} \end{cases} \tag{3}$$

The expectation $\mathbb{E}$ of the derivative of the trimming loss function is attained by

$$\mathbb{E}(L'_{tLoss}) = \frac{1}{2}Pr[\|\nabla_{kt}\|_2 < C_t] - \frac{1}{2}Pr[C_t \leq \|\nabla_{kt}\|_2]$$
$$= Pr[\|\nabla_{kt}\|_2 < C_t] - \frac{1}{2} \tag{4}$$

where the expectation $\mathbb{E} \in [-0.5, 0.5]$. Since the clipping loss function is convex and the gradient is bounded by 1, it is possible to obtain an estimated value of C_t with gradient descent. For a server that communicate with k clients, the average of the derivatives of the loss function $\bar{L}'_{tLoss}$ is computed by

$$
\begin{aligned}
\bar{L}'_{rLoss} = \frac{1}{2k}(&\sum_{i=1}^{k} Num((\|\nabla_{kt}\|_2)_i < C_t) \\
&- \sum_{i=1}^{k} Num(C_t \leq (\|\nabla_{kt}\|_2)_i))
\end{aligned}
\tag{5}
$$

where the $Num(\theta)$ function represents the sum of quantities when the condition θ establishes. During model training, it is expected that the clipped threshold C_t can change according to $\|\nabla_{kt}\|_2$. The more the mean value of the derivative of the loss function $\bar{L}'_{tLoss}$ tends to 0, the more appropriate the clipped threshold is. The average value of the derivative of the loss function $\bar{L}'_{tLoss} \in [-0.5, 0.5]$. But the variation of clipped threshold may span multiple orders of magnitude. However, linear function is not suitable for adjusting the clipped threshold. We use an exponential function for its flexibility. The initial clipped threshold C_0 is defined as the gradient norm of the first iteration of model training, which is attained by

$$
C_0 = \frac{1}{k} \sum_{i=1}^{k} (\|\nabla_{kt}\|_2)_i
\tag{6}
$$

The dynamically adjusted clipped threshold for the $t+1^{th}$ iteration is attained by

$$
C_{t+1} = e^{(\bar{L}'_{tLoss})} C_t
\tag{7}
$$

where C_{t+1} is the dynamically adjusted clipped threshold for the $t+1^{th}$ iteration.

4.3 Gradient Attenuation Clipped Threshold Based on Gradient Norm

The gradient norm always changes during the model training. When a single client trains the model with random gradient descent algorithm, the gradient norm decreases gradually.

Theorem 1. *When the changed gradient norm $\gamma_{min} \|\nabla f(x_{a-1})\|_2$ is relatively large or when the Gaussian deviation δ tends to 0, the gradient norm $\|\nabla f(x_a)\|_2$ must be reduced in the next iteration.*

Proof. Assuming that the training model $f : \mathbb{R} \mapsto \mathbb{R}$ is a quadratic differentiable strictly convex function, the second-order Taylor approximation expansion of the model f at x_{a-1} is shown below

$$
\begin{aligned}
f(x) =& \nabla f(x_a - 1)(x - x_{a-1}) + \frac{1}{2}(x - x_{a-1})^T H(x - x_{a-1}) \\
&+ f(x_a) + o(\|x - x_a\|_2^2)
\end{aligned}
\tag{8}
$$

where $o(\|x - x_a\|_2^2)$ is a higher order infinitesimal. Since model f is a strictly convex function, the Hessian matrix H is not only a symmetric matrix but also a positive definite matrix. In addition, the stochastic gradient descent algorithm with noise updates x_a based on the result of the previous iteration, which is obtained with

$$x_a = x_{a-1} - \eta \left(\nabla f \left(x_{a-1} \right) + N(\delta, n) \right) \tag{9}$$

where η is the learning rate and $N(\delta, n)$ is the noise that follows(observes) the Gaussian distribution. The gradient of x_a is computed by

$$\begin{aligned} \nabla f \left(x_a \right) &= \nabla f \left(x_{a-1} \right) - \eta H \left(\nabla f \left(x_{a-1} \right) + N(\delta, n) \right) \\ &= (I - \eta H) \nabla f \left(x_{a-1} \right) - \eta H N(\delta, n) \end{aligned} \tag{10}$$

The Eq. 11 is yielded by taking 2 norms on both sides of Eq. 10,

$$\|\nabla f \left(x_a \right)\|_2 \le \|(I - \eta H)\|_2 \|\nabla f \left(x_{a-1} \right)\|_2 - \|\eta H N(\delta, n)\|_2 \tag{11}$$

φ_{min} and $varphi_{max}$ represent the minimum and maximum eigenvalues of the Hessian matrix H. $\|(I - \eta H)\|_2$ is equivalent to $1 - \eta \gamma_{min}$ when $\eta \le 1/\gamma_{max}$. Equation 12 can be obtained as follows:

$$\|\nabla f(x_a)\|_2 - \|\nabla f(x_{a-1})\|_2 \le \eta \left(\|H N(\delta, n)\|_2 - \gamma_{min} \|\nabla f(x_{a-1})\|_2 \right) \tag{12}$$

It is sufficient to prove that the Eqn. $\|H N(\delta, n)\|_2 - \gamma_{min} \|\nabla f \left(x_{a-1} \right)\|_2 \le 0$ establishes. By specifying $\Psi = H^T H$, the deformation according to the multivariate Gaussian variable tail-bound estimation formula for any $t > 0$ yields

$$P(\delta \|H N(1, n)\|_2 \ge \sqrt{\operatorname{tr}(\Psi) + 2\sqrt{\operatorname{tr}(\Psi)t} + 2\|\Psi\|_2 t}) \le e^{-t} \tag{13}$$

$\|H N(\delta, n)\|_2$ can be equated to $\delta \|H N(1, n)\|_2$, the inequality $\|H N(\delta, n)\|_2 - \gamma_{min} \|\nabla f \left(x_{a-1} \right)\|_2 \le 0$ is always valid when $\gamma_{min} \|\nabla f \left(x_{a-1} \right)\|_2 \ge \operatorname{tr}(\Psi)$ is valid.

As a result, the gradient norm decreases almost with the number of training iterations during the training process. However, in federated learning, the aggregated model gradient increases in partly in some iterations due to the heterogeneity of data on each client. According to the proof in Eqs. 3–13, it is clear that the gradient vanes keep changing during the training procedure. The dynamic adjustment of the clipped threshold requires a global fit to control the change trend. A training approach with adaptively decaying gradient thresholds is designed: Firstly, in the early stage of training, using a larger clipped threshold makes the training model converge faster and allows more changes to be incorporated into the global model update. Secondly, in the later stage of training, a smaller clipped threshold increases the model training convergence robustness. Since the noise is scaled based on the clipped threshold, a smaller clipped threshold prevents excessive noise from destructing the original gradients.

4.4   Convergence Analysis

The most basic power function $F = x^a$ is chosen to construct the attenuation gradient threshold function. It can better conform to the training approach. The attenuation coefficient F_t is defined in the tth iteration compared to the first round according to the nature of the power function,

$$F_t = \frac{1}{\sqrt{t}} \tag{14}$$

For Eq. 14, the adaptive decay rate ρ_t for the tth iteration is obtained by mathematical induction,

$$\rho_t = \begin{cases} \sqrt{\left(1 - \frac{1}{t}\right)} & \text{if t} > 1 \\ 1 & \text{otherwise} \end{cases} \tag{15}$$

Since the number of training iterations $t \geq 1$ and $t \in [1, 2, 3, \dots]$, the minimum and maximum values of ρ_t are computed as follow:

$$\begin{aligned} (\rho_t)_{\min} &= \lim_{t \to 2} \rho_t = \sqrt{\tfrac{1}{2}} \\ (\rho_t)_{\max} &= \lim_{t \to \infty} \rho_t = 1 \end{aligned} \tag{16}$$

As can be seen from Eq. 16, ρ_t is a strictly bounded decay value. With the training going on, ρ_t tends to be 1. The DP-FedDGCN does not affect the convergence of the original model training and avoids a wide range of error fluctuations in the model training.

4.5   Algorithm

Gaussian noise is added during the model training procedure to achieve privacy protection. A good global model is trained by cropping the original gradients before adding the noise and scaling the additions. As described in Algorithm 1, the clipped loss function L_{tLoss} is computed on the client when training the model. The dynamic clipped threshold function C_t is calculated in each iteration based on the clipping loss function. The decay coefficients are constructed in the server. The adaptive decay rate ρ_t is obtained from the iterative information. Finally, the dynamic clipped threshold C_t is obtained in each iteration. The specific steps are as follows:

(1) Initialize the server, client model and the clipped threshold in the first round.
(2) In each iteration of communication, Each client conducts local model training using local privacy data, calculating the sum of the accumulated gradients.
(3) Crop the gradients and save the record Num between the gradient norm and the clipped threshold. Add scaled noise to the clipped gradients.
(4) Each client uploads the processed gradients value g and the record Num.

(5) The server aggregates the gradients of each client. According to the record Num, the server calculates the average value of the derivative of the loss function of the clipped threshold and the adaptive decay rate.

(6) Update the global model on the server based on the aggregated gradients and distribute them to each client to update the model.

Algorithm 1: Differential Privacy in Federated Dynamic Gradient Clipping based on Gradient Norm

---

Input: number of clients k, private dataset D_k in client k, model M_k in client k, size of local batch B, number of local iterations E, deviation of noise standard σ, relaxation factor δ, learning rate η, number of communication iterations R

Output: M

1 Initialize M, M_k, C ;
2 **for** *each communication round $r \in R$* **do**
3   **for** *each client k* **do in parallel**
4        $\theta_k \leftarrow M_k$;
5        $g_k, N_{k<}, N_{k\geq} \leftarrow$ clientTrain (θ_k, D_k, C) // `Train local models`
6   **end**
7     $\bar{L}'_{rLoss} \leftarrow \frac{1}{2k}\sum N_{k<}\sum N_{k\geq}$ // `Compute the average derivative of`
       // `the loss function of the clipped threshold`
8     $F_r \leftarrow \frac{1}{\sqrt{r}}, \rho_r \leftarrow \sqrt{\left(1-\frac{1}{r}\right)}$ // `Calculate adaptive decay rate`
9     $C \leftarrow e^{\left(\bar{L}'_{rLoss}\right)}\rho_r C$ // `Dynamic update clipped threshold`
10    $g \leftarrow \frac{1}{k}\sum g_k$ // `Aggregate gradients`
11    $\theta_{k+1} \leftarrow \theta_k - \eta g$ // `Update global model`
12 **end**
13 function *clientTrain*(θ, D, C) ;
14 **for** *each local epoch $i \in E$* **do**
15     $g_i \leftarrow \nabla L(\theta)$;
16     $\theta \leftarrow \theta - \eta g_i$;
17 **end**
18 $g \leftarrow \sum g_i$;
19 $g' \leftarrow \frac{g}{\max\left(1, \frac{\|g\|_2}{C}\right)}$ // `crop gradients`
20 **if** $\|g\|_2 \geq C$ **then**
21     $Num\left(\|g\|_2 \geq C\right) \leftarrow 1$ // `Num(`θ`) represents the sum of`
       // `quantities when the condition` θ `establishes.`
22 **else**
23     $Num\left(\|g\|_2 < C\right) \leftarrow 1$;
24 **end**
25 $g \leftarrow g' + N\left(0, \sigma^2 C^2 I\right)$ // `Add noise after scaling noise`
26 **return** $g, Num\left(\|g\|_2 < C\right), Num\left(\|g\|_2 \geq C\right)$

5   Performance Evaluation

5.1   Experimental Setup

Datasets and Models: The experiments are set up with a server and 100 clients involved in federated learning training with MNIST dataset, CIFAR-10 dataset, and CIFAR-100 dataset. We use CNN network model with different number of convolutional layers as well as full connectivity for model training. The results are implemented by Softmax normalization.

Hyperparameter Settings: Complex datasets require more iteration rounds compared to simple datasets, with a relatively slightly smaller learning rate. For the MNIST dataset, we set the iteration R to 200, the number of clients k to 100, and the learning rate η initialized to 0.1; For the CIFAR-10 and CIFAR-100 datasets, we set the iteration R to 500, the number of clients k to 100, and the learning rate η initialized to 0.05. The batch stochastic gradient descent algorithm is used for gradient computation (batch size was set to 128). The Dirichlet distribution parameter α is set to 1. The privacy budget ε was set to 1,2,4. The relaxation term δ was set to 0.001. 10 devices are involved in the training for each iteration of communication.

Baselines: The baselines are DP-FedAvg [10], DP-FedAGNC [21] and DP-FedDDC [22].

Evaluation Metrics: We measure the performance of the DP-FedDGCN from the following perspectives:

Performance of Differential Privacy Protection. We use the MIA [23], the ML-Leaks (adversary 1) [24], and the White-box method [25] to infer the trained model. In the evaluation of attacking models in this paper, the uncertainty of inference is maximized by using sets of the same size (i.e., equal number of members and non-members). The baseline accuracy is 0.5. We use the inference attack accuracy of the threat model on the test set to measure the privacy protection effect. The lower the inference attack accuracy, the better the privacy protection effect.

Global Model Availability. Model accuracy is used to assess the effectiveness of the DP-FedDGCN method. Higher model accuracy means higher model usability. We also discuss model accuracy under different privacy budgets.

Universality of Clipped Thresholds. The accuracy standard deviation is used to measure the universality of the clipped threshold. The standard deviation of accuracy SD is the standard deviation of the accuracy A_i of the global model and the average accuracy $\bar{A}$ of each client, which is computed by

$$SD = \sqrt{\frac{1}{i}\sum(A - \bar{A})^2} \tag{17}$$

The smaller the standard deviation SD, the closer the clipped threshold is to the optimum in each iteration.

5.2   Performance of Privacy Protection

The main purpose of adding differential privacy is to protect the privacy of local data in federated learning. The most intuitive way to measure the effectiveness of privacy preservation is to add inference attack methods to the training model. Since the privacy budget affects the accuracy of the inference attack, the experiments select three privacy budgets and three datasets. Comparative experiments are conducted with the basic MIA method, the ML-Leaks method, and the White-box method to analyze the privacy protection accuracy of the three inference attack methods. The experimental results are shown in Table 2. The experimental indicates:

Table 2. Attack accuracy of three attack models on three datasets.

Privacy Budget	Model training methods (DP-)	MNIST				CIFAR-10				CIFAR-100			
		Basic MIA	ML-Leaks	White-box	Average	Basic MIA	ML-Leaks	White-box	Average	Basic MIA	ML-Leaks	White-box	Average
1	FedAvg	**50.13%**	**50.56%**	**51.35%**	**50.68%**	**58.43%**	**62.01%**	67.84%	**62.76%**	**62.83%**	69.01%	74.44%	**68.76%**
	FedAGNC	50.39%	50.81%	51.62%	50.94%	59.05%	63.03%	68.30%	63.46%	63.16%	75.45%	78.29%	72.30%
	FedDDC	50.42%	50.63%	52.04%	51.03%	59.08%	62.47%	68.38%	63.31%	63.24%	69.88%	78.86%	70.66%
	FedDGCN	50.47%	50.61%	51.38%	50.82%	59.19%	62.42%	68.26%	63.29%	63.31%	69.36%	75.77%	69.48%
2	FedAvg	**50.22%**	**50.69%**	**51.64%**	**50.85%**	**59.41%**	**63.11%**	70.95%	64.49%	**64.51%**	75.09%	77.99%	**72.53%**
	FedAGNC	50.44%	50.96%	51.81%	51.07%	60.03%	64.06%	71.12%	65.07%	64.86%	76.82%	80.20%	73.96%
	FedDDC	50.48%	50.78%	51.98%	51.08%	60.06%	63.68%	71.20%	64.98%	64.90%	75.30%	80.39%	73.53%
	FedDGCN	50.52%	50.77%	51.77%	51.02%	60.16%	63.47%	71.07%	64.90%	65.02%	75.24%	78.11%	72.79%
4	FedAvg	**50.29%**	**50.89%**	52.06%	**51.08%**	**60.45%**	**63.54%**	76.02%	**66.67%**	**68.25%**	78.59%	**80.95%**	**75.93%**
	FedAGNC	50.51%	51.04%	52.11%	51.22%	61.17%	65.12%	75.88%	67.39%	68.87%	84.61%	85.32%	79.60%
	FedDDC	50.56%	50.98%	52.24%	51.26%	61.22%	64.21%	76.44%	67.29%	68.94%	79.07%	86.08%	78.03%
	FedDGCN	50.60%	50.96%	52.07%	51.21%	61.31%	63.68%	76.34%	67.11%	69.11%	78.95%	82.46%	76.84%

1) With different privacy budget settings and three inference attack methods, the DP-FedAvg achieves the best privacy protection. Since the DP-FedAvg uses a fixed clipped threshold to crop the gradients, the amount of added noise is the largest compared to the other three methods.

2) The DP-FedAGNC acquires the highest attacked accuracy under ML-Leaks inference attack. The defense against black-box attacks is poor. The attacker of ML-Leaks inference attack method can only access to the input and output of the model rather than the internal information of the model. The DP-FedAGNC can be obtained more information compared to the other three methods due to its more accurate input and output.

3) The DP-FedDGCN we proposed obtains the highest accuracy of being attacked under the basic MIA inference attack. The attacked accuracy under ML-Leaks inference attack and White-box inference attack is better than that of DP-FedAGNC and DP-FedDDC, which is comparable to the DP-FedAvg with similar privacy protection effects. Since the DP-FedDGCN adjusts the

clipped threshold according to the variation of the gradient norm, the gradients are cropped more accurately in each iteration. The DP-FedDGCN adds noise with lower perturbation resulting in a slightly higher success rate by MIA inference attack.

According to the average attack accuracy of the three inference attack methods in Table 2, the experimental results illustrates:

1) Due to the low complexity of the MNIST dataset, the average attack accuracy of four methods are not significantly different from each other. As the complexity of the training dataset increases, the inference attack accuracy of the various methods increases significantly.
2) With three datasets and different privacy budget settings, the average attack accuracy of DP-FedAvg are the lowest, which means the privacy protection is the best. Since the DP-FedAvg method only prevents the addition of larger noise and uses a fixed clipped threshold to crop the gradients, the amount of noise added is still the largest compared to the other three methods.
3) The average attacked accuracy of DP-FedDGCN outperforms that of DP-FedAGNC and DP-FedDDC with three datasets and different privacy budget settings. In particular, under high privacy budget, the privacy protection effect of the DP-FedDGCN improves compared to the DP-FedAGNC and DP-FedDDC methods, inferior to the DP-FedAvg by less than 1%.

In summary, the DP-FedDGCN can effectively defend against inference attack with high privacy budget, and the privacy protection performance is similar to that of DP-FedAvg, DP-FedAGNC, and DP-FedDDC. Black-box and white-box inference attacks can also be effectively defended by the DP-FedDGCN.

5.3 Global Model Availability

The availability of the DP-FedDGCN is evaluated by model accuracy. Table 3 and Fig. 3 show the global testing accuracy of four methods on three datasets, with privacy budget $\varepsilon = \{1, 2, 4\}$, respectively.

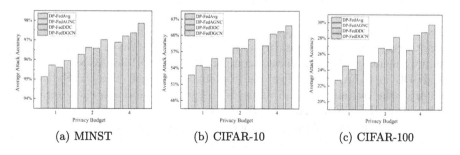

|(a) MNIST|(b) CIFAR-10|(c) CIFAR-100|

Fig. 3. Trend of global testing accuracy with different privacy budgets on three datasets.

Table 3. Global accuracy for different privacy budgets on three datasets.

Privacy Budget	Model training methodsDP-	MNIST	CIFAR-10	CIFAR-100
		Average Accuracy Rate	Average Accuracy Rate	Average Accuracy Rate
1	FedAvg	95.12%	52.73%	22.74%
	FedAGNC	95.72%	54.46%	24.52%
	FedDDC	95.61%	54.17%	24.09%
	FedDGCN	**95.94%**	**55.74%**	**25.81%**
2	FedAvg	96.27%	55.87%	24.97%
	FedAGNC	96.62%	57.69%	26.77%
	FedDDC	96.58%	57.61%	26.62%
	FedDGCN	**97.02%**	**59.30%**	**28.15%**
4	FedAvg	96.89%	58.10%	26.53%
	FedAGNC	97.21%	60.27%	28.45%
	FedDDC	97.37%	60.72%	28.76%
	FedDGCN	**97.87%**	**61.80%**	**29.74%**

As shown in Fig. 3(a), the average global accuracy of the DP-FedDGCN is better than DP-FedAvg, DP-FedAGNC, and DP-FedDDC by about 1.28%, 3.83%, and 0.30% with privacy budget $\varepsilon = \{1, 2, 4\}$ when using the MNIST dataset. The average global accuracy of the DP-FedDGCN outperforms DP-FedAvg, DP-FedAGNC, and DP-FedDDC by about 3.31%, 6.16%, and 0.15%, when applying the CIFAR-10 dataset with privacy budget $\varepsilon = \{1, 2, 4\}$, as shown in Fig. 3(b). As shown in Fig. 3(c), the average global accuracy of the DP-FedDGCN method exceeds DP-FedAvg, DP-FedAGNC, and DP-FedDDC by approximately 2.75%, 4.38%, and 0.43% with the privacy budget $\varepsilon = \{1, 2, 4\}$ when utilising the CIFAR-100 dataset. The experimental results indicates:

1) On the MNIST dataset, the global accuracy of the DP-FedDGCN outperforms the DP-FedAvg method by about 0.82% when the privacy budget is set to 1; and the global accuracy of the DP-FedDGCN method surpasses the DP-FedAvg by about 0.98% when the privacy budget is set to 4. Similarly, the global accuracy of the DP-FedDGCN with the privacy budget set to 4 is higher than that of the DP-FedAvg with the privacy budget set to 1 on the CIFAR-10 and CIFAR-100 datasets. Therefore, the DP-FedDGCN can improve the usability of the globally trained model under a high privacy budget.

2) The differences between the four methods are not significant on the MNIST dataset for its simplicity. For the CIFAR-10 and CIFAR-100 datasets, the DP-FedDGCN achieves the highest global accuracy by means of local gradient tracking and decay functions. In particular, the DP-FedDGCN obtains a large improvement in global accuracy compared to the DP-FedAGNC and DP-FedDDC in CIFAR-100, where the content is complex. The DP-FedDGCN controls the overall variation mode of the gradient norm in each iteration. The DP-FedDGCN incorporates the information of gradient change into the change pattern of the clipped threshold, so that the cropping is more accurate for each iteration in the complex data set.

5.4   Universality of Clipped Thresholds

The standard deviation of accuracy measures the local data universality of the global models. Figure 4 represents the experimental results of the local test accuracy standard deviation of the global model for four methods on the MNIST, CIFAR-10, and CIFAR-100 datasets with privacy budget $\varepsilon = \{1, 2, 4\}$.

As can be seen in Fig. 4, when the privacy budget $\varepsilon = \{1, 2, 4\}$, for the MNIST dataset, the local test accuracy standard deviations of the four methods are not significantly different due to the low complexity of the dataset. While for CIFAR-10 and CIFAR-100, where the complexity of the data is high, the DP-FedDGCN achieves the highest standard deviation of accuracy for local tests with three privacy budgets. The DP-FedAvg performs the worst for taking fixed threshold clipping. In contrast, the DP-FedDGCN compared with the DP-FedAGNC and DP-FedDDC methods, considers the relationship between the clipped threshold and the gradient norm and controls the overall trend of the clipped threshold in the training. In summary, the DP-FedDGCN exceeds the other three methods in terms of the universality of the clipped threshold.

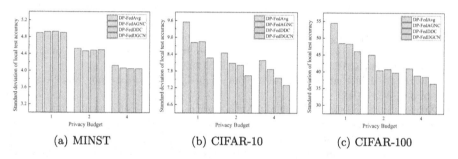

(a) MINST   (b) CIFAR-10   (c) CIFAR-100

Fig. 4. Standard deviation of local testing accuracy for different privacy budgets on three datasets.

6   Conclusions

In this paper, we proposed the differential privacy in federated dynamic gradient clipping based on gradient norm (DP-FedDGCN). There is no a priori knowledge of gradient clipped threshold in differential privacy method. The DP-FedDGCN can solve the problem that the imprecise clipped threshold affects the amount of gradient noise with high privacy budget, declining the model accuracy. The DP-FedDGCN method can adjust the gradient clipped threshold more precisely at each iteration to further control the amount of noise added by differential privacy, improving the problem of blindly setting the clipped threshold during client-side model training by analysis of the experimental results. The accuracy of model is improved without reducing the effectiveness of privacy protection, enhancing the usability of the trained model.

Acknowledgment. This work is supported by Key Research and Development Program of China, Grant/Award Number: 2022YFC3005401, Key Research and Development Project of Jiangsu Province of China (No. BE2020729), Science Technology Achievement Transformation of Jiangsu Province of China (No. BA2021002).

References

1. Ling, C., Zhang, W., He, H.: K-anonymity privacy protection algorithm for IoT applications in virtualization and edge computing. Cluster Comput. **26**, 1495–1510 (2020)
2. Mehta, B.B., Rao, U.P.: Improved l-diversity: scalable anonymization approach for privacy preserving big data publishing. J. King Saud Univ.-Comput. Inf. Sci. **34**(4), 1423–1430 (2022)
3. Gangarde, R., Sharma, A., Pawar, A., et al.: Privacy preservation in online social networks using multiple-graph-properties-based clustering to ensure k-anonymity, l-diversity, and t-closeness. Electronics **10**(22), 2877 (2021)
4. Li, R., Xiao, Y., Zhang, C., et al.: Cryptographic algorithms for privacy protection in online applications. Math. Found. Comput. **1**(4), 311–330 (2018)
5. Phong, L.T., Aono, Y., Hayashi, T., et al.: Privacy preserving deep learning via additively homomorphic encryption. IEEE Trans. Inf. Forensics Secur. **13**, 1333–1345 (2018)
6. Sayyad, S.: Privacy preserving deep learning using secure multiparty computation. In: 2020 Second International Conference on Inventive Research in Computing Applications (ICIRCA), pp. 139–142. IEEE (2020)
7. Dwork, C.: Differential privacy. In: Encyclopedia of Cryptography and Security, pp. 338–340 (2011)
8. Xu, Z., Shi, S., Liu, A.X., et al.: An adaptive and fast convergent approach to differentially private deep learning. In: IEEE INFOCOM 2020 - IEEE Conference on Computer Communications, pp. 1867–1876. IEEE (2020)
9. Wang, D., Xu, J.: Differentially private empirical risk minimization with smooth non-convex loss functions: a non-stationary view. In: Proceedings of the AAAI Conference on Artificial Intelligence, vol. 33, no. 01, pp. 1182–1189 (2019)
10. Abadi, M., Chu, A., Goodfellow, I., et al.: Deep learning with differential privacy. In: Proceedings of the 2016 ACM SIGSAC Conference on Computer and Communications Security, pp. 308–318 (2016)
11. Pan, Z., Hu, L., Tang, W., et al.: Privacy protection multi-granular federated neural architecture search: a general framework. IEEE Trans. Knowl. Data Eng. **35**(3), 2975–2986 (2021)
12. Tang, W., Li, B., Barni, M., et al.: An automatic cost learning framework for image steganography using deep reinforcement learning. IEEE Trans. Inf. Forensics Secur. **16**, 952–967 (2020)
13. Li, T., Li, J., Chen, X., et al.: NPMML: a framework for non-interactive privacy protection multi-party machine learning. IEEE Trans. Dependable Secure Comput. **18**(6), 2969–2982 (2020)
14. Wei, K., Li, J., Ding, M., et al.: Federated learning with differential privacy: algorithms and performance analysis. IEEE Trans. Inf. Forensics Secur. **15**, 3454–3469 (2020)
15. Guerraoui, R., Gupta, N., Pinot, R., et al.: Differential privacy and Byzantine resilience in SGD: do they add up? In: Proceedings of the 2021 ACM Symposium on Principles of Distributed Computing, pp. 391–401 (2021)

16. Yuan, Y., Zou, Z., Li, D., et al.: D-(DP)2SGD: decentralized parallel SGD with differential privacy in dynamic networks. Wirel. Commun. Mob. Comput. **6679453**, 1–14 (2021)
17. Huang, X., Ding, Y., Jiang, Z.L., et al.: DP-FL: a novel differentially private federated learning framework for the unbalanced data. World Wide Web **23**(4), 2529–2545 (2020)
18. Liu, J., Talwar, K.: Private selection from private candidates. In: Proceedings of the 51st Annual ACM SIGACT Symposium on Theory of Computing, pp. 298–309 (2019)
19. Augenstein, S., McMahan, H.B., Ramage, D., et al.: Generative models for effective ML on private, decentralized datasets. In: Proceedings of the 8th International Conference on Learning Representations (2020)
20. Jordon, J., Yoon, J., Schaar, M.: PATE-GAN: generating synthetic data with differential privacy guarantees. In: Proceedings of the 7th International Conference on Learning Representations (2019)
21. Lennart van der Veen, K., Seggers, R., Bloem, P., et al.: Three tools for practical differential privacy. In: Proceedings of the NeurIPS 2018 Workshop (2018)
22. Du, J., Li, S., Chen, X., et al.: Dynamic differential-privacy preserving SGD. arXiv preprint arXiv:2111.00173 (2021)
23. Gu, Y., Bai, Y., Xu, S.: CS-MIA: membership inference attack based on prediction confidence series in federated learning. J. Inf. Secur. Appl. **67**, 103201 (2022)
24. Salem, A., Zhang, Y., Humbert, M., et al.: ML-Leaks: model and data independent membership inference attacks and defenses on machine learning models. In: Network and Distributed Systems Security (NDSS) Symposium (2019)
25. Song, L., Shokri, R., Mittal, P.: Privacy risks of securing machine learning models against adversarial examples. In: Proceedings of the 2019 ACM SIGSAC Conference on Computer and Communications Security, pp. 241–257 (2019)

SolGPT: A GPT-Based Static Vulnerability Detection Model for Enhancing Smart Contract Security

Shengqiang Zeng[1,2], Hongwei Zhang[1,2(⊠)], Jinsong Wang[1,2], and Kai Shi[1,2]

[1] School of Computer Science and Engineering, Tianjin University of Technology,
Tianjin 300384, China
poilzero@stud.tjut.edu.cn, hwzhang@email.tjut.edu.cn,
{jswang,shikai0229}@tjut.edu.cn
[2] National Engineering Laboratory for Computer Virus Prevention and Control
Technology, Tianjin University of Technology, Tianjin 300457, China

Abstract. In this study, we present SolGPT, a novel approach to addressing the pivotal issue of detecting and mitigating vulnerabilities inherent in smart contracts, particularly those written in Solidity, the predominant language for smart contracts. Conventional deep learning methodologies largely rely on an abundant pool of labeled training instances, a resource that remains scarce in the domain, thereby limiting the efficacy of vulnerability detection. SolGPT seeks to counteract this limitation by employing an augmented GPT-2 architecture uniquely tailored for smart contract analysis. The model is enriched by Solidity Adaptive Pre-Training to amplify its feature extraction prowess, hence, reducing the reliance on copious amounts of labeled samples. SolGPT further enhances its field-specific adaptation via the introduction of SolTokenizer, a specialized tokenizer devised for smart contracts, thereby augmenting tokenization precision and efficiency. Subsequently, the model is refined to proficiently pinpoint known vulnerabilities in smart contracts, thereby offering real-time vulnerability detection and prescribing preventive countermeasures. Comprehensive evaluation demonstrates that SolGPT outperforms the state-of-the-art detection techniques in terms of accuracy, F1 score, and two other pertinent performance metrics. Notably, when compared to the best-performing alternative among the four vulnerabilities, SolGPT exhibits an average accuracy improvement of 12.85% and an average F1 score improvement of 18.55%. Consequently, the results underscore the potential of SolGPT in substantially advancing the security framework of the blockchain ecosystem.

Keywords: Smart Contract · GPT · Vulnerability detection ·
Adaptive Pre-Training · Blockchain

Supported by the National Natural Science Foundation of China (Grant No. 62072336), the Tianjin New Generation of Artificial Intelligence Science and Technology Major Project (Grant No. 19ZXZNGX00080).

Z. Tari et al. (Eds.): ICA3PP 2023, LNCS 14490, pp. 42–62, 2024.
https://doi.org/10.1007/978-981-97-0859-8_3

1   Introduction

With blockchain technology continuing to gain widespread adoption [1], it has exposed numerous security challenges. Smart contracts, a key component of blockchain applications [2], play a critical role in various industries. However, the pervasive presence of vulnerabilities in smart contracts, such as reentrancy, timestamp, delegate call, and integer overflow attacks, has hindered the pace of development and raised significant concerns regarding their security [3]. Consequently, the need for devising efficient techniques to detect and prevent these vulnerabilities in smart contracts has become a vital research direction.

Traditional smart contract vulnerability detection approaches, such as rule-based approaches [4], formal verification [5,6], symbolic execution [7,8], and fuzz testing [9,10], have been employed with varying degrees of success. However, they suffer from limitations including reliance on expert-defined rules, steep learning curves, path explosion, and the inability to exhaust all inputs. These drawbacks motivate the exploration of deep learning approaches to enhance the detection of vulnerabilities in smart contracts.

Deep learning has demonstrated remarkable potential across various security fields [11], encompassing malware detection and classification [12], vulnerability detection [13], intrusion detection [14], and privacy security [15]. In the smart contract vulnerability detection field, existing deep learning-based approaches are limited by the availability of labeled data, and they do not fully exploit the benefits of unsupervised pre-training and transfer learning [16–20]. The GPT series models already attained remarkable triumph in natural language processing (NLP) assignments due to their unsupervised pre-training and transfer learning capabilities [11,21,22]. These advantages make GPT models an attractive solution to address the challenges in smart contract vulnerability detection.

Inspired by these successes, SolGPT proposes a solidity adaptive pre-training approach for the smart contract security field using the GPT-based feature extracting structure. SolGPT also develop a specialized tokenizer, SolTokenizer, for smart contract-specific keywords. This evaluation across four smart contract vulnerability detections demonstrates SolGPT's superior performance compared to advanced benchmarks. SolGPT highlights the potential of unsupervised pre-training and transfer learning for improving smart contract vulnerability detection.

The major contributions of SolGPT can be delineated as follows:

1. Adapting the GPT architecture to the field of smart contract vulnerability detection, significantly enhancing performance metrics through the use of transfer learning, and addressing the challenge of limited labeled data.
2. Employing Solidity Adaptive Pre-Training and Vulnerability Detection Fine-tuning for the smart contract vulnerability detection, enabling the GPT-based model to better capture Solidity-specific features and further improve transfer learning performance.
3. Developing a specialized SolTokenizer, specifically designed for the smart contract field, resulting in improved tokenization performance.

2   Problem Formulation

In the context of blockchain technology and smart contract security, the fundamental problem addressed in this research can be formally defined as follows:

Given: A labeled dataset $\mathcal{L}$ containing pairs (C_i, y_i), where C_i represents a source smart contract code and y_i is a binary label where $y_i = 1$ represents C_i has a vulnerability of a certain type while $y_i = 0$ denotes C_i is safe.

Formulation: Given a smart contract C_i, SolGPT will produce a binary classification output $\hat{y}_i$ that predicts whether C_i is vulnerable or not. Mathematically, this can be expressed as:

$$\hat{y}_i = \text{SolGPT}(C_i) \tag{1}$$

where: $\hat{y}_i \in \{0, 1\}$ represents the predicted vulnerability status of the smart contract C_i. $\text{SolGPT}(C_i)$ denotes the prediction made by the SolGPT model for the input smart contract C_i.

Performance Metrics: The evaluation of SolGPT will be based on a set of performance metrics. These metrics, detailed in the experimental section, include but are not limited to accuracy, precision, recall, and F1-score.

The problem of smart contract vulnerability detection is inherently challenging due to the diversity of vulnerabilities and the limited availability of labeled data. SolGPT addresses these challenges by adapting the GPT architecture, employing Solidity Adaptive Pre-Training, and utilizing specialized tokenization techniques, ultimately aiming to achieve superior performance in identifying vulnerabilities in smart contracts.

3   Related Work

In this section, current deep learning-based approaches for smart contract vulnerability detection are discussed, and their limitations are analyzed. The motivation behind the proposed approach is introduced, drawing inspiration from the GPT models.

3.1   Deep Learning-Based Approaches for Smart Contract Vulnerability Detection

Deep learning-based approaches for smart contract vulnerability detection have been developed as an alternative to traditional approaches, aiming to improve performance by learning from labeled samples. Some of the notable works in this area include:

SaferSC [16] as an early application of deep learning to smart contract vulnerability detection, SaferSC utilized an LSTM (Long Short-Term Memory) network to enhance the Traditional tool Maian for more effective detection. This approach resulted in improved accuracy compared to traditional approaches, but it does not get rid of the drawbacks of traditional approaches.

ReChecker [17] **and ContractWard** [18] proposed a scheme for detecting vulnerabilities in smart contracts based on Word2Vec embeddings [23], improved recurrent neural networks, and attention mechanisms [24]. These schemes completely break away from the traditional approach for the first time, completely use the approach of deep learning, and achieve good results in some vulnerabilities.

DR-GCN [19] introduced the Graph Convolutional Neural Network (GCN) to detection and its improved information propagation, Temporal Message Passing (TMP). These approaches employed a contract graph representation and demonstrated excellent detection performance for certain vulnerabilities.

GCN with Expert Knowledge [20] proposes an approach for detecting smart contract vulnerabilities using graph neural networks and expert knowledge, which showed significant accuracy improvements on three types of vulnerabilities in Ethereum and VNT Chain platforms.

Notwithstanding the benefits of deep learning techniques in certain aspects, there remains significant room for improvement. One issue is that the neural network architectures employed in existing deep learning-based approaches may not effectively capture the contextual information in the code. This can lead to the potential loss of semantic information, contributing to the suboptimal performance of these approaches in vulnerability detection tasks.

Another key limitation of these deep learning approaches is their substantial reliance on the number of labeled samples. In the domain of smart contract vulnerability detection, the number of labeled samples is relatively limited compared to other domains, hindering the effectiveness of these approaches in improving performance.

Deep learning-based approaches also face challenges during the tokenization process when dealing with smart contract code. During tokenization, semantic blocks may be unintentionally split into different parts, causing a loss of complete semantic information. This issue could further impact vulnerability detection performance by reducing the model's understanding of the code structure and semantics.

3.2   GPT Models

To overcome these limitations, attention can be turned to the GPT (Generative Pre-trained Transformer) series of models [11,21,22], originally developed by OpenAI. GPT models have demonstrated significant success in various NPL assignments, such as sentiment analysis, text summarization, and text generation. The GPT architecture is built upon the Transformer model [25], which was introduced by Vaswani et al. in 2017 and has since become a cornerstone in the NLP field due to its outperforming scalability and high-level attention mechanism.

The GPT models employ unsupervised pre-training, where the model learns the statistical properties of the input data in unsupervised learning. And followed by transfer learning, where the pre-trained model is fine-tuned on a specific task using labeled data. This two-step process enables GPT models to utilize the knowledge learned from pre-training and adapt it to downstream assignments with relatively fewer labeled samples.

4   SolGPT: Solidity Adapative GPT

In this section, a novel deep learning-based approach called SolGPT is presented to improve vulnerability detection in Solidity smart contracts. SolGPT addresses the challenges faced by existing deep learning-based approaches by introducing the GPT-2 architecture, employing a solidity adaptive pre-training approach, and designing a specialized tokenizer. By these approaches, SolGPT aims to mitigate the dependency on a large quantity of labeled samples and further enhance the performance of smart contract vulnerability detection.

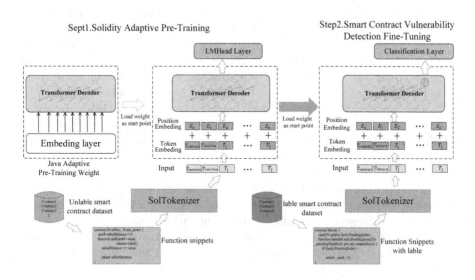

Fig. 1. Training Procedure.

The construction of SolGPT involves two training steps, Solidity Adapative Pre-Training, and Smart Contract Vulnerability Detection Fine-Tuning, with transfer learning applied between these two steps to transfer the model weights across different training tasks, as indicated by the dashed arrows in the Fig. 1. The feature extraction component of the model (enclosed by the dashed box in the Fig. 1) remains consistent across both two steps, leveraging the GPT-2 Base architecture (Parameters: 117M, Transformer Decoder Layers: 12, Hidden Size: 768) to extract relevant features from the smart contract code. The LMHead and ClassificationHead are task-specific heads, consisting of simple fully connected neural layers and training algorithms tailored to each training task. Furthermore, SolGPT introduce a specialized SolTokenizer to enhance the tokenization performance for smart contract segments. The details of this component are elaborated in Sect. 3.1. These two steps and the corresponding model structure improve the model's generalization capabilities and task performance, and the details of these two steps are discussed in Sects. 3.2 and 3.3.

In summary, Fig. 1 provides a comprehensive visualization of the entire Sol-GPT training process, showcasing the adaptation of GPT-2's feature extraction capabilities to the Solidity programming language.

4.1   SolTokenizer

In order to bolster SolGPT's ability to comprehend Solidity code, we employ a novel and specialized tokenizer, the SolTokenizer. This refinement ensures enhanced tokenization performance for Solidity, a distinct language with unique syntactic and structural properties that poses distinctive challenges for standard tokenization algorithms (Fig. 2).

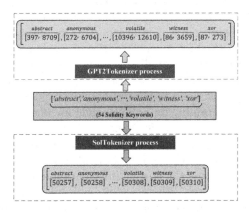

Fig. 2. Difference between GPT2Tokenizer and SolTokenizer.

Drawing inspiration from the Byte-Pair Encoding (BPE) [26] tokenization algorithm, a groundbreaking advancement in the field of Neural Machine Translation, SolTokenizer combines the technical expertise of BPE with a strategic approach to keyword tokenization. This amalgamation empowers SolTokenizer to deliver precise and refined tokenization results for smart contract codes, a challenging task due to their intricate nature and distinct characteristics.

Fundamentally, SolTokenizer leverages the BPE algorithm, enabling it to segment contract code not only at the character level but also into smaller sub-word sequences. This meticulous dissection uncovers subtle nuances and semantic information embedded within contract codes, thereby enhancing the performance of subsequent analytical tasks. In comparison to traditional character-level tokenization, SolTokenizer offers a more precise and comprehensive representation of the code.

Moreover, SolTokenizer introduces a novel approach by imposing tokenization restrictions on 54 crucial keywords commonly found within smart contracts. These keywords are treated as indivisible units, ensuring they remain intact without further subdivision into sub-words. This restriction strengthens the integrity and accuracy of these keywords, enabling the GPT network to intuitively comprehend and identify essential terminology within the contract codes, eliminating the need for additional semantic inference.

To summarize, SolTokenizer serves as an advanced tokenizer for smart contracts, merging the BPE algorithm with a strategic keyword tokenization approach to yield accurate, detailed, and semantically consistent tokenization results for GPT. This state-of-the-art tokenization technique not only enhances the understanding and analysis of smart contract code but also establishes a robust foundation for future research endeavors.

4.2   Solidity Adaptive Pre-training

To address the challenge posed by the limited number of labeled samples in smart contract vulnerability detection, the Solidity Adaptive Pre-training approach for SolGPT will be introduced, which incorporates an unsupervised pre-training step through modeling the high-level semantics of smart contracts. Specifically, SolGPT begins with CodeGPT-small-java, a GPT-2 pre-trained weights on a large Java code corpus by Microsoft, as a strong foundation. However, to account for the differences in syntax and semantics between Solidity and Java, SolGPT performs additional pre-training on Solidity code for better downstream smart contract vulnerability detection performance.

In this training step, the approach uses the same dataset as in the fine-tuning phase (detailed information will be provided in the specific experimental section), but without utilizing the labeled information indicating the presence of vulnerabilities in the dataset. Utilizing additional unlabeled datasets for training during this phase can further enhance generalization, resulting in improved performance on fine-tuning tasks. Moreover, in actuality, the vast majority of accessible data lacks labels, with only a limited fraction being labeled. Hence, unlabeled samples are plentiful and readily obtainable.

Prior to commencing the training process, weights from CodeGPT-java [27] are loaded as the initial state for the model's learning. This serves as the starting point for the subsequent training steps. Thereafter, the intelligent contract segment is introduced into the model for the extraction of features, resulting in

---

Algorithm 1. Solidity Adaptive Pre-training

1: Initialization: Load SolGPT with CodeGPT-small-java weights
2: Initialization: Load MPL ← Linear(hidden layer size, vocabulary size)
3: **for** each epoch in pre-training epochs **do**
4:     **for** each batch in dataset **do**
5:         tokens ← SolTokenizer(batch)
6:         labels ← tokens
7:         feature_vectors ← SolGPT(tokens)
8:     LMHead layer for predicting tokens:
9:         lm_logits ← MPL(feature_vectors)
10:        lm_logits ← Softmax(lm_logits)
11:    Calculate language modeling loss by comparing predicted tokens with real tokens:
12:        shift_logits ← lm_logits[:-1]
13:        shift_labels ← labels[1:]
14:        loss ← CrossEntropyLoss(shift_logits, shift_labels)
15:    Update model weights with gradient descent
16:    **end for**
17: **end for**
18: Save pre-trained SolGPT model
Input: Solidity dataset for pre-training
Output: Pre-trained SolGPT model

---

token-level feature vectors. Utilizing the LMHead layer, the nth feature vector is translated into its associated token within the vocabulary, signifying the prediction of the nth+1 token derived from this vector. In the concluding stage, the model computes the cross-entropy loss between the predicted token and the true nth+1 token present in the segment (with n ranging from 1 to $x - 1$ for every code segment incorporating x tokens), followed by backpropagation for the training process. The Solidity Adaptive Pre-training mechanism is delineated as Algorithm 1.

In Solidity Adaptive Pre-Training, SolGPT uses gradient descent optimization to update the model's parameters during the backpropagation. Specifically, SolGPT updates the vulnerability-specific parameter $\theta_{SolidityVulnerability}$ using the gradient of the cross-entropy loss function with respect to the model's parameters θ. The general domain parameter $\theta_{GeneralJavaDomain}$ is also used during the update process. The optimization formula is as follows:

$$\theta_{SolidityVulnerability} = \theta_{GeneralJavaDomain} - \alpha \nabla_{\theta} Lce \qquad (2)$$

Here, α is the learning rate, which determines the step size of each parameter update. The gradient of the cross-entropy loss function $\nabla_{\theta} Lce$ indicates the direction and magnitude of the change in the loss function with respect to each parameter. It measures how much each parameter affects the output of the model and is used to adjust the values of the parameters in order to minimize the loss function.

After completing the Solidity adaptive pre-training, SolGPT proceed with the fine-tuning process for smart contract vulnerability detection, as described in the "Vulnerability Detection Fine-tuning" subsection. The Solidity adaptive pre-training helps the model to better generalize to the vulnerability detection task, ultimately leading to improved performance in detecting vulnerabilities in Solidity smart contracts.

4.3   Vulnerability Detection Fine-Tunning

To assess the efficacy of SolGPT in detecting vulnerabilities in smart contracts, SolGPT chose four common types of smart contract vulnerabilities. To achieve this, SolGPT established a basic classification layer atop the pre-existing SolGPT model. This classification layer was composed of a fully connected layer and a softmax layer. By fine-tuning the SolGPT model through this process, SolGPT acquired a fine-tuned version of SolGPT that was optimized for this specific objective.

In vulnerability detection fine-tuning, SolGPT uses the hidden states produced by SolGPT, denoted as $SolFeature_i$, as inputs to an additional structure classification head. This layer is designed to map the high-dimensional hidden states to a lower-dimensional space for classification purposes. Specifically, SolGPT defines the logits of the classification head as a softmax function applied to the linear transformation of $SolFeature_i$, represented as:

$$logits = \text{softmax}(W_h SolFeature_i + b_h) \qquad (3)$$

Here, W_h and b_h are the weight matrix and bias vector of a multi-layer perceptron (MLP) layer that projects the 768-dimensional hidden layer feature vectors onto 2-dimensional vectors for vulnerability detection. The softmax function normalizes the output logits into a probability distribution over the existence of the specific smart contract vulnerability, allowing us to make a prediction for the input solidity code. The softmax function can be expressed as:

$$\text{softmax}(x_i) = \frac{e^{x_i}}{\sum_{j=1}^{n} e^{x_j}} \qquad (4)$$

where x_i represents the ith element of the logits, and n is the length of the logits.

During the fine-tuning phase, SolGPT used a dataset of smart contracts containing both vulnerable and non-vulnerable code. The corresponding code was first inputted into SolGPT with Classification Head to predict whether specific vulnerabilities existed. The predicted results were then compared with the actual labels, and the backpropagation algorithm was used to update the model parameters. The Solidity Fine-Tunning procedure can be described as Algorithm 2.

After completing the vulnerability detection fine-tuned, the fine-tuned Sol-GPT is evaluated by some experiments as described in Sect. 3 "Experiment And Analysis" which show significantly outperformed the baseline GPT-2 and cutting-edge smart contract vulnerability detection approaches in identifying vulnerabilities in smart contracts. The performance improvement is mainly due

Algorithm 2. Vulnerability Detection Fine-tuning

1: Initialization: Load pre-trained SolGPT model
2: Initialization: Load MPL ← Linear(hidden layer size, 2)
3: **for** each epoch in fine-tuning epochs **do**
4:     **for** each batch in dataset **do**
5:         tokens ← SolTokenizer(batch)
6:         labels ← vulnerability labels (0 or 1) for contracts in a batch
7:         feature_vectors ← SolGPT(tokens)
8:         lm_logits ← MPL(feature_vectors)
9:         lm_logits ← Softmax(lm_logits)
10:        Calculate vulnerability detection loss by comparing predicted labels with real labels:
11:            loss ← CrossEntropyLoss(logits, labels)
12:        Update model weights with gradient descent
13:    **end for**
14: **end for**
15: Save fine-tuned SolGPT model
Input: vulnerability detection dataset (vulnerable and non-vulnerable contracts)
Output: Fine-tuned SolGPT model for vulnerability detection

to the SolTokenizer, the Solidity Adaptive Pre-Training, and the vulnerability detection fine-tuning that SolGPT applied, which enables the SolGPT to better understand the unique syntax and structure of Solidity.

5   Experiment and Analysis

In this section, to evaluate the effectiveness of SolGPT in detecting various smart contract vulnerabilities, the fine-tuned SolGPT is constructed using the last section mentioned approach to detect four specific vulnerabilities and compare its performance with existing approaches. Additionally, ablation studies are conducted to analyze the contributions of different components of this proposed approach, including the specialized tokenizer, SolTokenizer, and Solidity Adaptive Pre-Training. Especially, the term "fine-tuned SolGPT" will be replaced by "SolGPT" in the following descriptions to avoid redundancy and improve readability.

Runtime Environment: In these experiments, the runtime environment can be described as Table 1.

Experimental Hyperparameters: This study utilizes cross-validation to evaluate the impact of various experimental hyperparameters on the outcomes. Through the comparison of diverse hyperparameter settings, the optimal parameters are eventually determined. The ultimate experimental parameters are outlined in Table 2.

Table 1. Fine-tuning runtime environment.

Category	Adaptive Pre-Training
Deep Learning Framework	Pytorch
System	Ubuntu22.04 LTS
CPU Processor	Intel Core i5-12400 CPU
GPU Processor	NVIDIA Tesla P40 24G
Memory Capacity	16 GB RAM

Table 2. Selection of fine-tuning hyperparameters.

Hyperparameter	Meaning	Value
Learning Rate	initial learning rate	5e−5
Batch Size	training sample amount per batch	4
Epoch	training rounds	50
Optimizer	learning rate optimization algorithm	AdamW
Dropout	random inactivation rate	0.1
Embed Size	word vector dimension	768

Datasets: Experiment utilized the recently released public Smart-Contract-Dataset [28]. In particular, the Resource2 dataset was meticulously chosen as the primary source for analysis due to its availability and suitability for evaluation purposes. While Resource1 comprises an extensive dataset, its unlabeled nature renders it unsuitable for direct assessment. Moreover, Resource3 was published subsequent to the experiment, prompting the deliberate selection of Resource2 for this investigation.

The flawed-marked dataset used in this study is derived from contract code crawled from Ethereum. The author of the dataset extracted code snippets related to vulnerabilities through automated control flow analysis and manually labeled the snippets. All experiments in this paper will be based on this dataset. The approach employed in the dataset to extract vulnerability-related code snippets through automated control flow analysis, as originally proposed in the 2018 paper "Vuldeepecker" [29], will not be further elaborated upon in this context.

In regard to the dataset size, we acknowledge that the selected dataset, particularly the Resource2 dataset, may have limitations in terms of the number of types and samples. The primary focus of our proposed method is to address the challenge of a scarcity of labeled samples, which is a common issue encountered in similar datasets during model training. Our approach incorporates techniques and strategies to effectively handle the limited dataset size, aiming to demonstrate robustness and generalization despite this constraint. We emphasize that the goal of this paper is to showcase the effectiveness and efficiency of our method, even under circumstances where the dataset is not extensive.

Table 3 presents the four subsets of the dataset, each corresponding to a specific vulnerability. It includes the number and percentage of positive samples (samples with the vulnerability) and negative samples (samples without the vulnerability) in each subset, as well as the total number of samples in each subset. The percentages are calculated based on the total number of samples in each subset. Prior to experimentation, the dataset was divided into a training set and a verification set according to a 7:3 ratio. The division ensured the balance of positive and negative samples proportion in both sets, mitigating issues that may arise from sample proportion imbalance caused by direct random division.

Table 3. Dataset sub-set positive and negative sample distribution for different vulnerability types.

Vulnerability Type	Positive Samples	Negative Samples	Total Samples
reentrancy	73 (26.74%)	200 (73.26%)	273
timestamp	179 (51.29%)	170 (48.71%)	349
delegatecall	62 (31.63%)	134 (68.37%)	196
integeroverflow	90 (32.73%)	185 (67.27%)	275

Evaluation Metrics: SolGPT employ commonly used evaluation metrics in the vulnerability detection field, including accuracy, precision, recall, and F1-Score. The formulas for these metrics are as follows:

- $Accuracy = (TP + TN)/(TP + TN + FP + FN)$.
- $Precision = TP/(TP + FP)$.
- $Recall = TP/(TP + FN)$.
- $F1 - Score = 2 \times (Precision \times Recall)/(Precision + Recall)$.

Here, TP represents true positives, TN represents true negatives, FP represents false positives, and FN represents false negatives.

5.1 Comparisons with Traditional Approaches

Initially, a comparison is conducted between the performance of SolGPT and that of traditional tools, such as Slither [4], Mythril [8], and Oyente [7], followed by an analysis of the obtained experimental data. For an in-depth examination of the results concerning each vulnerability type, the reader is directed to Fig. 3.

Reentrancy Vulnerability: SolGPT achieved an accuracy of 97.53%, outperforming Slither (89.37%), Mythril (62.63%), and Oyente (66.30%). In terms of precision, SolGPT achieved a perfect score of 100.00%, significantly higher than the other tools. The recall and F1 score were also superior, with SolGPT achieving 91.67% and 95.65%, respectively. This represents an improvement of 8.16% in accuracy and 12.81% in F1 score compared to existing traditional solutions for reentrancy vulnerability detection.

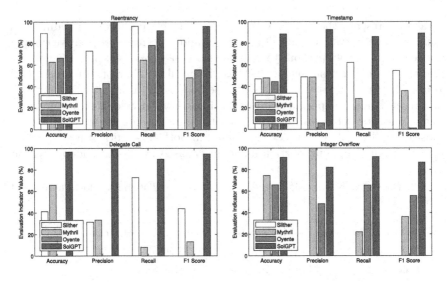

Fig. 3. Comparisons with traditional approaches.

Timestamp Vulnerability: For timestamp dependency vulnerabilities, Sol-GPT demonstrated an accuracy of 88.46%, substantially higher than Slither (46.99%), Mythril (47.85%), and Oyente (44.41%). The precision of SolGPT was 92.45%, while the recall and F1 score were 85.96% and 89.09%, respectively, outperforming the other tools by a wide margin. This marks a notable improvement of 40.61% in accuracy and 34.55% in F1 score compared to existing traditional solutions for timestamp vulnerability detection.

Delegate Call Vulnerability: In detecting delegate call vulnerabilities, Sol-GPT achieved an accuracy of 96.55%, compared to Slither's 41.32% and Mythril's 65.81%. Oyente did not provide results for this vulnerability type. This approach achieved perfect precision (100.00%) and outperformed other tools in terms of recall (90.00%) and F1 score (94.74%). The improvement over existing traditional solutions for delegate call vulnerability detection is substantial, with an increase of 30.74% in accuracy and 50.84% in F1 score.

Integer Overflow Vulnerability: For integer overflow vulnerabilities, SolGPT achieved an accuracy of 91.46%, while Mythril and Oyente achieved 74.54% and 65.81%, respectively. Slither did not provide results for this vulnerability type. SolGPT's precision was 82.14%, and its recall and F1 score were 92.00% and 86.79%, respectively, surpassing the other tools. This corresponds to a remarkable improvement of 16.92% in accuracy and 31.13% in F1 score compared to existing traditional solutions for integer overflow vulnerability detection.

This work observed that the three traditional tools generally exhibited low performance across various metrics in detecting the four types of vulnerabilities, especially in the case of Timestamp vulnerabilities. The paper speculates that with the increasing complexity of real-world vulnerabilities, traditional tools may struggle to effectively capture the evolving dependencies and related vulnerability patterns. In the case of Timestamp vulnerabilities, which involve dependencies on the contract's timestamp, more sophisticated analysis and reasoning are required to detect potential issues. This exacerbates the performance decline of traditional tools. Furthermore, our examination of the data reveals that this performance decline also affects SolGPT.

This phenomenon emphasizes the imperative of employing advanced methodologies such as SolGPT to address the escalating complexity of real-world vulnerabilities. SolGPT is designed to encompass the contextual framework within which code is composed. It possesses the capability to discern the interconnections between various components of the code and their interactions, a crucial aspect for the detection of vulnerabilities that might hinge on particular sequences or combinations of operations. As vulnerabilities continue to evolve, the demands for a more nuanced analysis intensify, and SolGPT stands out as an exceptional solution in this respect.

5.2   Comparisons with Deep Learning Approaches

In order to further validate the performance of the proposed SolGPT model, it is compared with various deep learning-based schemes such as ReChecker (based on Word2Vec and RNN, LSTM, BiLSTM, BiLSTM-att) [17] and DR-GCN(TMP) [20]. The results are shown in Fig. 4.

Based on the data provided in Table 4, SolGPT can analyze the performance of different deep learning schemes in detecting four types of vulnerabilities: Delegatecall Vulnerability, Integer Overflow Vulnerability, Reentrancy Vulnerability,

Table 4. Comparisons with deep learning approaches.

Vulnerability Type	Evaluation Metric	RNN	LSTM	BiLSTM	BiLSTM-ATT	TMP	SolGPT
Reentrancy Vulnerability	Accuracy	76.83%	73.17%	76.83%	78.05%	78.05%	97.53%
	Precision	80.00%	NaN	63.64%	70.00%	75.00%	100.00%
	Recall	18.18%	0.00%	31.82%	31.82%	27.27%	91.67%
	F1 Score	29.63%	NaN	42.42%	43.75%	40.00%	95.65%
Timestamp Vulnerability	Accuracy	64.76%	57.14%	68.57%	55.24%	71.43%	88.46%
	Precision	63.79%	59.52%	71.74%	57.50%	68.85%	92.45%
	Recall	69.81%	47.17%	62.26%	43.40%	79.25%	85.96%
	F1-score	66.67%	52.63%	66.67%	49.46%	73.68%	89.09%
Delegatecall Vulnerability	Accuracy	69.49%	67.80%	76.27%	69.49%	76.27%	96.55%
	Precision	66.67%	NaN	69.23%	60.00%	100.00%	100.00%
	Recall	10.53%	0.00%	47.37%	15.79%	26.32%	90.00%
	F1 Score	18.18%	NaN	56.25%	25.00%	41.67%	94.74%
Integer Overflow Vulnerability	Accuracy	66.27%	74.70%	78.31%	85.54%	66.27%	91.46%
	Precision	50.00%	57.14%	66.67%	76.67%	50.00%	82.14%
	Recall	25.00%	100.00%	71.43%	82.14%	53.57%	92.00%
	F1 Score	33.33%	72.73%	68.97%	79.31%	51.72%	86.79%

and Timestamp Vulnerability. The deep learning schemes include RNN, LSTM, BiLSTM, BiLSTM-ATT, TMP, and the proposed SolGPT model.

Reentrancy Vulnerability: The SolGPT model outshines other deep learning schemes, achieving the highest accuracy (97.53%), precision (100.00%), recall (91.67%), and F1 score (95.65%). In comparison to the second-best approach, our approach exhibits a remarkable improvement of 19.48% in accuracy and 51.9% in F1 score.

Timestamp Vulnerability: SolGPT outperforms other approaches, with the highest accuracy (88.46%), precision (92.45%), recall (85.96%), and F1 score (89.09%). In contrast to the second-best approach, TMP, our approach demonstrates a substantial improvement of 17.03% in accuracy and 15.41% in F1 score.

Delegatecall Vulnerability: Comparing the deep learning schemes, SolGPT outperforms other approaches, achieving the highest accuracy (96.55%), precision (100.00%), recall (90.00%), and F1 score (94.74%). In comparison to the second-best approach, BiLSTM, our approach shows a significant improvement of 20.28% in accuracy and 38.49% in F1 score.

Integer Overflow Vulnerability: SolGPT performs the best with an accuracy of 91.46%, precision of 82.14%, recall of 92.00%, and F1 score of 86.79%. When compared to the closest competitor, BiLSTM-ATT, our approach demonstrates a notable improvement of 5.92% in accuracy and 7.48% in F1 score.

In the comparative analysis of SolGPT against other deep learning approaches for smart contract vulnerability detection, several notable findings emerged. These findings can be attributed to various factors and provide insights into the strengths and limitations of SolGPT in this specific context.

Firstly, SolGPT demonstrates a substantial improvement in accuracy for three out of the four analyzed vulnerabilities, achieving nearly a 20% boost in accuracy when compared to the other methods. However, when examining the Integer Overflow vulnerability, the improvement in accuracy is comparatively modest, standing at 5.92%. This discrepancy can be elucidated by considering the inherent simplicity of the logic underlying the Integer Overflow vulnerability in contrast to the other three vulnerabilities. This simplicity may lead to higher detection rates by various tools, diminishing the relative improvement achievable by SolGPT in this case.

Moreover, this study unveils a remarkable enhancement in the Recall metric of SolGPT for detecting Reentrancy and Delegatecall vulnerabilities in comparison to the optimal solution. The Recall improvement for Reentrancy and Delegatecall vulnerabilities is 59.85% and 42.63%, respectively. This substantial boost in Recall signifies SolGPT's superior ability to correctly identify positive samples for these two vulnerabilities. This capability is particularly critical in vulnerability detection tasks where false negatives (missed vulnerabilities) can have severe consequences, often outweighing the cost of false positives.

Several factors contribute to these noteworthy results. Firstly, the control flow associated with Reentrancy and Delegatecall vulnerabilities is more intricate and challenging to model than that of other neural networks. SolGPT's extensive pre-training allows it to capture and understand these complex control flows effectively, contributing to its improved Recall. Secondly, the dataset used in the analysis contains a limited number of positive samples for Reentrancy and Delegatecall vulnerabilities. SolGPT's superior generalization capabilities, honed through its pre-training phases, enable it to maintain a high Recall rate even with a scarcity of positive samples. This indicates that SolGPT can leverage its comprehensive understanding of Solidity code, acquired during pre-training, to make accurate predictions even when faced with data scarcity.

In summary, the observed differences in performance between SolGPT and other deep learning approaches in the context of smart contract vulnerability detection can be attributed to the complexity of vulnerabilities, the nature of control flows, and SolGPT's ability to generalize effectively from pre-training. SolGPT's strengths in understanding intricate control flows and handling limited data make it particularly potent in identifying vulnerabilities like Reentrancy and Delegatecall, which are of paramount importance in security-critical applications.

5.3   Ablation Studies

In this section, we emphasize the importance of ablation studies as a crucial aspect of our evaluation. Ablation studies involve systematically disabling various components or modules within our proposed model to assess their individual contributions to the overall performance. By conducting these experiments, we aim to provide a clear understanding of the significance of each component in our approach and how it impacts the results. This analysis allows us to identify key mechanisms, optimize our model's architecture, and guide future developments in smart contract vulnerability detection. Ablation studies serve as a foundation for our performance evaluation, enabling readers to follow our rationale and insights effectively.

In this subsection, we investigate the influence of SolTokinzer and solidity adaptive pre-training on the task of detecting vulnerabilities in smart contracts. A suite of 16 models was trained, specifically for four distinct vulnerabilities in smart contracts, utilizing four distinct upstream models: GPT2-scratch (which involves no pre-training), GPT2-codeGPT (which leverages codeGPT weights as the pre-training initialization point), SolGPT-GPT2Tokenizer (which uses codeGPT weights as a starting point, followed by training with solidity adaptive pre-training), and SolGPT (which is identical to SolGPT-GPT2Tokenizer, but replaces GPT2Tokenizer with SolTokenizer). The hyperparameters and training procedures employed are consistent with those outlined earlier. The comparative performance of these diverse models under the four vulnerabilities is depicted in Fig. 4.

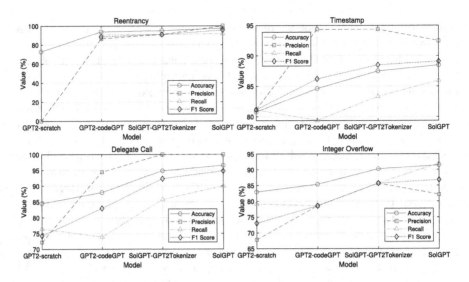

Fig. 4. Comparisons with traditional approaches.

Inspection of the results displayed in Fig. 4 corroborates the efficacy of the proposed SolGPT model. This is exemplified by the discernible gap between each procedure or module, which provides a clear comparison of their performance.

A clear trend is demonstrated in Fig. 4: the performance metrics for the four types of vulnerabilities consistently improve with the integration of the training steps or modules discussed in this paper. This enhancement is observable when compared to the baseline GPT2-scratch model, which has not undergone any pre-training. Notably, even the performance metrics of the GPT2-scratch model exhibit superiority in the Timestamp and Delegate Call vulnerabilities compared to the best alternative solutions evaluated in this study. The GPT2-scratch model's performance on the remaining two types of vulnerabilities is also highly competitive.

The transition from GPT2-scratch to GPT2-codeGPT demonstrated significant performance enhancement. Leveraging codeGPT as a pre-training initialization point capitalized on the commonalities between Java and Solidity, two programming languages. This allowed the model to learn shared features, leading to noticeable improvements in all performance metrics. On average, across the four types of vulnerabilities, the accuracy increased by a substantial 7.68%.

Moving from GPT2-codeGPT to SolGPT-GPT2Tokenizer, the introduction of solidity adaptive pre-training proved to be pivotal. This additional pre-training task enabled the model to grasp specific language features and keyword nuances unique to Solidity compared to Java. Consequently, the model's performance was further elevated, as reflected by a notable average F1-Score improvement of 5.34% across the four vulnerabilities.

The transition to SolGPT from SolGPT-GPT2Tokenizer marked a refinement in the tokenization process. This optimization ensured the semantic integrity of keywords, effectively emphasizing their semantic information during various training tasks. This nuanced emphasis resulted in a performance boost, as evidenced by an average F1-Score improvement of 2.21% across the four vulnerability types.

These results underscore the progressive impact of incorporating codeGPT-based pre-training and Solidity-specific pre-training on the model's ability to detect vulnerabilities in smart contracts. Additionally, it highlight the importance of optimizing the tokenization process, particularly in preserving and leveraging semantic information related to keywords, to further enhance the model's performance across various vulnerability detection tasks.

While there are a few data points of performance decline in our results, these are rather marginal and mainly concern indices with a specific class label bias, as opposed to the overall accuracy or F1-Score. Further analysis suggests that these declines could be attributed to dataset bias and should not be construed as evidence of overall performance decline.

Collectively, these results suggest that integrating SolTokenizer and solidity adaptive pre-training (SolGPT) substantially improves the model's ability to detect various vulnerabilities in smart contracts. The ablation studies underline the significant contribution of each component to the overall performance of the model. Future efforts should, therefore, focus on further optimizing these components and exploring additional techniques for enhancing the model's performance.

6   Conclusion

In today's era of distributed and network-based computing, ensuring the security and dependability of smart contracts has become a paramount concern. As the backbone of blockchain-based systems, smart contracts require robust protection against vulnerabilities that can lead to devastating consequences. To address this critical issue, this paper introduced SolGPT, a specialized GPT model for smart contract vulnerability detection in Solidity. Leveraging Solidity Adaptive Pre-Training, SolGPT boosts feature extraction capabilities while reducing reliance on labeled data. Additionally, SolGPT developed SolTokenizer for accurate and efficient tokenization in the smart contract field. Comparative Experimental Results show that SolGPT outperforms existing models in accuracy, F1 score, and other performance metrics, demonstrating its potential to improve smart contract security. In addition, we conducted ablation experiments to aid readers in gaining a better understanding of the impact of different modules/training methods on the final performance of SolGPT.

While this study constitutes a significant leap forward in enhancing smart contract security, it is not devoid of challenges. A prominent one stems from the inherent limitations in token length for large-scale contract processing in Large Language Models (LLMs) like GPT, used herein as GPT-2 base with a maximum token capacity of 512. To mitigate this, our methodology employs automated control flow analysis to extract vulnerability-related code segments, with the utilized dataset having already undergone this preprocessing step.

The controlled flow chains for the four types of vulnerabilities investigated in this study are not excessively long, hence minimizing the chances of surpassing the token limit. Nonetheless, an acknowledgment is in order that for vulnerabilities with inherently long control flow chains, the extracted code segments may still exceed the token limit, leading to truncation of the segment. This truncation could potentially impair semantic extraction and subsequent detection accuracy, a drawback warranting further investigation.

As we chart the course for future endeavors, our research focus includes refining the pre-training and fine-tuning processes of SolGPT to better encapsulate these challenges and to ensure more robust semantic extraction even with long control flow chains. Moreover, we aim to delve deeper into the implementation of advanced deep-learning techniques, designed to elevate the security of smart contracts further.

These endeavors are not merely academic pursuits but are essential to foster a more secure and reliable smart contract environment, an area of increasing importance in the rapidly advancing digital and decentralized world. We hope that our ongoing research will pave the way for innovative solutions, capable of overcoming the challenges faced by the smart contract security field.

References

1. Nakamoto, S.: Bitcoin: a peer-to-peer electronic cash system. Decentralized Bus. Rev. 21260 (2008)
2. Szabo, N.: Smart contracts: building blocks for digital markets. EXTROPY: J. Transhumanist Thought (16) **18**(2), 28 (1996)
3. Kushwaha, S.S., Joshi, S., Singh, D., Kaur, M., Lee, H.N.: Systematic review of security vulnerabilities in Ethereum blockchain smart contract. IEEE Access **10**, 6605–6621 (2022)
4. Feist, J., Grieco, G., Groce, A.: Slither: a static analysis framework for smart contracts. In: 2019 IEEE/ACM 2nd International Workshop on Emerging Trends in Software Engineering for Blockchain (WETSEB), pp. 8–15. IEEE (2019)
5. Grishchenko, I., Maffei, M., Schneidewind, C.: A semantic framework for the security analysis of Ethereum smart contracts. In: Bauer, L., Küsters, R. (eds.) Principles of Security and Trust: 7th International Conference, POST 2018, Held as Part of the European Joint Conferences on Theory and Practice of Software, ETAPS 2018, Thessaloniki, Greece, 14–20 April 2018, Proceedings 7, pp. 243–269. Springer, Cham (2018). https://doi.org/10.1007/978-3-319-89722-6_10
6. Amani, S., Bégel, M., Bortin, M., Staples, M.: Towards verifying Ethereum smart contract bytecode in Isabelle/HOL. In: Proceedings of the 7th ACM SIGPLAN International Conference on Certified Programs and Proofs, pp. 66–77 (2018)
7. Luu, L., Chu, D.H., Olickel, H., Saxena, P., Hobor, A.: Making smart contracts smarter. In: Proceedings of the 2016 ACM SIGSAC Conference on Computer and Communications Security, pp. 254–269 (2016)
8. Mueller, B.: Introducing Mythril: a framework for bug hunting on the Ethereum blockchain. https://medium.com/hackernoon/introducing-mythril-a-framework-for-bug-hunting-on-the-ethereumblockchain-9dc5588f82f6. Accessed 6 Mar 2020

9. Jiang, B., Liu, Y., Chan, W.K.: ContractFuzzer: fuzzing smart contracts for vulnerability detection. In: Proceedings of the 33rd ACM/IEEE International Conference on Automated Software Engineering, pp. 259–269 (2018)
10. He, J., Balunović, M., Ambroladze, N., Tsankov, P., Vechev, M.: Learning to fuzz from symbolic execution with application to smart contracts. In: Proceedings of the 2019 ACM SIGSAC Conference on Computer and Communications Security, pp. 531–548 (2019)
11. Lin, G., Xiao, W., Zhang, J., Xiang, Y.: Deep learning-based vulnerable function detection: a benchmark. In: Zhou, J., Luo, X., Shen, Q., Xu, Z. (eds.) Information and Communications Security: 21st International Conference, ICICS 2019, Beijing, China, 15–17 December 2019, Revised Selected Papers 21, pp. 219–232. Springer, Cham (2020). https://doi.org/10.1007/978-3-030-41579-2_13
12. Liu, Y., Tantithamthavorn, C., Li, L., Liu, Y.: Deep learning for android malware defenses: a systematic literature review. ACM J. ACM (JACM) (2022)
13. Guo, N., Li, X., Yin, H., Gao, Y.: VulHunter: an automated vulnerability detection system based on deep learning and bytecode. In: Zhou, J., Luo, X., Shen, Q., Xu, Z. (eds.) Information and Communications Security: 21st International Conference, ICICS 2019, Beijing, China, 15–17 December 2019, Revised Selected Papers 21, pp. 199–218. Springer, Cham (2020). https://doi.org/10.1007/978-3-030-41579-2_12
14. Ghazal, T.: Data fusion-based machine learning architecture for intrusion detection. Comput. Mater. Continua 70(2), 3399–3413 (2022)
15. Dong, Y., Chen, X., Shen, L., Wang, D.: Privacy-preserving distributed machine learning based on secret sharing. In: Zhou, J., Luo, X., Shen, Q., Xu, Z. (eds.) Information and Communications Security: 21st International Conference, ICICS 2019, Beijing, China, 15–17 December 2019, Revised Selected Papers 21, pp. 684–702. Springer, Cham (2020). https://doi.org/10.1007/978-3-030-41579-2_40
16. Tann, W.J.W., Han, X.J., Gupta, S.S., Ong, Y.S.: Towards safer smart contracts: a sequence learning approach to detecting security threats. arXiv preprint arXiv:1811.06632 (2018)
17. Qian, P., Liu, Z., He, Q., Zimmermann, R., Wang, X.: Towards automated reentrancy detection for smart contracts based on sequential models. IEEE Access 8, 19685–19695 (2020)
18. Wang, W., Song, J., Xu, G., Li, Y., Wang, H., Su, C.: ContractWard: automated vulnerability detection models for Ethereum smart contracts. IEEE Trans. Network Sci. Eng. 8(2), 1133–1144 (2020)
19. Zhuang, Y., Liu, Z., Qian, P., Liu, Q., Wang, X., He, Q.: Smart contract vulnerability detection using graph neural network. In: IJCAI, pp. 3283–3290 (2020)
20. Liu, Z., Qian, P., Wang, X., Zhuang, Y., Qiu, L., Wang, X.: Combining graph neural networks with expert knowledge for smart contract vulnerability detection. IEEE Trans. Knowl. Data Eng. 35(2), 1296–1310 (2021)
21. Radford, A., et al.: Language models are unsupervised multitask learners. OpenAI Blog 1(8), 9 (2019)
22. Brown, T., et al.: Language models are few-shot learners. Adv. Neural. Inf. Process. Syst. 33, 1877–1901 (2020)
23. Mikolov, T., Chen, K., Corrado, G., Dean, J.: Efficient estimation of word representations in vector space. arXiv preprint arXiv:1301.3781 (2013)
24. Zhou, P., et al.: Attention-based bidirectional long short-term memory networks for relation classification. In: Proceedings of the 54th Annual Meeting of the Association for Computational Linguistics (vol. 2: Short papers), pp. 207–212 (2016)
25. Vaswani, A., et al.: Attention is all you need. In: Advances in Neural Information Processing Systems, vol. 30 (2017)

26. Sennrich, R., Haddow, B., Birch, A.: Neural machine translation of rare words with subword units. arXiv preprint arXiv:1508.07909 (2015)

27. Asia, M.R.: CodeGPT-small-java-adaptedGPT2 model weights. https://huggingface.co/microsoft/CodeGPT-small-java-adaptedGPT2, date of publication not available. Accessed 2 Jan 2023

28. Qian, P.: Smart contract dataset (resource2) (2022). https://github.com/Messi-Q/Smart-Contract-Dataset. Accessed 28 Dec 2022

29. Li, Z., et al.: VulDeePecker: a deep learning-based system for vulnerability detection. arXiv preprint arXiv:1801.01681 (2018)

Neuron Pruning-Based Federated Learning for Communication-Efficient Distributed Training

Jianfeng Guan[1,2]([✉]) [iD], Pengcheng Wang[1], Su Yao[3], and Jing Zhang[1]

[1] School of Computer Science (National Pilot Software Engineering School), Beijing University of Posts and Telecommunications, Beijing 100876, China
{jfguan,wpc1021,jzhang2021}@bupt.edu.cn
[2] Key Laboratory of Networking and Switching Technology, Beijing University of Posts and Telecommunications, Beijing 100876, China
[3] Department of Computer Science and Technology, Tsinghua University, Beijing 100084, China
yaosu@tsinghua.edu.cn

Abstract. Efficient and flexible cloud computing is widely used in distributed systems. However, in the Internet of Things (IoT) environment with heterogeneous capabilities, the performance of cloud computing may decline due to limited communication resources. As located closer to the end, edge computing is used to replace cloud computing to provide timely and stable services. To accomplish distributed system and privacy preserving, Federated Learning (FL) has been combined with edge computing. However, due to the large number of clients, the amount of data transmitted will also grow exponentially. How to reduce the communication overhead in FL is still a big problem. As a major method to reduce the communication overhead, compressing the transmission parameters can effectively reduce the communication overhead. However, the existing methods do not consider the possible internal relationship between neurons. In this paper, we propose Neuron Pruning-Based FL for communication-efficient distributed training, which is a model pruning method to compress model parameters transmitted in FL. In contrast to the previous methods, we use dimensionality reduction method as the importance factor of neurons, and take advantage of the correlation between them to carry out model pruning. Our analysis results show that NPBFL can reduce communication overhead while maintaining classification accuracy.

Keywords: Edge Computing · Federated Learning · Privacy Preserving · Neuron Pruning · Internet of Things · Traffic Identification

This research was supported by the National Key R&D Program of China under Grant No. 2022YFB3102304 and in part by National Natural Science Foundation of China Grants (6222510562001057).

Z. Tari et al. (Eds.): ICA3PP 2023, LNCS 14490, pp. 63–81, 2024.
https://doi.org/10.1007/978-981-97-0859-8_4

1   Introduction

Nowadays, the number of mobile devices represented by Internet of Things (IoT) devices has increased dramatically, and the amount of data generated has also multiplied, which put forward a higher demand for reliable response to traditional central cloud computing [1, 2]. However, traditional central cloud computing still faces the problems of poor wireless communication channel stability, low bandwidth. Edge computing can provide flexible and stable services for users through the use of large-scale deployment of 5G technology. Therefore, in the edge network scenario, edge computing can be used to replace central cloud computing to provide users with more timely and stable services.

Apart from the communication problem, data privacy issues can not be underestimated. The IoT devices contain private data of businesses and factories that are critical to the proper functioning of the industry, thus placing high demands on the communication security [3]. Therefore, how to use IoT equipment and its generated data has become a hot topic in industry. Many consumers are willing to use their own devices for machine learning training to enjoy the performance improvement brought by using a large amount of data training. But they may not want their private information to be disclosed [4]. To solve this problem, Google has proposed a federated learning (FL) techniques that allows clients to cooperatively learn the model without sharing the raw training data [5]. In FL, all clients can work together to train a model. Specifically, each client has a local dataset and trains the model locally. They only exchanges parameters with the central server without exchanging privacy-sensitive raw data. Subsequently, the central server aggregates the model parameters. As a result, FL can train the model without transmitting a lot of training data.

In view of the above advantages, many scholars have conducted in-depth research on FL [6, 7] and widely integrated it with edge computing techniques, such as Internet of Vehicles (IoV) [8], smart city [9] and Industrial Internet of Medical Things (IoMT) [10]. However, the conventional FL still suffers from the huge communication overhead caused by the large number of clients and the complexity of neural network model. Most of the previous studies [11–21] have reduced the communication overhead by implementing some optimization algorithms on the clients or changing the communication mode.

For instance, optimization of communication scheme. Some works try to eliminate unnecessary duplicate communication by using the computing capability of the network node [12]. These works focus on optimizing communication schemes and reducing the number of participated node, which may have higher requirements on the performance of edge nodes.

There are also many studies using distillation [13], quantization [14, 17] and model pruning to compress the model, thus reducing the communication overhead. However, most studies focus on the relationship between convolution neurons and outputs, which includes applying sparse regularization on the filter weights [18], the scaling factors of each layer [19], the network channels [20] and evaluate the importance of each channel [21]. These works estimate the importance of a convolution neuron or channel separately, ignoring that the convolution neuron has multiple inputs, so there is a certain correlation between the convolution neuron or channel.

In Convolutional Neural Network (CNN), each layer will have many neurons that receive the same feature map as the input, so the parameters they trained are likely to have a certain correlation [22]. This correlation is more pronounced when the number of neurons in the same layer is large. If this correlation can be utilized to evaluate the importance factor of neurons and generate a set of linear combination primitives, representing all features of this layer with a small number of neurons, the number of parameters included in the model can be reduced on a large scale.

In order to fill the gap of using neurons correlation as an important factor in model pruning, this paper proposes a method using Principal Component Analysis (PCA), which is an example of dimensionality reduction, as the basis of model pruning. Using PCA-based pruning, the model can be efficiently compressed to reduce the amount of data transmitted. In particular, the PCA operation will not lose the knowledge of some nodes, and it is completely executed on the server, which is transparent to clients and will not put forward higher performance requirements for participated nodes.

Consequently, on the premise of reducing the communication overhead between mobile devices and central servers in the edge computing environment, the main contribution of this work is to develop a neuron Pruning-Based FL (NPBFL) for communication-efficient distributed training. The main contributions of this paper can be summarized as follows:

- Considering that mobile devices usually have weak communication capability in edge computing, we incorporate model pruning into the traditional FL process to perform model compression on the central server and reduce communication overhead.
- As for model pruning, considering that traditional methods usually only evaluate the importance of a single neuron or channel as an indicator, ignoring the correlation between neurons or channels, we introduce the dimensionality reduction method, taking PCA as an example, to reduce the dimension of neurons in each layer of the neural network, thus compressing the model and reducing the communication overhead.
- We implement a series of experiments to evaluate the performance of NPBFL on MNIST and CIFAR10. By comparing with two SOTA methods, we demonstrate that our scheme can perform better to effectively reduce communication overhead while achieving high accuracy. Moreover, we use a real network traffic dataset to verify the feasibility of NPBFL in practical applications.

The rest of this paper is organized as follows: Sect. 2 introduces the most related work about communication-efficient FL. Section 3 introduces our system and workflow. Section 4 describes the experimental settings and result analysis of the simulation experiment, and provides the application effect in the real network traffic classification scenario. Section 5 concludes this paper with further work.

2   Related Work

In this section, we will introduce the combination of FL and edge computing to show its feasibility in actual deployment. In addition, we will summarize the scheme of optimizing communication overhead in FL.

2.1   FL with Edge Computing

Due to the restrictions on communication overhead and privacy protection in edge scenario, traditional central cloud computing and machine learning will not be the primary options for actual application deployment. Edge computing is widely used in real-world deployment as a replacement for cloud computing, because it can provide more stable and timely services closer to the user side. FL, as a privacy-preserving distributed learning paradigm, can train a model without sharing the raw training data, which can perfectly adapt to edge scenarios.

Distributed edge clients and the central server collectively constitute a FL system. FL allows edge clients (e.g., mobiles and IoT devices) to cooperate in training a model without sharing their raw local data. In FL, there is a central server and distributed edge clients synchronize their model parameters updates through interaction with the central server [5]. In FL, the goal of model training is to find the optimal parameters ω' on the training datasets D that can minimize the global loss function, $F(\omega)$, of a given model. This objective can be expressed as Eq. (1) when $f(s)$ is the loss value computed from sample s.

$$\omega' = \arg\min F(\omega) = \arg\min \frac{1}{|D|} \sum\nolimits_{s \in D} f(s). \tag{1}$$

Many researches have combined edge computing and FL, such as Internet of Vehicles (IoV) [8], smart city [9] and Industrial Internet of Medical Things [10].

Zijia et al. [8] have used quantization and coding methods to reduce communication costs, and the federated distillation method is applied to the IoV to achieve real-time corresponding services. Kai et al. [9] have applied offloaded FL models in edge-cloud smart city to take intelligent analysis and data processing to deal with enormous network data. Tsang et al. have taken solder paste printing as a typical IIoT scenario, and introduced FL to build decision support system to ensure the automation and sustainable manufacturing management of the system. Otoum et al. [10] have utilized federated transfer learning to train intrusion detection system for IoMT to enhance the generalization ability of distributed IDS.

These researches show the effectiveness of combining edge computing and federated learning. However, the scale of the underlying Deep Neural Network (DNN) in FL is so large that it cannot be directly deployed in the edge environment with limited computing and communication resources.

2.2   Communication Overhead Reduction in FL

There have been some efforts to improve the communication efficiency in FL. Deng et al. [11] have proposed hierarchical federated learning framework to select a subset of distributed node as edge aggregators to conduct edge aggregations to reduce communication overhead. Luo et al. [12] have tried to eliminate unnecessary duplicate communication by using the computing capability of the network node. These kind of work focus on optimizing communication schemes, and reducing the number of participated nodes, which can indeed reduce the communication overhead in FL, but may have higher requirements on the performance of edge nodes.

Another type of work use the model compression to optimize the communication overhead. Liu et al. [13] have proposed Federated Distillation with active data sampling, which makes each client upload a subset of the logits with low entropy. Lian et al. [15] have reduced the communication time by layer-based parameter selection, which can select the valuable parameters for global aggregation to optimize the communication and training process. Qu et al. [14] have proposed FedDQ which achieves a reduction in communication overhead by continuously decreasing the range of parameters updated during training using quantization. Magnusson et al. [17] implements a general class of linearly convergent algorithms and uses quantization to compress the communication overhead (AQW). Saeed et al. [16] have used structured and unstructured model pruning to find a small subnetwork for each client to reduce communication overhead. These works have successfully reduced the communication overhead, but they either put forward higher requirements for participated nodes or ignored the knowledge owned by nodes. Therefore, our work will restrict the work of model compression to the central server, so as to release the computing pressure of mobile devices.

Model compression is mainly divided into tensor decomposition, quantization and pruning. We mainly introduce pruning because it is most relevant to our work.

Model pruning mainly uses the importance factors such as L_1/L_2 norm and geometric median to estimate whether a connection or channel is important, so as to decide how to prune. Most studies focus on the relationship between neurons and outputs, which includes applying sparse regularization on the filter weights [18], the scaling factors of each layer [19], the network channels [20] and evaluate the importance of each channel [21]. These methods apply sparse regularization filter weights so that the corresponding filter weights of unimportant channels are near zero. Then, these filters or channels could be safely pruned without affecting the output values. However, although these sparse means can avoid over-fitting, the difficulty in grasping the degree of regularization may also lead to the slow convergence of the model after pruning.

The importance of a neuron is estimated separately, thus ignoring the fact that neurons have multiple same inputs, which means they are heavily likely to have a certain correlation between them, especially when the number of neurons is very large [22]. By using a smaller number of neurons to represent the overall features through linear combination, we can reduce the number of parameters in the model.

Consequently, to fill the gap of using neurons correlation as an important factor in model pruning to achieve communication-efficient FL, we propose a method using PCA to replace sparse regularization pruning. We introduce PCA to compress the model structure, which makes full use of the data contained in each node and does not put forward higher performance requirements for nodes.

3   Algorithm Description

In this paper, we consider a traditional FL system which consists of one central server and multiple clients. We denote the set of clients as $A = \{A_1, ..., A_n\}$. Each client A_n has a dataset D_{A_n} stored locally. Under this circumstances, how to optimize the data transmitted between clients and central server becomes an essential problem. Model pruning has been used to solve the communication overhead problem, but the traditional sparse regularization method ignores the correlation between neurons. We considered the correlations among neurons in the same layer, using PCA for dimensionality reduction of neurons, reducing the number of channels, thus compressing the model.

The workflow of NPBFL is shown in Algorithm 1. NPBFL contains four main functions, namely global model distribution, local model training, global model aggregation and global model pruning. The clients are responsible for local model training while the rest three functions are performed on central server.

Global Model Distribution: The central server will select a basic global model that meets the client computing performance and distribute it to all participated clients.

Local Model Training: Distributed clients perform local training with received model based on the dataset saved locally, generates a new local model, and uploads it to the central server.

Global Model Aggregation: When the central server receives the returned local model, it invokes the model aggregation algorithm to generate a new global model and distributes the new global model down to the distributed clients.

Global Model Pruning: In order to ensure that the model parameters already reflect the importance of neurons during model compression, the central server only performs model compression when the number of clients exceeds 50% for the first time. The central server will use PCA as an example to prune the aggregated model, compress the model structure, reduce model parameters, and thus reduce communication costs.

Algorithm 1. NPBFL Procedure

Input:
    D_{A_i}: Dataset of distributed edge client A_i.
Output:
    M_g: Global model trained with distributed datasets.
1: Central server generates an initial model that meets the client computing performance and distributes it to participated distributed edge clients.
2: **for** $e = 1$ to E (Number of Epochs) **do**
3:    A_i trains received M_g using local dataset D_{A_i} to generate new local model M_{A_i}
      and upload it to the central server.
4:    **for** $n = 1$ to N (Number of paticipated distributed edge clients) **do**
5:       Central server receives local model M_{A_i}, invokes the model aggregation algo-
         rithm to generate a new M_g.
6:       **if** the number of clients exceeds 50% for the first time **then**
7:          Central server performs model pruning on M_g.
8:       **end if**
9:       Central server distributes new aggregated M_g down to edge clients.
10:    **end for**
11: **end for**

In this article, we mainly verify the optimization of communication overhead
for FL through model compression. Therefore, we simply choose to compress
when the number of clients exceeds 50% for the first time. The research on when
to compress and how to determine the compression timing is not the focus of
this article, and will be studied separately in the future.

3.1 Basic Model Selection

NPBFL can significantly reduce the communication overhead by using FL and
model pruning, but with the continuous development of deep learning, the CNN
has a dramatic growth in depth and breadth. Under this circumstances, the
limited computational and storage resources in edge clients motivate the central
server to generate a model with smaller scale that does not significantly degrade
accuracy.

We conducted a simple FL experiment in which LeNet-5, AlexNet, VGGNet
and GoogleNet were respectively applied as basic CNN model, generating
548 KB, 29.4 MB, 14.2 MB and 786 KB communication overhead. Table 1 and
Fig. 1 show the communication overhead and classification accuracy using different CNN models.

As shown in Fig. 1, there is no obvious difference in the accuracy of these
models after training while the difference in communication overhead during the
process of model aggregation is very conspicuous. The communication overhead
generated by LeNet-5 is one in sixty of that of AlexNet. This result means
that compared with other conventional CNN models, LeNet-5 can reduce the
communication overhead by at least 30%, which prompts us to use LeNet-5 as
the basic CNN model.

Table 1. Communication overhead for one epoch of different CNN models (Clients N = 2)

CNN	Communication overhead
LeNet-5	548 KB
AlexNet	29.4 MB
VGGNet	14.2 MB
GoogleNet	786 KB

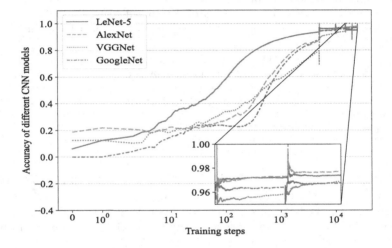

Fig. 1. Training Accuracy of different CNN models

3.2   Model Pruning

Model pruning is an important technical means to achieve model compression. As we mentioned above, the conventional pruning uses the importance factors such as L_1/L_2 norm and geometric median to estimate whether a neuron or channel is important, where the importance of a neuron or channel is estimated separately. However, the neuron has multiple inputs, which means there is a certain correlation between neurons that the conventional means ignores.

Specifically for a convolution layer, assuming that the size of the neurons is $k * k$, and a total of n neurons are used, the total number of parameters is $k * k * n$. The specific parameters in these neurons are trained through the same feature maps, so it is heavily likely to have a certain correlation between them, especially when the n is very large [22]. Hence, the importance factor can be evaluated based on the correlation between neurons, and a smaller number of neurons can be used to represent the overall features through linear combination.

In order to use model pruning to reduce the number of neurons, the dimensionality reduction method can be used to evaluate the importance of neurons. As a main way of dimensionality reduction, PCA can also perform matrix decomposition to compress the model. Therefore, to further reduce the communication overhead, we will use PCA to compress the structure of network models in central server.

For the convolution layer and full connection layer in CNN, each layer has a large number of neurons. When the number of neurons in one layer is too large, the similarity between them may increase significantly. Consider a layer in a neural network where the input is d feature maps and the output is n channels, so the parameter quantity of this layer is ndk^2. If we can reduce the number of neurons by 50%, then n will become $n/2$, and d is the number of neurons in the previous layer, which will also become $d/2$. Therefore, the overall parameter quantity will decrease to $ndk^2/4$.

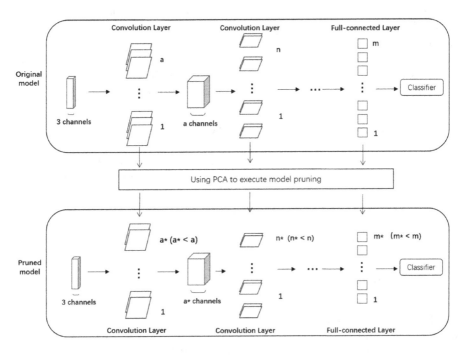

Fig. 2. Model pruning in NPBFL

Under such circumstances, if we can use a few base neurons to represent all neurons, the model structure will be reduced, thereby reducing communication overhead in FL. Therefore, we use the PCA method in CNN, and use 90% as the threshold to screen the neurons to be retained. PCA is performed layer by layer, thus reducing the number of neurons in each layer. The process of model pruning is shown in the Fig. 2.

As we mentioned above, the parameters of a layer can be expressed as a matrix of $d * k * k * n$. We then construct a matrix using neurons as rows, and construct a covariance matrix, and perform singular value decomposition to determine the contribution rate of each component. According to the contribution rate, retain the corresponding neurons to represent all neurons in this layer through linear combination.

Model pruning has been widely studied and implemented in various ways to compress model structure. In this paper, we only use PCA as an example to fill the gap of using the correlation between neurons as importance factor to determine how to prune model to reduce communication overhead in communication-efficient FL.

Taking FedDQ and AQW as examples, NPBFL differs from these existing advanced schemes in terms of design ideas. FedDQ achieves a reduction in communication overhead by continuously decreasing the range of parameters updated during training using quantization. AQW uses a general class of linearly convergent algorithms and uses quantization to compress the communication overhead. These methods reduce the communication overhead by compressing the internal parameter representation of the model, and do not start from the model itself. NPBFL directly considers the nature of the transmitted data, the model itself, and reduces the communication overhead by compressing the model structure and reducing the number of neurons contained in the model.

4   Simulation Results and Analysis

To assess the effectiveness and efficiency of NPBFL, we simulate the environment of FL and perform experimental validation. We also simulate the traditional FL, FedDQ [14] and Adaptive Quantization Weights (AQW) [17]. The datasets we use are MNIST and Cifar-10 to validate the effectiveness of NPBFL, and we use a network traffic dataset, USTC-TFC2016 [23], to build a distributed intrusion detection system to verify the availability of NPBFL in practical applications.

4.1   Experimental Setting

Considering that clients cannot have the same data distribution in reality, we use the non-independent identically distributed (Non-IID) scenario. There are quantity skew and label skew. Quantity skew means clients have same kind of samples but the number of samples is different. Label skew refers to clients having different types of samples, but the number of samples of the same type does not differ significantly. In our FL-based experiments, the dataset distribution of each client is different, which includes quantity and label. In extreme cases, some clients only have three or four kind of samples.

Experiment of NPBFL. NPBFL is used for this set of experiments. Each client have data samples with different distributions and the model will be trained locally. The training results of each client will be sent to central server to aggregate. When the number of clients exceeds 50% for the first time, the central server performs model pruning using PCA and compresses the model structure to reduce the subsequent communication overhead.

Experiment of Baseline. To prove the validity of NPBFL, the local datasets of edge clients are used to train the CNN model individually as a baseline, which means the clients can only use the knowledge contained in the local datasets. Except that the model will not be aggregated, the experiment is the same as that of NPBFL. In addition, we also implement conventional FL without model pruning in the same experimental setting with NPBFL to compare the availability of NPBFL.

Experiment of Other Methods. We implemented FedDQ [14] and AQW [17] to further evaluate the effectiveness of NPBFL in reducing the communication overhead. AQW implements a general class of linearly convergent algorithms and uses quantization to compress the communication overhead. FedDQ takes a step further to descend quantization scheme by reducing the range of updated model as the convergence rate decreases to reduce the communication overhead. To the best of our knowledge, FedDQ and AQW are the most effective in the current scenario, so we only compare NPBFL with these two state-of-the-art approaches.

4.2   Performance Analysis

In this section, we will analyze the experimental results in several dimensions.

Time Complexity. In real-world applications such as the Internet of things, users want to get the results quickly, so the time complexity of the algorithm should be as small as possible to meet the needs of fast response in the edge scene. In addition, time complexity can also represent the calculation pressure faced by edge client. The time complexity of NPBFL is related to the scale of CNN model, the number of simulated clients, the amount of training data, and other factors. Here, we assume that the time complexity of training a batch is $O(n)$, the size of the datasets of the ith client is D_i, the communication delay of the client is t, the maximum batch data is b, and the training round for convergence of the i client is k_i. Then the time complexity of the ith client training one round is as follows:

$$O(\frac{D_i}{b} * n + 2t) \tag{2}$$

In general, the time complexity required for all clients to complete the training convergence is as follows. T_{mp} represents the time spent in model pruning on the central server.

$$O(\sum k_i(\frac{D_i}{b} * n + 2t) + T_{mp}) \tag{3}$$

It can be seen from Eq. 3 that NPBFL only needs to prune the model when the number of clients exceeds 50% for the first time, so it only costs one T_{mp}. In FedDQ and AQW, the transmitted data needs to be compressed in each round, that is, the time spent needs to be accumulated, which means the time complexity is higher than NPBFL.

Classification Metrics. We use accuracy, recall and f1-score to evaluate the effect of NPBFL. In experiment of NPBFL, each client would learn its own datasets separately, then they upload the updated parameter to the central server to generate the global model. The central server aggregates a global model and then distributes it to the edge clients to upgrade the local model. In Experiment of individual client, each datasets is trained separately and the IIoT edge clients have no interaction with other nodes.

Table 2. Performance metrics

Experiment	Recall	F1-score	Accuracy
MNIST			
NPBFL	90.9%	91.7%	91.3%
FedDQ	91.7%	92.0%	91.8%
AQW	91.4%	92.7%	92.3%
Individual model	72.3%	73.1%	72.6%
Traditional FL	92.7%	93.1%	92.8%
Cifar-10			
NPBFL	72.3%	72.8%	72.6%
FedDQ	71.9%	72.2%	72.1%
AQW	72.9%	72.6%	72.8%
Individual model	56.5%	56.0%	56.2%
Traditional FL	75.2%	74.9%	74.9%

We use accuracy, recall and f1-score to evaluate the effect of NPBFL. MNIST and Cifar-10 are used for verification. Here we mainly analyze the results on MNIST, and the results on Cifar-10 are similar to those on MNIST, which are shown in Table 2. As shown in Fig. 3, Fig. 4(a) and Table 2, when the number of clients is 2, there is no obvious difference between the F1-score of NPBFL and centralized training model. Because the distribution of local datasets is Non-IID, NPBFL already has obvious advantages over individual model. When the number of clients is increased to 10, the gap between NPBFL and individual model increases to nearly 9%. When there are 50 clients in experiments, the gap increases to 19%. In terms of accuracy of classification, when the number of clients is 50, NPBFL also exceeds the individual model with an accuracy of 91.3%. The accuracy of NPBFL, FedDQ and AQW have no big difference, which is only slightly lower than that of traditional FL. In addition, NPBFL, FedDQ, and AQW each have their own advantages and disadvantages on different datasets, indicating that the model structure still has an impact on accuracy. NPBFL performs better than FedDQ on more complex Cifar-10 datasets through neural pruning, indicating that NPBFL has better adaptability to different scenarios.

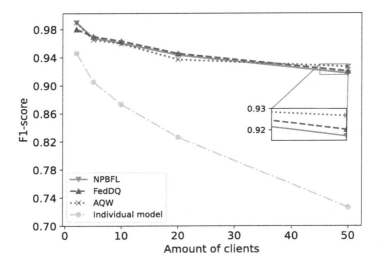

Fig. 3. F1-score of different models on MNIST

We can conclude that when the scale of datasets is very limited, the performance of individually model will have a great bottleneck. Under these circumstances, if combined with other clients, using NPBFL, this whole progress will greatly improve the model performance. Figure 4(a) shows the F1-score improvement in the results of NPBFL compared to the individual model when the number of clients is the same. In the range of 2 to 50, the number of clients in FL has a positive correlation with the improvement of F1-score performance, which means the greater the number of clients is, the more significant the performance improvement will be. As for the accuracy, it can be seen from Table 2 that the accuracy of NPBFL is similar to that of FedDQ and AQW, which is significantly better than the individual model.

The Impact of the Number of Clients on Time Consumption. The concept of relative time-consumption is also used to evaluate the model results. It represents the time spent on NPBFL's training (T_{NPBFL}) divided by the time spent on conventional training of individual edge client (T_{IEC}) in the case of the same number of clients, which can directly reflect the time comparison relationship between these two scenarios. In this section, we still use MNIST as

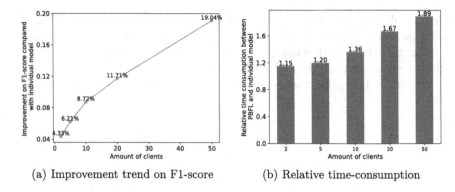

(a) Improvement trend on F1-score   (b) Relative time-consumption

Fig. 4. Performance on F1-score and time-consumption

an example to evaluate the results. The relative time-consumption is defines as following:

$$Relative\ time\text{-}consumption = \frac{T_{NPBFL}}{T_{IEC}} \quad (4)$$

The time consumed during 100 epochs of the model training is also counted in the process of Experiment of NPBFL and experiment of individual model to calculate the relative time-consumption.

In Fig. 4(b) when there are 2 clients, the training time of NPBFL is similar to individual model. When the number increases to 50, the training time of NPBFL is 1.89 times longer than that of the individual model. In addition, our experiments has eliminated the influence of communication delay theoretically.

The results show that the classification performance of NPBFL is similar to that of the two SOTA methods, but NPBFL will cause greater time consumption in the training process, which should be taken into account when it is used in actual scenarios.

Communication Overhead. In order to quantify the reduction of communication overhead, we propose the concept of *transmission packet load* (T_{PL}). A packet usually consists of a header and a payload. We define them as packet header (H_P) and packet loader (L_P). The composition of data packets (T_{PL}) can be expressed as follows without considering the need for fragmentation of data packets:

$$T_{PL} = H_P + L_P \quad (5)$$

The H_P generally does not exceed 100 bytes, which means it is negligible compared to L_P that represents the main load of the data. Therefore, the L_P can be approximated to represent the T_{PL}. T_{PL} represents the amount of exchanged data, which is communication overhead. To verify whether the NPBFL can reduce the communication overhead effectively, we measure the T_{PL} of NPBFL, conventional FL, FedDQ and AQW.

Table 3 shows the T_{PL} of different models. In centralized scenario, the edge clients are responsible for not only collecting but also transmitting raw data to the central server. We can see that the T_{PL} of centralized model is 1.5 GB, which means that the edge clients need to transmit a total of GB-level data, which is a very large communication overhead for edge networks.

Table 3. T_{PL} metric for experimental models

Experiment	T_{PL}
Conventional FL	48.7 MB
NPBFL	10.9 MB
FedDQ	25.4 MB
AQW	23.1 MB

The T_{PL} of conventional FL is 48.7 MB and 10.9 MB of NPBFL. The results show that only MB-level data is transmitted in NPBFL, which significantly reduces the communication overhead. Compared with FedDQ and AQW, their communication costs are 25.4 MB and 23.1 MB respectively. Although they have a significant advantage over traditional training methods, their communication overhead is still more than twice that of NPBFL.

4.3   Network Traffic Classification

To verify the feasibility of NPBFL in practical applications, we use a real network traffic dataset and built a distributed IDS to simulate the security system of the IoT environment. We set an organization of one central server, and 50 clients that represents IIoT edge clients. Clients A_i and A_j have two different datasets which are D_{A_i} and D_{A_j}. The feature space of the two datasets is the same, while the sample space is different.

We use USTC-TFC2016 which contains 10 kinds of benign traffic and 10 kind of malicious traffic to evaluate the performance of NPBFL. The distribution of USTC-TFC2016 datasets is shown in Table 4. In the aspect of data processing, we process the raw data into 28×28 pictures. The MAC address and IP address are replaced by randomly generated addresses, so as to reduce the possible interference of these addresses and protect the data privacy. The cleaned data is then divided by flow and processed in a uniform length of 784 bytes, where zeros are used to fill empty spaces, and converted to IDX format. We divide the dataset into 50 sub-datasets as the local data of each client. Some clients only have several types of data, and the data distribution of clients with all types of data is also different.

Table 4. Traffic distribution in USTC-TFC2016 datasets

	Labels	Samples		Labels	Samples
Benign	BitTorrent	7517	Malicious	Cridex	16995
	Facetime	6000		Geodo	37489
	FTP	60000		Htbot	7079
	Gmail	8629		Miuref	15512
	MySQL	60000		Neris	31019
	Outlook	7525		Nsis-ay	6425
	Skype	6322		Shifu	9522
	SMB	32662		Tinba	10964
	Weibo	24954		Virut	30890
	WorldOfWarcraft	7884		Zeus	11512
	Total				398900

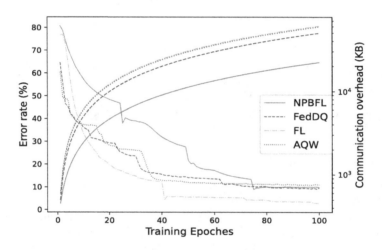

Fig. 5. Communication overhead and Error rate of NPBFL and existing approaches

Figure 5 shows the communication overhead of different methods, where the error rate is obtained by 1 minus accuracy. The conventional FL achieves the best accuracy, which is because there is no loss of model structure, and it has similar communication overhead with AQW and FedDQ. NPBFL achieves similar accuracy with FedDQ and AQW while significantly reducing communication overhead. As can be seen from the Fig. 5, the error rates of different methods have a sudden drop in the training process, that is, the accuracy rate has a sudden increase. This may be the learning rate value is too large. However, we have used an adaptive learning rate adjustment mechanism, so the overall error rate decline was still relatively flat. It is prudent to note that the optimization algorithms of AQW and FedDQ are executed at edge clients, which puts forward higher requirements on the performance of clients. As for NPBFL, the compression algorithm is executed at central server, which does not impose additional calculation burden on the edge clients. Compared with existing research, NPBFL can further reduce communication overhead and ensure accuracy without putting forward additional resource requirements for edge clients. Considering the actual deployment scenario, we can make a choice between accuracy and communication overhead according to the communication pressure of the IIoT environment.

5 Conclusion and Future Work

This paper proposed a neuron pruning-based federated learning for communication-efficient distributed training to reduce the communication overhead in edge network. In particular, NPBFL trains the model with FL and compresses the global model to reduce communication overhead. The results show that NPBFL effectively reduce the communication overhead while ensuring accuracy.

Further work will consider using other optimization and model compression methods to reduce the communication overhead, and investigate the performance of different model compression in non independent and identically distributed environments. Considering the existing method of optimizing the FL communication framework, we will also explore the possibility of combining framework optimization with model compression in the future. In addition, during the experimental phase, we raised the issue of compression timing. In future work, we will conduct further experiments and discuss the impact of selecting when to compress and when to end compression based on parameters such as training rounds and clients participated ratio on compression performance.

References

1. Peng, C., Hu, Q., Chen, J., Kang, K., Li, F., Zou, X.: Energy-efficient device selection in federated edge learning. In: 30th International Conference on Computer Communications and Networks, ICCCN 2021, Athens, Greece, 19–22 July 2021, pp. 1–9. IEEE (2021). https://doi.org/10.1109/ICCCN52240.2021.9522303
2. Yang, D., Cheng, Z., Zhang, W., Zhang, H., Shen, X.: Burst-aware time-triggered flow scheduling with enhanced multi-CQF in time-sensitive networks. IEEE/ACM Trans. Networking **31**(6), 2809–2824 (2023). https://doi.org/10.1109/TNET.2023.3264583
3. Tange, K., Donno, M.D., Fafoutis, X., Dragoni, N.: A systematic survey of industrial Internet of Things security: requirements and fog computing opportunities. IEEE Commun. Surv. Tutorials **22**(4), 2489–2520 (2020). https://doi.org/10.1109/COMST.2020.3011208
4. Al-Garadi, M.A., Mohamed, A., Al-Ali, A.K., Du, X., Ali, I., Guizani, M.: A survey of machine and deep learning methods for Internet of Things (IoT) security. IEEE Commun. Surv. Tutorials **22**(3), 1646–1685 (2020). https://doi.org/10.1109/COMST.2020.2988293

5. Sha, J., Basara, N., Freedman, J., Xu, H.: FLOR: a federated learning-based music recommendation engine. In: 31st International Conference on Computer Communications and Networks, ICCCN 2022, Honolulu, HI, USA, 25–28 July 2022, pp. 1–2. IEEE (2022). https://doi.org/10.1109/ICCCN54977.2022.9868921
6. Zhang, W., et al.: Optimizing federated learning in distributed industrial IoT: a multi-agent approach. IEEE J. Sel. Areas Commun. **39**(12), 3688–3703 (2021). https://doi.org/10.1109/JSAC.2021.3118352
7. Yang, D., et al.: DetFed: dynamic resource scheduling for deterministic federated learning over time-sensitive networks. IEEE Trans. Mobile Comput. (2023). Early Access, https://doi.org/10.1109/TMC.2023.3303017
8. Mo, Z., Gao, Z., Zhao, C., Lin, Y.: FedDQ: a communication-efficient federated learning approach for internet of vehicles. J. Syst. Archit. **131**, 102690 (2022). https://doi.org/10.1016/j.sysarc.2022.102690
9. Peng, K., Zhang, H., Zhao, B., Liu, P.: Edge-cloud collaborative computation offloading for federated learning in smart city. In: IEEE International Conference on Dependable, Autonomic and Secure Computing, International Conference on Pervasive Intelligence and Computing, International Conference on Cloud and Big Data Computing, International Conference on Cyber Science and Technology Congress, DASC/PiCom/CBDCom/CyberSciTech 2022, Falerna, Italy, 12–15 September 2022, pp. 1–7. IEEE (2022). https://doi.org/10.1109/DASC/PiCom/CBDCom/Cy55231.2022.9927848
10. Otoum, Y., Wan, Y., Nayak, A.: Federated transfer learning-based IDS for the internet of medical things (IoMT). In: IEEE Globecom 2021 Workshops, Madrid, Spain, 7–11 December 2021, pp. 1–6. IEEE (2021). https://doi.org/10.1109/GCWkshps52748.2021.9682118
11. Deng, Y., et al.: SHARE: shaping data distribution at edge for communication-efficient hierarchical federated learning. In: 41st IEEE International Conference on Distributed Computing Systems, ICDCS 2021, Washington DC, USA, 7–10 July 2021, pp. 24–34. IEEE (2021). https://doi.org/10.1109/ICDCS51616.2021.00012
12. Luo, S., Fan, P., Xing, H., Luo, L., Yu, H.: Eliminating communication bottlenecks in cross-device federated learning with in-network processing at the edge. In: IEEE International Conference on Communications, ICC 2022, Seoul, Korea, 16–20 May 2022, pp. 4601–4606. IEEE (2022). https://doi.org/10.1109/ICC45855.2022.9838381
13. Liu, L., Zhang, J., Song, S., Letaief, K.B.: Communication-efficient federated distillation with active data sampling. In: IEEE International Conference on Communications, ICC 2022, Seoul, Korea, 16–20 May 2022, pp. 201–206. IEEE (2022). https://doi.org/10.1109/ICC45855.2022.9839214
14. Qu, L., Song, S., Tsui, C.: FedDQ: communication-efficient federated learning with descending quantization. In: IEEE Global Communications Conference, GLOBECOM 2022, Rio de Janeiro, Brazil, 4–8 December 2022, pp. 281–286. IEEE (2022). https://doi.org/10.1109/GLOBECOM48099.2022.10001205
15. Lian, Z., Wang, W., Su, C.: COFEL: communication-efficient and optimized federated learning with local differential privacy. In: ICC 2021 - IEEE International Conference on Communications, Montreal, QC, Canada, 14–23 June 2021, pp. 1–6. IEEE (2021). https://doi.org/10.1109/ICC42927.2021.9500632
16. Vahidian, S., Morafah, M., Lin, B.: Personalized federated learning by structured and unstructured pruning under data heterogeneity. In: 41st IEEE International Conference on Distributed Computing Systems Workshops, ICDCSW 2021, Washington, DC, USA, 7–10 July 2021, pp. 27–34. IEEE (2021). https://doi.org/10.1109/ICDCSW53096.2021.00012

17. Magnússon, S., Ghadikolaei, H.S., Li, N.: On maintaining linear convergence of distributed learning and optimization under limited communication. IEEE Trans. Signal Process. **68**, 6101–6116 (2020). https://doi.org/10.1109/TSP.2020.3031073

18. Li, Y., Gu, S., Mayer, C., Gool, L.V., Timofte, R.: Group sparsity: the hinge between filter pruning and decomposition for network compression. In: 2020 IEEE/CVF Conference on Computer Vision and Pattern Recognition, CVPR 2020, Seattle, WA, USA, 13–19 June 2020, pp. 8015–8024. Computer Vision Foundation/IEEE (2020). https://doi.org/10.1109/CVPR42600.2020.00804, https://openaccess.thecvf.com/content_CVPR_2020/html/Li_Group_Sparsity_The_Hinge_Between_Filter_Pruning_and_Decomposition_for_CVPR_2020_paper.html

19. Zhao, K., Zhang, X., Han, Q., Cheng, M.: Dependency aware filter pruning. CoRR abs/2005.02634 (2020). https://arxiv.org/abs/2005.02634

20. Louizos, C., Welling, M., Kingma, D.P.: Learning sparse neural networks through L0 regularization. CoRR abs/1712.01312 (2017). http://arxiv.org/abs/1712.01312

21. Lin, M., et al.: HRank: filter pruning using high-rank feature map. In: 2020 IEEE/CVF Conference on Computer Vision and Pattern Recognition, CVPR 2020, Seattle, WA, USA, 13–19 June 2020, pp. 1526–1535. Computer Vision Foundation/IEEE (2020). https://doi.org/10.1109/CVPR42600.2020.00160, https://openaccess.thecvf.com/content_CVPR_2020/html/Lin_HRank_Filter_Pruning_Using_High-Rank_Feature_Map_CVPR_2020_paper.html

22. Chen, Y., Dai, X., Liu, M., Chen, D., Yuan, L., Liu, Z.: Dynamic convolution: attention over convolution kernels. In: Proceedings of the IEEE/CVF Conference on Computer Vision and Pattern Recognition (CVPR), June 2020

23. Wang, W., Zhu, M., Zeng, X., Ye, X., Sheng, Y.: Malware traffic classification using convolutional neural network for representation learning. In: 2017 International Conference on Information Networking (ICOIN), pp. 712–717 (2017). https://doi.org/10.1109/ICOIN.2017.7899588

A Novel Network Topology Sensing Method for Network Security Situation Awareness

Yixuan Wang[1(✉)], Bo Zhao[1], Zhonghao Sun[2], Zhihui Huo[3], Xueying Li[4], Yabiao Wu[4], and Jiao Li[4]

[1] School of Cyber Science and Engineering, Wuhan University, Wuhan 430072, China
{njwyx94,zhaobo}@whu.edu.cn
[2] The 28th Research Institute of China Electronics Technology Group Corporation, Nanjing 210007, China
[3] Zhejiang Dahua Technology Corporation, Hangzhou 310053, China
[4] Topsec Network Technology Inc., Beijing 100085, China
{li_xueying,wyb,li_jiao}@topsec.com.cn

Abstract. In Network Security Situation Awareness (NSSA), topology information of the monitored network constitutes the foundation of the whole NSSA process. This paper presents a novel method for network topology sensing in non-collaborative networks. The proposed method leverages trusted agents and Group Decision Making (GDM) policies to provide more accurate and complete topology information. To ensure the reliability of the proposed approach, the initial trusted agents are regarded as the experts and the GDM process is carried out solely under their control. Additionally, a core topology description ontology is employed to integrate detected information in a more efficient manner. Furthermore, the approach is exemplified through a comparative analysis in a practical network environment comprising of 20 subnets and over 400 nodes. The experimental results demonstrate that compared with previous network topology sensing methods, our method exhibits a relatively higher coverage rate and is more adept at selecting worker agents. Such outcomes lend credence to the possibility that our approach is a useful practice in detecting complex network environments, ultimately contributing to a security analyst's cognitive perspective of situation awareness.

Keywords: network security situation awareness (NSSA) ·
non-collaborative network · trusted agents · group decision making (GDM)

1 Introduction

The field of Situation Awareness (SA) has gained significant traction in recent research [1]. Network Security Situation Awareness (NSSA) is a subfield of SA that is concerned with environments where the core asset is a network. In such

Z. Tari et al. (Eds.): ICA3PP 2023, LNCS 14490, pp. 82–101, 2024.
https://doi.org/10.1007/978-981-97-0859-8_5

circumstances, a plethora of network attacks could be carried out by adversaries, potentially causing incalculable losses. The current aim of NSSA is to provide a higher-level view to monitor all aspects of the network and provide decision support to network security analysts once malicious behaviors are detected.

In this regard, the seminal challenge is that data collectors on one side may provide inaccurate information, particularly with respect to the topology information of the monitored network. With the rapid advancement of cloud computing and the Internet of Things, many mobile devices and ad hoc networks have emerged, thereby rendering network topology sensing an increasingly vital matter. Hence, it is imperative to develop novel approaches that can accurately gather network topology information and promptly respond to any changes within the network.

It is also worth noting that some networks may consist of devices from both sides, which grants these networks a non-collaborative nature.

Another challenging aspect of network topology sensing methods is the trustworthiness of the collected topology information, which is garnered from an untrustworthy network source. In traditional NSSA, agents equipped with various data collectors play a pivotal role in collecting network topology information. Yet, these agents may also be targeted by malicious adversaries, leading to hijacking or attacks. To address this issue, trusted agents can be employed to conduct the information gathering process in a logistically more secure manner. Besides, the efficiency and effectiveness of topology sensing is equally important.

Due to the challenges of the network topology sensing method, research questions of this paper are listed as below:

Research Question 1: In a non-collaborative network, how to obtain the topology information with high efficiency?

Research Question 2: In faced with potential malicious behaviors of adversaries, how to obtain reliable and correct topology information?

We study the research problems in depth and try to increase the efficiency of the detection process and the accuracy of the detection results with limited cost of some network bandwidth. In this paper, a novel network topology sensing method for network security situation awareness is presented. Trusted agents act as a useful mechanism for dealing with untrustworthy networks as they can utilize trusted network connection technology to transfer information securely among them. Furthermore, they possess a secure execution environment to process and handle sensitive information collected. There is a rising trend in the number of non-collaborative networks, and the specific number is contingent on the practical scenario under discussion.

During the initial state, more than one agent is required to initialize the topology information collection task. While agents are discovering the topology, they may encounter limitations on the scope they can reach, which necessitates an agent injection process to produce additional agents. In each detection round, determining which hosts need to be injected and for how long can pose a challenge. To overcome this issue, Group Decision Making (GDM) technology is employed to enhance the cooperative decision-making process among agents.

Trusted agents act as experts in the GDM process where they generate candidate agents and select an optimal subset for implementation.

This presented paper has the following advantages and novelties:

- Firstly, a novel network topology sensing method is proposed for NSSA with an aim of obtaining high-quality topology information in non-collaborative network. This is accomplished through the utility of GDM based on trusted agents for better organization of detection periods and improved trustworthiness of situation elements.
- In addition, two approaches are presented to support the GDM process. The first outlines the behavioral patterns to be employed by the worker agents in each new round of detection tasks, ensuring the collection of useful topology information and the generation of potential alternatives for negotiation. The second approach illuminates the mechanism used by the trusted agents to decide the next round of candidate agents and the termination of topology detection tasks. These two approaches are combined to implement the entire GDM process.
- Finally, we evaluate the impact of the proposed design and performance of the topology detection tasks within a specialized non-collaborative network scenario, and draw conclusions based on our observations and evaluations.

The paper is structured as follows: Sect. 2 outlines the preliminaries and related works necessary to depict the proposed network topology sensing process in detail. Section 3 describes the proposed network topology sensing method in a comprehensive manner. Section 4 presents an experimental example that provides a better understanding of the method. Furthermore, the experimental results are presented and thoroughly discussed. Lastly, in Sect. 5, we draw several conclusions based on our findings and observations.

2   Preliminaries and Related Works

In this section, the basis required to comprehend the proposed method is presented.

2.1   Network Security Situation Awareness (NSSA)

Currently, NSSA has become a popular topic of research amongst scholars [2]. The origins of NSSA can be traced back to Situation Awareness (SA), which was first proposed by Endsley in 1988 [3]. Subsequently, Bass et al. proposed the functional model of NSSA in 1999, which aimed to recognize and comprehend the environmental factors of the monitored network within a specific time and space and predict future trends [4]. The process of NSSA is divided into three stages, while the network posture has three aspects [5]. These stages include the situation elements extraction stage, the situation evaluation stage, and the situation prediction stage, respectively. During the first stage, the original data is fused, and active events within the network are recognized. In the second stage, active events that could pose a threat are captured and evaluated [6]. Subsequently, situation evaluation values are generated for the monitored network. Finally, in the last stage, predictions based on the ongoing network situation are conducted to support users of NSSA system. It is important to note that these stages are not straightforward, and the results of the prediction stage can feed back to the state of the environment and change the elements of the current situation. Therefore, NSSA is an iterative process [7].

In order to capture the situation elements of the three network postures and obtain an accurate security situation of the monitored network, many supportive technologies are utilized and integrated into the whole NSSA system. As the topology posture provides the structure and assets of the monitored network, which are vital for the NSSA system, an appropriate topology information extraction policy is required [8]. To the best of our knowledge, most research on NSSA focuses on the situation evaluation and situation prediction stages and pays little attention to the situation elements collection stage, particularly, neglecting topology information. The reason for this is that the topology of a traditional network is a collaborative network, where the nodes are assets belonging to us. However, once the monitored network is a fluctuating and non-collaborative environment that consists of nodes from more than two sides, the changes in network topology may significantly affect the situation evaluation and predication stages [9].

2.2   Group Decision Making (GDM)

GDM is an activity that requires a group of decision makers to choose the optimal alternative. These decision makers can either be experts, individuals, or agents with the authority to make decisions. The purpose of GDM is to bring together the knowledge, expertise, and opinions of multiple individuals to make a decision that has higher quality and efficiency than decisions made individually. The different phases of solving GDM problems consist of comparing different alternatives and communicating the group opinion.

Figure 1 illustrates the four phases of the GDM process: (i) problem analysis; (ii) rank policy; (iii) weight calculation; and (iv) determining the optimum solution. Firstly, decision makers generate their own solutions by analyzing the given problem. Secondly, an aggregation method is applied to the decision makers' opinions, and a rank policy is then applied to generate the alternatives, which represent the relatively appropriate solutions. Thirdly, weight calculation is conducted to compare the alternatives. Finally, the optimum solution is determined as the final decision.

In recent years, GDM has gained widespread adoption in addressing complex and challenging real-world problems [10]. Zhao et al. [11] utilized a multi-attribute group decision-making method to tackle incomplete probabilistic linguistic problems. Moreover, GDM has also proven to be effective in addressing practical decision making problems. Zhou et al. [12] proposed a large-scale group decision making based method to select the best-performing solar water heater and demonstrated the availability of the proposed method. Additionally, various

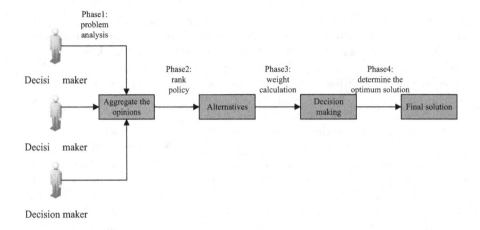

Fig. 1. Process of GDM.

efforts have been made to enhance the performance of GDM methods in certain scenarios. For example, to combat potential stagnation in GDM, Morente-Molinera et al. [13] introduced fuzzy ontologies into the GDM process. Focusing on fairness in GDM, Gong et al. [14] developed a maximum fairness consensus model with limited cost to achieve a balanced consensus solution. Furthermore, Liu et al. [15] proposed a GDM-based model to resolve consensus problems in social networks and demonstrated its efficiency.

It can be inferred that GDM provides an effective mechanism for addressing multi-party decision-making problems, and can be applied to many practical optimal selection scenarios. In the context of topology sensing in non-cooperative networks, optimal node selection and topology discovery represent critical selection problems.

3   Network Topology Sensing Method for NSSA

3.1   Overview

There are various circumstances where network detection proves to be untrustworthy for NSSA, particularly when it encompasses assets from two sides. In other words, the target network is comprised of assets (hosts, routers, switches, mobile devices, etc.) from two sides, with only a portion of the aforementioned nodes belonging to us while others may potentially be malicious entities, the sources of which are unknown. Obtaining the entirety of the target network's topology is problematic, and traditional network detection methods may produce wholly inaccurate results. To obtain dependable and dynamic network topology information, we propose a novel method for network topology sensing. In Fig. 2, several trusted agents are appointed and initialized [16], whose communications are conducted under Trusted Network Connections (TNC) [17]. By this way, the trusted agents authenticate each other to ensure that they are trustworthy. The GDM module runs on the trusted agents, determining which agents will be injected in each decision round. The outcomes of each round then serve as input for the agent injection module. The agent injection module is deemed somewhat invasive since its responsibility entails taking control of the target node within the target network and transforming it into an agent via the injection of detection and communication modules. The target network is composed of various subnets, each containing nodes that are initially unknown to us. Once the injection process is completed, some of the nodes become agents of this round and fall under the control of trusted agents. These agents then react and perform tasks in accordance with the commands issued by the trusted agents, detecting and collecting useful topology information that they subsequently convey backward to the trusted agent. Following this, trusted agents exchange the gathered information and generate a new set of agent alternatives for the next round. The mechanisms of the modules will be thoroughly discussed in Sect. 3.2 and Sect. 3.3.

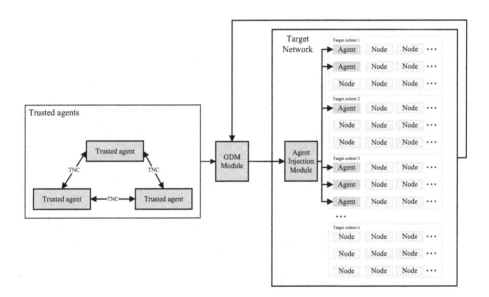

Fig. 2. Overview of the proposed method.

The target network is one that comprises nodes from two or more sides, with those nodes belonging to us requiring monitoring within NSSA. To obtain an understanding of asset node elements, collecting topology information from the network they're in is crucial. With topology information in hand, we can acquire four valuable situation elements: (i) the distinction between my asset nodes and those of others; (ii) the dynamic topology of the target network; (iii) the active traffic of the target network; and (iv) the adversarial situation of the target network. Our primary countermeasure is to gain control of the target host through agent injection. The target nodes result from the GDM in one round, and they also serve as the targets for injection. If the target node belongs to our side, we can directly control it and make it an agent. If it is from the adversary, the agent injection module completes injection and gains control of it, thereby transforming it into an agent. After multiple rounds involving probing and decision-making, the number of working agents increases until there are enough worker agents in every reachable subnet of the target network. However, trusted agents will determine new worker agents to cope with changes in topology when the detection results vary again, as this could be due to changes in topology. Ultimately, the shifted topology information is promptly relayed back to the NSSA system.

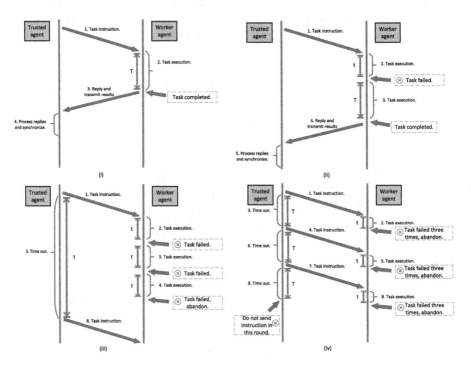

Fig. 3. Typical communication scenarios between trusted agent and worker agent: (i) a successful communication; (ii) task execution fails once but finally succeeds; (iii) task execution fails three times and retransmit the task instruction command; (iv) task instructions have been sent three times and get no reply.

3.2   Agent Model

In the Agent Model (AM), there are two typical types of agents: trusted agents and worker agents.

Trusted agents play a dual role of serving as experts within the GDM module and the commanders of specific worker agents. The chief functions of trusted agents include: (i) collecting and reporting topology information from worker agents; (ii) collaborating to make decisions through the GDM module; and (iii) data fusion and aggregation. Conversely, worker agents are subordinate to trusted agents, and their principal functions are: (i) constructing and sending probes to implement active detection; (ii) sensing and monitoring traffic to acquire topology information while implementing passive detection; (iii) performing SNMP topology aggregation if possible; and (iv) receiving instructions from trusted agents and conveying detection results to them.

The communication mechanism between worker agents and trusted agents is as follows: each worker agent is controlled by a trusted agent that manages it. Upon receiving a task command from the trusted agent, the worker agent executes the task promptly. Once task execution is completed, the result is transmitted to the trusted agent. Various typical scenarios are depicted in Fig. 3. We elucidate the state transition process of trusted agents and worker agents by explicitly specifying their Finite State Machine (FSM) in Fig. 4 and Fig. 5. Figure 4 lists 11 typical states of a trusted agent. Each state has a corresponding state transition event. Once the corresponding event occurs, the state of the trusted agent changes. Similarly, Fig. 5 lists four typical states of worker agents.

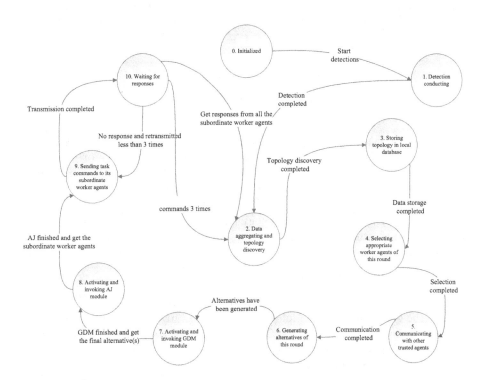

Fig. 4. FSM of trusted agents.

In a given round, trusted agents receive topology information from their respective worker agents. Subsequently, they exchange this information with each other and complete information aggregation and data fusion. This process is accomplished under the predefined data fusion module, leading to the current detection of the dynamic topology structure. The data fusion process is guided by the security core ontology [18, 19], which effectively aids in directing the fusion process while also describing the topology information thoroughly. The primary fragments of the core topology description ontology are displayed in Fig. 6.

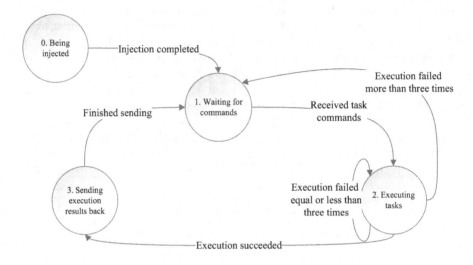

Fig. 5. FSM of worker agents.

During the data fusion implementation, concepts and data properties in the ontology can be mapped to relational tables and table attributes of relational databases, and object properties in the ontology can be mapped to constraint relations between tables in relational databases. Because the data fusion module is executed by the trusted agent, the corresponding database is also installed on the trusted agent, and the saved topology restore results are passed to the NSSA system by the trusted agent.

3.3   Group Decision Making Based on Trusted Agents

The process-level overview of our proposed GDM module is shown in Fig. 7. In general, the group decision module can be divided into three steps.

Step I: data preprocessing. The trusted agent uses the information collected by the worker agent to integrate and extract attributes such as network transmission rate, number of open ports, operating system, RTT and packet loss rate. The mentioned attributes are used to construct the worker agent decision matrix, which is denoted as A. In fact, attribute information can be adjusted,

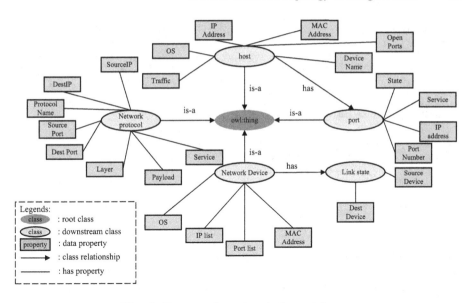

Fig. 6. Core topology descriptive ontology.

and in order to discover injectable nodes, it is sometimes necessary to pay attention not only to how many ports the node is open, but also to which ports are open, since the injection process also requires corresponding ports to be open. So-called best performance agents are also nodes that can be easily injected and hijacked as worker agents.

Assume there are m alternative worker agents and initialize matrix A_m.

$$A_m = \begin{pmatrix} a_{11} & \cdots & a_{15} \\ \vdots & \ddots & \vdots \\ a_{m1} & \cdots & a_{m5} \end{pmatrix} \tag{1}$$

In formula (1), each column indicates a specific attribute and each row indicates the quantitative metric. Then, use the deviation minimization method to standardize the decision matrix and map each attribute to the interval $[0, 1]$.

$$b_{ij} = (b_{ij} - b_j^{max})/(b_j^{max} - b_j^{min}) \tag{2}$$

$$b_{ij} = (b_j^{max} - b_{ij})/(b_j^{max} - b_j^{min}) \tag{3}$$

$$b_j^{max} = max\{b_{1j}, b_{2j}, b_{3j}, \cdots, b_{mj}\} \tag{4}$$

$$b_j^{min} = min\{b_{1j}, b_{2j}, b_{3j}, \cdots, b_{mj}\} \tag{5}$$

The formula (2) is used to measure the beneficial attributes (network transmission rate, number of device open ports, operating system version), and the formula (3) is used to measure the intrinsic attributes (round-trip delay, packet

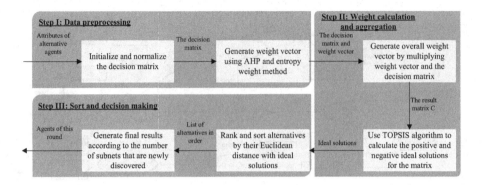

Fig. 7. Process-level overview of the proposed GDM approach.

loss rate). The formula (4) and (5) are supplementary explanations to (2) and (3), representing the maximum and minimum values of attributes respectively.

Step II: weight calculation and aggregation. Generate the agent decision matrix and calculate the weight of multiple attributes to complete the decision of this round based on the alternative worker agents. According to the characteristics of the current detection network, the importance of each attribute is measured and given different weights. In the process of weight calculation, the subjective weight algorithm and objective weight algorithm are combined to get more comprehensive weight calculation results.

As Fig. 8 depicts, we use Analytic Hierarchy Process (AHP) method [20] to obtain the subjective weight. The problem of worker agent selection is divided into four layers. The top layer is the worker agents that have best performance. In the criterion layer, important agent attributes are listed. In the sub-criterion layer, typical operating systems are listed because the detection module has different performance under different operating systems. The lowest layer is the scheme layer that contains the final worker agents out of the alternatives. We use entropy weight method to obtain the objective weight vector of the alternatives. Considering entropy and standard deviation at the same time, the specific calculation process is in formula (6).

$$w_j = ((\delta_j + E_j) \sum_{i=1}^{n} (1 - \varepsilon_{ij})) / (\sum_{j=1}^{m} (\delta_j) \sum_{i=1}^{n} (1 - \varepsilon_{ij})) \tag{6}$$

In formula (6), w_j denotes the weight of the jth attribute, ε_{ij} indicates the correlation between attributes i and j, δ_j is the standard deviation of the jth attribute, E_j is the entropy value of the jth attribute, n is the number of attributes, m is the number of alternatives.

We use deviation minimization method to get the comprehensive weight vector by combining subjective weight and objective weight. In deviation minimization method, assume the number of attributes is n and there exists L ways to calculate weight, then we can get L weight vectors. We get a linear combination

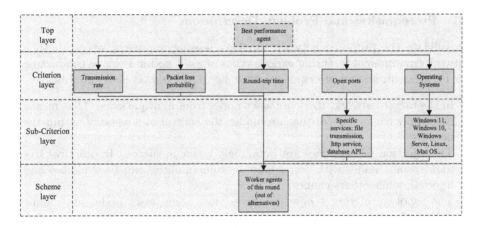

Fig. 8. Layer architecture of AHP model in the proposed method.

of weight vectors as W. As formula (7) illustrates, α_k is the weight coefficient, μ_k is the weight vector obtained by the kth method.

$$W = \sum(\alpha_k \mu_k^T), \alpha_k > 0 \qquad (7)$$

After getting the integrated weight of alternatives, we use TOPSIS algorithm [21] to calculate positive and negative ideal solutions.

Step III: sort and decision making. Set the aggregation operator and perform the aggregation calculation. After aggregation, several alternatives of this round are decided. Also, alternatives are sorted according to whether they are newly discovered in the previous round or key nodes in the topology. Firstly, calculate the Euclidean distance between ideal solutions. Secondly, calculate the alternative agent correlation measure between the positive ideal solution. Finally, compare the correlation measure of each alternative agents and sort them in descending order. Then, pick the top n of them as the result of this round. The value of n equals to the number of newly discovered subnets. If there are no newly discovered subnet in the previous one round, set the value of n equal to or less than one fifth of the number of subnets already discovered.

4   Experimental Studies and Analysis

As mentioned in Sect. 3, the agent model and GDM serve as the fundamental framework for network topology sensing. In this section, we proceed to evaluate the proposed approach's availability by addressing two distinct problems. Additionally, we present an illustrative case study to showcase how the proposed approach improves network topology sensing in NSSA.

4.1 Prerequisites and Problem Descriptions

To validate the proposed network topology sensing approach, we establish a network environment featuring various types of asset nodes. Prior to conducting the experiment, several prerequisites must be considered, as elaborated below:

- In the target network, there are asset nodes from multiple sides. That means there may exists devices that are under the control of adversary within the network.
- The injection process does not unfailingly result in success. In practical scenarios, some nodes with corresponding vulnerabilities can be exploited and injected, while others cannot.
- The topology of target network varies since some asset nodes are mobile devices that can easily switch among subnets.

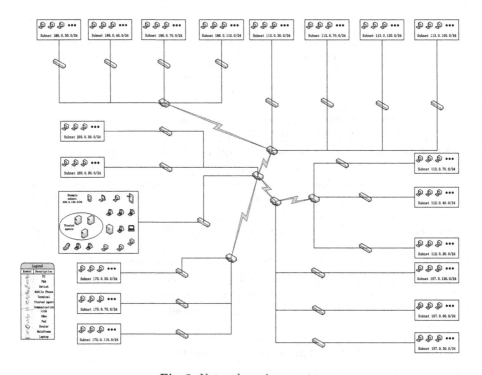

Fig. 9. Network environment.

To streamline the experimental process, we provide detailed descriptions below.

Goal: Detect and measure the target network while sensing changes within a limited time frame. The proposed network topology sensing method must surpass existing methods in terms of evaluation metrics.

Evaluation Metrics: To assess the reliability of the proposed approach in this study, we consider coverage rate and efficiency.

Experimental Environment: Our experimental environment is a hybrid network comprising asset nodes from different sides, as depicted in Fig. 9 as a sample network environment. The simulation network environment comprises 20 subnets, with trusted agents being injected and initialized by satellite signals in the example subnet "202.0.120.0/24". For the sake of simplicity, we ignore the subnets among the six core routers. Of these six core routers, only half of them enable the SNMP configuration. Terminal devices connected to the target network can communicate with the example subnet either directly or indirectly. We assume that each subnet has at least two vulnerable nodes that can be exploited via a multi-step injection, allowing the agent injection modules to attain root privilege of the nodes. Furthermore, all subnet addresses from the experimental environment are self-defined private addresses.

Table 1 presents fundamental quantization data of the subnets in the experimental environment. Connected nodes pertain to the nodes connected to the given subnet. Vulnerable nodes denote nodes that can be injected and controlled, while unreachable nodes pertain to nodes that do not generate any traffic in the network. Since we need to simulate a varying environment, the quantization information can change from time to time. For better demonstration and comparison, we set up at least one vulnerable node in each subnet. The agent injection process to subnets where there are no vulnerable nodes tends to fail. Once there is no agent in the target subnet, the active detection task cannot be conducted. Moreover, the unreachable nodes are set up to avoid 100% coverage rate.

4.2   Experimental Process and Numerical Example

To better elucidate the issue at hand, we offer a comprehensive account of a complete execution and illustrate how the proposed method enables network topology sensing. Figure 10 depicts an instance of the experimental process, which requires 7 rounds to identify all the 20 subnets in the experimental network environment. The GDM module generates new worker agents in each round, which are then injected by trusted agents or worker agents from the preceding round. In this way, the process of network topology sensing dynamically adapts to the changing conditions of the targeted network as time elapses. To provide a concrete example, when at least two worker agents have already been injected into each subnet, the GDM module will not generate new worker agents in subsequent rounds. Nevertheless, existing worker agents remain poised to execute detection tasks as directed by trusted agents.

Firstly, determine the n attributes $e_1, e_2, \cdots,$ and e_n of the alternatives. We determine the elements by evaluating the results of different selections. The considered elements are: (1) the version of running OS; (2) number of the open ports; (3) transmission rate; (4) RTT; (5) packet loss rate. These elements are selected due to the fact that they reflect how active the nodes are, and generally

Table 1. Detailed subnet quantization information.

Subnet	Connected nodes	Vulnerable nodes	Unreachable nodes
186.0.30.0/24	25	2	4
186.0.40.0/24	20	3	0
186.0.70.0/24	20	2	0
186.0.110.0/24	21	4	0
113.0.30.0/24	24	2	7
113.0.70.0/24	20	2	0
113.0.120.0/24	20	2	0
113.0.150.0/24	29	2	3
203.0.30.0/24	26	2	0
203.0.90.0/24	20	2	0
203.0.120.0/24	23	1	0
173.0.50.0/24	20	2	3
173.0.70.0/24	18	2	0
173.0.110.0/24	24	3	2
157.0.130.0/24	20	5	0
157.0.60.0/24	23	2	5
157.0.30.0/24	20	1	8
112.0.70.0/24	21	2	0
112.0.40.0/24	21	2	1
112.0.20.0/24	20	2	0
Total	435	45	33

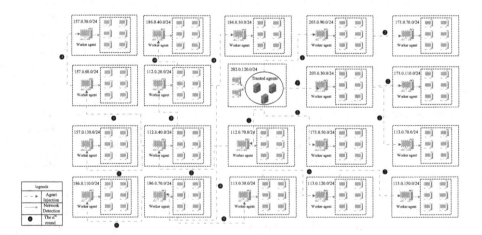

Fig. 10. Example experimental process.

a node that takes more bandwidth plays a more important role in the whole network. And in our experimental environment, nodes with potential vulnerabilities tend to provide more services and generate more traffic. Secondly, initialize the decision matrix A, as formula (1) depicts. Thirdly, use formula (2) and (3) to map the value of each element into the interval [0, 1]. Next, initialize the decision matrix of the selection of worker agents, as is shown in Table 2. In Table 2, the terms are abbreviations: Transmission Rate (TR), Operating System (OS), Round-Trip Time (RTT), Open Ports (OP), Packet Loss probability (PL), Consistency Index (CI), Consistency Ratio (CR). The weight vector is denoted as ω.

Table 2. Decision matrix of the optimal agent selection problem.

	TR	OS	RTT	OP	PL
TR	1	1/2	4	3	3
OS	2	1	7	5	5
RTT	1/4	1/7	1	1/2	1/3
OP	1/3	1/5	2	1	1
PL	1/3	1/5	3	1	1

$\lambda_{max} = 5.073, CI = 0.018, CR = 0.016, \omega = (0.263, 0.475, 0.055, 0.090, 0.110)^T$

Also, initialize the decision matrix of the OS in targeted devices, as is shown in Table 3. In Table 3, six operating systems are denoted as specific attributes.

Table 3. Decision matrix of the OS detected in the target device.

	Windows	Ubuntu	Kali	CentOS	Kylin	MacOS	Others
Windows	1	2	3	4	3	1	1
Ubuntu	1/2	1	3	7	4	1/2	1/2
Kali	1/3	1/3	1	5	3	1/3	1/3
CentOS	1/4	1/7	1/5	1	1/5	1/7	1/7
Kylin	1/3	1/4	1/3	5	1	1/3	1/3
MacOS	1	2	3	7	3	1	1
Others	1	1/2	3	7	3	1	1

$CI = 0.049, CR = 0.037, \omega = (0.577, 0.175, 0.103, 0.014, 0.073, 0.041, 0.017)^T$

4.3 Comparisons and Discussions

We employed a test network environment to experimentally evaluate the efficacy of the group decision-based network topology sensing method. In addition, we conducted comparative tests on three other approaches, specifically:

1) single-point network detection without GDM; 2) active network detection; and 3) passive network detection. Both the active and passive network detection approaches utilize the GDM policy to select worker agents in each new round. In contrast, the single-point network detection approach with GDM policy only considers the most recently discovered subnet when selecting worker agents in each subsequent round. This implies that in every round, only nodes from the previously identified subnet are chosen as the newly generated worker agents.

The basic descriptive information of these methods is shown in Table 4. As a supplementary note, active prober means that agents can take active detection in their subnet, using tools such as NMAP. Passive sniffer means that agents can take passive detection in their subnet with tools like p0f or TCPDUMP. SNMP topology collector means that the routers are equipped with SNMP protocol. Recently discovered policy means that trusted agents select nodes that are recently discovered as worker agents in the next round. Trusted agents means that the topology sensing method uses trusted agents as the center agents that controls the whole process of topology discovery. In order to better illustrate the effectiveness of the collaborative topology awareness method based on GDM proposed in this paper, we control whether active and passive detection cooperation is adopted and whether GDM strategy is adopted. All the methods use the same task duration limit.

Table 4. Decision matrix of the optimal agent selection problem.

	Simple point decision	Active detection	Passive detection	Proposed method
Active prober	✓	✓	✗	✓
Passive sniffer	✓	✗	✓	✓
SNMP topology collector	✓	✓	✓	✓
Recently discovered policy	✓	✗	✗	✗
Proposed GDM policy	✗	✓	✓	✓
Trusted agents	✓	✓	✓	✓
Round duration limit	5 min	5 min	5 min	5 min

We present in Fig. 11 a comparative analysis of the time consumption and the number of discovered nodes, using four different techniques. As the duration of network topology sensing is variable, we performed 10 experiments for each methodology. In Fig. 11 (i), the single point decision approach employs a novel policy to determine the optimal worker agent in each round. Although it has a higher coverage rate, it requires more time. In Fig. 11 (ii), the active detection method uses the GDM policy to select multiple worker agents in each round of decision-making, resulting in high and stable coverage rates with reduced time consumption. In Fig. 11 (iii), the passive detection method also employs the GDM policy to select more than one worker agent in each decision round; however, it exhibits an unpredictable coverage rate and varied time consumption due to its inability to detect nodes with no network traffic during a specific time frame. Finally, our proposed technique, as shown in Fig. 11 (iv), combines the active and passive detection methods, delivering a higher coverage rate compared to the rest while surpassing other methods in terms of stability.

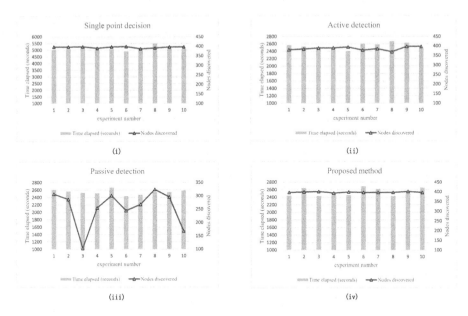

Fig. 11. Time cost and nodes discovered of different methods: (i) single point decision method without GDM; (ii) active detection method with GDM; (iii) passive detection method with GDM; (iv) our proposed method with GDM.

Figure 12 (i) illustrates the comparative performance of all experimental methods, with our proposed technique displaying the most significant coverage rate. In our testing environment, the proposed method discovers nearly all the reachable nodes with comparable time consumption to the active and passive detection methods. Figure 12 (ii) illustrates the duration of network topology sensing for all four methods. We collected the average transmission rate using the SAR tool in the trusted agent subnet "203.0.120.0/24." This allowed us to observe the transmission rate's impact attributed to different topology discovery techniques. As presented in Fig. 13, the passive detection method exhibits the lowest value since it does not actively create probes into the network apart from communication between trusted and worker agents. The active detection approach delivers a marginally reduced transmission rate compared to our proposed technique, which actively forms probes. It is observed that the single point decision technique leads to the highest values and consumes the most bandwidth resources. However, there are still certain limitations: (i) the communication between trusted agents and worker agents incurs some bandwidth costs; (ii) worker agents are needed to obtain topology information. Despite these limitations, our work represents a meaningful attempt, and further research is warranted.

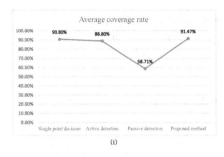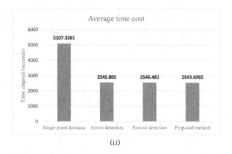

(i)   (ii)

Fig. 12. Average coverage rate and time cost.

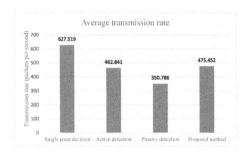

Fig. 13. Average transmission rate.

5   Conclusions

In this paper, we address the challenge of network situation collection in non-cooperative networks where a comprehensive view of the topology and detailed information is unavailable. We propose a network topology sensing technique based on GDM.

Acknowledgements. We like to thank to Yujun Li, Minrong Xie, and Yiqiang Zhou for their guidance. The work presented in this paper is supported by the Innovation Funding Plan by Topsec Network Technology Inc.

References

1. Håvard, J.O., Sokratis. K.: Understanding situation awareness in SOCs, a systematic literature review. Comput. Secur. **126**, 103069 (2023)
2. Chen, C., Ye, L., Yu, X.Z., Ding, B.L.: A survey of network security situational awareness technology. In: International Conference on Artificial Intelligence and Security. LNCS, pp. 101–109. Springer, New York (2019)
3. Endsley, M.R.: Design and evaluation for situation awareness enhancement. In: Proceedings of the Human Factors Society Annual Meeting, pp. 97–101. (1988)
4. Bass, T.: Multi-sensor data fusion for next generation distributed intrusion detection systems. In: 1999 IRIS National Symposium Draft (1999)

5. Gutzwiller, R., Dykstra, J., Payne, B.: Gaps and opportunities in situational awareness for cybersecurity. Digital Threats: Res. Pract. **1**(3), Article 18. (2020)
6. Li, D., Hu, Y.K., Xiao, G.Q., Duan, M.X., Li, K.L.: An active defense model based on situational awareness and firewalls. Concurr. Comput.: Pract. Exper. **35**(6), e7577 (2023)
7. Endsley, M.R.: Toward a theory of situation awareness in dynamic systems. Human Factors: J. Human Factors Ergonom. Society. **37**(1), 32–64 (1995)
8. Liu, Z.T., Ding, G.R., Wang, Z., Zheng, S.L., Wu, Q.H.: Cooperative topology sensing of wireless networks with distributed sensors. IEEE Trans. Cogn. Commun. Network. **7**(2), 524–540 (2021)
9. Liu, Z.T., Wang, W., Ding, G.R., Wu, Q.H., Wang, X.B.: Topology sensing of non-collaborative wireless networks with conditional granger causality. IEEE Trans. Netw. Sci. Eng. **9**(3), 1501–1515 (2022)
10. Mohd, W.R.W., Abdullah, L.: Aggregation methods in group decision making: a decade survey. Informatica **41**(1), 71–86 (2017)
11. Zhao, M., Kou, D., Li, L., Lin, M.W.: An incomplete probabilistic linguistic multi-attribute group decision making method based on a three-dimensional trust network. Applied Intelligence. **53**, 5029-5-047 (2023)
12. Zhou, Y.Y., Zheng, C.L., Zhou, L.G., Chen, H.Y.: Selection of a solar water heater for large-scale group decision making with hesitant fuzzy linguistic preference relations based on the best-worst method. Appl. Intell. **53**, 4462–4482 (2023)
13. Morente-Molinera, J.A., Morfeq, A., AI-Hmouz, R., Ashary, E.B., Su, J.F., Herrera-Viedma, E.: Introducing disruption on stagnated group decision making processes using fuzzy ontologies. Appl. Soft Comput. **132**, 109868 (2023)
14. Gong, G.C., Li, K., Zha, Q.B.: A maximum fairness consensus model with limited cost in group decision making. Comput. Indust. Eng. **175**, 108891 (2023)
15. Liu, X., Zhang, Y.Y., Xu, Y.J., Li, M.Q., Herrera-Viedma, E.: A consensus model for group decision-making with personalized individual self-confidence and trust semantics: a perspective on dynamic social network interactions. Inf. Sci. **627**, 147–168 (2023)
16. Bhattacharyya, S., Kalaimani, R.K.: Resilient dynamic average consensus based on trusted agents. arXiv:2303.09171 (2023)
17. Sangster, P., Narayan, K.: PA-TNC: a posture attribute protocol compatible with trusted network connect. RFC 5792 (2010)
18. Mountasser, I., Ouhbi, B., Hdioud, F., Frikh, B.: Semantic-based big data integration framework using scalable distributed ontology matching strategy. Distrib. Parallel Databases. **39**, 891–937 (2021)
19. Júnior, P.S.S., Barcellos, M.P., Falbo, R.D.A., Almeida, J.P.A.: From a scrum reference ontology to the integration of applications for data-driven software development. Inf. Softw. Technol. **136**, 106570 (2021)
20. Grošelj, P., Dolinar, G.: Group AHP framework based on geometric standard deviation and interval group pairwise comparisons. Inf. Sci. **626**, 370–389 (2023)
21. Fan, R.L., Zhang, H.L., Gao, Y.: The global cooperation in asteroid mining based on AHP, entropy and TOPSIS. Appl. Math. Comput. **437**, 127535 (2023)

Distributed Task Offloading Method Based on Federated Reinforcement Learning in Vehicular Networks with Incomplete Information

Rui Cao[1], Zhengchang Song[1], Bingxin Niu[1($\boxtimes$)], Junhua Gu[1], and Chunjie Li[2]

[1] School of Artificial Intelligence & Hebei Province Key Laboratory of Big Data Calculation, Hebei University of Technology, Tianjin, China
`niubingxin666@163.com`
[2] Hebei Expressway Group Limited, Shijiazhuang, China

Abstract. With the development of Internet of Vehicles, a large number of different types of applications have emerged. However, given the limited computing power of vehicles, many tasks cannot be completed within specified time. The task offloading method based on mobile edge computing (MEC) effectively solves the problem. However, the existing research does not fully consider user needs and has the disadvantages of limited application scope and slow training speed, so it is not suitable for high-speed vehicle scenarios. Considering the heterogeneity of tasks and the different needs of users, this paper first designs a dynamic weighting method for delay and vehicle energy consumption, and then proposes a distributed algorithm FDQN based on Deep Q-Network (DQN) and federated learning. The algorithm does not require any information about the base station and can utilize user collaboration to achieve fast convergence. The experimental results show that our algorithm can achieve best optimization effect under different vehicle energy and speeds.

Keywords: Task offloading · Federated learning · Vehicular network · Deep reinforcement learning

1 Introduction

The advancement of connected vehicle technology has spurred the creation of various types of applications, such as autonomous driving and virtual reality. However, these applications often require tasks to be completed within strict time constraints, which is often unfeasible due to the limited computing power of vehicles [1]. In the past, delivering tasks to cloud servers for computation was a common approach. However, this method is no longer sustainable due to the exponential increase in data volume, leading to unpredictably long transmission delays and placing significant demands on bandwidth resources. Task offloading method based on MEC has been an effective approach to overcome these challenges by offloading tasks to nearby edge servers [2]. This method can satisfy computing delay requirements while ensuring the reliability of results. Nevertheless, task offloading is still a challenging problem due to the following reasons:

(1) Uncertainty of the edge servers' computational capabilities: Users often have several optional nodes for task offloading, but the computing power of these nodes is unknown to the user. If the information is obtained by means of communication, it will bring additional delay.

(2) Environmental dynamics: The time-varying nature of the channel, the mobility characteristics of the vehicle, and the heterogeneity of the task will make it difficult to make offloading decisions.

(3) Limitations of vehicle energy consumption: Low-carbon travel is the development trend of the automotive industry in the future. For new energy vehicles, limited battery capacity and endurance should also be considered in the algorithm. For emergent tasks, priority should be given to ensuring that tasks can be completed within the specified time. For secondary tasks, it is necessary to balance delay and energy consumption as much as possible.

Firstly, different applications have different delay tolerances and different users have different travel needs. Therefore, simply minimizing the delay and energy consumption cannot satisfy the Quality of Service (QoS) of each user. Secondly, most of the existing methods require the information of edge servers, but this information is difficult to obtain accurately and timely. And these methods based on deep reinforcement learning are only effective within a limited range. Once a vehicle enters a new resource scenario, the algorithm needs to relearn the characteristics of the environment, and the previous knowledge information cannot be utilized, resulting in a decrease in algorithm performance. To address these issues, this article proposes a distributed task offloading strategy based on DQN and federated learning, aiming to minimize vehicle energy consumption while meeting task success rates. This method combines deep reinforcement learning and federated learning methods, utilizing the DQN algorithm to learn information about the current environment from historical feedback, and utilizing federated learning to share the knowledge of each vehicle user, allowing multiple users to participate in model training. At the same time, each region saves a copy of the latest model parameters for new vehicles to reference and use. The major contributions of our paper are as follows:

(1) We design a dynamic reward function for delay and energy consumption to guide agents to implement flexible offloading strategies.

(2) We propose FDQN algorithm based on federated reinforcement learning, which can converge quickly without relying on any environmental knowledge.

(3) Simulation results show that FDQN algorithm can effectively reduce task delay and vehicle energy consumption.

2   Related Work

The difference between the vehicular networks and other networks is the dynamic network topology caused by vehicle mobility. Therefore, most of the current research focuses on the mobility of vehicles and the dynamics of resources. In addition, the development of vehicular networks has also created a large number of service requirements. While designing the algorithm, we must also consider the various needs of users.

Reference [3] proposes a collaborative multi-agent deep reinforcement learning algorithm from the perspective of mobile network operators. Reference [4] studied the joint optimization of computation offloading and resource allocation in dynamic multi-user MEC systems, and proposed a reinforcement learning (RL) method based on value iteration. Reference [5] considered the task offloading problem in the data-driven scenario. Based on the idea of asynchronous advantage actor-critic (A3C) and DQN, the author proposed an asynchronous ADQN algorithm for task offloading. The model uses asynchronous training to make the global model converge quickly. Reference [6] considered how to make decisions in ultra-dense networking scenarios with time-varying channel conditions. The author takes delay and energy consumption as optimization objectives and proposed a masked-DQN algorithm. Reference [7] considered the privacy problem of users in partial task offloading, and proposed an adaptive task offloading algorithm based on DRL and differential privacy technology. Reference [8] considered the dynamic optimization requirements of different applications between service delay and quality loss of results, and proposed an event-triggered dynamic task allocation framework based on linear programming optimization and binary particle swarm optimization. This scheme can adjust the service delay sensitivity and quality sensitivity according to the actual needs. In order to defend security threats (such as sniffing and tampering) in task offloading, Reference [9] proposed a security and cost-aware task offloading strategy, and used DQN to find the optimal solution of the problem. Reference [10] studied the offloading mechanism of cloud edge collaborative mobile computing for industrial networks, and proposed a cloud-edge collaborative mobile computing method based on DRL. In order to make full use of vehicle resources, Reference [11] divided the task into multiple small tasks, and realized distributed task offloading based on DQN algorithm. Reference [12] proposed a cloud based layered vehicle edge computing offloading framework, and the author designed an optimal multi-stage offloading scheme using Stackelberg game theory. Reference [13] considered the issue of task workload and proposed a tuple scheduler to reduce queue backlog. Reference [14] proposed a new performance metric-coflow age (CA) for coflows generated by distributed streaming applications. Reference [15] focused on the problem of jointly considering endpoint placement and coflow scheduling to minimize the average CCT of coflows across geo-distributed datacenters. Reference [16] studied the problem of optimizing deadline and non-deadline coflows simultaneously. The author presented a new optimization framework to schedule deadline coflows to minimize and balance their bandwidth footprint, such that non-deadline coflows can be scheduled as early as possible.

In summary, the current research considers multiple optimization requirements such as delay, energy consumption, cost, and result quality, but generally adjusts the weight in a constant weighted manner, which ignores the heterogeneity of tasks and cannot accurately meet the optimization needs of users. Some scholars consider adjusting the weight of delay by task priority [17], but there is no clear standard for the division of task priority. Secondly, most algorithms require the relevant information of the environment node as the input of the algorithm, but this information is difficult to obtain in the actual scene. Reference [18] proposed a task offloading method based on DRL, which does not require any instantaneous channel state information or prior knowledge of the computing power of the base station. Reference [19] considered the task offloading problem in the scenario of unknown edge information and user coexistence, and used the multi-armed bandit method to solve it. But these methods are only effective in a limited area and both require a long iterative process, so they are not suitable for high-speed vehicle scenarios.

In order to solve above problems, this paper proposes a dynamic weighting method to adjust the impact of task delay and vehicle energy consumption, and then implements a distributed offloading algorithm FDQN based on DQN and federated learning. This algorithm does not require information about the channel or computing power of the base station, and can reduce delay and energy consumption effectively.

3   System Model

We assume that a number of small-cell base stations (SBSs) and macro-cell base stations (MBSs) are deployed in an area. These base stations are connected by optical fibers. Each base station is equipped with an edge server. The SBS is responsible for task offloading and the MBS is the manager of the region, as shown in Fig. 1. Suppose that the set of SBS servers is $\mathcal{M} = \{1, 2, \ldots M\}$, each server is defined as m and $m \in \mathcal{M}$. The characteristics of each server can be described by the tuple (f_m, b_m), which represents the computing power and wireless bandwidth of the edge server m, respectively. We discretize time into a set of equal-interval time slots, denoted as $\mathcal{T} = \{1, 2, \ldots, T\}$, and each time slot $t \in \mathcal{T}$.

It is assumed that the vehicle u runs on the road at a speed of v and its position is (x_t, y_t) at time slot t. The vehicle carries a number of tasks and can choose to calculate its tasks locally or offload partial task to the edge server. However, due to the limitation of communication range, the vehicle can only choose the SBS within the signal coverage. The set of tasks is $\mathcal{R} = \{1, 2, \ldots, R\}$. Each task $r \in \mathcal{R}$ and can only be processed in order. Task r has three characteristics (D_r, C_r, ϕ_r). D_r represents the data size of task r, C_r represents the required CPU cycles of task r, and ϕ_r is the maximum tolerable delay of task r. The offloading action corresponding to task r is defined as a_r. When the task r is offloaded to server m, there is $a_r = m$, and $a_r^m = 1$, otherwise $a_r^m = 0$. If the task is calculated locally, the variable $a_r = 0$ and $a_r^l = 1$.

The total delay of the task includes three parts: transmission time, computation time and additional time. If the task is completed at the server m and

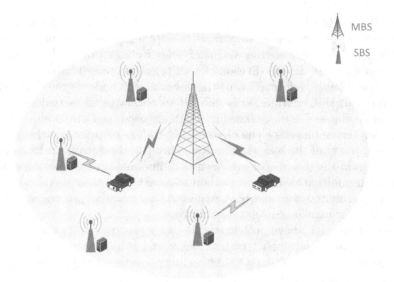

Fig. 1. Illustration of computation offloading scenario

the vehicle has left the coverage of the SBS, several relay MBSs are required to transmit the calculation results of the task, and additional processing time is required in this process. The total energy consumption of the task includes the transmission energy consumption sent to the base station and the local computing energy consumption. We model the local computing and edge computing respectively.

3.1 Local Computing

When the task r is calculated locally, assuming that f_{local} is the CPU frequency of the vehicle, the calculation time is

$$T_{local} = \frac{C_r}{f_{local}} \tag{1}$$

According to [20], the local computing energy consumption in this process is

$$E_{local} = \kappa f_{local}{}^2 C_r \tag{2}$$

where κ is the effective switched capacitance depending on chip architecture of the device [21].

3.2 Edge Computing

When the task r is offloaded to server m and the offloading ratio is θ_r, the transmission time can be defined as T_{trans}, which can be calculated as:

$$T_{trans} = \frac{\theta_r \cdot D_r}{R_{mu}(t)} \tag{3}$$

$R_{mu}(t)$ is the channel data rate between the vehicle u and m, which can be calculated as:

$$R_{mu}(t) = b_m \log_2 \left[1 + \frac{\rho_0 p_r}{\sigma^2 [d_m(t)]^\beta} \right] \qquad (4)$$

where ρ_0 is the channel power gain at this distance, β is the path-loss index, p_r is the transmit power, σ^2 is the power noise, and $d_m(t)$ is the distance between the vehicle and m at time t.

Define the computing resource allocated to the task r by server m as f_m^r, and the calculation time T_{cmp} can be calculated as:

$$T_{cmp} = \frac{\theta_r \cdot C_r}{f_m^r} \qquad (5)$$

If the vehicle has driven out of the communication range of the m when the task calculation is completed, there is a variable $z_r = 1$, otherwise $z_r = 0$. At this time, the calculation results need to be returned by several MBSs as relay points. Assuming that the number of relay points is k, the additional communication time is:

$$T_{ext} = k \cdot \lambda \qquad (6)$$

λ is the turnaround time of data in the MBS, and its value is a fixed constant. Since the calculation results of the task are often much smaller than the size of the original data, the delay of the calculated results transmitted back to the vehicle is not considered [22,23].

The offloading delay of the task can be expressed as the maximum value of the local computing delay and the offloading delay:

$$T_{off} = max \left(T_{trans} + T_{cmp} + z_r \cdot T_{ext}, \frac{C_r \cdot (1 - \theta_r)}{f_{local}} \right) \qquad (7)$$

Offloading tasks to the edge server will bring energy loss. Energy consumption is related to transmission time and transmission power. Assuming that the vehicle is equipped with multiple transmitters with different powers, the transmit power of the selected transmitter is p_r, the energy consumed by task r can be calculated as:

$$E_{off} = p_r T_{trans} + \kappa \cdot f_{local}^2 \cdot C_r \cdot (1 - \theta_r) \qquad (8)$$

Combining the above two cases, the total offloading delay of task r is:

$$T_r = \sum_{\forall m \in \mathcal{M}} a_r^m T_{off} + a_r^{local} T_{local} \qquad (9)$$

The total energy consumption is:

$$E_r = \sum_{\forall m \in \mathcal{M}} a_r^m E_{off} + a_r^{local} E_{local} \qquad (10)$$

Assuming that the residual energy of the vehicle is E_{res}, the energy consumption of task r can't exceed the residual energy, so the constraint condition can be expressed as:

$$E_r < E_{res} \tag{11}$$

The vehicle updates the remaining energy at the moment after each task offloading:

$$E_{res} \leftarrow E_{res} - E_r \tag{12}$$

4   Problem Formulation

Due to the high dynamic characteristics of the vehicular network, the impact of delay and energy consumption cannot be weighted simply. We need to adjust the weight according to the environmental characteristic [24,25].

Suppose that the vehicle's task to be calculated is r,and the maximum tolerable delay is ϕ_r. We use the vehicle speed and the maximum tolerable delay of the task to prioritize the task. The weight of delay can be calculated as:

$$W_r^l = \frac{\alpha}{\phi_r} \cdot v \tag{13}$$

where α is a scaling factor to adjust the weight. Next, we define the weight of energy consumption. Taking the new energy vehicle as an example, we assume that the maximum mileage of the vehicle u under full power is L_u, the remaining distance from the destination is S_u, the expected power consumption is roughly expressed as E_p, and the power used for task offloading is E_{ext}, which can be calculated as:

$$E_p = \frac{S_u}{L_u} \tag{14}$$

$$E_{ext} = E_{res} - E_p \tag{15}$$

When E_{ext} is relatively sufficient, we use a constant N to weight the energy consumption. At this time, the delay is the main optimization target. When E_{ext} is less than the threshold H, we weight it in logarithmic form. At this time, the weight of energy consumption is increased rapidly. The weight of energy consumption can be calculated as:

$$W_r^e = \begin{cases} \omega \cdot N, & E_{ext} > H \\ -\omega \cdot \ln(E_{ext}), & E_{ext} \leq H \end{cases} \tag{16}$$

where ω is the weight coefficient, N is a constant, H is the energy consumption threshold and satisfies $H = e^{-N}, H \in (0,1)$. Users can flexibly adjust the size of H and ω to achieve personalized energy consumption bias. Therefore, the optimization objective OV can be expressed as:

$$OV = \sum_{\forall r \in \mathcal{R}} W_r^l \times T_r + W_r^e \times E_r \tag{17}$$

The final optimization problem is defined as:

$$\min_{a_r^l, a_r^m, \theta_r} \frac{OV}{|\mathcal{R}|}$$

$$
\begin{aligned}
s.t. \quad & C1: && E_r < E_{res}, \forall r \in \mathcal{R} \\
& C2: && a_r^l + \sum_{\forall m \in \mathcal{M}} a_r^m \leq 1, \forall r \in \mathcal{R} \\
& C3: && a_r^l \in \{0,1\}, \forall r \in \mathcal{R} \\
& C4: && a_r^m \in \{0,1\}, \forall r \in \mathcal{R} \\
& C5: && \theta_r \in \{\theta_1, \theta_2, \ldots, \theta_n\} \\
& C6: && T_r \leq \phi_r, \forall r \in \mathcal{R}
\end{aligned}
\tag{18}
$$

The C1 is the energy constraint. The C2 is the offloading action constraint and the vehicle can only select one offloading action. The C3 and C4 constrain the range of the action, and C5 indicates that the offloading ratio can only take the optional discrete value to reduce the complexity of the problem. The constraint condition C6 indicates that the task must be completed within the specified time.

5   Algorithm Design

This section first introduces the principle of DQN algorithm, and then introduces the basic components of the FDQN algorithm, which including state space, action space, reward function, communication mask and the training process.

5.1   Deep Q-Network

Traditional dynamic programming method or heuristic algorithms can solve the above problem, but these methods have great complexity and can not meet the real-time requirements of vehicle applications. With the continuous development of artificial intelligence, deep reinforcement learning has gradually become a popular method to solve such problems.

The most typical reinforcement learning algorithm is the Q-learning algorithm. The Q-learning algorithm selects the optimal action by learning the value function in each state. However, as the dimension of the state space and the action space increase, the algorithm will be very time-consuming. In order to solve this problem, scholars have begun to use neural networks instead of Q-table, so the DQN algorithm was born.

The DQN algorithm includes four parts: agent, environment, action and reward. Firstly, the agent obtains the state space s_t according to the environment, inputs s_t into the neural network Q, and then selects the most valuable action a_t according to the ϵ-greedy strategy. The environment will feedback a reward r_t according to the action a_t, and then update the state s_{t+1}. Finally, (s_t, a_t, r_t, s_{t+1}) is saved to the experience pool D, and the DQN model parameter weights are updated by experience. The training process of the DQN algorithm can be expressed as Algorithm 1.

Algorithm 1. DQN Algorithm

Input: Prediction network Q with weights w, target network Q' with weights w', environment state s_t, experience pool D;

Output: Prediction network and target network;

1: for each episode e do;
2: Initial time slot $t = 0$;
3: Initializing the environment state s_t;
4: Taking s_t as the input of the current Q network, and using the ϵ-greedy method to select the action a_t;
5: Perform action a_t under state s_t, get new state s_{t+1}, reward r_t and end flag *done*;
6: Put $\{s_t, a_t, r_t, s_{t+1}\}$ into the experience pool D;
7: Sampling q samples from D, then calculating the target value y_j.

$$y_j = \begin{cases} r_j & done = True \\ r_j + \gamma \max_{a'} Q'\left(s_{j+1}, a'; w'\right) & done = False \end{cases}$$

8: Constructing loss function $\frac{1}{q}\sum \left(y_j - Q(s_j, a_j; w)\right)^2$, using back propagation algorithm to update the parameter w of Q network;
9: If $t \% C = 0$, update the target Q network weights $w' \leftarrow w$;
10: If $done = True$, $e \leftarrow e + 1$, go to step 2;
11: $s_t \leftarrow s_{t+1}$, $t \leftarrow t + 1$, go to step 4;
12: end for

5.2 State Space

In order to avoid dependence on global information, we learn environmental knowledge through historical offloading feedback of the vehicle. In order to guide the agent to choose a better decision, we use the offloading information and feedback results of the two latest historical tasks as a reference, and define the state space as $S_u(t)$, which consists of $S_u^1(t), S_u^2(t), S_u^3(t), S_u^4(t)$. The specific definition is as follows:

(1) Historical task information $S_u^1(t) = \{D_{r-2}, D_{r-1}, C_{r-2}, C_{r-1}, W_{r-2}^l, W_{r-1}^l, W_{r-2}^e, W_{r-1}^e\}$

(2) Current task Information $S_u^2(t) = \{D_r, C_r, W_r^l, W_r^e\}$

(3) Current vehicle information $S_u^3(t) = \{x_t, y_t, S_u, L_u, E_{res}\}$

(4) Historical action and feedback $S_u^4(t) = \{a_{r-2}, a_{r-1}, T_{r-2}, T_{r-1}, E_{r-2}, E_{r-1}\}$

5.3 Action Space

The action space is defined as A, which includes four parts: local offloading action, edge offloading action, optional sending devices and offloading ratios. Assuming that the vehicle has π different power sending devices and τ optional offloading ratios, then $A = \{local\} + \{a_{m1}, a_{m2}, \ldots, a_{m|\mathcal{M}|}\} \times \{dev_0, dev_1, \ldots, dev_\pi\} \times \{ratio_0, ratio_1, \ldots, ratio_\tau\}$.

5.4 Reward Function

The agent performs action $a'_r \in A$ according to the state $S_u(t)$, and the environment will feedback a reward $R_u(t)$ to evaluate the action and guide the network to update the parameter weights. When the task is completed within the specified time, we use the negative number of the target value as the reward, otherwise return a smaller constant $-G$ as the penalty, so there is

$$R_u(t) = \begin{cases} -(W^l_r \times T_r + W^e_r \times E_r), & \text{if success} \\ -G, & \text{otherwise} \end{cases} \tag{19}$$

5.5 Communication Mask

Because of the limited communication range and computing power of the base station, MBS will issue a mask to each vehicle according to the change of the environment, and the vehicle u cannot offload tasks to the base station outside the communication range. Therefore, the mask can be defined as $M_u(t) = \{m_1, m_2, \ldots, m_{|A|}\}$, where the mask corresponding to the optional action is 1, and the rest is 0. In addition, the stronger the computing power of the node, the more prone to resource competition. So the corresponding action masks of the optional node $i (i \in \mathcal{M})$ can also be set to 0 according to a certain probability P_i, which can be defined as:

$$P_i = \frac{\delta_i \cdot f_i}{\sum_{\forall q \in \mathcal{M}} f_q} \tag{20}$$

δ_i is a scaling factor whose value can be set according to the vehicle density in the actual environment. Assuming that the action-value function is $Q(t)$, so its output after the mask operation is $Q'(t) = \{Q_1 m_1, Q_2 m_2, \ldots, Q_{|A|} m_{|A|}\}$.

5.6 FDQN Algorithm

Due to the special definition of state space and the limited communication range of the vehicle, the training of the algorithm requires frequent interaction with the environment, which leads to extremely low learning efficiency. Federated Learning is a new technology of artificial intelligence. It can use multiple computing nodes to carry out efficient machine learning under the premise of protecting personal data privacy [26]. Inspired by this idea, we propose an offloading algorithm FDQN based on DQN and federated learning, as shown in Fig. 2.

We integrate the mechanism of federated learning into the DRL algorithm to improve its effectiveness. The specific process is as follows: we assume that each MBS acts as a manager within its communication range and is responsible for the aggregation and distribution of the global prediction network. Each user can freely download the global network or upload the local network parameter weights. Every user uses action feedback and the experience pool to train local prediction networks based on their own environment. When a certain number of training steps are met, the user uploads the local network parameter weights and the latest reward value to the MBS. The MBS will regularly aggregate the collected networks according to the reward value uploaded by the user as a weight and update the global network parameters. When there is a new global network version, the user can choose to download the global network parameters to update the local prediction network and continue training based on this. In the subsequent process, when new users come to this range, they can download the latest global network and continue training based on previous work, thus effectively improving the training speed of the DRL algorithm and achieving "hot start". In summary, the FDQN algorithm procedure can be expressed as Algorithm 2.

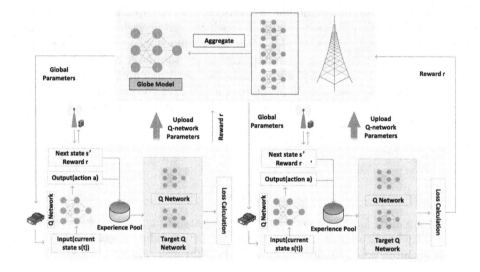

Fig. 2. FDQN procedure.

In addition, in order to make the model adapt to the environment changes (such as a node is unavailable or hardware changes) in a timely manner, the entire federated learning process is lifelong, and each client must regularly upload local network parameters to update the global network.

6   Performance Evaluation

In this section, we first introduce the simulation setup and baselines. Then we evaluate the performance of our algorithm through a large number of simulations.

6.1   Simulation Setup

We simulate a 2 km × 2 km area with 36 SBSs and 1 MBS. The computing power of the edge server obeys the uniform distribution of [5,15] GHz and each

Algorithm 2. FDQN algorithm

Input: Global prediction network parameters Q_g, local prediction network parameters Q_u, local target network parameters Q'_u, vehicle set $\mathcal{U}$, current vehicle $u \in \mathcal{U}$, experience pool D, upload threshold $Th1$ and update threshold $Th2$;

Output: Local prediction network and target network;

1: When the vehicle u enters the MBS range, download Q_g as the initial weights of the local network, $Q_u = Q_g, Q'_u = Q_g$;

2: Initialization $step = 0$;

3: For r in $\mathcal{R}$;

4: The best action a'_r is obtained by using the local prediction network Q_u, mask $M_u(t)$ and state space $S_u(t)$;

5: The user offloads the task according to the best action, and the edge server returns the result after completing the calculation;

6: Get the reward value R_u according to the reward function, then save and update the experience pool D, $step \leftarrow step + 1$;

7: Using experience playback method to update local network models Q_u and Q'_u.

8: If $step \% Th1 = 0$, the user uploads the latest reward R_u and local prediction network parameter Q_u. The MBS aggregates multiple local models according to the reward value as a weight every certain time slot. The calculation method is as follows:

$$w_u = \frac{R_u^{-1}}{\sum\limits_{\forall u \in \mathcal{U}} R_u^{-1}}$$

$$Q_g = \sum\limits_{\forall u \in \mathcal{U}} w_u \cdot Q_u$$

9: If the global model is updated, then $Q_u \leftarrow Q_g$;

10: If $step \% Th2 = 0$, update target network $Q'_u \leftarrow Q_u$;

11: Update the state space $S_u(t) \leftarrow S_u(t+1)$;

12: End for;

---

server allocates resources equally according to the number of tasks received. The turnaround time in the MBS is 50 ms, the path-loss index $\beta = 4$, the channel bandwidth of the base station is set to 10 MHz, and the noise power $\sigma^2 = 2 \times 10^{-13} W$.

For the user's transmit device, the available transmit power is 150 mW or 300 mW. We define three levels of offloading ratios: 0.33, 0.66 and 1. The battery capacity E_{max} is 100% and the threshold H is 0.37. The input data size $D_r \in [0.3, 4]$Mbits and the required CPU cycles $C_r \in [2, 10]$GHz/task, the maximum tolerable delay $\phi_r \in [200, 1000]$ ms.

In addition, we set the size of the experience pool to 2000, the batch size to 128, the discount factor to 0.9, the learning rate to 0.01, the exploration probability to 0.05, the update frequency of the target network parameters to 200. The number of user clients is 20, and their models are trained for 600 episodes in parallel. After each episode, the vehicle's speed, location, initial power, remaining mileage and other information are changed randomly to cover different vehicle states, and the global model(FDQN) are updated and evaluated. The baseline algorithm includes greedy algorithm(Greedy), random offloading algorithm(Random), and local computing algorithm(Local). The greedy algorithm will choose to offload all tasks to the nearest edge server. The random offloading algorithm uses a completely random method to offload the task, and the local computing algorithm will process all tasks locally. Considering the heterogeneity of vehicles and tasks, we use weighted delay and energy consumption to measure the effectiveness of the algorithm.

6.2 Evaluation Results

We choose Greedy, FDQN, DQN and A3C-MEC [18] to compare their convergence effects. The experimental training curve is shown in Fig. 3.

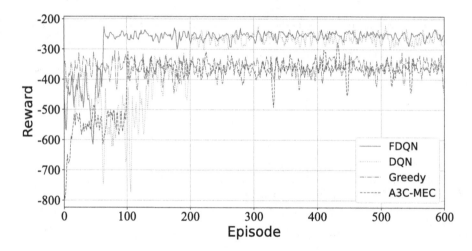

Fig. 3. Convergence performance.

It can be seen that the convergence speed of the FDQN are better than the DQN algorithm. When the global model converges, subsequent vehicle users can download the global model directly, thus avoiding the tedious initial training process. The Greedy algorithm is superior to the FDQN algorithm at the beginning, but the effect is not as good as the converged FDQN algorithm. The A3C-MEC algorithm has good performance in ultra dense networking environments, but its performance is poor in simulation environments with sparse networking in this article. Its optimization effect is basically the same as that of Greedy. Therefore, the training method of multi vehicle collaborative learning is significantly superior to single vehicle independent learning. When the initial energy of the vehicle changes, the delay and energy consumption of different algorithms are shown in Fig. 4 and Fig. 5.

Figure 4 compares the delay of the algorithm under different initial energy. It can be found that our algorithm is obviously better than the other three

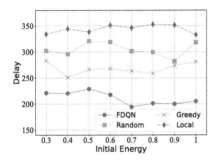

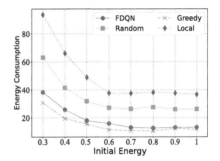

Fig. 4. Task offloading delay under different vehicle energy.

Fig. 5. Energy consumption under different vehicle energy.

offloading algorithms. Figure 5 compares the energy consumption of different algorithms. As the vehicle energy increases, the weight of energy consumption gradually decreases, so the energy consumption of all algorithms gradually decreases. The Greedy will send all the tasks to the nearest base station for calculation, so the vehicle energy consumption is the lowest. However, the nearest base station may not be the most suitable node, so the total delay of tasks cannot be guaranteed to be the shortest. However, the FDQN algorithm can achieve the lowest delay, and the energy consumption is only slightly higher than the Greedy.

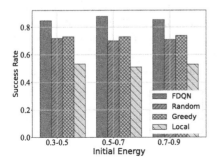

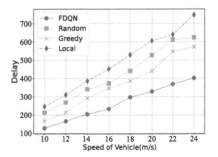

Fig. 6. Number of successful tasks under different vehicle energy.

Fig. 7. Task offloading delay at different vehicle speeds.

Figure 6 compares the task success rate of different algorithms and the FDQN algorithm has the most successful tasks. Due to the limitation of the vehicle's own computing power, the performance of the local strategy is the worst. After that, we tested the delay and energy consumption comparison at different vehicle speeds. The experimental results are shown in Fig. 7 and Fig. 8.

When the speed is faster, the weight of the delay also increases, so the overall curve is on the rise, but the energy consumption is not affected by the speed, so the curve is parallel. The number of successful tasks at different speeds is shown

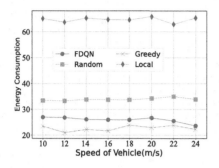

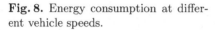

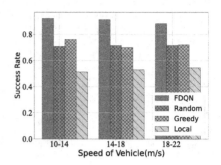

Fig. 8. Energy consumption at different vehicle speeds.

Fig. 9. Number of successful tasks at different vehicle speeds

in Fig. 9. It can be seen that the offloading success rate of our method is still the highest at different speeds.

7   Conclusion

In this paper, we propose a distributed task offloading method for delay and energy consumption optimization in vehicular networks with incomplete information. We first construct a task offloading scenario based on mobile edge computing and implement a dynamic weighting method for delay and energy consumption. Then we propose an offloading algorithm FDQN based on DQN and federated learning. This algorithm can cross the limitation of geographical scope and does not require any prior knowledge of the computational capabilities of edge nodes. We conduct a large number of simulation experiments to verify the effectiveness of our algorithm. The results indicate that our method can achieve fast convergence and optimal offloading effect regardless of the vehicle speed and energy status.

Acknowledgements. This work was supported in part by the Hebei Province Innovation Capability Enhancement Plan Project under Grant 22567603H, in part by S&T Program of Hebei under Grant 20310801D.

References

1. Liu, J., et al.: Rl/drl meets vehicular task offloading using edge and vehicular cloudlet: A survey. IEEE Internet Things J. **9**(11), 8315–8338 (2022)
2. Li, Z., Qi, W., Yifan, C., Guoqi, X., Li, R.: A survey on task offloading research in vehicular edge computing. Chinese J. Comput. **44**(05), 963–982 (2021)
3. Li, K., Wang, X., He, Q., Yi, B., Morichetta, A., Huang, M.: Cooperative multiagent deep reinforcement learning for computation offloading: a mobile network operator perspective. IEEE Internet Things J. **9**(23), 24161–24173 (2022)
4. Zhou, H., Jiang, K., Liu, X., Li, X., Leung, V.C.: Deep reinforcement learning for energy-efficient computation offloading in mobile-edge computing. IEEE Internet Things J. **9**(2), 1517–1530 (2021)

5. Dai, P., Hu, K., Wu, X., Xing, H., Yu, Z.: Asynchronous deep reinforcement learning for data-driven task offloading in mec-empowered vehicular networks. In: IEEE INFOCOM 2021-IEEE Conference on Computer Communications, pp. 1–10. IEEE (2021)
6. Zhang, Y., Liu, T., Zhu, Y., Yang, Y.: A deep reinforcement learning approach for online computation offloading in mobile edge computing. In: 2020 IEEE/ACM 28th International Symposium on Quality of Service (IWQoS), pp. 1–10. IEEE (2020)
7. Pang, X., Wang, Z., Li, J., Zhou, R., Ren, J., Li, Z.: Towards online privacy-preserving computation offloading in mobile edge computing. In: IEEE INFOCOM 2022-IEEE Conference on Computer Communications, pp. 1179–1188. IEEE (2022)
8. Zhu, C., et al.: Folo: latency and quality optimized task allocation in vehicular fog computing. IEEE Internet Things J. 6(3), 4150–4161 (2019)
9. Huang, B., et al.: Security and cost-aware computation offloading via deep reinforcement learning in mobile edge computing. Wirel. Commun. Mob. Comput. **2019**, 1–20 (2019)
10. Chen, S., Chen, J., Miao, Y., Wang, Q., Zhao, C.: Deep reinforcement learning-based cloud-edge collaborative mobile computation offloading in industrial networks. IEEE Trans. Signal and Inform. Process. Over Netw. 8, 364–375 (2022)
11. Chen, C., Zhang, Y., Wang, Z., Wan, S., Pei, Q.: Distributed computation offloading method based on deep reinforcement learning in ICV. Appl. Soft Comput. **103**, 107108 (2021)
12. Zhang, K., Mao, Y., Leng, S., Maharjan, S., Zhang, Y.: Optimal delay constrained offloading for vehicular edge computing networks. In: 2017 IEEE International Conference on Communications (ICC), pp. 1–6. IEEE (2017)
13. Li, W., Liu, D., Chen, K., Li, K., Qi, H.: Hone: Mitigating stragglers in distributed stream processing with tuple scheduling. IEEE Trans. Parall. Distrib. Syst. **32**(8) (2021)
14. Li, W., et al.: Efficient coflow transmission for distributed stream processing. In: IEEE INFOCOM 2020-IEEE Conference on Computer Communications, pp. 1319–1328. IEEE (2020)
15. Li, W., Yuan, X., Li, K., Qi, H., Zhou, X., Xu, R.: Endpoint-flexible coflow scheduling across geo-distributed datacenters. IEEE Trans. Parallel Distrib. Syst. **31**(10), 2466–2481 (2020)
16. Xu, R., Li, W., Li, K., Zhou, X., Qi, H.: Scheduling mix-coflows in datacenter networks. IEEE Trans. Netw. Serv. Manage. **18**(2), 2002–2015 (2020)
17. Klaimi, J., Senouci, S.M., Messous, M.A.: Theoretical game approach for mobile users resource management in a vehicular fog computing environment. In: 2018 14th International Wireless Communications & Mobile Computing Conference (IWCMC), pp. 452–457. IEEE (2018)
18. Lin, Z., Gu, B., Zhang, X., Yi, D., Han, Y.: Online task offloading in udn: A deep reinforcement learning approach with incomplete information. In: 2022 IEEE Wireless Communications and Networking Conference (WCNC), pp. 1236–1241. IEEE (2022)
19. Wang, X., Ye, J., Lui, J.C.: Decentralized task offloading in edge computing: a multi-user multi-armed bandit approach. In: IEEE INFOCOM 2022-IEEE Conference on Computer Communications, pp. 1199–1208. IEEE (2022)
20. Wu, C., Huang, Z., Zou, Y.: Delay constrained hybrid task offloading of internet of vehicle: a deep reinforcement learning method. IEEE Access **10**, 102778–102788 (2022)

21. Burd, T.D., Brodersen, R.W.: Processor design for portable systems. J. VLSI Signal Process. Syst. Signal, Image Video Technol. **13**(2), 203–221 (1996). https://doi.org/10.1007/BF01130406
22. Li, M., Gao, J., Zhang, N., Zhao, L., Shen, X.: Collaborative computing in vehicular networks: a deep reinforcement learning approach. In: ICC 2020–2020 IEEE International Conference on Communications (ICC), pp. 1–6. IEEE (2020)
23. Liu, Y., Yu, H., Xie, S., Zhang, Y.: Deep reinforcement learning for offloading and resource allocation in vehicle edge computing and networks. IEEE Trans. Veh. Technol. **68**(11), 11158–11168 (2019)
24. Nath, S., Wu, J.: Deep reinforcement learning for dynamic computation offloading and resource allocation in cache-assisted mobile edge computing systems. Intell. Converged Netw. **1**(2), 181–198 (2020)
25. Ale, L., Zhang, N., Fang, X., Chen, X., Wu, S., Li, L.: Delay-aware and energy-efficient computation offloading in mobile-edge computing using deep reinforcement learning. IEEE Trans. Cogn. Commun. Netw. **7**(3), 881–892 (2021)
26. Li, T., Sahu, A.K., Talwalkar, A., Smith, V.: Federated learning: challenges, methods, and future directions. IEEE Signal Process. Mag. **37**(3), 50–60 (2020)

Carbon Trading Based on Consortium Chain: Building, Modeling, and Analysis

Chaoying Yan[1] , Lijun Sun[1]([✉]) , Shangguang Wang[2] , and Shuaiyong Li[3]

[1] Qingdao University of Science and Technology, Qindao 266000, China
lijunsun@qust.edu.cn

[2] School of Computer Science, Beijing University of Posts and Telecommunications, Beijing 100000, China

[3] Chongqing University of Posts and Telecommunications, Chongqing 400065, China

Abstract. Traditional carbon trading suffers from poor interoperability, low transparency and reliance on manual drawbacks. In this paper, we analyze the combination of carbon trading and blockchain technology to design a trusted process for carbon trading. Then, we implement relevant smart contracts and build a DC-chain based on Hyperledger Fabric for efficient carbon trading. For larger-scale applications, analyzing blockchain protocols is an essential and necessary task. We propose a queuing network-based modeling approach to analyze the blockchain endorsement, ordering, and validation stages, and explore the impact of blockchain parameter configuration on system performance through theoretical modeling and experiments. To demonstrate the effectiveness of the proposed model, we utilize Caliper, a blockchain performance benchmarking framework, to compare the theoretical and experimental value. The results show that our proposed queuing model can fit the blockchain performance well.

Keywords: Consortium Blockchain · Carbon Trade · Queue Theory · Performance Analysis

1 Introduction

With the growing problem of global climate change, environmental protection and sustainable development have become important issues in today's society [1]. The dramatic increase in carbon emissions poses a great threat to the Earth's ecosystem and human society, and there is an urgent need to take effective measures to reduce the release of greenhouse gases. In this context, carbon trading

Supported by Open Foundation of State key Laboratory of Networking and Switching Technology (Beijing University of Posts and Telecommunications) SKLNST-2022-1-11; Key Laboratory of Industrial Internet of Things & Networked Control, Ministry of Education 2022FF08; Natural Science Foundation of Qindao City under Grant No.23-2-1-164-zyyd-jch.

Z. Tari et al. (Eds.): ICA3PP 2023, LNCS 14490, pp. 119–130, 2024.
https://doi.org/10.1007/978-981-97-0859-8_7

[2] as a market mechanism is widely regarded as one of the important means to achieve carbon emission reduction and promote sustainable development.

However, traditional carbon trading platforms have many challenges and limitations [3–5]. They lack transparency in the trading process, leading to information asymmetry and trust issues. In addition, the reliance on manual verification and processing leads to inefficient and error-prone transactions. To address these issues and promote the sustainable development of carbon trading, blockchain technology has attracted much attention in recent years as an innovative solution.

Blockchain technology offers new possibilities for carbon trading with its decentralized, transparent and tamper-proof nature. Blockchains can be divided into public and consortium chains, depending on the access restrictions of the blockchain network. In recent years, public chains [6] have been successful in the financial sector, but public chains are only suitable for scenarios where information interaction is entirely on the chain and decentralization is quite required. Consortium chains are usually managed by a group of related organizations, and nodes need to be licensed and authenticated before joining the network, which can prevent attacks and deal with malicious nodes in a timely manner. Consortium chains have greater advantages in practical scenarios where privacy and performance requirements are high and decentralization requirements are not as stringent. In carbon trading, we take carbon emission quota (CEQ) and country carbon emission reductions (CCER) as traceable and tradable on-chain digital assets, build a coalition chain as the underlying infrastructure for trading, use smart contracts [7] as business logic carriers to proxy manual operations, and combine smart edge devices for credible data collection to ensure the authenticity of source data and achieve traceability and transparency throughout the carbon trading cycle.

Due to its distributed collaboration paradigm, the performance problem of blockchain has been a long-standing pain point. In general, system performance is determined by the combination of communication model, consensus protocol [8,9], number of nodes, and block parameters. Due to the impossibility triangle constraint of blockchain [7], any attempt to improve system performance will correspondingly reduce its decentralization or security. Consortium chain platforms, such as Hyperledger Fabric [10], divide system processing into three phases: endorsement (execution), sequencing, and validation. Different phases are handled by different nodes (endorsement, sorting, and commit) to further separate transaction execution from consensus and improve the performance and scalability of the system. To analyze and improve the performance of the blockchain, we establish a queueing network model for the three-phase processing process of the consortium chain to guide and optimize the performance parameter settings, achieving carbon trade more efficient.

In this paper, we propose a DC(double carbon)-chain for distributed carbon trading, and then perform a theoretical performance analysis using a queueing theory model and an experimental validation of the established performance model. The main contributions of this paper can be summarized as follows:

- We have developed a credible and transparent double carbon trading framework that removes the trust barrier between carbon emitters and carbon reducers to enable reliable carbon trading between them. Using smart edge devices to detect and record carbon reduction data in real time, which ensure the authenticity of the data in the chain.
- Based on the above transaction process, we build the DC-chain and realize the complete business process through node management, carbon emission quota allocation, carbon emission reduction certification and trading and swap smart contracts. By combining on-chain contracts and off-chain processing, we improve execution efficiency and reduce storage pressure.
- Queueing theory is used to analyze the theoretical performance of the system. Appropriate queueing models are developed for its three processing stages, which are connected in series to form an overall queueing network. The average transaction confirmation time of each processing stage is evaluated theoretically and experimentally, and the results show that the established queuing model is well adapted.

The rest part of this paper is organized as follows. In Sect. 1, we introduce previous studies related with this paper. Section 2 presents carbon trade process and implement of DC-chain. Section 3 describes the queueing model for blockchain in detail. Experimental setup and evaluation results are given in Sect. 4. At last, we conclude our work in Sect. 5.

2 DC-Chain for Carbon Trading

2.1 Role Assignment

The DC-Chain framework consists of three main participants: CE P(Carbon Emissions Party), CRP (Carbon Reduction Party), and Regulatory Agency (RA). We introduce their roles as follows.

CEP: It is noted that its main characteristic is to produce a large amount of carbon emissions in its production and operation process. Such as conventional thermal power plants, etc. There is a cap on the amount of carbon they can emit in an emission cycle. Their production behavior will gradually depletes its CEQ.

CRP: They reduce greenhouse gases in a direct or indirect way. For example, small and medium-sized distributed photovoltaic power generation. Smart terminals are deployed locally on their side and used to calculate emission reduction data. The calculated carbon reductions must be certified to CCER before they can be traded on the carbon market.

RA: They collect and analyze historical emissions data from major emitters for statistical purposes, set regional emissions caps for the current period, and assign CEQ for various CEPs at the beginning of the compliance cycle. They also certify carbon reductions for CRPs.

2.2 DC-Chain Construction and Smart Contract Design

We use Fabric as carbon trading basic infrastructure to construct DC-chain. Fabric provides a pluggable consensus mechanism, allowing developers to freely choose the consensus and authentication scheme that suits them. Membership service is achieved through identity management. Scalable network architecture that allows users to scale the network by adding new components and nodes. Next, we describe the basic building blocks of the Fabric base network in the following.

Channel: Channel is a private communication path created in a Fabric network to pass transactions and data between a specific set of participants. Channels are designed to provide isolation and privacy, allowing some participants to transact and share information on the channel while other participants cannot access or view those transactions and information.

Ledger: Ledger has two parts: a world state and a blockchain ledger. The world state is a KV database that keeps the up-to-date state of the account. The blockchain ledger is maintained by all peers, and is an append-only data structure that uses a hash chain to record all transactions.

Endorsement Policy: Endorsement policy specifies the validity and legitimacy of a client-side transaction request. A transaction is legitimate only if the client receives a certain number (typically chosen to be greater than $1/2$) or a particular endorsing node's signature.

Chaincode: Chaincode is a smart contract that defines and executes business logic running in the Fabric blockchain network.Chaincodes are applications developed and deployed on the Fabric platform that control the behavior of assets and transactions on the network,whcih can be written in Go, Java and other programming languages.

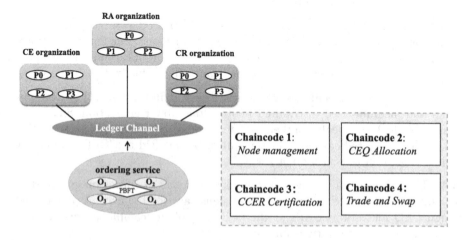

Fig. 1. DC-Chain network configuration and chaincode

We constructed a Fabric network with three organizations of CE, CR, and RA, which together form a ledger channel as shown in Fig. 1. Each organization has peer nodes representing a CEP or CRP, which are registered by CA. Practical Byzantine Fault Tolerance (PBFT) can resist to Byzantine fault even in worse network condition. Due to the weak trust of distributed energy trading, we choose it as the consensus algorithm in the ordering service to maintain the consistency of transaction order. First, we create a channel ledger and initialize the block configuration of block size equals to 10, and block interval equals to 1 s. Nodes in RA organization are various sector regulators and service providers.

Four main chaincode are designed to support carbon trading by using Fabric-go-contract SDK, as also shown in Fig. 1, they are installed on different peers to provide different services. The details of chaincode are following.

Node Management: This chaincode is only installed on peers in the RA organization and is responsible for approving new nodes to join and leave the channel, as well as defining peer read/write permissions and ledger data visibility.

CEQ Allocation: This chaincode is installed on peers in both CE and RA organizations. CE nodes send transaction $<t, D_l, \sigma_e>$ to one of RA nodes, where t is timestamp, D_l represent the remaining CEQ allowance in last period, σ_e is digital signature. RA nodes first verify σ_e, making sure it belongs to a correct organization, calculate the CEQ for a CE peer at this period, and finally start quota allocation transaction $<t, D_c, \sigma_a>$.

CCER Certification: This chaincode is installed on peers in both CR and RA organization. CR nodes send transaction $<$cetification, t, n, T, $\sigma_r>$ every once in a interval T, That means it produces some carbon emission reduction during time T, where n is the carbon reduction volume not yet certified, σ_r is the digital signature. RA node issues a digital asset certificate for the CR peer.

Trade and Swap: This chaincode is installed on peers in both CE and CR organizations. Each clients registered can lauch MPT and LPT request to swap their carbon asset.

3   Modeling Blockchain Performance

In this section, we will model the system process of DC-chain through queueing theory and analyze the relationship between the system performance and the influence of its relevant parameters, as a guide for the performance optimization of the blockchain. In DC-chain, each transaction goes through three stages of sequential execution before it is recorded on the ledger, as shown in Fig. 2. First, transactions are executed by the endorsing node. Second, the transactions signed by the endorsing node will be packaged into unverified blocks by the ordering node. The ordering nodes reach consensus on the order of the transactions in a block using a specified consensus algorithm. Finally, commit nodes independently validate these blocks. Transactions that pass validation are added to the channel ledger, and transactions that fail validation will be discarded, and the ledger world state will be updated with valid transactions. In the following, we model three stages using queueing theory and perform a theoretical performance analysis.

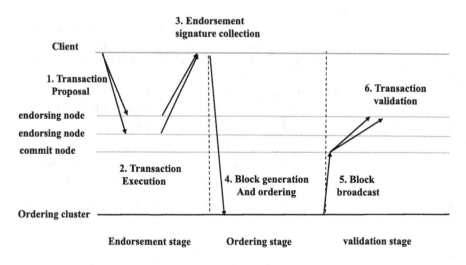

Fig. 2. Transaction process flow in single organization

The typical queuing model consists of the following elements: customer source, service station, service rules and service station capacity. In our model, the customer source is the arriving transactions, the service station is the various nodes responsible for endorsement, ordering and validation, and the service rule is FCFS (First Come First Service), assuming that the capacity of the service station is infinite. We consider the transaction processing of the consortium blockchain as three tandem queuing service processes: stage 1, stage 2, and stage 3, which represent the endorsement, ordering, and validation processes respectively. The ordering stage uses the PBFT consensus algorithm. The established queuing network model is shown in Fig. 3.

In the endorsement stage, clients send requests to the endorsing nodes. The endorsing nodes invoke the corresponding smart contract according to the request, and perform a simulated execution. (The results are not recorded in the channel ledger at this stage). Finally, they return the read/write set generated by the chaincode and node's signature to the client. We model the process of executing a smart contract by endorsing nodes as the queuing model of $M/M/1$, i.e., the transaction requests arrive at the endorsing node with the arrival rate λ_e, and the service time follows the exponential distribution with the parameter μ_e. Let the idle probability of the system be ρ_e at stage1, and its expression is

$$\rho_e = 1 - \frac{\lambda_e}{\mu_e} \tag{1}$$

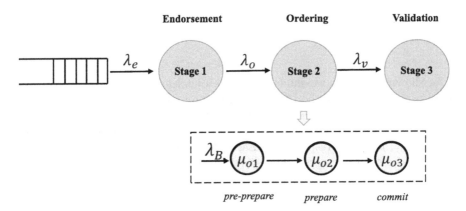

Fig. 3. Queue network model

According to the Little's Formula, the average waiting time w_e of the transaction request at stage1 can be obtained, and its expression is

$$w_e = \frac{1}{\mu_e - \lambda_e} \tag{2}$$

In the ordering stage, the PBFT algorithm is a classical Byzantine fault-tolerant algorithm that can resist Byzantine attacks when the total number of network nodes is n and the number of Byzantine nodes is f, satisfying $n > 3f$. Due to the complexity of the ordering service, its service time cannot simply be measured by exponential distribution. To simplify the analysis, we assume that the number of nodes involved in the ordering service is four, and there is one primary node and the rest are replica nodes. They are connected by reliable communication way. The difference in computing power will cause the speed of processing message at each node to be inconsistent. Let μ_1, μ_2, μ_3, μ_4 represent the service rate of these four nodes respectively. The complete PBFT consensus algorithm consists of five phases: $request$, $pre-prepare$, $prepare$, $commit$ and $reply$. In Fabric, the request phase is for clients to send the endorsed transaction proposal to the primary node. The primary node packs the received transactions into block, which then goes through the three-phase message dissemination of $pre-prepare$, $prepare$ and $commit$ to reach the consensus on the order of transactions in the block. Finally, the $reply$ phase is passing block to commit nodes, transactions enter the validation stage. The following is a detailed description of modelling the three main phases of PBFT message propagation.

Endorsement stage transactions arrive at the primary node with transaction arrival rate $\lambda_o = \mu_e$, and wait to be packed into block. Since stage 2 is based on the block as a whole, rather than a single transaction for consensus. That is to say, the ordering service processes multiple transactions in bulk. It is necessary to first convert the transaction arrival rate into the block arrival rate λ_B, where

B represents the setting block size.

$$\lambda_B = \frac{\lambda_e}{B} \tag{3}$$

As shown in Fig 3, we model it as the queueing model of $M/E_r/1$. The arrival process abides by the Poisson distribution with the arrival rate λ_B, and the service time is the distribution of Erlang-r, which represents r continuous exponential service process, specifically here $r = 3$, and the service rate of each phase is μ_{o1}, μ_{o2}, μ_{o3}, respectively. Each block passes sequentially these three phases of the service process. The system state of stage 2 can be represented by (k,j), where $k(0 < k <= B)$ is the number of transactions, in this case specifically the number of transactions contained in the block, and j $(j = 0,1,2)$ represents the block in the ordering subphase of pre-prepare, prepare, commit. For example, state $(5,2)$ means 5 transactions in the prepare phase. The transition rate matrix Q has the typical block tridiagonal form:

$$Q = \begin{pmatrix} B_{00} & B_{01} & 0 & 0 & 0 & \cdots \\ B_{10} & A_1 & A_2 & 0 & 0 & \cdots \\ 0 & A_0 & A_1 & A_2 & 0 & \cdots \\ 0 & 0 & A_0 & A_1 & A_2 & \cdots \\ 0 & 0 & 0 & A_0 & A_1 & \cdots \\ \vdots & \vdots & \vdots & \vdots & \vdots & \ddots \end{pmatrix} \tag{4}$$

the submatrix can be obtained by

$$B_{00} = -\lambda_B, B_{01} = \begin{pmatrix} \lambda_B & 0 & 0 \end{pmatrix}, B_{10} = \begin{pmatrix} 0 \\ 0 \\ r\mu \end{pmatrix} \tag{5}$$

$$A_0 = \begin{pmatrix} 0 & 0 & 0 \\ 0 & 0 & 0 \\ r\mu & 0 & 0 \end{pmatrix} A_1 = \begin{pmatrix} -\lambda_B - r\mu & r\mu & 0 \\ 0 & -\lambda_B - r\mu & r\mu \\ 0 & 0 & 0 \end{pmatrix} A_2 = \lambda_B I = \begin{pmatrix} \lambda_B & 0 & 0 \\ 0 & \lambda_B & 0 \\ 0 & 0 & \lambda_B \end{pmatrix} \tag{6}$$

Let the stationary probability vector $\pi = (\pi_0, \pi_1, \pi_2, ..., \pi_k)$, which is the probability that the number of transactions in a block is k when the system is stable, satisfy the following linear equation $\pi Q = 0$, using a matrix geometric solution [26]. Here π_0 is a vector of length 1 (its value is equal to the probability that the system is idle) and $\pi_k(k=1,2)$ is a vector of length 3. Its ith element represents the probability of being in phase i, when there are k transactions in the system. π satisfies $\pi_{i+1} = \pi_i * R$, where R is called Neuts' rate matrix.

$$R_{l+1} = -(V + R_l^2 W) \tag{7}$$

where $V = A_2 A_1^{-1}$, and $W = A_0 A_1^{-1}$ Next, we need to compute the initial vectors π_0 and π_1 using the matrix geometry method. from $\pi Q = 0$,

$$\begin{cases} \pi_0 B_{00} + \pi_1 B_{00} = 0 \\ \pi_0 B_{01} + \pi_1 A_1 + \pi_1 A_2 = 0 \end{cases} \tag{8}$$

Let e denote a column vector of length r whose components are all equal to 1, then

$$\pi_0 + \pi_1 \left(I - R \right)^{-1} e = 1 \tag{9}$$

Due to $\pi_0 = 1 - \rho$, the average waiting time for the stage $2w_o$ can be obtained by

$$w_o = \pi_1 \left(I - R \right)^{-2} \tag{10}$$

In the validation stage, blocks ordered are passed to the commit nodes of each organization. Upon receiving a block, it calls the VSCC (Validation System Chaincode) associated with that transaction chaincode as part of the transaction validation process to determine the validity. Transactions in the block that are validated as legitimate are committed to the channel ledger, and then the world state is updated. The validation stage is similar to the stage 1 and can be modeled as a $M/M/1$ queuing model. According to Burke's theorem [11], the transaction arrival rate of the validation stage satisfies $\lambda_v = \mu_o$, the average transaction wait time w_v in stage 3 can also be obtained by:

$$w_v = \frac{1}{\mu_v - \lambda_v} \tag{11}$$

Lately, we demonstrate the applicability of our established queuing network through comparison of theoretical modeling values and actual test results in Sect. 5.

4   Analytical and Experimental Results

Based on the queueing network model built in Sect. 4, we apply three organizations with four peer nodes each, and an ordering cluster service with four ordering nodes on Fabric v2.4. Each node is launched as a Docker container and then connected to the network using the Docker Swarm with Ubuntu 22.04.3. We use Caliper v0.5.1 to generate transaction requests by calling our *Trade* and *Swap* chaincode in Sect. 4 with the Fabric Go SDK, and test the average wait time of each stage.

4.1   Transaction Confirmation Time

First, we compared the transaction confirmation time under different transaction arrival rate. Block size is set to default value of 10, block interval is 1 s. The total number of transaction requests is 5000 in one test. The analytical and experimental results are shown below.

Using the queuing model developed in Sect. 4, we calculated the average transaction confirmation time for each stage in the steady state and compared it with the actual experimental results. Figure 4 shows the average transaction confirmation time for each stage for different transaction arrival rates. It can be seen that as the transaction arrival rate increases, the transaction confirmation time for each stage increases monotonically, and the experimental results agree well with the results of the modeling analysis. This indicates that our queuing model is appropriate. Figure 4 d) compares the confirmation times of each stage and it can be seen that when the transaction arrival rate is small, the confirmation times of the three stages increase moderately and the total average confirmation time is relatively low. However, when the transaction arrival rate is greater than 85, the time of the ordering stage increases sharply, which is due to the fact that when the block size is fixed, most of the transactions are jammed in the ordering stage and cannot be packed into the block in time.

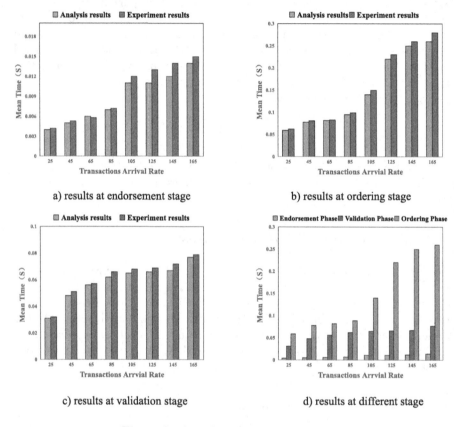

a) results at endorsement stage

b) results at ordering stage

c) results at validation stage

d) results at different stage

Fig. 4. Analytical vs experimental results

4.2 Block Size

We also measure the mean transaction confirmation time under block size ranging from 5,20,50,100,200,300 in block interval equals to 0.5,2,4 s at the fix transaction arrival rate 50. The result shows in Fig. 5.

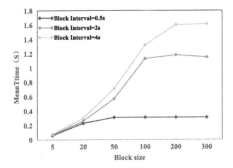

Fig. 5. Mean confirmation time with different block size

With respect to different block intervals, the average transaction confirmation time becomes longer as the block interval increases. This is due to the fact that when the transaction send rate and block size are fixed. The first transaction is ordered and then filled into the block waiting for final packing. Generally, the block packing strategy is divided into two cases, one is when the number of transactions reaches the block size limit and the other is when the waiting time reaches the block interval. In the small block size situation, the block is easily filled up resulting in block packing. As shown above, when the block size is less than 20, the average confirmation time does not make much difference. This is because the block interval does not have enough effect on the transaction confirmation time. But with the block size increasing, the block is become difficult to filled up right away, where the block interval is dominant to the speed of block packing time. When the block size exceeds 20, the gap in the average transaction confirmation time starts to increase sharply. For the same block interval, the average transaction confirmation time first increases rapidly with increasing block size and then reaches a steady state. The block size no longer affects the average transaction confirmation time. From the experimental results, we can analyze that the average transaction confirmation time is affected by a combination of block size, block interval and transaction arrival rate. In general, the transaction arrival rate is determined by the participating clients and the speed is unknown. In order to keep the latency of the blockchain system low, the block interval and block size should be reduced as much as possible.

5 Conclusion

In this paper, we propose a carbon trading framework based on consortium blockchain. Blockchain can empower carbon neutrality and solve the problems of difficult data traceability and audit difficulties through carbon emission data and emission reduction data validation, while enhancing data security and transparency. Smart contracts can achieve distributed and credible automated execution. To ensure the availability in large-scale distributed scenarios, we analyzed the transaction execution flow of blockchain, building a suitable queuing theory model for each stage. It helps theoretically analyze and evaluate the performance metrics of the consortium chain system. In future work, we will investigate the improvement of the consensus mechanism to enhance the performance of the system from a more fundamental level.

Acknowledgements. This work was spported by Open Foundation of State key Laboratory of Networking and Switching Technology (Beijing University of Posts and Telecommunications) (SKLNST-2022-1-11); Key Laboratory of Industrial Internet of Things & Networked Control, Ministry of Education,2022FF08; Natural Science Foundation of Qindao City under Grant No.23-2-1-164-zyyd-jch.

References

1. Mallapaty, S.: How China could be carbon neutral by mid-century. Nature **586**, 482–483 (2020)
2. Zhang, S., Chen, W.: Assessing the energy transition in China towards carbon neutrality with a probabilistic framework. Nat Commun. **13**(1) (2022)
3. Bai, Y., Song, T., Yang, Y.: Construction of carbon trading platform using sovereignty blockchain. In: International Conference on Computer Engineering and Intelligent Control, pp. 149–152 (2020)
4. Richardson, A., Xu, J.: Carbon trading with blockchain. In: Pardalos, P., Kotsireas, I., Guo, Y., Knottenbelt, W. (eds.) Mathematical Research for Blockchain Economy. SPBE, pp. 105–124. Springer, Cham (2020). https://doi.org/10.1007/978-3-030-53356-4_7
5. Mengelkamp, E., Notheisen, B., Beer, C.: A blockchain-based smart grid: towards sustainable local energy markets. Comput. Sci. Res. Dev. **33**, 207–214 (2018)
6. Nakamoto, S.: Bitcoin: A Peer-to-Peer Electronic Cash System (2008)
7. Wood, G.: Ethereum: A secure decentralized generalized transaction ledger (2014)
8. Ongaro, D., Ousterhout, J.: In search of an understandable consensus algorithm. In: 2014 USENIX Conference on USENIX Annual Technical Conference, pp. 305–320 (2014)
9. Castro, M., Liskov, B.: Practical Byzantine fault tolerance and proactive recovery. ACM Trans. Comput. Syst. **20**(4), 398–461 (2002)
10. Androulaki, E., Barger, A., Bortnikov, V.: Hyperledger fabric: a distributed operating system for consortium blockchains.In: EuroSys'18: Thirteenth EuroSys Conference, pp. 1–15 (2018)
11. Burke, P.: The output process of a stationary M/M/s queueing system. Ann. Math. Statist. **39**(4), 1144–1152 (1968)

CAST: An Intricate-Scene Aware Adaptive Bitrate Approach for Video Streaming via Parallel Training

Weihe Li[1], Jiawei Huang[2], Yu Liang[3], Jingling Liu[2(✉)], Wenlu Zhang[2], Wenjun Lyu[4], and Jianxin Wang[2]

[1] University of Edinburgh, Edinburgh EH8 9YL, UK
[2] Central South University, Changsha 410083, China
jinglingliu@csu.edu.cn
[3] Lancaster University, Lancaster LA1 4YW, UK
[4] Rutgers University, Newark, NJ 08901, USA

Abstract. Adaptive Bitrate (ABR) algorithms have become increasingly important for delivering high-quality video content over fluctuating networks. Considering the complexity of video scenes, video chunks can be separated into two categories: those with intricate scenes and those with simple scenes. In practice, improving the quality of intricate chunks can lead to more significant improvements in Quality of Experience (QoE) than improving simple chunks. However, current schemes either assign equal priority to all chunks or optimize using a fixed linear-based reward function, making them inadequate for meeting real-world requirements. To tackle these limitations, this paper introduces a novel ABR approach that explicitly considers bitrate adaptation as the primary objective. The proposed approach, CAST (Complex-scene Aware bitrate algorithm via Self-play reinforcemenT learning), leverages the power of parallel computing with multiple agents to train a neural network, aiming to achieve superior video playback quality for intricate scenes while minimizing frequent freezing events. The extensive trace-driven evaluation and subjective test results demonstrate that CAST outperforms existing off-the-shelf schemes.

Keywords: Video streaming · Bitrate adaption · Parallel computing · Self-play reinforcement learning · Scene complexity

This work was conducted during the pursuit of a Master's degree by Weihe Li at Central South University. This work was supported in part by the National Natural Science Foundation of China (62302524, 62132022); in part by the Key Research and Development Program of Hunan under Grant 2022WK2005; in part by the Natural Science Foundation of Hunan Province, China, under Grant 2021JJ30867; and in part by using computing resources at the High Performance Computing Center of Central South University.

Z. Tari et al. (Eds.): ICA3PP 2023, LNCS 14490, pp. 131–147, 2024.
https://doi.org/10.1007/978-981-97-0859-8_8

1   Introduction

The proliferation of smart mobile devices and the widespread accessibility of wireless connectivity have precipitated a substantial surge in network traffic [1], particularly in the context of video streaming. HTTP Adaptive Bitrate Streaming, also known as Dynamic Adaptive Streaming over HTTP (DASH) [2], has become the preferred method for delivering video content over fluctuating networks. In the DASH system, video content is pre-encoded and pre-chunked at different quality levels (bitrates) on the server side. The player on the client side dynamically selects the appropriate bitrate for each chunk based on an estimation of network capacity and measured buffer occupancy, intending to provide a high QoE for viewers.

The two primary encoding methods for video content delivery in the DASH system are Constant Bitrate (CBR) and Variable Bitrate (VBR). With CBR, the entire video is encoded using a fixed bitrate for a quality level, resulting in uniform bit allocation across all video chunks and causing inconsistent quality across chunks with different scenes [3]. In contrast, VBR allocates more bits to intricate scenes characterized by high dynamics while giving fewer bits to low-motion scenes, resulting in a more uniform quality for chunks with the same bitrate level. The benefits of VBR, like the ability to achieve the same quality with a lower bit budget, have led to a shift in content providers' encoding strategies from CBR to VBR in recent years [4].

Despite the efforts of Variable Bitrate (VBR) encoding to maintain consistent quality across video chunks with varying scene complexity, research has shown that the quality of chunks with intricate scenes (herein referred to as *"intricate"* or *"complex"* or *"dynamic"* chunks) remains considerably lower than that of chunks with simple scenes (herein referred to as *"simple"* chunks) due to limitations in existing encoding techniques [3]. However, since intricate chunks play a critical role in determining viewing quality, enhancing the quality of such chunks can result in more significant QoE improvements [5,6]. Unfortunately, existing ABR algorithms seldom consider the impact of scene complexity, leading to subpar quality for intricate chunks and subsequent degradation in QoE. Although state-of-the-art approaches that account for scene complexity aim to optimize a score using a linear-based formula comprising weighted sum metrics, precise calculation of the optimization function is, however, challenging [17]. A poorly designed function may mislead ABR algorithms to make inappropriate bitrate decisions, ultimately compromising users' viewing experience.

Furthermore, many existing learning-based ABR algorithms train the neural network with a single agent [19], while employing multiple agents in parallel for reinforcement learning training offers notable benefits. Parallel training enables simultaneous learning, effectively utilizing computational resources and accelerating the learning process. Additionally, independent agents explore the environment differently, facilitating comprehensive exploration of the state space and the discovery of various optimization paths.

Motivated by these challenges, we aim to harness parallel training to introduce a new ABR algorithm. In practice, the task of bitrate adaption can be regarded as a straightforward goal or rule. For instance, the primary aim of most ABR schemes is to minimize rebuffering time while ensuring high playback quality [34]. In this paper, we propose CAST, which utilizes multi-agent self-play reinforcement learning to train the neural network in parallel for delivering high-quality video chunks with intricate scenes without compromising the quality of simple chunks and minimizing stall time on variable networks. By training on the actual goal, CAST can accurately fulfill explicit requirements. Our experimental results indicate that CAST surpasses existing ABR algorithms, improving the quality of intricate chunks by 8.69% to 40.03%.

2   Background and Motivation

2.1   ABR Algorithms

Recent research in ABR approaches can be broadly categorized into heuristic-based and learning-based methods. The heuristic-based approaches include FES-TIVE [30], which estimates available bandwidth based on past throughput and selects the highest bitrate not exceeding the estimated capacity. BBA [31] and BOLA [32] employ bitrate selection based on playback occupancy to drive buffer occupancy to a proper value, thereby reducing rebuffering events. MPC [6] maximizes an optimization problem over a horizon of several chunks ahead by fusing the signal of rate and buffer. QUETRA [7] transforms the video streaming task into a queuing model and utilizes the queuing theory in conjunction with the bandwidth prediction to select the bitrate for each chunk by maintaining the buffer occupancy fluctuates around half of the maximum buffer size. PIA [8] strategically harnesses the Proportional-Integral-Derivative (PID) control concepts and incorporates several novel strategies for various requirements of ABR streaming to ameliorate the overall QoE. MSPC [9] leverages the Kalman filter to predict network bandwidth and adopts multi-step prediction to provide responsive adaptation and smooth playback for mobile video applications. Nevertheless, these hybrid heuristic methods are mainly built on a series of assumptions, such as negligible bandwidth variations during a short period, rendering them insufficient for fluctuating networks, especially the cellular network.

On the other hand, learning-based approaches have gained significant attention in recent years. Pensieve [10] uses Deep Reinforcement Learning (DRL) to train a neural network for bitrate adaptation from scratch to maximize a linear QoE function. Stick [18] employs DRL to find optimal parameter settings for BBA under different network environments by maximizing the QoE function. BayesMPC [12] leverages the Bayesian neural network model to enhance the accuracy of bandwidth predictions, and subsequently applies MPC for bitrate selection for the upcoming video segment. Fugu [13] employs online learning techniques to estimate the download time for video chunks at each level. Subsequently, it applies MPC to select the optimal bitrate for the succeeding video chunk. RAV [19] utilizes deep reinforcement learning techniques to acquire bitrate selection policies for both audio and video content, significantly mitigating the bitrate gap for the audio and video chunks with the same index. PRIOR [14] introduces an accurate network bandwidth prediction method with the attention mechanism and utilizes the DRL to learn a bitrate selection policy via maximizing the QoE value. However, these approaches have limitations as they fail to consider the difference in scenes between diverse chunks, resulting in poor quality for intricate chunks. Other learning-based methods, such as [11,16,20–23], overlook scene differences and lack appropriate methods for scene-aware bitrate adaptation.

2.2   Scene Complexity

The content of video chunks can vary in terms of scene complexity, with some chunks featuring low-motion and simple scenes (see Fig. 1(a)) while others have high-motion and intricate scenes (see Fig. 1(b)). In order to investigate the impact of chunk size on quality, we conducted experiments using the video "Big Buck Bunny" [25] as our test case. The video was encoded into six different tracks, ranging from 144p to 1080p, using two different encoders: H.264 and H.265. Each track was then divided into chunks with a fixed duration of 2 s.

(a) Chunk with simple scene.

(b) Chunk with intricate scene.

Fig. 1. Chunk with different scenes [24].

When assessing the scene complexity, several metrics such as Spatial Information (SI) and Temporal Information (TI) [27] may be utilized. Nevertheless, implementing these metrics in commercial streaming services is challenging due to the computationally intensive content-level analysis they require, which would necessitate significant modifications to the streaming pipeline [3]. Fortunately, in VBR encoding, the size of a chunk can effectively indicate relative scene complexity, with intricate chunks allocated more bits than simple ones. Given that scene complexity is constant across quality levels at a given playback point, we designate a middle-quality track (480p) as a reference track and classify chunks based on their size, with larger chunks in the first half classified as intricate and the remaining chunks classified as simple [5]. It is important to note that the classification method is not fixed, and other strategies, such as using four classes, can be employed. In this study, we utilize the Video Multi-method Assessment Fusion (VMAF) metric [33] to evaluate the quality of each chunk, as it is known to accurately reflect users' subjective viewing experience [11,15,28,29].

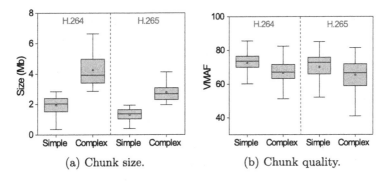

(a) Chunk size.   (b) Chunk quality.

Fig. 2. Analysis of chunk size and quality. (a) Comparison of the size of simple and intricate chunks. (b) Comparison of the quality of simple and intricate chunks.

As depicted in Fig. 2(a), we can observe that intricate chunks are considerably larger than simple ones, with an average size of around $1.18\times$ and $1.10\times$ higher under the H.264 and H.265 codecs, respectively. However, despite utilizing more bits for encoding, the quality of intricate chunks is still inferior to that of simple ones, as shown in Fig. 2(b). Specifically, under the H.264 and H.265 formats, the average quality of intricate chunks is respectively 7.88% and 6.73% lower than that of simple chunks. These findings have prompted us to consider different treatment strategies for intricate and simple chunks to achieve optimal QoE for users. It is critical to allocate bandwidth to prioritize the transmission of intricate chunks while ensuring that the quality of simple chunks is not significantly compromised. Additionally, minimizing stall time is of utmost importance, as this can significantly impact the overall QoE.

3   Design

3.1   CAST Overview

This paper proposes a new approach, CAST, that leverages self-play reinforcement learning to learn a bitrate selection policy that considers scene complexity.

Unlike the widely adopted deep reinforcement learning (DRL) that trains via a linear reward function, CAST transforms the ABR problem into an explicit objective and views the learning task as a competition between different trajectories collected by itself. By training the neural network towards the gradient of approaching the predefined goal, the converged policy can nicely satisfy the actual demands [35]. The training process is illustrated as follows:

The training process begins by sampling D distinct trajectories, denoted as $T_d = \{s_0^d, a_0^d, s_1^d, a_1^d, \cdots, s_t^d, a_t^d\}$, from the same environment and starting point using a deep neural network. We record these trajectories in a collection M, where D is the total number of trajectories collected, $d \in D$, s_t represents the environment status at the t-th time, and a_t denotes the action for video bitrate selection at the t-th time. Further details about the status and action can be found in Section III.B.

Afterward, CAST compares each trajectory's performance to evaluate the current policy's performance. Specifically, given two different samples T_p and T_q from the sample set M, CAST first computes their average quality for intricate chunks C_p, C_q, average rebuffering time B_p, B_q, and average quality for simple chunks S_p, S_q. Then, CAST compares their performances via a deterministic rule consisting of two cases.

Case 1: Given three thresholds α, β and η, if the absolute difference $|C_p - C_q| < \alpha$ and $|B_p - B_q| < \beta$ and $|S_p - S_q| < \eta$, we consider the competition between T_p and T_q a *draw* game.

Case 2: Otherwise, the competition process is based on a priority (Tables 1 and 2). In detail, we first compare the average rebuffering time between T_p and T_q. If the absolute difference $|B_p - B_q|$ exceeds the threshold β, the sample with a shorter rebuffering time is the *winner*, and the competition is over. Otherwise, the competition compares the absolute difference in intricate chunks' quality, followed by comparing the absolute difference in simple chunks' quality. Here, we set the threshold α, β, and η as 1, 0.01, and 1, respectively.

After the battle completes, each trajectory can obtain how many times it won A_i in all sampled trajectories D. Thus, the win rate of each sample is $\frac{A_i}{|D|-1}$ because there is no battle between the trajectory and itself.

Table 1. Rule: reducing rebuffering time (1st priority)

Rule	$< \beta$	$\geq \beta$
$|B_p - B_q|$	TABLE 2	**Winner: min(B_p, B_q)**

Table 2. Rule: improving quality for intricate chunks (2nd priority)

Rule	$< \alpha$	$\geq \alpha$
$|C_p - C_q|$	**Winner: max(S_p, S_q)**	**Winner: max(C_p, C_q)**

The last step of CAST involves training a neural network using the collected trajectory samples and their win rates. To achieve this, the neural network is trained using the state-of-the-art Dual-clip Proxy Policy Optimization (Dual-PPO) algorithm [36], which owns better stability and convergence than existing learning methods, such as the original PPO algorithm. The neural network includes a policy network and a value network. The policy network is trained to increase the probability of winning samples and decrease the probability of losing samples by optimizing a loss function. The loss function of the policy network can be expressed as follows.

$$\mathcal{L} = \begin{cases} \hat{\mathbb{E}}_t[\max(\min(p_t(\theta)\hat{A}_t, clip(p_t(\theta), 1 - \varepsilon, 1 + \varepsilon)\hat{A}_t), c\hat{A}_t)], \hat{A}_t < 0, \\ \hat{\mathbb{E}}_t[\min(p_t(\theta)\hat{A}_t, clip(p_t(\theta), 1 - \varepsilon, 1 + \varepsilon)\hat{A}_t)], \hat{A}_t \geq 0, \end{cases}$$

The loss function of the policy network includes the advantage function $\hat{A}_t$ and the policy ratio $p_t(\theta)$ between the current policy π_θ and the old policy $\pi_{\theta_{old}}$. The hyper-parameters ε and c are set to the same values as in [36]. On the other hand, the value network is trained to minimize the estimated error of the advantage function, which is expressed as $\mathcal{L}^V = \frac{1}{2}\hat{\mathbb{E}}_t[\hat{A}_t]^2$. Furthermore, an entropy term is added to the loss function to encourage exploration. Therefore, the overall loss function is given by

$$\nabla\mathcal{L}^{CAST} = -\nabla_\theta\mathcal{L} + \nabla_{\theta_p}\mathcal{L}^V + \nabla_\theta\gamma H^{\pi_\theta}(s_t).$$

where γ denotes the entropy weight.

3.2 Neural Network Structure

We will now delve into the neural network in greater detail, focusing on its inputs, outputs, and network architecture.

Inputs. The inputs for CAST are divided into three categories: (1) player environment features, (2) intricate chunk features, and (3) simple chunk features. Player environment features are represented as $P = \left(\vec{c_k}, \vec{b_k}, e_t\right)$, where $\vec{c_k}$ represents the network capacity for downloading the last k chunks, $\vec{b_k}$ denotes the buffer occupancy of the last k chunks, and e_t is the quality of the last chunk. To maintain consistency with previous research [10], the value of k is set to 8. Intricate chunk features are represented as $C = \left(\vec{Q_t}, \vec{G_t}, I_t, R_t\right)$, where $\vec{Q_t}$ denotes the quality of each track for the next intricate chunk, $\vec{G_t}$ is the size of each track for the next intricate chunk, I_t is the index of the next intricate chunk, and R_t indicates the number of unretrieved intricate chunks in the video. Simple chunk features are denoted as $S = (\vec{q_t}, \vec{g_t}, r_t)$, where $\vec{q_t}$ represents the quality of each track for the next simple chunk, $\vec{g_t}$ denotes the size of each track for the next simple chunk, and r_t indicates the number of unretrieved simple chunks in the video. It is important to note that the classification of intricate and simple chunks is based on their size distribution, which is identical to that employed in Sect. 2.2.

Outputs. The output is a v-dimensional vector that indicates the probability of selecting each video bitrate under the current environment. For CAST, we set v to 6, corresponding to the bitrate levels used in [10].

Network Architecture. As shown in Fig. 3, the neural network employed in CAST comprises six Conv1D layers with a feature number of 128, a kernel size of 4, and four fully connected layers with 128 neurons. These layers are used to extract the player environment feature, intricate chunk feature, and simple chunk feature. The network is then separated into the policy and value networks by aggregating all these features via a concatenated layer. The policy network outputs a v-dimensional vector that indicates the likelihood of each bitrate being selected, while the value network outputs a scalar. The activation function *ReLU* is utilized for each layer in the policy network, with *softmax* used for the last layer. Meanwhile, the value network employs *tanh* as the activation function.

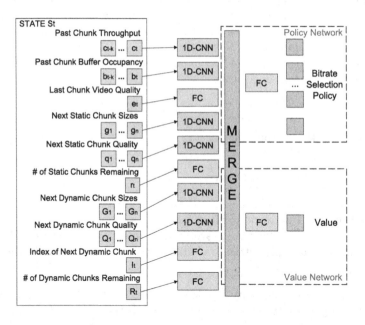

Fig. 3. Neural network architecture.

3.3 Implementation

The implementation of CAST is carried out using TensorFlow [37], and the neural network is constructed using the TFlearn [38] library. The policy network and value network share the same network structure and hyperparameters. To achieve convergence, we set the number of sampled trajectories D as 16 and the learning rate as 10^{-4}. We leverage **eight parallel agents** instead of a single process on a workstation equipped with an Intel Xeon W-2255 Processor and an NVIDIA RTX3090 GPU card to expedite the training process. The training lasts approximately 12 h, after which a stable convergence is reached.

4   Evaluation

4.1   Methodology

Virtual Player. To train CAST, we employ a faithful virtual player with a maximum buffer size of 60 s that accurately simulates the dynamics of video streaming using network traces and video information [10,11]. This approach is more efficient than training on an actual streaming platform and can significantly reduce the training time.

Network Traces. To evaluate CAST's performance on various real-world network conditions, we generate bandwidth traces from two widely used public datasets, namely HSDPA [39] and FCC [40]. The HSDPA dataset comprises throughput measurements collected by a mobile device while in motion, such as on a tram or car, within a mobile wireless network in Norway. On the other hand, the FCC dataset is a broadband dataset that contains throughput traces gathered in the United States of America. To ensure that our evaluation avoids exceptional circumstances where bitrate adaptation is constrained, such as when the capacity is too low to support the lowest bitrate or too high that the highest bitrate is always the optimal choice, we restrict our analysis to traces whose average capacity falls between the minimum and maximum available bitrates. We partition the dataset by randomly selecting 80% of the traces for training and reserve the remaining 20% for testing, as described in Pensieve [10].

Video Parameters. We choose the commonly employed test video, "Envivo-Dash3" [5], which is encoded with six distinct bitrates, namely 300 Kbps, 750 Kbps, 1200 Kbps, 1850 Kbps, 2850 Kbps, and 4300 Kbps, with each chunk having a duration of 4 s. Following the methodology outlined in Sect. 2.2, we categorize video chunks into intricate and simple segments based on their size distribution. For this study, we designate the 1850 Kbps bitrate as the reference track.

Benchmarks. To assess the effectiveness of our proposed CAST algorithm, we have chosen to compare its performance against several established ABR schemes, encompassing both heuristic-based and learning-based approaches. Specifically, we have selected the following representative ABR schemes for evaluation:

- Rate-Based (RB) [26]: This is a basic approach that selects the maximum available bitrate below the estimated capacity.
- FESTIVE [30]: This approach utilizes the harmonic mean of the past five throughput measurements as the capacity prediction value and selects the highest available bitrate below the predicted capacity.
- BOLA [32]: This approach reformulates the ABR task as a utility maximization problem, which is addressed through the Lyapunov function.

- RobustMPC (RMPC) [6]: This hybrid-based method combines both the rate and buffer signal and chooses the bitrate by maximizing a linear-based reward function using information on estimated throughput and buffer occupancy.
- Pensieve [10]: This approach leverages deep reinforcement learning to derive a bitrate selection policy without making any prior assumptions about the environment, which has been shown to provide noteworthy improvements compared to heuristic-based algorithms.

Evaluation Metrics. In this study, we adopt the Elo rating system [41], a well-established method for determining the relative skill levels of players in zero-sum games, as our primary evaluation metric. The Elo rating system computes a player's skill level as a numerical value that changes depending on the outcome of each game. Specifically, after each game, the winner gains points from the loser, and the number of points gained or lost is determined by the difference between the ratings of the two players [42]. For this experiment, we set the initial Elo score to 1000, which is consistent with previous studies [16,17].

QoE Representation. To assess the overall performance of different ABR schemes, we adopt the following QoE function. In practice, quality switching is often considered negligible and thus omitted in the QoE metric, as reported in prior studies [16,34]. Here, V represents VMAF, T denotes rebuffering time, and φ, ζ, ψ are set as 3, 1, and 100, respectively [43]. *It is worth noting that we use this QoE metric only for evaluation and not for training the CAST model.*

$$QoE = \sum_{h \in Complex} \varphi * V(R_h) + \sum_{h \in Simple} \zeta * V(R_h) - \psi \sum_{i=1}^{N} T_i.$$

4.2   Comparison with Existing ABR Algorithms

We first adopt the Elo rating system to evaluate the performance of various ABR methods. Specifically, the Elo ratings of all learning-based approaches are obtained through their trained models.

As presented in Table 3, our results show that CAST outperforms other ABR techniques, exhibiting a performance improvement of 25.84% (Pensieve), 6.71% (RobustMPC), 33.85% (BOLA), 101.49% (FESTIVE), and 161.44% (RB), respectively. The superiority of CAST can be attributed to two factors: (i) its ability to favor intricate chunks without excessively degrading the quality of simple chunks; and (ii) its reliance on actual goals instead of inaccurate linear reward functions.

4.3   CAST vs. Existing ABR Approaches

Figure 4 presents a comparative analysis of the performance of various ABR schemes in terms of the quality of intricate/simple chunks and rebuffering time over the HSDPA dataset. The results show that CAST achieves the highest

Table 3. Elo Rating for Different ABR Algorithms.

Scheme	Elo Rating
CAST	**1514.76**
Pensieve	1203.74
RobustMPC	1419.45
BOLA	1131.7
FESTIVE	751.77
RB	579.39

quality for intricate chunks among the existing schemes, with an improvement ranging from 6.69% to 19.13% (Fig. 4(a)). Moreover, CAST performs similarly to other approaches in terms of simple chunk quality (Fig. 4(b)) and also maintains a lower stall time (Fig. 4(c)[1]). The overall QoE results show that CAST

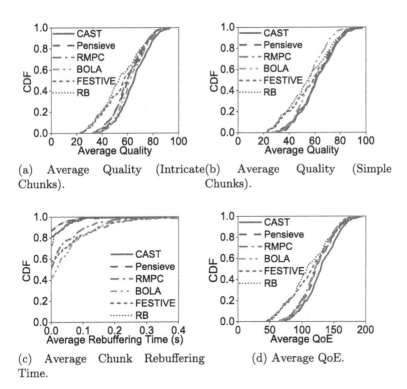

(a) Average Quality (Intricate Chunks).

(b) Average Quality (Simple Chunks).

(c) Average Chunk Rebuffering Time.

(d) Average QoE.

Fig. 4. Performance comparison of CAST vs existing algorithms over the HSDPA dataset.

---

[1] Here, we employ the average chunk rebuffering time as an indicator of the rebuffering experience. This measure is computed by dividing the total rebuffering time by the total number of chunks [19].

outperforms other schemes with an improvement ranging from 4.91% to 17.46% (Fig. 4(d)).

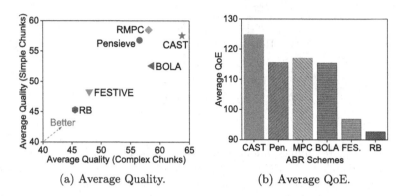

(a) Average Quality.   (b) Average QoE.

Fig. 5. Comparing CAST with existing schemes across FCC traces (in sub-figure (b), Pen. represents Pensieve, MPC indicates RobustMPC, FES. denotes FESTIVE).

Furthermore, we evaluate CAST across the FCC traces and observe that it improves the quality of intricate chunks by 8.69% to 40.03% compared to the other schemes (Fig. 5(a)). CAST achieves a similar quality for simple chunks as the learning-based method Pensieve (Fig. 5(a)). Finally, we also observe that CAST outperforms other approaches and enhances the overall QoE by up to 34.65% (Fig. 5(b)), thereby validating its superiority.

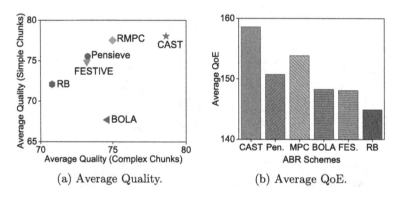

(a) Average Quality.   (b) Average QoE.

Fig. 6. Comparing CAST with existing schemes across LTE traces (in sub-figure (b), Pen. represents Pensieve, MPC indicates RobustMPC, FES. denotes FESTIVE).

4.4   Performance Under Different Network Environments

In the experiments reported earlier, CAST was trained and evaluated on the HSDPA and FCC datasets. However, in practice, CAST may encounter diverse

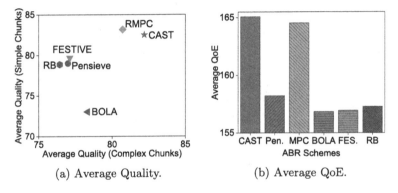

(a) Average Quality.   (b) Average QoE.

Fig. 7. Comparing CAST with existing schemes across Oboe traces (in sub-figure (b), Pen. represents Pensieve, MPC indicates RobustMPC, FES. denotes FESTIVE).

network conditions. To evaluate the generalization ability of CAST, we test it on two additional datasets: LTE traces [44] collected in China and a dataset collected by Oboe [45] from wired, WiFi, and cellular networks. As shown in Fig. 6 and Fig. 7, we find that CAST's performance in improving the quality of intricate chunks remains optimal even under new network conditions, achieving improvements of up to 9.46% and 5.25% across the LTE and Oboe datasets, respectively. Moreover, CAST exhibits a high quality for simple chunks and low rebuffering time, demonstrating its robustness and superiority in diverse network settings.

4.5 Performance Under Different Player Settings

Here, we conduct a sensitivity analysis of CAST under different player configurations with the HSDPA traces by varying the maximum buffer size from 10 s to 50 s. Figure 8 illustrates the results, indicating a significant improvement in the performance of all algorithms as the maximum buffer capacity increases from 10 s to 30 s. This is because a larger buffer size enables the player to store more video content to handle network capacity variations and network breakdowns effectively. However, once the buffer size is sufficient, further increase in the buffer size does not significantly improve the overall performance. Even under the tight buffer size of 10 s, CAST still achieves superior performance compared to existing methods, with an improvement of up to 69.68%, verifying its robustness and effectiveness.

4.6 Subjective Tests from Real Users

To evaluate the performance of CAST in practice, we set up a testbed deployment consisting of a client player using Dash.js [46], an ABR server, and a content server for storing video content. CAST is implemented on the local ABR server to reduce the computational burden on the video client [10]. We

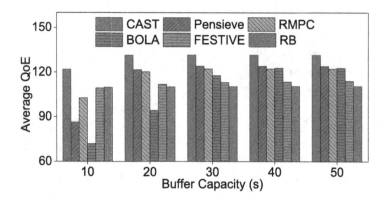

Fig. 8. Average QoE under different buffer capacities over the HSDPA dataset.

invited 20 volunteers to watch online videos using different ABR schemes in an HSDPA network environment. Since assessing viewers' subjective feelings is complex, we asked them to complete a questionnaire and rate their responses rather than providing a quantitative score for the overall viewing quality [19,47]. The questionnaire results are summarized in Table 4. Our analysis of the subjective evaluation data reveals that CAST provides users with an enjoyable viewing experience, with 16 out of 20 volunteers agreeing. Additionally, 19 out of 20 volunteers agreed that CAST offered them good quality for the dynamic chunks. These results confirm the effectiveness of CAST in practice.

Table 4. Subjective tests from 20 real users via watching videos with different ABR schemes.

Questions	Disagree	Neutral	Agree
CAST provides you the optimal viewing experience	2	2	16
CAST delivers a good overall playback quality	1	2	17
CAST offers you a satisfactory quality for chunks with dynamic scenes	0	1	19
CAST incurs few playback freezing events	1	4	15

5   Conclusions

This paper presents a new self-play reinforcement learning-based bitrate adaptation method, CAST, incorporating scene complexity and training with multiple agents in parallel. Through explicit goal training, CAST can effectively meet the actual requirements of the system. Experimental evaluations demonstrate that CAST outperforms existing ABR approaches regarding quality for intricate chunks across diverse network environments. Moreover, CAST achieves comparable quality for simple chunks as prior methods without inducing excessive rebuffering incidents.

References

1. Hu, J., Huang, J., Li, Z., Wang, J., He, T.: A receiver-driven transport protocol with high link utilization using anti-ECN marking in data center networks. IEEE Trans. Netw. Serv. Manage. **20**(2), 1898–1912 (2023)
2. Sodagar, I.: The MPEG-DASH standard for multimedia streaming over the internet. IEEE Multimed. **18**(4), 62–67 (2011)
3. Qin, Y., et al.: ABR streaming of VBR-encoded videos: characterization, challenges, and solutions. In: Proceedings of ACM CoNEXT, pp. 366–378 (2018)
4. Qin, Y., Hao, S., Pattipati, K.R., Qian, F.: Quality-aware strategies for optimizing ABR video streaming QoE and reducing data usage. In: Proceedings of ACM MMSys, pp. 189–200 (2019)
5. Li, W., Huang, J., Wang, S., Wu, C., Liu, S., Wang, J.: An apprenticeship learning approach for adaptive video streaming based on chunk quality and user preference. IEEE Trans. Multimed. **25**, 2488–2502 (2023)
6. Yin, X., Jindal, A., Sekar, V., Sinopoli, B.: A control-theoretic approach for dynamic adaptive video streaming over HTTP. In: Proceedings of ACM SIGCOMM, pp. 325–338 (2015)
7. Yadav, P.K., Shafiei, A., Ooi, W.T.: QUETRA: a queuing theory approach to DASH rate adaptation. In: Proceedings of ACM MM (2017)
8. Qin, Y., Jin, R., Hao, S., Pattipati, K.R., Qian, F.: A control theoretic approach to ABR video streaming: a fresh look at PID-based rate adaptation. In: Proceedings of IEEE INFOCOM (2017)
9. Wang, B., Ren, F.: Towards forward-looking online bitrate adaptation for DASH. In: Proceedings of ACM MM (2017)
10. Mao, H., Netravail, R., Alizadeh, M.: Neural adaptive video streaming with pensieve. In: Proceedings of ACM SIGCOMM, pp. 197–210 (2017)
11. Huang, T., et al.: Quality-aware neural adaptive video streaming with lifelong imitation learning. IEEE J. Sel. Areas Commun. **38**(10), 2324–2342 (2020)
12. Kan, N., Li, C., Yang, C., Dai, W., Zou, J., Xiong, H.: Uncertainty-aware robust adaptive video streaming with bayesian neural network and model predictive control. In: Proceedings of ACM NOSSDAV, pp. 17–24 (2021)
13. Yan, F.Y., et al.: Learning in situ: a randomized experiment in video streaming. In: Proceedings of USENIX NSDI, pp. 495–511 (2020)
14. Yuan, D., Zhang, Y., Zhang, W., Liu, X., Du, H., Zheng, Q.: PRIOR: deep reinforced adaptive video streaming with attention-based throughput prediction. In: Proceedings of ACM NOSSDAV (2022)
15. Zuo, X., Yang, J., Wang, M., Cui, Y.: Adaptive bitrate with user-level QoE preference for video streaming. In: Proceedings of IEEE INFOCOM, pp. 1279–1288 (2022)
16. Huang, T., Zhang, R., Sun, L.: Zwei: a self-play reinforcement learning framework for video transmission services. IEEE Trans. Multimed. **24**(1), 1350–1365 (2022)
17. Huang, T., Yao, X., Wu, C., Zhang, R.X., Pang, Z., Sun, L.: Tiyuntsong: a self-play reinforcement learning approach for ABR video streaming. In: Proceedings of IEEE ICME, pp. 1678–1683 (2019)
18. Huang, T., Zhou, C., Zhang, R.X., Wu, C., Yao, X., Sun, L.: Stick: a harmonious fusion of buffer-based and learning-based approach for adaptive streaming. In: Proceedings of IEEE INFOCOM, pp. 1967–1976 (2020)
19. Li, W., Huang, J., Lyu, W., Guo, B., Jiang, W., Wang, J.: RAV: learning-based adaptive streaming to coordinate the audio and video bitrate selections. IEEE Trans. Multimed. (2022). https://doi.org/10.1109/TMM.2022.3198013

20. Li, W., Huang, J., Liang, Y., Liu, J., Gao, F.: Synthesizing audio and video bitrate selections via learning from actual requirements. In: Proceedings of IEEE ICME (2022)
21. Li, W., Huang, J., Liu, J., Jiang, W., Wang, J.: Learning audio and video bitrate selection strategies via explicit requirements. IEEE Trans. Mob. Comput. (2023). https://doi.org/10.1109/TMC.2023.3265380
22. Huang, T., Zhou, C., Zhang, R., Wu, C., Sun, L.: Learning tailored adaptive bitrate algorithms to heterogeneous network conditions: a domain-specific priors and meta-reinforcement learning approach. IEEE J. Sel. Areas Commun. **40**(8), 2485–2503 (2022)
23. Qiao, C., Li, G., Ma, Q., Wang, J., Liu, Y.: Trace-driven optimization on bitrate adaptation for mobile video streaming. IEEE Trans. Mob. Comput. **21**(6), 2243–2256 (2022)
24. Lipa, D.: Physical (Official Music Video) (2020). https://www.youtube.com/watch?v=9HDEHj2yzew
25. Xiph. Org, Xiph.org Video Test Media (2016). https://media.xiph.org/video/derf/
26. Liu, C., Bouazizi, I., Gabbouj, M.: Rate adaptation for adaptive HTTP streaming. In: Proceedings of ACM MMSys, 2011, pp. 169–174 (2011)
27. ITU-T P. 910, Subjective Video Quality Assessment Methods for Multimedia Applications (2008)
28. Huang, T., Zhang, R., Zhou, C., Sun, L.: QARC: video quality aware rate control for real-time video streaming based on deep reinforcement learning. In: Proceedings of ACM MM, pp. 1208–1216 (2018)
29. Huang, J., et al.: Opportunistic transmission for video streaming over wild internet. ACM Trans. Multimed. Comput. Commun. Appl. **18**(140), 1–22 (2023)
30. Jiang, J., Sekar, V., Zhang, H.: Improving fairness, efficiency, and stability in HTTP-based adaptive video streaming with FESTIVE. In: Proceedings of ACM CoNEXT (2012)
31. Huang, T.Y., Johari, R., McKeown, N., Trunnell, M., Watson, M.: A buffer-based approach to rate adaptation: evidence from a large video streaming service. In: Proceedings of ACM SIGCOMM, pp. 187–198 (2014)
32. Spiteri, K., Urgaonkar, R., Sitaraman, R.K.: BOLA: near-optimal bitrate adaptation for online videos. In: Proceedings of IEEE INFOCOM, pp. 1–9 (2016)
33. Li, Z., Aaron, A., Katsavounidis, I., Moorthy, A., Manohara, M.: Toward a practical perceptual video quality metric. https://techblog.netflix.com/2016/06/toward-practical-perceptual-video.html
34. Huang, T.Y., et al.: Hindsight: evaluate video bitrate adaptation at scale. In: Proceedings of ACM MMSys, pp. 86–97 (2019)
35. Bai, Y., Jin, C., Yu, T.: Near-optimal reinforcement learning with self-play. In: Proceedings of NeurIPS, pp. 1–12 (2020)
36. Ye, D., et al.: Mastering intricate Control in MOBA Games with Deep Reinforcement Learning (2019). arXiv preprint arXiv:1912.09729
37. Abadi, M., Barham, P., Chen, J., et al.: TensorFlow: a system for large-scale machine learning. In: Proceedings of USENIX OSDI, pp. 265–283 (2016)
38. Yuan, T.: TF.Learn: TensorFlow's High-level Module for Distributed Machine Learning (2017). https://tflearn.org/
39. Riiser, H., Vigmostad, P., Griwodz, C., Halvorsen, P.: Commute path bandwidth traces from 3G networks: analysis and applications. In: Proceedings of ACM MMSys, pp. 114–118 (2013)

40. Federal Communications Commission, Raw Data - Measuring Broadband America (2016). https://www.fcc.gov/reports-research/reports/measuring-broadband-america/raw-data-measuring-broadband-america-2016
41. Coulom, R.: Whole-history rating: a bayesian rating system for players of time-varying strength. In: Proceedings of Springer International Conference on Computers and Games (2008)
42. Elo Rating System. https://en.wikipedia.org/wiki/Elo_rating_system
43. Li, W., Huang, J., Wang, S., Liu, S., Wang, J.: DAVS: dynamic-chunk quality aware adaptive video streaming using apprenticeship learning. In: Proceedings of IEEE GLOBECOM, pp. 1–6 (2020)
44. The ACM Multimedia 2018 Live Video Streaming Grand Challenge, LTE/WiFi Dataset (2018). https://www.aitrans.online/competition_detail/competition_id=2
45. Akhtar, Z., et al.: Oboe: auto-tuning video ABR algorithms to network conditions. In: Proceedings of ACM SIGCOMM, pp. 44–58 (2018)
46. Dash.js (2017). https://github.com/Dash-Industry-Forum/dash.js
47. Qiao, C., Wang, J., Liu, Y.: Beyond QoE: diversity adaption in video streaming at the edge. IEEE/ACM Trans. Networking **29**(1), 289–302 (2021)

A Hybrid Active and Passive Cache Method Based on Deep Learning in Edge Computing

Zhengchang Song[1], Rui Cao[1], Bingxin Niu[1(✉)], Junhua Gu[1], and Chunjie Li[2]

[1] School of Artificial Intelligence & Hebei Province Key Laboratory of Big Data Calculation, Hebei University of Technology, Tianjin, China
niubingxin666@163.com
[2] Hebei Expressway Group Limited, Shijiazhuang, China

Abstract. In recent years, the rapid development of micro-video in the multimedia field has brought about a huge increase in Internet traffic. Ensuring consumer quality of experience (QoE) has become a major challenge for Internet service providers (ISPs). To alleviate the burden of Internet traffic, we propose a hybrid active-passive cache update strategy. For newly released micro-videos, the multimodal transformer popularity prediction model (MTPP) is used to actively predict the popularity of micro-videos. For micro-video files in the base station, a dynamic cache based on the popularity prediction algorithm (DCPP) is used to update the local popularity through changes in local requests. Extensive experiments on public datasets demonstrate that our proposed popularity prediction method outperforms traditional prediction models in the field of micro-video prediction. In the simulation experiment, our proposed dynamic cache algorithm based on popularity prediction outperforms the traditional cache replacement algorithm.

Keywords: active-passive cache · multimodal popularity prediction · dynamic caching · deep learning

1 Introduction

With the increasing popularity of social media, more and more users are sharing information and interacting online, which has greatly changed the way people entertain and communicate with each other. Micro or online short videos have become a new trend in user-generated content, as streaming video traffic continues to surge with the growing number of mobile users. By 2023, video traffic alone will account for 82% of global internet usage [1]. During video playback, initial buffering delay and stalling are key factors that hinder users from watching videos. The requirement for quality of experience (QoE) can be defined as meeting the initial delay requirement and ensuring minimal stalling during playback [2]. Deploying caching of popular content in cellular networks helps compensate for the sustained network latency resulting from the huge increase

Z. Tari et al. (Eds.): ICA3PP 2023, LNCS 14490, pp. 148–159, 2024.
https://doi.org/10.1007/978-981-97-0859-8_9

in mobile video consumption. As cache capacity is limited, it is crucial to determine which videos should be cached. Furthermore, only a few very popular videos are accessed by a large number of users, which is the reason for the increase in traffic, while most content remains rarely accessed. Therefore, predicting the popularity of video content in advance is critical. Different studies on the popularity of video content show that they follow a Zipf distribution [3], where 10% of videos contribute to nearly 80% of views, while the remaining 90% of videos only contribute to 20% of total views. Therefore, predicting the popularity of content is a crucial factor, as cache storage size is limited. As an emerging research field, short video caching poses unique challenges to the design of caching strategies. First, in traditional video-on-demand platforms, the length of videos ranges from a few minutes to several hours, the daily output of new videos is usually not high, and users usually request several videos per day. But for short videos, the video length is usually no more than one minute, and the daily release volume of short videos is huge, and users often request dozens of videos in a short period of time. Therefore, if the newly released popular short video is not cached on the edge node in time, it will cause significant network delay. Finally, in the short video network, popular topics change rapidly. Researchers have statistically analyzed the most popular videos and found that the number of visits dropped from around 1,500 after just 50 min to under 100, illustrating that the rapidly changing popularity of videos makes designing efficient caching strategies even more challenging [4]. The contributions of this article can be summarized as follows:

1) We propose a multimodal transformer popular prediction model (MTPP) to explore knowledge from different modalities, where in order to address data loss and dimensional differences between different modalities, considering the need to learn the interaction between different modes in multi-modal prediction tasks. On the basis of using multi-modal deep learning models for feature extraction, we introduce the Transformer model into our framework. This enables short videos with high predictive heat to deploy to the edge cache before the user request arrives, effectively solving the cold start of short video cache.

2) We propose a dynamic caching algorithm based on popularity prediction (DCPP), which combines active prediction of newly released short videos with local request changes to address the characteristics of difficulty in predicting popularity in the early stages of short video publishing and rapid changes in popularity after publishing.

2   Related Work

Existing cache decision-making can be divided into two categories: I) Reactive caching II) Proactive caching In reactive caching, the edge node makes a caching decision (whether to store the requested content) only when a request for specific content arrives [5] and [6]. By caching a large amount of dynamic content in the edge cache, the average response time can be shortened, which benefits from higher cache hit rates. Effective strategies have been proposed in [7] and [8] that focus on utilizing historical or statistical data of user demands. In [9], authors predict the popularity score of future content based on input historical request records.

In proactive caching, the edge node proactively predicts the popularity of content and makes caching decisions before any user request [10] and [11] transforms popularity prediction in deep learning into a classification problem by discretizing the popularity of content into multiple categories. Therefore, improving prediction accuracy is essential for improving cache hit rates. The accuracy of the popularity prediction model affects the cache hit rate. The popularity of content changes dynamically based on different factors such as events, content type, and content lifespan. Deep learning has achieved significant success in areas such as speech recognition and computer vision due to its high prediction accuracy. Based on deep learning popularity prediction models have been widely used for predicting video content popularity. For instance. Bielski et al. [12] predict the popularity of videos before their release using soft self-attention mechanism to mine the spatiotemporal features of videos. McParlane et al. [13] adopt a content-based strategy and only use visual appearance and user context to predict the popularity of Flickr images. Chen et al. [14] first use four types of heterogeneous features, namely visual, acoustic, textual, and social features to characterize short videos and propose an inductive multi-modal learning model that finds an optimal common space for all modalities. Later, Jing et al. [15] discuss the adverse effects of internal noise on micro-video analysis research, but few works consider integrating the use of deep learning to predict popularity into caching. Taking into account the above two situations, this article proposes a method that combines active caching and passive caching.

Due to the large number of new short videos being released in each time slot, we use deep learning-based short video popularity prediction to address the proactive caching problem for newly released short videos. For videos in the base station, a strategy based on updating the local popularity of the base station according to user requests is used to address the dynamic changes in short video popularity.

3   System Model

This chapter studies a simple MEC scenario consisting of an application provider, a cloud center, an SBS, and a MEC server. We assign a unique index for each video content stored in the cache, which serves as the unique content ID. We assume that in each time slot, the users in the SBS coverage area can request a fixed number of contents, which can be represented as $F \in \{f1, f2, \ldots, fn\}$. The network operates in a time slot manner, and a group of time slots can be defined as $T \in \{t1, t2, \ldots, tn\}$. After the user's request reaches SBS, the content in the MEC server cache is first retrieved. If the requested content is retrieved from the MEC server, it will be directly transmitted to the user. Otherwise, the user request will be sent to the cloud and the required content will be retrieved and sent from the content server provider. Finally, the central cloud delivers content to users through SBS (Fig. 1).

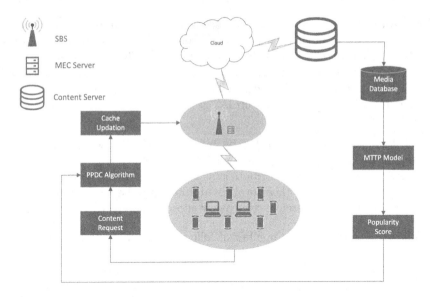

Fig. 1. Caching system architecture

4 Proposed Methodology

Existing research has recognized that the key factors affecting predicting the popularity of online videos are content and social factors. Specifically, we apply visual, auditory, and textual modalities to represent content factors, and social modalities to represent uploader factors. Therefore, we use a combination of extracted features to represent each video.

4.1 Feature Extraction from Four Modalities

Visual Modality. Based on the calculation method of viewing times on online video platforms and its relatively short length nature, a single video's theme is usually singular. This allows us to use keyframes to represent the intrinsic theme of the entire video. Therefore, we adopt a mean pooling strategy to extract fixed-length vector features from certain keyframes and obtain deep visual features of micro-videos through pre-trained ResNet-50 [16] on ImageNet [17].

Acoustic Modality. As we know, acoustic information and visual information are inseparable and interconnected in conveying information and enhancing atmosphere, such as movie review videos that narrate the speaker's ideas and perspectives, and entertaining clips with popular background music on streaming services. These features have a great impact on the popularity of videos. Therefore, we extract auditory features using VGGish [18] pre-trained on AudioSet [19].

Text Modality. Textual information is a common condition that has a considerable impact on popularity prediction. Specifically, the textual modality is usually revealed through video titles and descriptions. Through textual keyword and sentiment reflection, the video will be disseminated to vertical segmented fields. We adopt pre-trained Glove [20]. The designed model that is more suitable for short texts to extract text features.

Social Factors. Videos uploaded by internet celebrities with high follower counts and platform authentication will become trendier after publishing. For example, the view counts contributed by fans will statically converge to the number of followers due to their loyalty. Additionally, a larger audience size will attract more viewers. Therefore, this paper encodes four types of social clues:

- Follower count: given the number of followers and the followed accounts of the publisher;
- View count: the number of views after short video uploading and the total view count of all short videos posted by the publisher;
- Post count: the total number of posts of each publisher;
- Twitter verification: a binary value that reflects whether the publisher is a verified user or not.

4.2 Multimodal Prediction Model

For edge caching scenarios, predicting the popularity of newly published short videos in advance is necessary to solve the cold start problem. However, predicting the popularity of new online videos is challenging in multimedia. In the task of multimodal short video prediction, Most existing prediction models only consider the complementarity between different modules. In order to better solve the problems of missing data and inconsistent feature dimensions between original features, as well as processing position information in input sequences and learning the correlation between different modalities. Our prediction model mainly adopts Transformer [21], which is currently an important model in the field of natural language processing with both transformer network and self-attention mechanism network as its core.

First, A self attention mechanism in transformer can be used to encode original features to observe more effective feature representations, which fully solve problems such as missing data between different modal features, significant dimensional differences, and modularity interference Second, the transformer network introduces position encoding, which enables the model to better handle position information in the input sequence. The transformer network can simultaneously handle input data from different modalities, such as video images and audio. By using a multi-head attention mechanism, the transformer network can effectively integrate information from different modalities and better learn the correlation between them (Fig. 2).

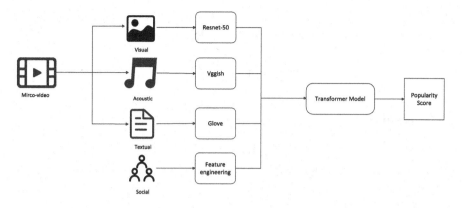

Fig. 2. MTPP Model

5 Experiment Setup

5.1 Dataset Setup

The micro video popularity prediction dataset used in this paper was built by the Media Lab of the National University of Singapore[1]. This dataset contains a total of 303,242 user-generated micro videos collected from the online short video sharing website Vine, uploaded by 98,166 users. All micro videos are no longer than 8 s, with approximately 75% of them being 6 to 7 s long. Because the popularity is related to the high degree of social networking, when calculating the final popularity score of each micro video, you need to consider the average of the four statistical data of comments, sharing, like, and browsing. In the experiments of this paper, only 10,235 pieces of data were downloaded from the dataset. 10 rounds of random experiments were conducted on the dataset, using 90% of the samples for training and the remaining 10% for testing. The average result of the 10 tests was ultimately taken. Both training and testing were conducted on a GPU with a GeForce RTX 3090 configuration. The model framework operating environment is as follows: python $= 3.6$; pytorch $= 1.7.1$; numpy $= 1.19.5$. The model uses stochastic gradient descent (SGD) with a learning rate set to 0.01.

5.2 Evaluating Indicator

This article uses normalized mean square error to measure the consistency between predicted and true values

$$\text{nMSE} = \frac{1}{n} \sum_{i=1}^{n} \frac{(y_i - \hat{y}_i)^2}{\sigma_y^2}. \tag{1}$$

---

[1] https://acmmm2016.wixsite.com/micro-videos.

In addition, we measure the Spearman's Rank Correlation (SRC) coefficient as a complementary metric, which is defined as follows:

$$\text{SRC} = 1 - \frac{6 \sum d_i^2}{n(n^2 - 1)} \tag{2}$$

5.3 Baselines

We compared the proposed algorithm framework with existing methods,

- Multiple social network learning (MSNL) [22] solves incomplete data in terms of source credibility and source consistency by simultaneously modeling source credibility and source consistency.
- Multi-view discriminant analysis (MvDA) [23] is a multi-view learning model that searches for a potential common space by enhancing the view consistency of multi-linear transformations.
- Transductive multi-modal learning (TMALL) [14] is a multi-modal learning model for predicting the popularity of micro-videos, which unifies and preserves different modal features in a latent common space to address the problem of information insufficiency.
- Transductive low-rank multi-view regression (TLRMVR) [15] sets a new low-rank constraint on the learned short video embedding in the model, retaining only the main components in the feature space in the final feature representation.
- The multimodal variational encoder- decoder (MMVED-REG) [24] encodes and fuses four modalities of micro- videos to obtain a high-level hidden stochastic representation, where the popularity score could be decoded. The suffixes -REG represents the decoding structure for using MMVED for popularity prediction tasks (Tables 1 and 2).

It can be seen that compared to models that only consider the complement tarity between different modules and MMVED based on automatic encoding methods, our proposed MTPP framework shows better performance when evaluated using nMSE and SRC as metrics. This is because the transformer network used in our model has stronger representation ability. The transformer model has a multi-layer self attention mechanism and positional encoding, enabling it to capture long-range dependencies in input sequences, enabling the model to better understand and represent complex patterns in multimodal data. In addition, the transformer network also improves the model's generalization ability for samples with large modal differences.

Table 1. Comparison of prediction result between proposed MTPP and several baselines.

Methods	nMSE
MSNL	1.098
MvDA	0.982
TMALL	0.979
TLRMVR	0.934
MMVED-REG	0.914
MTPP	0.911

Table 2. Comparison of prediction result between proposed MTPP and several baselines in terms of SRC.

Methods	SRC
MSNL	0.356
TMALL	0.370
MvDA	0.391
MMVED-REG	0.762
MTPP	0.768

Lower nMSE and higher SRC indicate better performance.

6   Cached Updates

This chapter introduces a video cache update method called dynamic cache based on the popularity prediction. Specifically, the algorithm determines whether to add it to the cache based on the popularity of each new micro video, and first provide a video from the cache when the user request. At the same time, when a video's global popularity exceeds a certain threshold, it will be added to the cache to replace less popular files.

The core part of the cache update algorithm predicts the popularity of new upload videos based on predictive models, and determines whether it is added to the cache based on the popularity score. For cached videos, if they are present in the cache when a user requests them, they are immediately provided with the video, and the local popularity is updated. Otherwise, the video will be downloaded from the core network, and its global popularity will be updated. Additionally, if a video's global popularity exceeds a certain threshold, it will replace the least popular file in the current cache.

Algorithm 1: DCPP algorithm

Input: Video request vi, add video vj
Output: updated popular files of v in Cache C

1 **for** *each add video v_j* **do**
2   **if** *predicted popularity of v_j > threshold* **then**
3   cache C_s updation

4 **for** *each video request v_i* **do**
5   **if** *v_i belongs to smallcell cache C_s* **then**
6   serve from cache C_s
7   update P_f local of video v_i in C_s

8   **else**
9   download via backhaul
10   update P_f global of video v_i
11   **if** *P_f global > threshold* **then**
12   cache updation

---

7   Evaluation Results

In this section, we first introduce the setup of the experimental simulation and the baseline used for comparison. Then, we evaluate the performance of the proposed algorithm through simulation experiments.

7.1   Simulation Setup

In the simulation experiment, it is assumed that the size of each video is the same, and there are five thousand requests generated in each time slot t. In order to simulate the fast-changing characteristics of short video popularity, the popularity of each video fluctuates randomly by 10% in each time slot t. Each request is determined to access a file through the Zipf distribution. The transmission delay from MEC server to user is defined as 10 ms, and the transmission delay from content server to user is 135 ms.

7.2   Evaluation Results

We chose FIFO caching algorithm, random caching algorithm, and LFU caching algorithm to compare the cache hit rate and average latency with our algorithm under different cache capacities (Figs. 3 and 4).

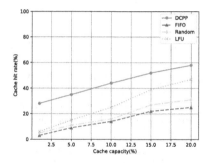 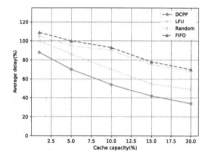

Fig. 3. The impact of cache capacity on cache hit rate

Fig. 4. The impact of cache capacity on average latency

We compared our proposed DCPP caching algorithm with FIFO, Random, and LFU caching algorithms to evaluate the cache hit rate and average latency under different cache capacities. Compared with traditional caching policies, our proposed DCPP caching algorithm significantly improves both cache hit rate and average latency.

Since FIFO algorithm always replaces the most popular video in the base station every time, its caching performance is the poorest. LFU algorithm replaces the least accessed video in each time slot t, so it achieves better performance. Random algorithm randomly replaces videos in the base station, so its performance is between that of FIFO and LFU. Since DCPP algorithm predicts the popularity of new videos each time and places highly popular videos in the cache, while dynamically updating their popularity based on the number of accesses, it achieves better performance than the other algorithms.

8   Conclusion

In this paper, we propose a multimodal transformer prevalence prediction model and a dynamic cache based population prediction algorithm. The effectiveness of the algorithm is verified by simulation experiments. The results show that the algorithm can effectively improve the cache hit rate and reduce the average delay in consideration of predicting the popularity of new short videos and changes in the popularity of short videos.

Acknowledgements. This work was supported in part by the Hebei Province Innovation Capability Enhancement Plan Project under Grant 22567603H, in part by S&T Program of Hebei under Grant 20310801D.

References

1. Cisco Visual Networking Index: Forecast and Methodology 2016–2021. CISCO White Paper (2017)
2. Martin, A., et al.: Network resource allocation system for QOE-aware delivery of media services in 5g networks. IEEE Trans. Broadcast. **64**(2), 561–574 (2018)
3. Chen, Z., He, Q., Mao, Z., Chung, H.M., Maharjan, S.: A study on the characteristics of Douyin short videos and implications for edge caching. In: Proceedings of the ACM Turing Celebration Conference, China, pp. 1–6 (2019)
4. Zhang, Y., Li, P., Zhang, Z., Bai, B., Zhang, G., Wang, W., Lian, B.: Challenges and chances for the emerging short video network. In: IEEE Conference on Computer Communications Workshops (INFOCOM WKSHPS), IEEE INFOCOM 2019, pp. 1025–1026. IEEE (2019)
5. Thar, K., Tran, N.H., Ullah, S., Oo, T.Z., Hong, C.S.: Online caching and cooperative forwarding in information centric networking. IEEE Access **6**, 59679–59694 (2018)
6. Ndikumana, A., et al.: In-network caching for paid contents in content centric networking. In: 2017 IEEE Global Communications Conference, GLOBECOM 2017, pp. 1–6. IEEE (2017)
7. Abolhassani, B., Tadrous, J., Eryilmaz, A.: Achieving freshness in single/multi-user caching of dynamic content over the wireless edge. In: 2020 18th International Symposium on Modeling and Optimization in Mobile, Ad Hoc, and Wireless Networks (WiOPT), pp. 1–8. IEEE (2020)
8. Kam, C., Kompella, S., Nguyen, G.D., Wieselthier, J.E., Ephremides, A.: Information freshness and popularity in mobile caching. In: 2017 IEEE International Symposium on Information Theory (ISIT), pp. 136–140. IEEE (2017)
9. Thar, K., Oo, T.Z., Tun, Y.K., Kim, K.T., Hong, C.S., et al.: A deep learning model generation framework for virtualized multi-access edge cache management. IEEE Access **7**, 62734–62749 (2019)
10. Chen, M., Saad, W., Yin, C., Debbah, M.: Echo state networks for proactive caching in cloud-based radio access networks with mobile users. IEEE Trans. Wirel. Commun. **16**(6), 3520–3535 (2017)
11. Liu, W.X., Zhang, J., Liang, Z.W., Peng, L.X., Cai, J.: Content popularity prediction and caching for ICN: a deep learning approach with SDN. IEEE Access **6**, 5075–5089 (2017)
12. Bielski, A., Trzcinski, T.: Understanding multimodal popularity prediction of social media videos with self-attention. IEEE Access **6**, 74277–74287 (2018)
13. McParlane, P.J., Moshfeghi, Y., Jose, J.M.: "Nobody comes here anymore, it's too crowded"; predicting image popularity on Flickr. In: Proceedings of International Conference on Multimedia Retrieval, pp. 385–391 (2014)
14. Chen, J., Song, X., Nie, L., Wang, X., Zhang, H., Chua, T.S.: Micro tells macro: predicting the popularity of micro-videos via a transductive model. In: Proceedings of the 24th ACM International Conference on Multimedia, pp. 898–907 (2016)
15. Jing, P., Su, Y., Nie, L., Bai, X., Liu, J., Wang, M.: Low-rank multi-view embedding learning for micro-video popularity prediction. IEEE Trans. Knowl. Data Eng. **30**(8), 1519–1532 (2017)
16. He, K., Zhang, X., Ren, S., Sun, J.: Deep residual learning for image recognition. In: Proceedings of the IEEE Conference on Computer Vision and Pattern Recognition, pp. 770–778 (2016)

17. Russakovsky, O., et al.: ImageNet large scale visual recognition challenge. Int. J. Comput. Vision **115**, 211–252 (2015)
18. Hershey, S., et al.: CNN architectures for large-scale audio classification. In: 2017 IEEE International Conference on Acoustics, Speech and Signal Processing (ICASSP), pp. 131–135. IEEE (2017)
19. Gemmeke, J.F., et al.: Audio set: an ontology and human-labeled dataset for audio events. In: 2017 IEEE International Conference on Acoustics, Speech and Signal Processing (ICASSP), pp. 776–780. IEEE (2017)
20. Pennington, J., Socher, R., Manning, C.D.: Glove: global vectors for word representation. In: Proceedings of the 2014 Conference on Empirical Methods in Natural Language Processing (EMNLP), pp. 1532–1543 (2014)
21. Vaswani, A., et al.: Attention is all you need. In: Advances in Neural Information Processing Systems 30 (2017)
22. Song, X., Nie, L., Zhang, L., Akbari, M., Chua, T.S.: Multiple social network learning and its application in volunteerism tendency prediction. In: Proceedings of the 38th International ACM SIGIR Conference on Research and Development in Information Retrieval, pp. 213–222 (2015)
23. Kan, M., Shan, S., Zhang, H., Lao, S., Chen, X.: Multi-view discriminant analysis. IEEE Trans. Pattern Anal. Mach. Intell. **38**(1), 188–194 (2015)
24. Xie, J., Zhu, Y., Zhang, Z., Peng, J., Chen, Z.: A multimodal variational encoder-decoder framework for micro-video popularity prediction. In: WWW 2020: The Web Conference 2020 (2020)

We Will Find You: An Edge-Based Multi-UAV Multi-Recipient Identification Method in Smart Delivery Services

Yi Xu[1]([✉]), Ruyi Guo[1], Jonathan Kua[2], Haoyu Luo[3], Zheng Zhang[1], and Xiao Liu[2]

[1] School of Computer Science and Technology, Anhui University, Hefei, China
08055@ahu.edu.cn, guory@stu.ahu.edu.cn, e21201136@stu.ahu.edu
[2] School of Information Technology, Deakin University, Geelong, Australia
{jonathan.kua,xiao.liu}@deakin.edu.au
[3] College of Mathematics and Informatics, South China Agricultural University, Guangzhou, China
luohy@whu.edu.cn

Abstract. Unmanned aerial vehicle (UAV) is increasingly becoming a promising solution for last-mile delivery in smart logistics, and multi-UAV scenarios have become increasingly common. In multi-UAV delivery services, the ability to accurately and efficiently identifying multiple target recipients is a critical issue. Face recognition has been widely used in UAV delivery services but since the UAVs need to passively wait for the target recipient to arrive at the designated location, the efficiency of identification is undesirable. Furthermore, as one UAV is intended only for one recipient, there is no collaboration between multiple UAVs at the same location which could otherwise reduce the computational and communication time for some common tasks. To address the aforementioned issues, in this paper, we propose an active multi-UAV multi-recipient identification method (named MM4ID) in edge-based smart delivery services, based on person re-identification (ReID) technology. Specifically, multiple UAVs scan the recipient's activity area and transmit their video streams to a nearby edge server. In the meantime, a multiple object tracking (MOT) algorithm is performed at the edge server to track the recipients and obtain one image for each recipient from the video streams to reduce duplicated data. We perform person re-identification algorithm on the obtained recipient images to determine the location of the recipient, and then we use the face recognition algorithm to confirm the identity of the recipient. Finally, successful identification results including the target recipients and their locations are sent to all UAVs via broadcast so that each UAV can complete their final delivery task. Experimental results based on a real-world edge-based smart UAV delivery service system successfully demonstrate the effectiveness of our proposed method and yield better performance compared with other representative solutions.

Keywords: Face Recognition · UAV Delivery · Person Re-identification · Multiple Object Tracking

Z. Tari et al. (Eds.): ICA3PP 2023, LNCS 14490, pp. 160–173, 2024.
https://doi.org/10.1007/978-981-97-0859-8_10

1   Introduction

In recent years, smart logistics has become an effective means to improve delivery efficiency and customer satisfaction [1,2]. UAVs delivery service has become a promising solution for last-mile delivery in smart logistics, and scenarios of multiple UAVs for logistics delivery are becoming increasingly common [3–6]. In multiple UAVs delivery systems, fast identification of target recipients with high accuracy is a key challenge for satisfactory user experience [7]. A more effective and cost-efficient solution is to use face recognition to confirm the identity of the recipient [8].

At present, the target recipient identification method based on face recognition algorithm still has some challenges, particularly in the UAV logistics scenario [9,10]. On the one hand, although recognition can be achieved with only face recognition method, its efficiency is low because it requires the recipient to be at a designated location. Until then the UAV needs to passively hover at the designated location and wait for the target recipient to arrive at the designated location, which will incur time costs, energy consumption and reduce the quality of service (QoS). On the other hand, the recipient should be able to receive the delivery from the UAV as soon as it arrives at the designated location. The identification should be done actively by the UAV before the recipient arrives at the designated location. Although there are some existing research works, they all identify the recipient after the recipient has arrived at a designated location, which can be very inefficient. Face recognition is performed after the recipient location is determined, which undoubtedly increased the practicality of UAVs to identify the recipient [11,12]. Pose recognition algorithm is used to proactively find the target recipients' location, but the algorithm can be time-consuming [11]. Person re-identification algorithm is used to proactively find the target recipients' location [12], but it does not consider the highly redundant data generated by the terminal device. In the frame-by-frame data processing, there are several duplicated data between frames, and the data of one pedestrian is processed several times.

To address the aforementioned issues, in this paper, we propose an active multi-UAV multi-recipient identification method (named MM4ID) in edge-based smart delivery services, based on person re-identification. First, the UAVs capture video streams in the recipient's activity area and transmit the data to a nearby edge server. Second, for the video streams captured by UAVs, the multiple object tracking (MOT) [13] algorithm named YOLOv5-DeepSORT is used to track the person and obtain one image for each recipient to reduce duplicated data. Third, we perform person ReID algorithm [14] on the obtained recipient image to determine the location of the recipient. Fourth, we use the face recognition algorithm to confirm the recipient. Finally, successful identification results are sent to all UAVs via broadcast and the corresponding UAV completes the logistics delivery. Experimental results in a simulation of a real smart UAV delivery system demonstrate the effectiveness of the method.

The rest of this paper is organized as follows. Section 2 presents some related work. Section 3 provides an overview of the framework and describes the detailed models and algorithms used in our proposed framework. Section 4 presents the experimental results of the real-world UAV delivery system. Finally, Sect. 5 concludes the paper and indicates future work.

2   Related Work

In the process of UAV delivery with edge computing, the ability to efficiently identify the recipient is a popular research topic [15–17]. There are already some methods based on face recognition algorithms to solve the target recipient identification problem in the UAV logistics delivery scenario. Shao et al. used a face detection algorithm to detect the recipient's faces and then input the detected faces into the face recognition network for identity confirmation [18]. However, since recipients often keep moving and their positions are mostly random in the real world, this approach cannot ensure satisfactory face recognition in UAV delivery scenarios. Shen et al. proposed a face scoring mechanism and a UAV flight control algorithm that enables the UAV to adjust its orientation according to the recipient's pose [19], thus enabling automatic capture of the recipient's frontal for face recognition. Since this method ensures that the UAV always captures the frontal aspects of the recipient, it can improve its effectiveness in UAV delivery scenarios. However, the UAV requires a lot of time and energy to adjust its flight direction to capture the frontal aspect of the recipient. Therefore, this method is not energy efficient for UAVs, which makes it impractical to use in realistic scenarios. Although the recognition accuracy can be achieved in the aforementioned methods, the efficiency of these methods is low because face recognition algorithm requires the recipient to be in a designated location. The UAV needs to passively wait for the target recipient to arrive at a designated location, which will bring time as well as energy consumption and reduce the quality of service (QoS). To solve this problem, Gao et al. proposed a method to first filter some recipients' image by processing the video frame by frame through pose recognition to proactively find the target recipients' location, and then complete the identity confirmation by capturing the frontal face [11]. This method improves the applicability of face recognition, but the time complexity of the pose recognition algorithm is usually high. Therefore, the efficiency of the whole process cannot be guaranteed and powerful computing hardware is required. Zhang et al. performed target recipient identification by person ReID algorithm to complete logistics delivery [12], but the accuracy of single-person ReID is low in practical scenarios, which is difficult to meet the high precision requirements of logistics delivery. In addition, it does not consider the highly redundant data generated by the terminal device.

Therefore, there is a need to design a method for the UAV to actively and efficiently find the target recipient thereby improving the efficiency of target recipient identification.

3   Framework and Algorithms

3.1   Framework

The UAV logistics delivery system in edge computing consists of three main components: the cloud server, the edge server, and the UAV. The cloud server stores all users' information (including their registered face images and full body images) and continuously provides intelligent services to the whole system. The edge server has sufficient computing resources, and it is closer to the end device than the cloud server, which can effectively reduce the data transmission time. AI models and algorithms (such as MOT, person ReID and face recognition algorithms) can be deployed on the edge server. UAVs are used as end devices for transportation systems, such as sensors, cameras, computing devices, batteries, and other resources on board, mainly for flight and cargo-related tasks.

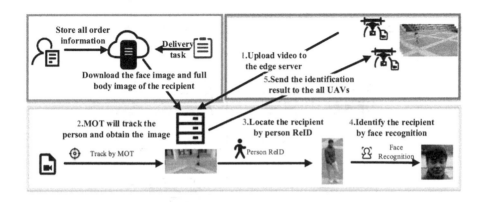

Fig. 1. Processing flow of MM4ID.

The processing flow of MM4ID is shown in Fig. 1. When there is a delivery task, the edge server will request the data of the target recipient on the cloud server and arrange the UAV to deliver the goods. After the UAV arrives in the recipient's activity area, the video captured by the UAV is analyzed frame by frame to determine the location and identify the recipient. The recipient identification process is divided into four main steps as follows.

Step 1: The UAV captures the original video frames in the recipient's activity area and sends the video data to the edge server.
Step 2: The video data captured by the UAV will then be processed on the edge server. The edge server first assigns an identity document (ID) to each pedestrian in the video and tracks and locates it by MOT. For each ID's pedestrian, obtain one image to reduce duplicate data.
Step 3: The recipient is identified using the person ReID with the recipient's picture to determine the recipient's location. After determining the existence

of a pedestrian matching the picture of the recipient by the person ReID, the UAV approaches this pedestrian and shoots the front face images of the pedestrians.

Step 4: The face recognition algorithm is used to confirm whether the face of the pedestrian determined by the person ReID matches the image of the recipient's face downloaded from the cloud server.

Step 5: The edge server broadcasts the identification result to all UAVs after successful matching. After the corresponding UAV receives the identification result, the UAV will land near the target recipient and complete the delivery.

3.2   Algorithms

Multiple Object Tracking (MOT). The MOT used in this paper is YOLOv5-deepsort [20]. YOLOv5 is used as the recipient detector, and deepsort is used to track the detected target recipient. As an object detection algorithm, YOLOv5 can efficiently detect pedestrians and YOLOv5 network structure is shown in Fig. 2 [21], which consists of four components: Input, Backbone, Neck, and Prediction.

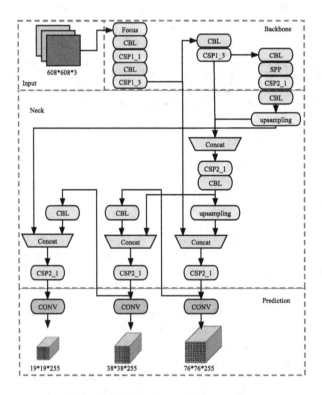

Fig. 2. YOLOv5 network structure.

The Input component completes the preprocessing of the image, including Mosaic data enhancement, adaptive anchor frame calculation, and adaptive image filling. The Backbone component adopts a focused structure. It is responsible for the feature extraction of the input image. The Neck component is the feature pyramid network (FPN) and path aggregation network (PAN) structure. It uses a CSP2 structure borrowed from the cross stage partial network (CSP-net) design, which enhances the ability of network feature fusion. The Prediction adopts the GIoU loss function to achieve faster and better convergence. The loss function is defined as follows:

$$Loss_{GIoU} = 1 - \frac{A^p \bigcap A^g}{A^p \bigcup A^g} + \frac{C - A^p \bigcup A^g}{C} \tag{1}$$

Where A^p denotes the prediction box, A^g denotes the ground truth box, and C denotes the smallest outer rectangle of the two boxes. The main problem involved in detection-based tracking is the problem of prediction, matching, and updating after detection. As a tracking algorithm, deepsort also has good target tracking performance in the presence of occlusion and the process is shown in Fig. 3 [22].

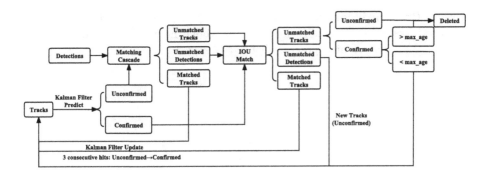

Fig. 3. Processing flow of deepsort.

The target is predicted by Kalman filtering, and the Mahalanobis distance between the obtained predicted target and the current target is calculated. When its value is less than the specified threshold, it means that the predicted target is correct. The Mahalanobis distance is calculated as follows:

$$d^{(1)}(i,j) = (d_j - y_i)^T S_i^{-1} (d_j - y_i) \tag{2}$$

$d^{(1)}(i,j)$ denotes the motion match between the i-th detection target and the j-th trajectory. S_i is the covariance matrix of the observation space of the current moment predicted by the trajectory after using the Kalman filter, y_i denotes the predicted observations of the trajectory at the current moment, and d_j is the position information of the j-th detection target. To prevent the ID from changing constantly, the minimum cosine value between the predicted target and the feature vector of the predicted target contained in the trajectory is used as the apparent matching degree to measure the prediction accuracy. The expression is calculated as follows:

$$d^{(2)}(i,j) = min \left\{ 1 - r_j^T r_k^{(i)} \mid r_k^{(i)} \in R_i \right\} \tag{3}$$

Where r_j denotes one feature vector corresponding to each detection block d_j, $|r_j| = 1$, and r_k is the feature vector of the most recent successful association stored for each tracking target. Finally, the final measure is as follows:

$$c(i,j) = \lambda d^{(1)}(i,j) + (1 - \lambda) d^{(2)}(i,j) \tag{4}$$

Where λ is the weight parameter and finally the Hungarian algorithm is introduced to detect whether a target in the current frame, is the same as a target in the previous frame. To avoid a track being obscured for a period of time or after losing frames for a long time, cascade matching is used to ensure that the predicted target is the target before being obscured or before losing frames for a long time.

Person Re-identification (ReID). In this paper, person ReID is used as a method for recipient localization. The backbone network of person ReID is ResNet50 and the training process is shown in Fig. 4 [14].

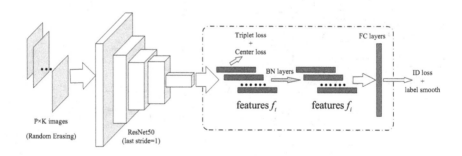

Fig. 4. Training process of person ReID.

Random Erasing Augmentation (REA) is used in the training process to reduce the degree of model overfitting, so it can improve the performance of the model, and the last stride down sampling operation in the backbone is removed to increase the size of the feature map. A batch normalization (BN) layer is added after the features f_t. The features before BN layer are f_t and after BN is f_i, which makes the ID loss easier to converge.

Face Recognition. The face recognition model is trained by the deep convolutional network [8], and triplet loss is used as the loss function. The loss function is as follows:

$$\varphi(A, P, N) = \max\left(\|f(A) - f(P)\|^2 - \|f(A) - f(N)\|^2 + \alpha, 0\right) \quad (5)$$

Where $f(A)$ denotes the face recognition model we need to train. The 128 feature vectors of image A can be calculated through $f(A)$. A is an image of a person, P is another image of the person, N is an image of another person, and α is the spacing distance. The specific training process is to divide the training picture into triples, make A and P as close as possible, and A and N as far as possible. Through the trained face recognition model, the feature values F_i can be extracted, which is composed of 128 feature vectors x_{in} ($n = 1, 2, \ldots, 128$) in the face image. For face matching, the ED (Euclidean distance) is used as the similarity measure between the face feature values. The ED is calculated as follows:

$$ED(F_i, F_j) = \sum_{n=1}^{128} \sqrt[2]{(x_{in} - x_{jn})^2} \quad (6)$$

F_i and F_j represent the face feature values of two images. If the calculated $ED(F_i, F_j)$ is less than the threshold value we set, it is determined that the same person is identified, otherwise, it is not.

4   Evaluation

In this section, we present a case study and experimental results to verify the effectiveness of our proposed method. The videos used in the experiments were collected by using the UAV (DJI Matrice 600 Pro) and all experiments were run on the computer with an 11th generation Intel (R) Core (TM) i5-11400 processor and 16 GB of RAM, with a 3060Ti 8G GPU and Windows 10 (64-bit) operating system. The data and algorithms used in our experiments are made available online on GitHub: https://github.com/GRY-Code/MM4ID.

4.1   Case Study

We introduce MM4ID with a real-world case study. We suppose that there are two delivery tasks T_1 and T_2. UAV_1 and UAV_2 denote the UAVs corresponding to T_1 and T_2. U_1 and U_2 denote the target recipient corresponding to T_1 and T_2, $Image\_Face_1$ and $Image\_Face_2$ denote the face images of U_1 and U_2, $Image\_Body_1$ and $Image\_Body_2$ denote the body images of U_1 and U_2, ES denotes the edge server, and CS denotes the cloud server. j denotes the ID of the pedestrian we assign, and P_j denotes the pedestrian whose ID is j. ES in the area will first launch a data request to the CS to download the $Image\_Face_1$, $Image\_Face_2$, $Image\_Body_1$ and $Image\_Body_2$. The corresponding processing flow is as follows.

Firstly, the ES arranges UAV_1 and UAV_2 to fly to the recipient's activity area to deliver the goods. The UAV_1 and UAV_2 send the captured video streams to the ES as shown in Fig. 5. Secondly the ES first uses the MOT to track and locate the pedestrian in the video and assign ID as shown in Fig. 6.

Fig. 5. UAV capture data.

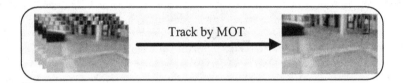

Fig. 6. Track by MOT.

Thirdly each numbered pedestrian will be matched once with $Image\_Face_1$ and $Image\_Face_2$ by person ReID. If P_m captured by UAV_1 is matched with $Image\_Body_2$ as shown in Fig. 7. UAV_1 will move to the vicinity of P_m and capture the positive frontal face.

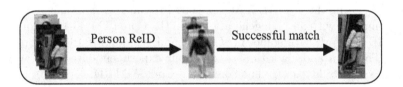

Fig. 7. Locate by person ReID.

Fourthly, U_2 is matched by face recognition as shown in Fig. 8. If the face picture of P_m taken by UAV_1 matches with $Image\_Face_2$ through face recognition, the P_m is determined to be the U_2.

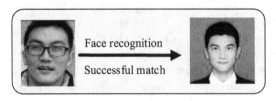

Fig. 8. Face recognition to determine identity.

Finally, the ES will broadcast the successful identification result is sent to UAV_1 and UAV_2 as shown in Fig. 9. When UAV_2 receives the broadcast message, it will fly to the location corresponding to P_m and lands to complete the logistics delivery. UAV_1 continues to use the above steps to find U_1.

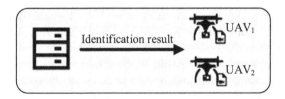

Fig. 9. ES broadcasts the result to UAV_1 and UAV_2

If the target recipient is not found when the recipient's activity area search has been completed, it is determined that the recipient is not within the receiving area. The corresponding UAV will then fly back to the logistics station. When the recipient initiates the next delivery request, another delivery will be made.

4.2 Experiments

We perform three sets of comparison experiments. The first set of experiments evaluates the efficiency of the task performed in cloud server and edge server. In the second set of experiments, we compare the efficiency of our method with two other methods, a baseline method using only face recognition [18] and Edge4Sys which combines pose recognition and face recognition [11]. The third set of experiments is the ablation experiments.

In the first set of experiments, we compare the time for recognition tasks processed in the cloud server and in the edge server as the number of target recipients increases when processing seven video frames per second using MM4ID.

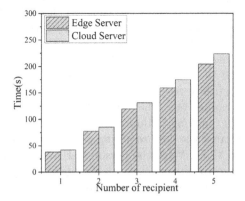

Fig. 10. Comparison between edge server and cloud server.

As shown in Fig. 10, the time for processing tasks in the edge server is less than the time of processing tasks in the cloud server. This is because when data is processed in the cloud server, the data captured by the UAV needs to be transferred to the cloud server, and the time of the transfer affects the overall task processing rate. The edge server is located near the UAV so the time for data transfer is low, making it more suitable for UAV logistics delivery scenarios.

In the second set of experiments, we compare the efficiency of these three methods. The baseline method is a passive target recipient recognition approach using only face recognition algorithm. Edge4Sys is using tiny-YOLO [23] to capture pedestrian images and pose recognition to locate recipients before face recognition algorithm. We compare the time consumption of the three methods for processing the number of video frames per second of 3, 5, and 7 for the number of target recipients of 1, 2, 3, 4, and 5.

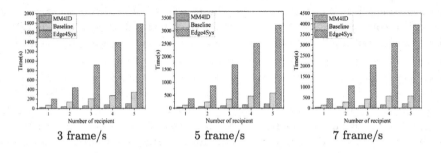

Fig. 11. Comparison of the efficiency of different methods.

The efficiency comparison results of the three methods are shown in Fig. 11. Edge4Sys has the lowest efficiency because pose recognition takes more time to process the data, and there is also the case where the data of the same person is processed multiple times. These factors significantly affect task processing efficiency. The baseline method takes less time because it only uses face recognition and does not use some other methods of recipient localization such as pose recognition. However, to only use face recognition in a practical delivery scenario, the recipient needs to stand at a designated location and wait, the UAV passively comes to find the recipient. If the location is not designated, or if the recipient is moving, and face recognition has certain requirements for the pixel density of the photographed face, the UAV will need to fly over a wide area to perform face matching, making it impractical. The efficiency of our proposed method is optimal because the data requiring person ReID is effectively reduced subsequently through MOT. With the person ReID, our proposed method can accurately locate the recipient, which is highly efficient in UAV logistics delivery. The final face recognition also further ensures the accuracy of identifying the recipient.

The accuracy of existing face recognition algorithm reached 99.68% [8], and their extremely low error rate made them widely used for mobile phone unlocking, access control, etc. The methods we compare all use face recognition as the final means of identification. The accuracy of different methods is the same. Therefore, in this experiment, we only consider the case of successful delivery, and for the scenario of extremely low probability of logistics delivery errors, we will consider it in our future work.

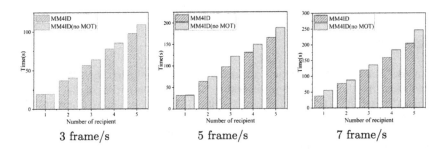

3 frame/s   5 frame/s   7 frame/s

Fig. 12. Comparison of the efficiency of with and without MOT.

In the third set of experiments, we compared the time consumption of MM4ID and MM4ID without MOT to process 3, 5, and 7 video frames per second with the number of target recipients of 1, 2, 3, 4, and 5. As shown in Fig. 12, using MOT is more efficient than without using MOT. This is because the use of MOT can effectively remove duplicated data and improve recognition efficiency, and the effect will be more obvious when the number of recipients increases or the number of images to be processed per second increases.

5   Conclusions and Future Work

Face recognition has been widely used for person identification in many intelligent Internet of Things systems, including intelligent UAV delivery systems for smart logistics. However, the efficiency of face recognition methods in edge-based smart UAV delivery services are still open issues. In this paper, we proposed an active multi-UAV multi-recipient identification method named MM4ID. Specifically, our proposed method can effectively coordinate the computation and communication between multiple UAVs and the edge server, and take full advantage of the multiple object tracking algorithm, person re-identification model, and the face recognition method to improve the overall efficiency of the recipient identification process. Experimental results based on real images captured by UAVs demonstrated that our proposed method can achieve the best results in recognition efficiency compared with other representative solutions.

There are still some future research directions in this area. For example, in this paper, we only considered the case in which the recipient is correctly identified, but we have not discussed the rare but complex cases where some errors may occur. This could potentially be handled as exceptions with human intervention. In addition, we will consider integrating data security measures for personal ReID (such as biometric information protection) to further improve the identification process.

Acknowledgements. This work was supported by the National Natural Science Foundation of China (Nos. 62076002, 61402005, 61972001), and the Natural Science Foundation of Anhui Province of China (No. 2008085MF194).

References

1. Yu, B., Zhao, A.: Development of internet finance industry with the core of e-commerce platform services optimised by the edge computing of the internet of things based on artificial intelligence. Int. J. Ad Hoc Ubiquit. Comput. **39**(4), 192–200 (2022)
2. Shee, H.K., Miah, S.J., De Vass, T.: Impact of smart logistics on smart city sustainable performance: an empirical investigation. Int. J. Logist. Manag. **32**(3), 821–845 (2021)
3. Sah, B., Gupta, R., Bani-Hani, D.: Analysis of barriers to implement drone logistics. Int. J. Log. Res. Appl. **24**(6), 531–550 (2021)
4. Škrinjar, J.P., Škorput, P., Furdić, M.: Application of unmanned aerial vehicles in logistic processes. In: Karabegović, I. (ed.) NT 2018. LNNS, vol. 42, pp. 359–366. Springer, Cham (2019). https://doi.org/10.1007/978-3-319-90893-9_43
5. Song, B.D., Park, K., Kim, J.: Persistent UAV delivery logistics: MILP formulation and efficient heuristic. Comput. Ind. Eng. **120**, 418–428 (2018)
6. Shahzaad, B., Bouguettaya, A., Mistry, S.: A game-theoretic drone-as-a-service composition for delivery. In: 2020 IEEE International Conference on Web Services (ICWS), pp. 449–453. IEEE (2020)
7. Shahzaad, B., Bouguettaya, A., Mistry, S.: Robust composition of drone delivery services under uncertainty. In: 2021 IEEE International Conference on Web Services (ICWS), pp. 675–680. IEEE (2021)
8. Schroff, F., Kalenichenko, D., Philbin, J.: FaceNet: a unified embedding for face recognition and clustering. In: Proceedings of the IEEE Conference on Computer Vision and Pattern Recognition, pp. 815–823 (2015)
9. Adjabi, I., Ouahabi, A., Benzaoui, A., Taleb-Ahmed, A.: Past, present, and future of face recognition: a review. Electronics **9**(8), 1188 (2020)
10. Ali, W., Tian, W., Din, S.U., Iradukunda, D., Khan, A.A.: Classical and modern face recognition approaches: a complete review. Multimedia Tools Appl. **80**, 4825–4880 (2021)
11. Gao, H., et al.: Edge4Sys: a device-edge collaborative framework for MEC based smart systems. In: Proceedings of the 35th IEEE/ACM International Conference on Automated Software Engineering, pp. 1252–1254 (2020)
12. Zhang, C., et al.: An edge based federated learning framework for person re-identification in UAV delivery service. In: 2021 IEEE International Conference on Web Services (ICWS), pp. 500–505. IEEE (2021)
13. Luo, W., Xing, J., Milan, A., Zhang, X., Liu, W., Kim, T.K.: Multiple object tracking: a literature review. Artif. Intell. **293**, 103448 (2021)
14. Luo, H., Gu, Y., Liao, X., Lai, S., Jiang, W.: Bag of tricks and a strong baseline for deep person re-identification. In: Proceedings of the IEEE/CVF Conference on Computer Vision and Pattern Recognition Workshops (2019)

15. Xu, J., Liu, X., Li, X., Zhang, L., Jin, J., Yang, Y.: Energy-aware computation management strategy for smart logistic system with MEC. IEEE Internet Things J. **9**(11), 8544–8559 (2021)
16. Alkouz, B., Bouguettaya, A., Lakhdari, A.: Density-based pruning of drone swarm services. In: 2022 IEEE International Conference on Web Services (ICWS), pp. 302–311. IEEE (2022)
17. Shahzaad, B., Bouguettaya, A.: Service-oriented architecture for drone-based multi-package delivery. In: 2022 IEEE International Conference on Web Services (ICWS), pp. 103–108. IEEE (2022)
18. Shao, Y., Zhang, D., Chu, H., Zhang, X., Chang, Z.: Aerial photography pedestrian target recognition based on yolo and face net. Manuf. Autom. (56–60) (2020)
19. Shen, Q., Jiang, L., Xiong, H.: Person tracking and frontal face capture with UAV. In: 2018 IEEE 18th International Conference on Communication Technology (ICCT), pp. 1412–1416. IEEE (2018)
20. Gai, Y., He, W., Zhou, Z.: Pedestrian target tracking based on deepsort with YOLOv5. In: 2021 2nd International Conference on Computer Engineering and Intelligent Control (ICCEIC), pp. 1–5. IEEE (2021)
21. Nepal, U., Eslamiat, H.: Comparing YOLOv3, YOLOv4 and YOLOv5 for autonomous landing spot detection in faulty UAVs. Sensors **22**(2), 464 (2022)
22. Veeramani, B., Raymond, J.W., Chanda, P.: Deepsort: deep convolutional networks for sorting haploid maize seeds. BMC Bioinform. **19**, 1–9 (2018)
23. Redmon, J., Farhadi, A.: YOLO9000: better, faster, stronger. In: Proceedings of the IEEE Conference on Computer Vision and Pattern Recognition, pp. 7263–7271 (2017)

Multi-objective Optimization for Joint Handover Decision and Computation Offloading in Integrated Communications and Computing 6G Networks

Dong-Fang Wu[1], Chuanhe Huang[1]([✉]), Yabo Yin[1], Shidong Huang[1], and Hui Gong[2]

[1] School of Computer Science, Wuhan University, Wuhan, China
{wudongfang,huangch}@whu.edu.cn
[2] Unit No. 92192, Ningbo, China

Abstract. Mobile edge computing (MEC) deploys the edge computing servers (ECSs) to the network edge and alleviates the problems of limited computational resources and power for mobile users' equipment (UE). Thus, MEC supports the massive computation-intensive applications in the integrated communications and computing 6G network (CCN). However, in mobility management of CCN, users' mobility triggers the new handover between not only two BSs but also two ECSs. When we select the optimal BS, we also need to consider whether the co-located ECS has the sufficient computational resources and low queuing delay. To obtain the lower offloading delay, the existing offloading methods produce extra handover in the decision of the optimal ECS. What's more, the existing handover decision methods ignore the problem of limited computational resources of ECS. In this paper, to meet the demands of communication and computation services, we propose a joint decision method based on multi-objective optimization method (JD-MOO) to solve the joint handover decision and computation offloading problem. We define the services satisfaction degree functions to evaluate the quality of two services. Simulation results show that the proposed JD-MOO method has good performance of handover and offloading.

Keywords: Integrated Communications and Computing 6G
Networks · Handover Decision · Computation Offloading

1 Introduction

Mobile edge computing [1] pushes the computational resources and storage to the network edge (e.g., BS), which can reduce the offloading delay and alleviate the shortage of computational resources and energy of UE [2]. Therefore, combining with MEC and 6G, CCN can provide mobile UEs with the high quality communication services and low delay computation services. In CCN, mobile

Supported by the National Natural Science Foundation of China (No. 61772385).

users connect the nearby BS and access the real-time communication services. Meanwhile, when UEs produce the computing demands, the computing tasks of UEs are offloaded to the co-located ECSs [2]. In the process of offloading, the transmission delay is low and the energy cost of local computing is reduced by the offloading. Thus, CCN is one promising next generation wireless networks, which supports the computation-intensive and latency-sensitive applications.

However, in the mobility management, users' mobility triggers the new handover between not only two BSs but also two ECSs [2]. When we determine the optimal BS, we also need to consider the computational resources allocation problem of ECS. The existing handover decision methods based on deep reinforcement learning (DRL) [3,4], optimization method [5], and threshold [2] solve the handover decision problem in cellular network [4] and IoT. Most offloading methods based on DRL [6,7], optimization method [8], mobility-aware [9], and event-driven method [10] solve the offloading and migration problems in IoT [7] and vehicle network [6], which tradeoff the offloading delay and energy costs [2]. But, the handover and communication traffic are ignored, which lead to the extra handover and frequent handover. To sum up, the new research issues of handover are described as follow: 1) This new handover is between not only two BSs but also two ECSs. The decision of the optimal BS and ECS should be considered jointly. 2) The recent handover decision methods ignore the demands of computation offloading of UEs [2]. And the existing offloading methods produce the extra handover in the decision of the optimal ECS. 3) How to efficiently manage the limited computational resources of ECSs in handover decision?

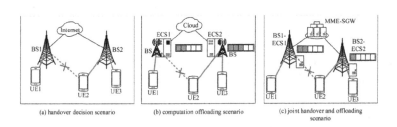

(a) handover decision scenario   (b) computation offloading scenario   (c) joint handover and offloading scenario

Fig. 1. The comparison with different research scenarios.

Moreover, the research scenario of the new handover in CCN is different from the traditional handover and computation offloading. Figure 1(a) shows the handover decision in cellular network. Because of user's movement and the changes in wireless link quality, UE2 finishes the handover from BS1 to BS2. And the handover decision factors (e.g., SINR, link quality) are used to determine the optimal BS. In handover decision, the optimal BS provides mobile user with high quality communication services. Figure 1(b) shows the handover of mobile user between two ECSs in MEC. MEC mainly focuses on the computation services. UE offloads the computing tasks to the proper ECS by the connected BS. Considering the available computing resources, the queuing waiting delay, and energy costs, UE2 executes the handover from ECS1 to ECS2. But, the decision of the optimal ECS produces the extra handover between two BSs, which affects the performance of communication services [9]. Figure 1(c) shows the joint handover decision and computation offloading problem in CCN. Different from the MEC, CCN provides mobile users with the communication services and computation services. Mobile users can connect the nearby BS and access the communication services. Meanwhile, mobile users can offload the computing tasks to the co-located ECSs. In this research, to provide mobile users with high quality communication services and low delay computation services, we propose a JD-MOO method to solve the joint handover decision and computation offloading problem. The main contributions in this research are as follows:

- Considering that the handover precedes the computation offloading, we innovatively add the offloading factors in handover decision to solve the joint handover decision and computation offloading problem in CCN. Specifically, we select the throughput and computing delay as the joint decision factors.
- To balance the performance of handover and offloading, we define the services satisfaction degree functions to express the quality of services. Then, we formulate the joint decision problem as a multi-objective optimization problem (MOP) to maximize the total services satisfaction degree.
- We propose a JD-MOO method. By the objective function method, the MOP is transformed into a single-objective optimization problem. And we utilize the Lagrange dual decomposition method and the gradient descent method with the constant step to solve this optimization problem.
- We select the traditional method, optimization method, and threshold-based method as the benchmarks to prove the performance of JD-MOO method.

The rest of this paper is structured as follows. Section 2 reviews the latest research works. Section 3 mainly presents the framework of CCN. The details of the proposed JD-MOO method are described in Sect. 4. Section 5 discusses the experimental results. Section 6 concludes this paper.

2   Related Work

Computation Offloading Method. The computation offloading methods mostly balance the energy costs and computing delay in MEC systems. The DRL method [6,8], optimization method, and threshold-based method are used to solve the computation offloading problem. In [5], to minimize the execution latency, a two-step algorithm is proposed to solve the service matching game and the optimal bandwidth allocation problem. In [8], a DQN-based method is used to solve the joint task offloading and resources allocation problems [9,11]. Considering the CSI outdated problem in the data collection process, the reinforcement learning (RL) is used to predict the accurate CSI. In [9], considering that the handover between two BSs in MEC system can make the computation offloading not smooth, channel unstable and reduce the network performance, a mobility-aware offloading and resource allocation scheme is proposed to maximize the utility. In [10], different from the existing time-driven schemes, the event-driven computation offloading scheme based on the double deep Q-network (DDQN) is proposed to solve the offloading problem. In [7], a twin-delayed deep deterministic policy gradient (DDPG) method is proposed to address the user selection and task allocation problems. In [12], a mobility virtual reality (VR) system based on task offloading method is designed. In [13], a joint cost-effective and resource-aware decision method is proposed to solve the task offloading problem. In [14], the computation offloading problem in unmanned aerial vehicle (UAV) network is formulated as a two-side matching problem.

Handover Decision Method. The handover decision methods mainly focus on the decision factors and methods. The user's movement and changes in wireless link quality mostly lead to the handover between two BSs for UE. In [4], the prioritized experience replay-based DDQN [15] handover decision method is proposed to solve the frequent handover and ping-pong effect. In [16], considering the channel characteristics, a RL-based handoff method is proposed to solve the redundant handoffs in HetNets. In [17], an event-triggered adaptive handover method solves the handover problem and balances the user access of BSs. In [18], the frequent handover problem in UDN is solved by the movement aware handover method. The Q-learning-based handover method is proposed to solve the frequent handover problem in UDN [19]. In [20], the user-centric cooperative transmission-based handover method is proposed to solve the frequent handover problem. What's more, the sample additive weighting method [21], TOPSIS [22], optimization method [16], and RL [15, 19, 23] are used to solve the handover problem in 5G UDN, satellite network [24], and internet of vehicles [23].

Joint Handover and Offloading Method. The optimization method [1], DRL method [3], and threshold-based method [2] are used to solve the joint handover and offloading problem in MEC system or wireless network. Meanwhile, the deployment of MEC in cellular network needs to solve the mobility management problem of mobile users [1, 2]. Therefore, to tackle the joint migration and handover problem, the relaxation-and-rounding-based solution approach [1] jointly manages the computation and radio resources in the MEC enabled wireless network. In [2], a joint handover and offloading decision algorithm based on user's priority and threshold is used to solve the joint radio and computation offloading issues. The user's priority is based on the speed, deadline, and traffic conditions. But, the design of decision threshold depends on the prior knowledge and cannot guarantee the adaptability of this method. In [3], a DQN-based algorithm is proposed to solve the joint server selection, cooperative offloading and handover problem in MEC wireless network. Different from the references [2, 3], the proposed JD-MOO selects the offloading delay factors in the handover decision process and solves the extra handover problem. The designed handover and offloading mechanisms enable the adaptability. And the comparing of handover and offloading methods is shown in Table 1.

Table 1. Comparing of several handover and offloading methods

Ref.	Scenarios	HO	CO	Problems	Methods
[2]	MEC network	✓	✓	Joint radio offloading and computation offloading problem	A joint handover and offloading decision algorithm based on user's priority and decision threshold
[3]	MEC network	✓	✓	Joint server selection, cooperative offloading and handover problem	A DQN method based on recursive decomposition of the action space
[6]	MEC network	✗	✓	Tasks Offloading problem	A DRL-based solution to minimize the energy consumption
[9]	MEC network	✗	✓	Joint offloading and computing resource allocation problem	A mobility-aware offloading and re-source allocation scheme
[10]	MEC IoT	✗	✓	Computation offloading problem	An event-driven schemes based on DDQN and task priority
[17]	HetNet	✓	✗	Handover problem	An event-triggered adaptive handover method
[15]	UDN	✓	✗	Handover failure and ping-pong effect	A DDQN-based handover parameters optimization method minimize the handover failure rate
[13]	MEC network	✗	✓	Task offloading and energy cost problem	Joint cost-effective and resource-aware decision method
This paper	CCN	✓	✓	Joint handover decision and computation offloading problem	JD-MOO method selects the throughput and offloading delay as the joint decision factors

Handover decision: HO, Computation offloading: CO, HetNet: heterogeneous network, UDN: ultra dense network.

3 System Model and Problem Formulations

3.1 System Model

As Fig. 2 shown, CCN includes N macro base stations (BSs), in which A user's equipment (UEs) follow the random walk model. One ECS with limited computational resources is deployed in each BS, which provide mobile users with communication and computing services. UEs connect the nearby BS and offload the computing tasks to the co-located ECS by the same access link. The optimal BS with ECS provides UE with the high quality access link, available computational resource, and low queuing delay. The computing tasks of each UE are divided into M sub-tasks and ECS without further offloading tasks to other ECSs or cloud. The state data of UE includes SINR, set of candidate BSs, and task tuple. BSs record the list of connected users. The task list of ECS records the offloaded tasks. The waiting queue follows first-in first-out (FIFO) rule. The mobility management entity (MME) collects the real-time data by control link, which is responsible for handover and offloading. Table 2 shows the comparison analysis of MEC system and CCN from three vital aspects. Different from the MEC system, CCN integrates the computing, communication, and storage resources, which provides mobile users with communication and computation services. But, MEC mainly focuses on the performance of computation services and balances the energy cost and offloading delay. The most important difference is that the offloading decision produces the extra handover between two BSs which degrades the network performance. And CCN jointly solves the computation offloading problem in the handover process.

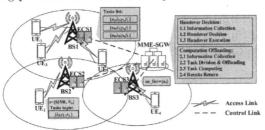

Fig. 2. An example system model in CCN.

Table 2. The comparison analysis of MEC system and CCN.

Differences	MEC System	CCN
System Framework	Push edge servers to the network edge	Resources integration, collaborative management
Service Content	Computation services	Computation and communication services
Mobility management	Offloading, migration	Offloading, migration, handover

3.2 Communication Model

The path loss model of 6G is under research. Therefore, for the convenience of experiments, the path loss model of macro BS for urban area in 5G is adopted [25]. The path loss between UE a and BS n is defined as,

$$PL_{a,n} = 28.0 + 20lg\left(f\right) + 22lg\left(d_{a,n}\right) \tag{1}$$

where f is the carrier frequency, $d_{a,n}$ is the distance between UE and BS. The throughput rate between UE a and BS n is expressed as,

$$r_{a,n} = W_a \log_2\left(1 + \frac{P_{a,n}H_{a,n}}{\sigma^2 + I_a}\right) \tag{2}$$

$$I_a = \sum_{j\in N, j\neq n} P_{a,j} \cdot H_{a,j} \tag{3}$$

where W_a is the sub-channel bandwidth. The transmit power is $P_{a,n}$. σ^2 is the noise power and I_a is the co-channel interference. $H_{a,n}$ is the channel gain.

3.3   Computation Model

During user movement, each UE randomly generated several computing tasks which can be divided into M sub-tasks. And the task tuple is expressed with $(\eta_{a,m}, \tau_{a,m}, \gamma_{a,m})$, where $\eta_{a,m}$ is the amount of input data of task, $\tau_{a,m}$ is the deadline of task, $\gamma_{a,m}$ is the computing intensity of task. And the total delay of computation offloading includes transmission delay, queuing delay, computing delay, and the delay of results return. Because the size of computation results is small, the results return time is very low which can be regarded as a constant [1]. The computing capability of UE depends on its available CPU clock speed f_a. The local computing delay of UE is defined as,

$$du_{a,m} = \frac{\eta_{a,m} \cdot \gamma_{a,m}}{f_a} \tag{4}$$

where the higher CPU resources f_a the lower computing delay $du_{a,m}$. For the local computing, the energy cost is linearly dependent on the CPU clock speed. The energy cost of UE in local computing is expressed as,

$$eu_{a,m} = H\_coef \cdot \eta_{a,m} \cdot \gamma_{a,m} \cdot (f_a)^2 \tag{5}$$

where H_coef is a constant which is related to the CPU hardware. Before the offloading, the computing tasks should be transmitted to the BS and ECS. The transmission delay and energy cost of task in UE are expressed as,

$$dtr_{a,m} = \frac{\eta_{a,m}}{r_{a,n}} \tag{6}$$

$$etr_{a,m} = \frac{P_{a,n} \cdot \eta_{a,m}}{r_{a,n}} \tag{7}$$

The computing delay in ECS includes queuing and computing delay, which is,

$$ds_{a,m} = dq_{a,m} + \frac{\eta_{a,m} \cdot \gamma_{a,m}}{f_{n,a}} \tag{8}$$

where $f_{n,a}$ is the available CPU resource for UE a in ECS n. When the computational resource is not available, the offloaded tasks enter the waiting queue of ECS, and the waiting queue adopts the FIFO order. Assume that the order of computing task m in waiting queue is k, the queuing delay is expressed as,

$$dq_{a,m} = \sum_{i=1}^{k-1} \frac{\eta_{i,m} \cdot \gamma_{i,m}}{f_{n,i}} \tag{9}$$

where $dq_{a,m}$ is the sum computing delay of former k-1 offloaded tasks in waiting queue. ECSs have the abundant power and not consider the energy cost.

3.4   Problem Formulation

To tradeoff the quality of communication and computation services, we define the service satisfaction degree functions related to handover decision and computation offloading. We select the network throughput and offloading delay to evaluate the quality of communication and computation services. The satisfaction degree function of communication services is expressed as

$$I_{HO} = \frac{r_t}{r_c} \tag{10}$$

where r_t and r_c are the throughput rate of target BS and current serving BS, respectively. I_{HO} is the ratio between two throughput rates, which is positive. When $I_{HO} \succ 1$, the target BS provides the better quality wireless links. Otherwise, the current serving BS provides the better quality wireless links. The larger the function value, the better the quality of wireless link of target BS. The satisfaction degree function of computation services is expressed as

$$I_{MEC} = \frac{2\tau_m}{du_{a,m} + dtr_{a,m} + ds_{a,m}} \tag{11}$$

where τ_m is the deadline of task m. I_{MEC} is the ratio between deadline and the sum of local computing delay and ECS computing delay, which is positive. When $I_{MEC} \succ 1$, the local computing delay or ECS computing delay meet the deadline of task. The larger the function value, the smaller the computing delay of task. In CCN, mobile users connect the nearby BSs and offload tasks to ECSs. UE can access the communication and computation services at the same time. Meanwhile, the user movement can lead to the handover and computation offloading problems. To jointly solve the handover decision and computation offloading problems in mobility management, we formulate this joint decision problem as a multi-objective optimization problem (MOP) which is expressed,

$$(P1) \ \max \left(\frac{\sum_{a=1}^{A} \sum_{n=1}^{N} x_{a,n} \cdot I_{HO}}{\sum_{a=1}^{A} \sum_{n=1}^{N} x_{a,n} \cdot I_{MEC}} \right) \tag{12}$$

$$s.t. \begin{cases} C1: \sum_{a=1}^{A} x_{a,n} \leq B_n, \forall n = 1, \dots, N \\ C2: \sum_{n=1}^{N} x_{a,n} = 1, \forall a = 1, \dots, A \\ C3: x_{a,n} \in \{0,1\}, \forall a = 1, \dots, A, n = 1, \dots, N \end{cases}$$

where the objective function is to maximize the services satisfaction degree of two services for all the mobile users. Constraint C1 expresses the available capacity of BS n. Constraint C2 and C3 ensures that one UE at most access one BS at a time. The value of indicator function $x_{a,n}$ is 0 or 1. When $x_{a,n} = 1$, user a connects the BS n. To solve this MOP, we transform the MOP (P1) to a single-objective optimization problem (SOP) (P2) via the objective function method. The proof of problem transformation is shown as Theorem 1. The total service satisfaction degree function is defined by the linear weighting method.

$$I_{a,n} = w_1 \cdot I_{HO} + w_2 \cdot I_{MEC} \tag{13}$$

where w_1 and w_2 are positive. And the weighted coefficients w_1 and w_2 balance the performance of handover decision and computation offloading. I_{HO} and I_{MEC} express the throughput rate and computing delay of target BS and the co-located ECS, which are normalized. The larger the $I_{a,n}$ value, the better the quality of services. The SOP of joint decision problem is expressed as:

$$(P2) \; \max \sum\nolimits_{a=1}^{A} \sum\nolimits_{n=1}^{N} x_{a,n} \cdot I_{a,n} \tag{14}$$

$$s.t. \, C1 \sim C3 \, of \, (P1)$$

where all the constraints of optimization problem (P1) without change. The object of optimization problem (P2) is to maximize the total service satisfaction degree of communication and computation services.

Theorem 1: The optimal solution of SOP (P2) is the Pareto's efficient solution for the MOP (P1). (Proof by Contrapositive)

Proof: Given that the optimal solution of (P2) is n^*. Assume that n^* is not the Pareto's efficient solution of (P1). There is a feasible solution $\overline{n} \in N$ which is better and meets the inequality constraint of $I_{HO}(\overline{n}) \geqslant I_{HO}(n^*)$ and $I_{MEC}(\overline{n}) \geqslant I_{MEC}(n^*)$, $w_1, w_2 \geqslant 0$. We obtain $w_1 \cdot I_{HO}(\overline{n}) \geqslant w_1 \cdot I_{HO}(n^*)$ and $w_2 \cdot I_{MEC}(\overline{n}) \geqslant w_2 \cdot I_{MEC}(n^*)$. Sum two inequalities can obtain $w_1 \cdot I_{HO}(\overline{n}) + w_2 \cdot I_{MEC}(\overline{n}) \geqslant w_1 \cdot I_{HO}(n^*) + w_2 \cdot I_{MEC}(n^*)$. To sum up, the feasible solution $\overline{n}$ is better than n^* in (P2), which contradicts the known conditions, and n^* is not the optimal solution of (P2). Therefore, the assumption is not true. Namely, the optimal solution n^* of (P2) is the Pareto's efficient solution of (P1).

4   Joint Handover Decision and Computation Offloading Method Based on MOO

4.1   The Flowchart of Mobility Management in CCN

As Fig. 3 shown, the mobility management technique in CCN includes three steps: collect data, handover decision, and offloading decision. During user movement, UEs connect the nearby BSs and offload the computation-intensive and delay-sensitive tasks to the available ECSs. Obviously, users first connect the BSs and then offload computing tasks to ECSs. Handover decision precedes computation offloading. Even if handover and offloading don't happen at the same time, we comprehensively consider the throughput and offloading delay in the handover decision. In collect data, the real-time data are collected and transmitted to MME. When the designed handover mechanism is triggered, the optimal target BS is determined in handover decision. In offloading decision, when the designed offloading mechanism is triggered, the tasks of UE can be offloaded to ECS or computed locally. If the available power of UE don't support transmission and local computing, the computing of tasks fails. At each time slot, this flowchart runs one time. The trigger mechanisms and the proposed JD-MOO method are deployed in MME. And the collected data expresses the real-time state of CCN and support the joint decision in mobility management. As Algorithm 1 shown, the mobility management is introduced.

Fig. 3. The flowchart of mobility management technique in CCN.

Algorithm 1. Mobility management technique

Require: BS set N, UE set A, tasks tuple $(\eta_{a,m}, \tau_{a,m}, \gamma_{a,m})$.
Ensure: Handover and offloading matrix X^*, Y^*.
1: Initialize the simulation time T, handover threshold $RSSI_{th}$, offloading gain $Gain_{th}$, available power of UE for computing E_{at}, tasks list in ECS.
2: **for** $t = 1, 2, \ldots, T$ **do**
3:     Collect the real-time data and transmit the collected data to MME.
4:     **if** $RSSI_{a,n} \leqslant RSSI_{th} - Hys$ **then**
5:         Execute the JD-MOO in **Algorithm 2**. Obtain the handover matrix X^*.
6:     **end if**
7:     **if** $du_{a,m} - dtr_{a,m} - ds_{a,m} \geqslant Gain_{th}$ and $E_{a,t} \geqslant etr_{a,m}$ **then**
8:         Offload user task to ECS.
9:     **else if** $E_{a,t}$ support local computing **then**
10:         Compute user task locally.
11:     **else**
12:         Task computing failure.
13:     **end if**
14:     Determine the offloading matrix Y^*, update $E_{a,t}$, $t = t + 1$, and tasks list.
15: **end for**
16: Return handover and offloading matrix X^*, Y^*

4.2 Trigger Mechanisms of Handover and Offloading

We design the trigger mechanisms of handover and offloading to monitor the mobility of users and computing demands. Therefore, A2 event is considered in handover. The trigger mechanism of handover relays on the signal strength, which is defined as

$$RSSI_{a,n} \leqslant RSSI_{th} - Hys \tag{15}$$

where $RSSI_{a,n}$ and $RSSI_{th}$ denote the signal strength and handover threshold, respectively. Hys is a hysteresis parameter, which reduces the redundant handover and ping-pong effect of mobile users.

For the operation of computation offloading, the total offloading delay need to meet the tasks deadline. And the available power of UE is enough to support the energy costs of transmission or local computing. The trigger mechanism of offloading is expressed as,

$$du_{a,m} - dtr_{a,m} - ds_{a,m} \geqslant Gain_{th} \tag{16}$$

$$E_{a,t} \geqslant etr_{a,m} \tag{17}$$

where $Gain_{th}$ is the offloading threshold, which expresses the saved computing time. And $E_{a,t}$ is the remaining power of UE a at time slot t, which should be larger than the energy cost of transmission in offloading $etr_{a,m}$. Obviously, the energy cost of transmission is less than the energy cost of local computing. If the available power $E_{a,t}$ cannot support the transmission or local computing, the task computing failures.

4.3   Joint Decision Method Based on MOO

The joint handover decision and computation offloading problem (P2) is also a generalized assignment problem (GAP) [16], which is a NP-hard problem. We adopt Lagrange dual decomposition method to tackle this optimization problem. After the relaxation operation of $x_{a,n} \in [0, 1]$, (P2) becomes a linear problem. The Lagrange function of (P2) is expressed as

$$L(x, \lambda) = \sum_{a=1}^{A} \sum_{n=1}^{N} x_{a,n} \cdot I_{a,n} - \sum_{n=1}^{N} \lambda_n \cdot \left(\sum_{a=1}^{A} x_{a,n} - B_n \right) \quad (18)$$

$$= \sum_{a=1}^{A} \sum_{n=1}^{N} x_{a,n} \cdot (I_{a.n} - \lambda_n) + \sum_{n=1}^{N} \lambda_n \cdot B_n$$

where λ_n is the Lagrange multiplier. The constraint of BS capacity is reduced by the Lagrange dual operation. When the vector λ is fixed, the Lagrange dual function is expressed as

$$d(x, \lambda) = \max_x L(x, \lambda) \quad (19)$$

$$s.t. \begin{cases} C1 : \sum_{n=1}^{N} x_{a,n} = 1, \forall a = 1, \ldots, A \\ C2 : 0 \le x_{a,n} \le 1, \forall a = 1, \ldots, A, n = 1, \ldots, N \end{cases}$$

where the object of Lagrange dual function is to maximize the Lagrange function. Constraint C1 expresses the mobile user can connect at most one BS at a time. Constraint C2 describes the relaxation operation of $x_{a,n}$. For the original joint decision problem in mobility management, the dual problem is $\min_{\lambda} d(x, \lambda)$. The Lagrange dual function doesn't have cross-term of $x_{a,n}$. And, the maximum value is unrelated to the user index. The computing order of Lagrange dual function is changed to

$$d(x, \lambda) = \max_x \sum_{a=1}^{A} \sum_{n=1}^{N} x_{a,n} \cdot (I_{a,n} - \lambda_n) + \sum_{n=1}^{N} \lambda_n \cdot B_n \quad (20)$$

$$= \sum_{a=1}^{A} \max_{x_{a,n}, n \in N} \sum_{n=1}^{N} x_{a,n} \cdot (I_{a,n} - \lambda_n) + \sum_{n=1}^{N} \lambda_n \cdot B_n$$

where B_n is the available capacity of BS. For the second item of Lagrange dual function, B_n is fixed for the Lagrange dual function. The first item of the Lagrange dual function consists of A sub-problem, which search for the optimal BSs. Therefore, we can solve the sub-problem for each user a,

$$d_a(x, \lambda) = \max_{x_{a,n}, n \in N} \sum_{n=1}^{N} x_{a,n} \cdot (I_{a,n} - \lambda_n) \quad (21)$$

where $I_{a,n}$ is the total services satisfaction degree of communication and computation, which evaluates the performances of handover and offloading. When vector λ is fixed, the sub-problem $d_a(x, \lambda)$ changes to

$$\max_{x_{a,n}, n \in N} \sum_{n=1}^{N} x_{a,n} \cdot (I_{a,n} - \lambda_n) = \max (I_{a,n} - \lambda n) \qquad (22)$$

where $I_{a,n}$ is the service satisfaction degree. Thus, for user a, we choose the optimal BS n from the candidate BS set N_a to maximize the first item of Lagrange dual function [16]. Thus, when λ is fixed, Lagrange dual function (19) can be solved. Then, we minimize the $d(x, \lambda)$ over λ to obtain the optimal Lagrange multiplier λ^* for the dual problem. We adopt the gradient descent method with the constant step to update λ with positive value.

$$\lambda_n^{k+1} = \lambda_n^k - \delta^k \cdot g^k \qquad (23)$$

$$\delta^k = \frac{\gamma}{\|g^k\|_2} \qquad (24)$$

$$g^k = \left(B_n - \sum_{a=1}^{A} x_{a,n} \right) \qquad (25)$$

where k is the iteration number. δ^k is the update step size, and the value of δ^k is positive. g^k is the gradient of Lagrange function $d(x, \lambda)$ over λ. $\gamma = \|x^k - x^{k-1}\|$ expresses the difference of decision matrix between two iterations. The proposed JD-MOO algorithm is described in Algorithm 2. The time complexity of JD-MOO method is $O(K \cdot N \cdot A)$, where K, N, and A are the iteration number, the BS number, and user number. According to the experiments, the running time of the proposed JD-MOO method for one user at each time slot are 0.01, 0.03, 0.05, and 0.1 ms under 50, 100, 200, and 300 mobile users, respectively. The running time meets the requirement of millisecond time cost of handover. The proposed JD-MOO method is deployed in MME, which collects the real-time data of BSs, users, and ECSs. And the energy cost of JD-MOO method is supported by the energy system of MME. For the computation offloading, the transmission of task data from user to ECS consumes energy of users.

The Algorithm 2 return the handover decision matrix X^* to Algorithm 1. The Algorithm 2 runs once per time slot t. When the difference between handover matrixes of k-th iteration and k-th iteration meets the convergence requirements, the JD-MOO method stop the iterative operation. ε_1 is obtained from the experiments.

4.4   The Convergence Analysis of JD-MOO Method

The proposed JD-MOO adopts the gradient descent method with constant step to search the approximate optimal solution of dual problem over the dynamically changing variable λ_k. To prove the convergence of JD-MOO, we present the Proposition 1. Assume that d^* is the optimal solution of dual problem.

Algorithm 2. Joint decision method based on multi-objective optimization

Require: BS set N, UE set A, tasks tuple $(\eta_{a,m}, \tau_{a,m}, \gamma_{a,m})$, Lagrange multiplier λ_k.
Ensure: Handover decision matrix X^*.
1: Initialize the handover decision matrix x^0, maximum iteration number K, initial iteration number $k = 0$.
2: In collected data, the candidate BSs set N_a, and BS capacity B_n are obtained.
3: Service satisfaction degree of communication and computation I_{HO}, I_{MEC} are obtained by (10, 11). Compute the total service satisfaction degree $I_{a,n}$ by (13).
4: **while** $k \leqslant K$ **do**
5:    Update iteration number with $k = k + 1$.
6:    According to (21, 22), search the maximum value $(I_{a,n} - \lambda_n)$, solve the subproblem for each user a, and obtain the handover matrix x_k in $k - th$ iteration. Update the handover decision matrix X^*.
7:    According to (23, 24, 25), the Lagrange multiplier λ_k is updated by the gradient descent method with constant step. Update the capacity of BS B_n.
8:    **if** $\left\| x^k - x^{k-1} \right\| \leq \varepsilon_1$ **then**
9:       STOP the program.
10:   **end if**
11: **end while**

Proposition 1: Assuming that step size δ^k is computed by (24, 25). When k approaches infinity, we have

$$\lim_{k \to \infty} d\left(x, \lambda_{best}^k\right) - d^* \leq \frac{G\gamma}{2} \tag{26}$$

where $d\left(x, \lambda_{best}^k\right) = \min \left\{ d\left(x, \lambda^k\right), k = 1, 2, \ldots, K \right\}$, which is the locally minimum solution of Lagrange dual function (19) after K times iterations. $\gamma = \left\| x^k - x^{k-1} \right\|$, and $\left\| g^k \right\|_2 \leq G$.

Proof: According to (23), we have $\left\| \lambda^{k+1} - \lambda^* \right\|_2^2 = \left\| \lambda^k - \delta^k \cdot g^k - \lambda^* \right\|_2^2$, where λ^* is the optimal Lagrange multiplier. By the perfect square expression and triangle inequality, we have

$$= \left\| \lambda^k - \lambda^* \right\|_2^2 - 2\delta^k \cdot g^k \cdot \left(\lambda^k - \lambda^*\right) + \left(\delta^k\right)^2 \cdot \left\| g^k \right\|_2^2$$

$$\leq \left\| \lambda^k - \lambda^* \right\|_2^2 - 2\delta^k \cdot \left(d\left(x, \lambda^k\right) - d^*\right) + \left(\delta^k\right)^2 \cdot \left\| g^k \right\|_2^2$$

For the K times iterations, after the summation, we get

$$2\sum_{k=1}^{K} \delta^k \cdot \left(d\left(x, \lambda^k\right) - d^*\right) - \sum_{k=1}^{K} \left(\delta^k\right)^2 \cdot \left\| g^k \right\|_2^2 \leq \left\| \lambda^1 - \lambda^* \right\|_1^1$$

$$2\sum_{k=1}^{K} \delta^k \cdot \left(d\left(x, \lambda_{best}^k\right) - d^*\right) \leq 2\sum_{k=1}^{K} \delta^k \cdot \left(d\left(x, \lambda^k\right) - d^*\right)$$

After the simplification, we have

$$d\left(x,\lambda_{best}^{k}\right) - d^* \le \frac{G\left\|\lambda^1 - \lambda^*\right\|_2^2}{2K\gamma} + \frac{G\gamma}{2}$$

When k approaches infinity, we have

$$\lim_{k\to\infty} d\left(x,\lambda_{best}^{k}\right) - d^* \le \frac{G\gamma}{2}$$

We obtain the Proposition 1 and prove the convergence of JD-MOO method.

5   Performance Evaluation

5.1   Experiments Setting

We consider $N = 25$ BSs which are evenly deployed in a 2.3 km×2.3 km urban area, as Fig. 4 shown. The radius of BS is 300 m. Besides, the bandwidth of BS is 500 MHz, which includes 100 sub-channels. The carrier frequency is $f = 3.5$ GHz. Then, users' mobility follows the random walk model, which generates the user mobility trajectory. The simulation time T is 600 s, and time slot t is 1 s. The number of UE is $A = 50, 100, 200, 300$. The velocity range of UE is [0,120] km/h. The transmit power of UE is 0.5 W. The CPU frequency of UE is $f_a = 2$ GHz. The sum frequency of ECS is $F_n = 5$ GHz. The number of sub-tasks of UE is $M = 60$. For the user tasks, $\eta_{a,m} \in [0.5, 1.5]$ Mbits, $\gamma_{a,m} \in [500, 1000]$ cycles/bits, and $\tau_{a,m} \in [800, 1000]$ ms. And we consider five benchmarks for comparison. **Q-learning** [3]: The total computation delay forms the negative reward function. **SMART** [16]: Use the Lagrange dual decomposition method to maximize the volume of transmitted data. **RBH** [16]: The rate-based handover decision method selects the optimal BS with maximum transmission data rates. **SBH** [16]: The SINR-based handover selects the optimal BS with maximum SINR. **JD-Threshold** [2]: Use the multi-level thresholds and users' priority to perform handover and offloading.

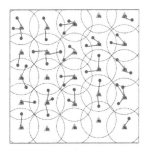

Fig. 4. The simulated scenarios of CCN.

5.2   Performance of Handover Decision

As Fig. 5(a) shown, the average handover number increases as the user number increases. From the comparison of average handover numbers, the proposed JD-MOO method obtains the optimal performance with 5.3, 5.7, 11.4, and 29.8. RBH and SBH methods are classified as the multi-attributes-based handover decision, and the optimal decision attributes lead to the redundancy handover. SMART method adopts the iterative optimization method. And the mobile users connect the optimal edge computing servers, which produce the extra handover and sacrifice the performance of handover in CCN. JD-Threshold method defines the multi-level thresholds and trigger mechanisms to solve the handover and offloading. But it depends on the prior knowledge, which cannot guarantee the performance of handover and offloading. As Fig. 5(b) shown, when the number of users increases, the average throughput increases. From the analysis of average throughput of users, the JD-MOO method has a good performance of communication. The average throughput of JD-MOO method are 10.4 Mbits, 10.5 Mbits, 10.8 Mbits, and 11.3 Mbits, respectively. And the RBH method obtains the best performance of communication. RBH method based on throughput rate selects the optimal BS with the optimal throughput rate. And the throughput is the cumulative sum of throughput rate in time. And the proposed JD-MOO defines the services satisfaction degree function to describes the throughput rate of the target BS. Therefore, the JD-MOO method also obtain the good performance.

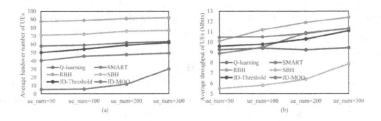

Fig. 5. (a) Average handover number of UE, (b) Average throughput of mobile users

5.3   Performance of Computation Offloading

As Fig. 6(a) shown, the average offloading delay increases as the user number increases. From the comparison of average offloading delay, the proposed JD-MOO method obtains the minimum offloading delay. The average offloading delay of JD-MOO method are 146 ms, 151.3 ms, 169.9 ms, and 181.9 ms, respectively. As Fig. 6(b) shown, the average energy cost increases as the number of users increases. According to the defined trigger mechanism of offloading, when the offloading delay increases, more computing tasks of mobile users are completed locally. The local computing increases the energy cost of UE. From the

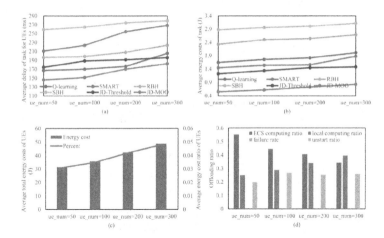

Fig. 6. (a) Average delay of tasks for mobile users, (b) Average energy costs of task, (c) Total energy costs of task for UE, (d) Offloading ratio of mobile users by JD-MOO.

analysis of average energy cost, the proposed JD-MOO method obtains the minimum energy costs. When the user number are 50, 100, 200, and 300, the average energy costs of JD-MOO methods are 0.55 J, 0.61 J, 0.72 J, and 0.82 J, respectively. Because the SBH method ignores the offloading demands of users, which leads to more handover and high average energy costs. As Fig. 6(c) shown, for the proposed JD-MOO method, the total energy costs and energy costs ratio increase as the number of users increases. The total energy costs are 31.2 J, 35.6 J, 42.2 J, and 48.7 J, respectively. And the corresponding energy cost ratios are 0.031, 0.035, 0.042, and 0.048, respectively. The offloading ratio includes ECS computing ratio, local computing ratio, failure rate, and unstart ratio. The sum ratio of above four ratios is 1. As Fig. 6(d) shown, when the user number increases, the ECS computing ratio decreases and the local computing ratio increases. With the defined trigger mechanism of offloading, the deadline of task and the gain of computation offloading lead to the increase of local computing. As Fig. 6(c) shown, every UE has 1000 J energy to support the local computing and transmission of tasks, so the failure rate is 0. And the unstart ratio is equal to the difference between 1 and the sum of ECS computing ratio and local computing ratio. Table 3 shows the comparison of offloading ratio. ECR is the ECS computing ratio. LCR is the local computing ratio. The greater the ECR, the lower the average energy cost of each task. And the sum of ECR and LCR is a fixed value. The ECR and LCR of the proposed JD-MOO method obtain the optimal value, the JD-MOO method obtains the lower energy costs. By the designed handover and offloading mechanisms, the offloading performance is good than the threshold-based joint decision method.

Table 3. The comparison of offloading ratio

Method\user number		50	100	200	300
Q-learning	ECR	0.46	0.385	0.349	0.303
	LCR	0.341	0.347	0.397	0.437
SMART	ECR	0.477	0.397	0.363	0.307
	LCR	0.324	0.335	0.383	0.433
RBH	ECR	0.402	0.346	0.311	0.279
	LCR	0.399	0.386	0.435	0.461
SBH	ECR	0.356	0.322	0.287	0.264
	LCR	0.445	0.41	0.459	0.476
JD-Threshold	ECR	0.496	0.407	0.368	0.321
	LCR	0.305	0.325	0.378	0.419
JD-MOO	ECR	0.553	0.445	0.406	0.344
	LCR	0.248	0.287	0.34	0.396

5.4 Analysis of Services Satisfaction Degree

As Fig. 7(a) shown, when the weighted coefficient w_1 of communication services increases, the weighted coefficient w_2 of computation services decreases. And $w_1 + w_2 = 1$. This is because that the normalized services satisfaction degree of communication I_{HO} is less than the normalized services satisfaction degree of computation I_{MEC}. The differences of the services satisfaction degree functions of communication and computation lead to the differences of two normalized services satisfaction degree. What's more, when the user number increases, for the fixed w_1 and w_2, the total service satisfaction degree decreases. This is because that the increases of user number lead to the decreases of quality of communication and computation services. As Fig. 7(a) shown, the increases of user number lead to the increases of offloading delay. Because the services satisfaction degree of computation consists of the estimated computing delay, the increases of the estimated computing delay lead to the decreases of services satisfaction degree of computation. Meanwhile, when the user number increases, the competition among mobile users become more intense, which further lead to the decreases of services satisfaction degree of communication. According to the experiments, we set the $w_1 = 0.2$, $w_2 = 0.8$. As Fig. 7(b) shown, when the offloading delay increases, the normalized services satisfaction degree decreases. We designed three services satisfaction degree of computation, $f_1 = \tau_m/(dtr_{a,m} + ds_{a,m})$, $f_2 = du_{a,m}/(dtr_{a,m} + ds_{a,m})$, $f_3 = 2\tau_m/(du_{a,m} + dtr_{a,m} + ds_{a,m})$. f_1 evaluates the performance of offloading delay in target ECS. f_2 is the ratio between the delay of ECS computing and the delay of local computing, which presents the offloading performance. f_3 is the ratio between the deadline of tasks and the estimated computing delay, which evaluates the performance of ECS computing and local computing. Obviously, f_2 meet the demands of objective function. And,

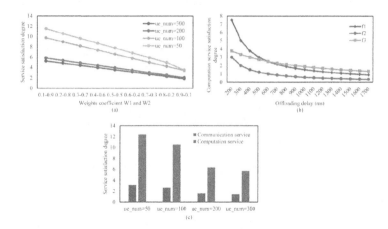

Fig. 7. (a) Analysis of weight coefficients, (b) Analysis of service satisfaction degree functions, (c) Comparison of services satisfaction degree ($w_1 = 0.2$, $w_2 = 0.8$).

handover decision precedes the offloading decision, which don't happen at the same time. As Fig. 7(c) shown, for the JD-MOO method ($w_1 = 0.2$, $w_2 = 0.8$), when the number of users increases, the services satisfaction degrees decrease. This is because that the increases of user number lead to the increases of throughput rate and offloading delay. The increases of offloading delay lead to the decrease of service satisfaction degree of computation. And when the user number increases, the resources competition among mobile users becomes intense and the throughput rate of target BS degrades. So the services satisfaction degree of communication decreases. And the services satisfaction degree is a relative value. The services satisfaction degrees of communication are 3.1, 2.6, 1.5, and 1.4, respectively. And the corresponding services satisfaction degrees of computation are 12.4, 10.5, 6.3, and 5.6, respectively. As Fig. 6(d) shown, when the user number increases, more tasks select the local computing which further leads to the decreases of services satisfaction degree of communication.

5.5   Analysis of Running Time and Convergence

As Fig. 8(a) shown, when the user number increases, the running time of different methods increases. From the analysis of running time, the proposed JD-MOO method has the minimum running time of a single user running once. The running time of JD-MOO method are 0.01 ms, 0.02 ms, 0.05 ms, and 0.1 ms, respectively. And Q-learning method needs to search the state-action table, which costs much search time. And the RBH and SBH methods adopt the multi-attributes-based method and decision threshold, which are more efficient than Q-learning. As Fig. 8(b) shown, when the user number increases, the iterative steps of JD-MOO method also increase. This is because that the increases of user number lead to the numbers of handover decision and offloading decision, which further lead to that the iterative optimization of target BS and ECS

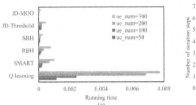 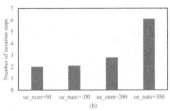

Fig. 8. (a) Running time, (b) Iterative steps for JD-MOO method.

become more complex and the iterative steps also increase. The iterative steps of JD-MOO method are 2.01, 2.08, 2.78, and 6.1, respectively.

6   Conclusion

In this research, we studied the joint handover decision and computation offloading problem in CCN. We designed the effective trigger mechanisms and the service satisfaction degree function to balance the quality of services. Then, we proposed a JD-MOO method to solve this joint decision problem. Simulation results show that the proposed method outperforms the baselines and provides UE with the high quality communication and computation services.

References

1. Liang, Z., Liu, Y., Lok, T., Huang, K.: Multi-cell mobile edge computing: joint service migration and resource allocation. IEEE Trans. Wirel. Commun. **20**(9), 5898–5912 (2021)
2. Nasrin, W., Xie, J.: A joint handoff and offloading decision algorithm for mobile edge computing (MEC). In: 2019 IEEE Global Communications Conference, GLOBECOM 2019, Waikoloa, HI, USA, 9–13 December 2019, pp. 1–6. IEEE (2019)
3. Ho, T.M., Nguyen, K.K.: Joint server selection, cooperative offloading and handover in multi-access edge computing wireless network: a deep reinforcement learning approach. IEEE Trans. Mob. Comput. **21**(7), 2421–2435 (2022)
4. Wu, D.F., Huang, C., Yin, Y., Huang, S., Guo, Q., Zhang, L.: State aware-based prioritized experience replay for handover decision in 5G ultradense networks. Wirel. Commun. Mob. Comput. **2022**, 1–16 (2022)
5. Zeng, H., Li, X., Bi, S., Lin, X.: Delay-sensitive task offloading with D2D service-sharing in mobile edge computing networks. IEEE Wirel. Commun. Lett. **11**(3), 607–611 (2022)
6. Kazmi, S.M.A., et al.: Computing on wheels: a deep reinforcement learning-based approach. IEEE Trans. Intell. Transp. Syst. **23**(11), 22535–22548 (2022)
7. Chen, Y., Sun, Y., Wang, C., Taleb, T.: Dynamic task allocation and service migration in edge-cloud IoT system based on deep reinforcement learning. IEEE Internet Things J. **9**(18), 16742–16757 (2022)
8. Ai, L., Tan, B., Zhang, J., Wang, R., Wu, J.: Dynamic offloading strategy for delay-sensitive task in mobile-edge computing networks. IEEE Internet Things J. **10**(1), 526–538 (2023)

9. Xia, C., Jin, Z., Su, J., Li, B.: Mobility-aware offloading and resource allocation strategies in MEC network based on game theory. Wirel. Commun. Mob. Comput. **2023**, 1–12 (2023)
10. Wei, Z., Zhao, B., Su, J.: Event-driven computation offloading in IoT with edge computing. IEEE Trans. Wirel. Commun. **21**(9), 6847–6860 (2022)
11. Yan, Z., Cheng, P., Chen, Z., Vucetic, B., Li, Y.: Two-dimensional task offloading for mobile networks: an imitation learning framework. IEEE/ACM Trans. Netw. **29**(6), 2494–2507 (2021)
12. Sun, Y., Chen, J., Wang, Z., Peng, M., Mao, S.: Enabling mobile virtual reality with open 5G, fog computing and reinforcement learning. IEEE Netw. **36**(6), 142–149 (2022)
13. Tout, H., Mourad, A., Kara, N., Talhi, C.: Multi-persona mobility: joint cost-effective and resource-aware mobile-edge computation offloading. IEEE/ACM Trans. Netw. **29**(3), 1408–1421 (2021)
14. Wang, Y., et al.: Task offloading for post-disaster rescue in unmanned aerial vehicles networks. IEEE/ACM Trans. Netw. **30**(4), 1525–1539 (2022)
15. Huang, W., Wu, M., Yang, Z., Sun, K., Zhang, H., Nallanathan, A.: Self-adapting handover parameters optimization for SDN-enabled UDN. IEEE Trans. Wirel. Commun. **21**(8), 6434–6447 (2022)
16. Sun, Y., Feng, G., Qin, S., Liang, Y., Yum, T.P.: The SMART handoff policy for millimeter wave heterogeneous cellular networks. IEEE Trans. Mob. Comput. **17**(6), 1456–1468 (2018)
17. Narmanlioglu, O., Uysal, M.: Event-triggered adaptive handover for centralized hybrid VLC/MMW networks. IEEE Trans. Commun. **70**(1), 455–468 (2022)
18. Sun, W., Wang, L., Liu, J., Kato, N., Zhang, Y.: Movement aware comp handover in heterogeneous ultra-dense networks. IEEE Trans. Commun. **69**(1), 340–352 (2021)
19. Khosravi, S., Ghadikolaei, H.S., Petrova, M.: Learning-based handover in mobile millimeter-wave networks. IEEE Trans. Cogn. Commun. Netw. **7**(2), 663–674 (2021)
20. Kibinda, N.M., Ge, X.: User-centric cooperative transmissions-enabled handover for ultra-dense networks. IEEE Trans. Veh. Technol. **71**(4), 4184–4197 (2022)
21. Ndashimye, E., Sarkar, N.I., Ray, S.K.: A multi-criteria based handover algorithm for vehicle-to-infrastructure communications. Comput. Netw. **185**, 107652 (2021)
22. Hu, Q., Gan, C., Gong, G., Zhu, Y.: Adaptive cross-layer handover algorithm based on MPTCP for hybrid LiFi-and-WiFi networks. Ad Hoc Netw. **134**, 102923 (2022)
23. Tan, K., Bremner, D., Kernec, J.L., Sambo, Y.A., Zhang, L., Imran, M.A.: Intelligent handover algorithm for vehicle-to-network communications with double-deep Q-learning. IEEE Trans. Veh. Technol. **71**(7), 7848–7862 (2022)
24. Wang, F., Jiang, D., Wang, Z., Chen, J., Quek, T.Q.S.: Seamless handover in LEO based non-terrestrial networks: service continuity and optimization. IEEE Trans. Commun. **71**(2), 1008–1023 (2023)
25. 3GPP: Study on channel model for frequencies from 0.5 to 100 GHz. Technical report (TR) 38.901, 3rd Generation Partnership Project (3GPP), December 2019, version 16.1.0

A Joint Optimization Scheme in Heterogeneous UAV-Assisted MEC

Tian Qin, Pengfei Wang[(✉)], and Qiang Zhang[(✉)]

School of Computer Science and Technology, Dalian University of Technology,
Dalian 116024, China
tianqin@mail.dlut.edu.cn, {wangpf,zhangq}@dlut.edu.cn

Abstract. Mobile Edge Computing (MEC) is considered as a promising technology to meet the high-quality service requirements of emerging applications in mobile intelligent terminals. It can effectively handle computation-intensive and latency-sensitive tasks in the Internet of Things (IoT). However, location-fixed edge servers in MEC cannot efficiently handle time-varying tasks in the hot-spot area. Therefore, it is necessary to utilize the Unmanned Aerial Vehicle (UAV) with communication and computation resources making on-demand network deployment for handling the time-varying tasks above. In this paper, we build a novel MEC system based on heterogeneous multi-UAV, in which we take both the UAV scheduling problem and the task allocation problem into consideration. What's more, in order to minimize the system energy consumption, we propose a joint optimization method, named **JoSA**, for the two problems mentioned above. To be specific, we first regard the UAV scheduling problem as a knapsack problem. Based on this, we then divide the tasks in the hot-spot area according to geographic location and allocate them in different situations. Finally, compared with the other two benchmarks, the simulation experiments show that our method demonstrates good generalization ability and makes better performance with a reduction of 8% and 11% in system energy consumption, and 3% and 4% in system time cost, respectively.

Keywords: Mobile edge computing (MEC) · heterogeneous multi-Unmanned Aerial Vehicle (multi-UAV) · scheduling · task allocation

1 Introduction

With the continuous expansion of the scale of cellular IoT users and the overall growth of mobile intelligent terminals, more and more new mobile applications are making significant contributions to improve the quality of human life, such as Augmented Reality (AR) [23], Virtual Reality (VR) [8] and Mixed Reality (MR) [6]. However, the computing resource of these applications are often insufficient due to the limited storage and computing capabilities of mobile devices. Although cloud computing [26] can somehow enhance the data storage

© The Author(s), under exclusive license to Springer Nature Singapore Pte Ltd. 2024
Z. Tari et al. (Eds.): ICA3PP 2023, LNCS 14490, pp. 194–216, 2024.
https://doi.org/10.1007/978-981-97-0859-8_12

and management capabilities of mobile devices, it cannot meet the requirements of latency-sensitive tasks due to the massive response time caused by the long transmission distance between cloud servers and mobile devices. Therefore, how to meet the above requirements as well as improve the service quality is an urgent problem to be solved.

Compared with cloud computing described above, MEC is also a promising technology to provide computing resources for mobile devices through deploying servers at the edge of the network [18,22]. However, MEC consumes less transfer time and energy as these mobile devices can offload their tasks to nearby edge servers for execution. It has to be noticed that, the computation resources and processing coverage of location-fixed edge servers in MEC are limited. Thus, the dynamic demand of users will lead to uneven distribution of tasks and location-fixed edge servers cannot be flexibly deployed according to the needs of mobile users, which will cause some edge servers to be overloaded while others are idle. To mitigate this, existing techniques such as task migration [4] can balance the load among edge servers but causing additional communication and time cost which can be solved by on-demand network deployment. On-demand network deployment is a valuable proposal for catering to dynamic hot-spots in the face of large events or disasters. Simultaneously, it also offers improved utilization of computing resources compared with Base Station (BS) sleep technology [28] by providing on-demand access.

In recent years, UAV has been widely used in many aspects of military and civilian fields, e.g., real-time monitoring, automatic tracking and relay transmission [24], due to its small size, low cost and easy deployment. Adopting a UAV-assisted MEC system, in which the edge servers are installed in the UAV. This kind of system offers two key advantages. Firstly, one of which is a higher altitude of the edge servers in UAVs allowing for improved line-of-sight connections to mobile applications and increasing the likelihood of seamless links. The other is a flexible deployment of UAVs enabling a reduction in transmission distance [3,40]. As a result, this system enhances the overall service quality provided to mobile applications. Compared with a single UAV, the coordination of multiple UAVs [38] is more effective, fast and flexible, which is also the key point of our research. In emergency and complex scenarios, the utilization of multiple UAVs can significantly enhance the applications of MEC by efficiently supporting a larger number of user tasks within a shorter time. Note that, in our research, the characteristic parameters of each UAV are not the same, that is, heterogeneous.

In general, we propose a MEC system with the assist of heterogeneous multi-UAV to determine the number and location of UAVs (UAV scheduling) as well as the offloading decision and resource assignment (task allocation), among them, the offloading decision is to decide which UAV the users' tasks will be executed on, and the resource assignment is to consider how many resources should be allocated to the users' tasks. This system jointly optimizes the UAV scheduling and task allocation problems to minimize system energy consumption and ensure that all tasks can be executed in parallel.

To sum up, our work contributions are as follows:

(1) We propose a novel heterogeneous UAV-assisted MEC system, which makes an improvement on the performance of traditional MEC system. Considering to minimize the system energy consumption, we propose a joint optimization scheme of UAV scheduling and task allocation, named JoSA, which can improve the Quality of user Experience (QoE).

(2) We innovatively regard the UAV scheduling problem as a knapsack problem. Based on this, we also propose a UAV scheduling decision algorithm which can obtain an optimal solution for minimizing the energy consumption.

(3) We also propose a parallel computed task allocation algorithm, which trades off the UAV memory capacity and users' task size and realizes the on-demand deployment of the network.

(4) To evaluate the effectiveness of our scheme, we design a series of simulation experiments to compare with the baselines under our settings. Experimental results verify the effectiveness of our method. To be specific, our method makes a reduction of 8% and 11% in system energy consumption, and 3% and 4% in system time cost, respectively, compared with the other two benchmarks.

The rest of the paper is organized as follows. We present the related work in Sect. 2. We provide an overview of our proposed heterogeneous UAV-assisted MEC system and the system model in Sect. 3. We describe the UAV scheduling algorithm and task allocation algorithm in the system respectively in Sect. 4. We evaluate the performance of the proposed joint optimization method JoSA in the heterogeneous UAV-assisted MEC system in Sect. 5. Finally, we make a summary of this paper.

2   Related Work

2.1   MEC

By deploying servers at the edge of the network, MEC enables mobile devices to offload their tasks to nearby edge servers, resulting in less transmission time and energy consumption due to the shorter transmission distance. Jiang et al. [13] traded off the time-varying network state and placement cost in the problem of dynamic edge servers placement, based on this, they proposed an intelligent long-term dynamic decision process. Qin et al. [25] introduced a groundbreaking framework named user-centric MEC by integrating the user-centric network with the MEC computing service. Wang et al. [29] developed a hierarchical network-assisted architecture for task offloading in green IoT with the objective of reducing energy consumption and carbon emission. Li et al. [16] designed a model based on the multi-user MEC scenario which realized the trade-off between the task offloading delay and energy consumption. Yang et al. [34] divided the task offloading problem in MEC into two steps and proposed an efficient offloading framework. This framework altered the order of affloading decision and resource assignment to achieve the purpose of reducing the dimension of the state space and action space simultaneously.

2.2   UAV Scheduling

In the scene of UAV-assisted MEC, UAV scheduling refers to determining the number and location of UAVs. Keshavamurthy et al. [14] deduced an optimal control strategy for a single relay using the semi-Markov decision process formula. Based on this, they regarded multiple UAVs as a competitive group and optimized the overall trajectory. Long et al. [19] jointly optimized the UAV path planning and scheduling strategy to minimize the long-term information age in wireless communication. Fan et al. [9] considered a communication scenario supported by UAVs. In this scenario, UAVs need to select the optimal communication location and avoid obstacles under the constraints of known ground eavesdropping nodes to achieve safe communication with the ground. In order to minimize the weighted sum of the task transmission energy consumption and UAV hovering energy consumption, Diao et al. [7] considered the UAV as a relay node to assist the MEC system. Zhang et al. [36] defined different computing strategies for terminal devices and proposed a new optimization method for path planning, power allocation and bit allocation of UAVs. This method ensured the minimum energy consumption of the system. As to the traffic overload problem of edge servers on UAVs, Lai et al. [15] proposed a UAV deployment scheme which were both fair and self-adaptive. Luna et al. [20] analyzed the multi-UAV rapid coverage path planning problem from two aspects of software framework and algorithm, and mainly adopted heuristic algorithm and minimization technology to optimize the planning task. However, all the above literatures do not analyze the impact of UAV heterogeneity on UAV scheduling, which is taken fully consideration by us.

2.3   Task Allocation

Task allocation can be divided into two parts, i.e., affloading decision and resource assignment, in UAV-assisted MEC. The former is to determine which edge server the users' tasks are executed on, and the latter is to consider how many resources should be allocated to the users' tasks. Chen et al. [5] designed a joint task scheduling, routing, and charging decision-making scheme, which removed the limitation of UAV battery capacity. Inspired by the simulated annealing algorithm, Liu et al. [17] proposed a multi-UAV cooperative task allocation algorithm. Wei et al. [32] divided the tasks into different types based on the detection value, and then proposed a task allocation method by improving the multiverse optimization algorithm. Wu et al. [33] introduced a fusion genetic algorithm that integrates improved simulated annealing to enhance the task execution efficiency of multiple UAVs. Akter et al. [1] conducted research on the task offloading in a multi-layer UAV-assisted MEC system based on the hybrid genetic algorithm. Zhang et al. [35] investigated the task assignment problem for the cooperative multi-UAV system with temporal coupling constraints, aiming to minimize the equivalent distance of all tasks. Zhang et al. [39] realized the task allocation in the process of multi-UAV combat based on the clone selection algorithm, which optimized multiple targets at the same time. However, the aforementioned literatures do not take the importance of UAV parallel process-ing tasks on QoE into consideration in the problem of task allocation.

3 System Model

In this section, we introduce a main scenario of the UAV scheduling problem and the task allocation problem. Based on this scenario, we propose a heterogeneous UAV-assisted MEC system and its main notations. What's more, we also adopt three models, i.e., the UAV scheduling model, the task executing model and the UAV hovering model in this system. The final optimization goal of the system is also stated.

3.1 Problem Statement

We depict a scenario that includes the hot-spot area and the UAV area. To be specific, there are several randomly distributed computing tasks generated by users within the hot-spot area. Note that the properties of these tasks are different from each other. Within a certain range outside the hot-spot area, i.e., the UAV area, several UAVs are randomly distributed. It should be noted that each of the UAVs also has different attributes which can also be regarded as heterogeneity. This scenario mainly has the following features:

(1) The scale of the hot-spot area is reasonable, neither too small so that multiple UAVs can hover over it, nor too large, at least guarantee that the UAVs can reach a specific location on time.
(2) The users' task requests in the hot-spot area change over time, e.g., a large number of requests may be generated in a sudden.
(3) Similar to (1), the scale of the UAV area also cannot be too small or too large, which ensures that the distribution of the UAVs are not too dense and the time of scheduling the UAVs is reasonable.

Therefore, we can abstract the scenario described above into the following problem: Given a hot-spot area and a UAV area whose scale and location is relatively fixed. Several computing tasks generated by users and heterogeneous UAVs are randomly distributed in the hot-spot area and the UAV area, respec-tively. We first need to schedule the UAVs to fly from the UAV area to the hot-spot area, and then allocate the computing tasks in the hot-spot area to the scheduled UAVs, which is the UAV scheduling problem and the task allocation problem.

3.2 System Overview

Under the above scenario, we propose a heterogeneous UAV-assisted MEC sys-tem. The main definitions of parameters used in this system are summarized in Table 1. The system contains N heterogeneous UAVs available for schedul-ing and M computing tasks requested by users, where the UAVs are denoted

Table 1. List of notation.

Notation	Explanation
N	The number of UAVs available for scheduling
M	The number of computing tasks
R	The radius of the UAV area
L	The side length of the hot-spot area
$\mathcal{U}$	The set of UAVs available for scheduling
$\mathcal{U}_{N'}$	The set of scheduled UAVs
$\mathcal{T}$	The set of computing tasks
$\mathcal{C}_N$	The set of initial coordinates of all UAVs
$\mathcal{C}_{N'}$	The set of hovering coordinates of scheduled UAVs
$\mathcal{C}_M$	The set of coordinates of all users' tasks
m_i	The memory size of u_i
q_i	The CPU processing power of u_i
p_i^h	The hovering power of u_i
p_i^f	The flight power of u_i
s_i	The flight speed of u_i
b_i	The battery power of u_i
r_i	The communication radius of u_i
d_j	The data size of t_j
c_j	The CPU cycles of t_j
a_{ij}	The binary variable: t_j is served by $u_i(a_{ij} = 1)$ or not$(a_{ij} = 0)$
W_i^{UH}	The distance from u_i to the hot-spot area
D_{ij}^{UT}	The distance between u_i and t_j
R_{ij}^{UT}	The uplink data rate of t_j on u_i
D_{ik}^{UU}	The distance between u_i and u_k

as $\mathcal{U}=\{u_1, u_2, ..., u_N\}$ and the tasks are represented by $\mathcal{T}=\{t_1, t_2, ..., t_M\}$. As shown in Fig. 1, the UAVs are heterogeneous and the k-th of UAVs can be denoted as $u_k=\{m_k, q_k, p_k^h, p_k^f, s_k, b_k, r_k\}$, where $m_k, q_k, p_k^h, p_k^f, s_k, b_k$ and r_k respectively represent the memory size, the CPU processing power, the hovering power, the flight power, the flight speed, the battery power and the communication radius. The initial three-dimensional coordinates of these UAVs are seemed as $\mathcal{C}_N=\{(X_k, Y_k, h), \forall u_k \in \mathcal{U}\}$. We assume to schedule N' of UAVs from $\mathcal{U}$ to the hot-spot area, denoted as $\mathcal{U}_{N'}=\{u_1, u_2, ..., u_{N'}\}$, in which the i-th UAV is denoted as u_i and the three-dimensional coordinates of UAVs in the area can be expressed as $\mathcal{C}_{N'}=\{(x_i, y_i, h), \forall u_i \in \mathcal{U}_{N'}\}$. It has to be noticed that $\mathcal{U}_{N'}$ and $\mathcal{C}_{N'}$ cannot be obtained in advance. As to computing tasks $\mathcal{T}$, the j-th task of it is denoted as $t_j=\{d_j, c_j\}$, where d_j and c_j respectively represent the task size and the CPU cycles. The three-dimensional coordinates of tasks $\mathcal{T}$ are represented as $\mathcal{C}_M = \{(x_j, y_j, 0), \forall t_j \in \mathcal{T}\}$. Note that $\mathcal{U}, \mathcal{T}, \mathcal{C}_N$ and $\mathcal{C}_M$ are provided in priori.

In our heterogeneous UAV-assisted MEC system, each task in $\mathcal{T}$ may be executed by any of UAVs in $\mathcal{U}_{N'}$. What's more, we define $a_{ij} \in \{0,1\}$ to indicate whether task t_j is allocated to UAV u_i, where $a_{ij} = 1$ means task t_j is allocated to UAV u_i and vice versa. In this paper, we stipulate that all UAVs are equipped with directional antennas with fixed beamwidth, serving all tasks through frequency division multiple access.

There are three models in our system: (1) The UAV scheduling model. (2) The task execution model. (3) The UAV hovering model.

3.3 UAV Scheduling Model

When the UAVs available for scheduling in the UAV area are scheduled to the hot-spot area, the scheduling time of any scheduled UAV u_i in $\mathcal{U}_{N'}$ is defined as:

$$T_i^F = \frac{W_i^{UH}}{s_i}, \forall u_i \in \mathcal{U}_{N'} \tag{1}$$

where W_i^{UH} is the distance between the initial position of u_i represented in a form of three-dimensional coordinate (X_i, Y_i, h) to the hot-spot area. It should be noted that it is necessary to ensure that the battery power of any scheduled UAV is greater than the minimum battery power, so the following constraint need to be satisfied:

$$C1 : b_i \geqslant b_{min}, \forall u_i \in \mathcal{U}_{N'} \tag{2}$$

where b_{min} is the minimum power required to satisfy our UAV-assisted MEC system. The flight energy consumption of u_i is:

$$E_i^F = p_i^f T_i^F, \forall u_i \in \mathcal{U}_{N'} \tag{3}$$

We can get the scheduling energy consumption of the scheduled UAVs from the Eqs. (1)–(3) above.

3.4 Task Execution Model

When a computing task generated by users is executed on a UAV, the following three steps are required: First, the task needs to be offloaded to the UAV. Then the edge server on the UAV processes the task. Finally, the processing results are transmitted back to the terminal device. The transmission distance between the UAV u_i and the task t_j can be calculated as:

$$D_{ij}^{UT} = \|(x_i, y_i, h) - (x_j, y_j, 0)\|$$
$$= \sqrt{(x_i - x_j)^2 + (y_i - y_j)^2 + h^2} \tag{4}$$
$$\forall u_i \in \mathcal{U}_{N'}, \forall t_j \in \mathcal{T}$$

Obviously, if the task t_j is executed on the UAV u_i, t_j must be within the coverage of the communication radius of u_i, that is, the following constraint need to be met as [2]:

$$C2 : a_{ij}D_{ij}^{UT} \leqslant r_i, \forall u_i \in \mathcal{U}_{N'}, \forall t_j \in \mathcal{T} \tag{5}$$

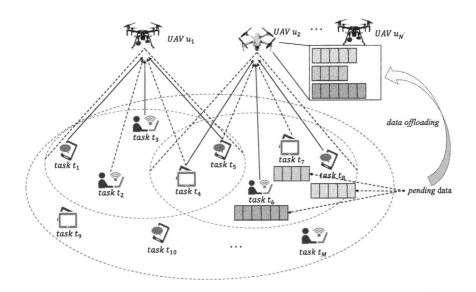

Fig. 1. Overview of heterogeneous UAV-assisted MEC system.

Considering the limited memory capacity of the edge server on the UAV. To ensure the service quality of the edge server, the total task size of the tasks executed simultaneously on each UAV should meet certain constraint:

$$C3 : \sum_{j=1}^{M} a_{ij}d_j \leqslant \alpha m_i, \forall u_i \in \mathcal{U}_{N'}, \forall t_j \in \mathcal{T} \tag{6}$$

where α is a normal number in the range of $(0, 1)$, which represents the maximum memory usage for optimal computing performance of the UAV. When the task size in the UAV exceeds the threshold αm_i, it is considered that the load capacity of the UAV has been exceeded. When the task t_j is offloaded to the UAV u_i, the uplink transmission rate is defined as [31]:

$$R_{ij}^{UT} = B_0 log_2(1 + \frac{P^{TE}G_0C_1}{N_0B_0(D_{ij})^2})$$
$$\forall u_i \in \mathcal{U}_{N'}, \forall t_j \in \mathcal{T} \tag{7}$$

where B_0 is the channel bandwidth, P^{TE} is the transmission power of the terminal device, G_0 is the channel gain, N_0 is the noise power, and C_1 is a positive constant.

The execution time of the task t_j includes the transmission time for offloading t_j to the UAV u_i and the computing time for u_i to process the task [30]:

$$T_{ij}^C = T_{ij}^{Comm} + T_{ij}^{Comp}$$
$$= \frac{d_j}{R_{ij}^{UT}} + \frac{c_j}{q_i} \tag{8}$$
$$\forall u_i \in \mathcal{U}_{N'}, \forall t_j \in \mathcal{T}$$

Therefore, the energy consumption of executing task t_j includes the transmission energy consumption of t_j offloaded to UAV u_i and the computing energy consumption of u_i processing t_j [31]:

$$E_{ij}^C = P^{TE} \frac{d_j}{R_{ij}^{UT}} + \varsigma(q_i)^{C_2} c_j, \forall u_i \in \mathcal{U}_{N'}, \forall t_j \in \mathcal{T} \tag{9}$$

where ς is the effective switched capacitance and C_2 is a positive constant. We can get the task executing energy consumption of all users' tasks from Eqs. (4)–(9).

It should be noted that when the task is returned to the terminal device after being processed on the UAV, the return transmission delay is negligible.

3.5   UAV Hovering Model

When the UAVs are executing the users' tasks, they will hover at a fixed position over the hot-spot area. To avoid collision, the distance between any two UAVs cannot be less than the minimum distance D_{min}^{UU}. Thus, the following constraint should be hold:

$$C4: D_{ik}^{UU} \geq D_{min}^{UU}, \forall u_i, u_k \in \mathcal{U}_{N'}, i \neq k \tag{10}$$

The hovering energy consumption of UAV u_i is expressed as:

$$E_i^H = p_i^h T_i^H, \forall u_i \in \mathcal{U}_{N'} \tag{11}$$

where T_i^h is the hovering time of u_i. We can get the hovering energy consumption of all scheduled UAVs from Eqs. (10)–(11).

3.6   Problem Formulation

In our heterogeneous UAV-assisted MEC system, the key problems to be solved are UAV scheduling and task allocation. We jointly optimize these two problems in order to minimize the total energy consumption of the system. Specifically, the total energy consumption of the system is the sum of the scheduling energy consumption, the task executing energy consumption and the hovering energy consumption in the above three models. Our final optimization goal can be expressed as:

$$\min_{N',x_i,y_i,a_{ij}} \sum_{i=1}^{N'}(E_i^F + E_i^H) + \sum_{i=1}^{N'}\sum_{j=1}^{M} a_{ij}E_{ij}^C$$

$$
\begin{aligned}
s.t. \quad & C1 : b_i \geqslant b_{min}, \forall u_i \in \mathcal{U}_{N'} \\
& C2 : a_{ij}D_{ij}^{UT} \leqslant r_i, \forall u_i \in \mathcal{U}_{N'}, \forall t_j \in \mathcal{T} \\
& C3 : \sum_{j=1}^{M} a_{ij}d_j \leqslant \alpha m_i \\
& \forall u_i \in \mathcal{U}_{N'}, \forall t_j \in \mathcal{T}, \alpha \in (0,1) \\
& C4 : D_{ik}^{UU} \geq D_{min}^{UU}, \forall u_i, u_k \in \mathcal{U}_{N'}, i \neq k \\
& C5 : \sum_{i=1}^{N} a_{ij} = 1, \forall u_i \in \mathcal{U}_{N'}, \forall t_j \in \mathcal{T} \\
& C6 : a_{ij}T_{ij}^C \leqslant T_i^H, \forall u_i \in \mathcal{U}_{N'}, \forall t_j \in \mathcal{T}
\end{aligned}
\tag{12}
$$

where $C1$, $C2$, $C3$ and $C4$ have been explained above. In addition, $C5$ guarantees that every task can be executed by the scheduled UAVs and one task can only be executed by one UAV. $C6$ is a constraint for the latency constraints of executing each task, that is, the task execution time cannot exceed the UAV hovering time.

4   Proposed Approach

In this section, we elaborate on the proposed UAV scheduling algorithm and task allocation algorithm. We first regard the UAV scheduling problem as a knapsack problem with geographical constraints and UAV attribute constraints. Based on this, we propose a heuristic UAV scheduling decision algorithm which can obtain an optimal solution for minimizing the energy consumption. Then, we cluster the users' tasks based on the geographical location. By analyzing the results of clustering, we propose a task allocation algorithm, which trades off the UAV memory capacity and users' task size. We finally obtain the number and location of scheduled UAVs as well as the offloading decision and resource assignment of users' tasks.

4.1   UAV Scheduling Algorithm

The requests of users' tasks in the hot-spot area have obvious time-variability, resulting the number and the location of users' tasks changing with time. Therefore, we need to flexibly deploy the number and location of the UAVs according to the size and the distribution of users' tasks. Our goal is to guarantee that the scheduled UAVs can execute all users' tasks in parallel as low as possible for energy consumption. Although we prioritize the UAVs with large memory for scheduling in a relatively small number, it still cannot guarantee the minimization of the scheduling energy consumption as it is limited by the initial position of the scheduled UAVs. At the same time, if we prioritize to schedule the UAVs near the hot-spot area, we cannot guarantee the number of the scheduled UAVs, which may also lead to higher scheduling energy consumption.

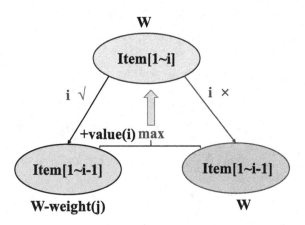

Fig. 2. The state transition process of the knapsack problem.

To make the trade-off between the above two situations, we regard the UAV scheduling problem as a knapsack problem [21]. As a NP-hard problem, the essence of the knapsack problem is a sort of constrained optimization problem whose goal is to maximize the total value of the items in the knapsack under a certain capacity limit. The state transition process of this problem is shown in Fig. 2. To be specific, We consider all the users' tasks as a whole and regard it as a knapsack. In addition, we regard all the UAVs as several items. Thus, the sum of the task size is regarded as the knapsack capacity, and the memory capacity of the UAV is regarded as the weight of the item. What's more, we regard the flight energy consumption of each UAV flying to the hot-spot area as the value of the item, and the minimum scheduling energy consumption can be expressed as:

$$\min_{N',a_{ij}} \sum_{i=1}^{N'} E_i^P = \max_{N',a_{ij}} \sum_{i=1}^{N'} \frac{1}{E_i^P} \tag{13}$$

$$s.t. \quad C1, \ C2 \ and \ C3$$

It should be noted that the goal of the knapsack problem is to maximize the value of the items in the knapsack, while the goal of the UAV scheduling problem is to minimize the scheduling energy consumption. Therefore, it is necessary to invert the scheduling energy consumption. The state transition equation of this method is:

$$F(i, \mathbf{S}) = \begin{cases} F(i-1, \mathbf{S}), & \mathbf{S} < \alpha m_i \\ max\{F(i-1, \mathbf{S}), & \\ F(i-1, \mathbf{S} - \alpha m_i)\} , & \mathbf{S} \geq \alpha m_i \end{cases} \tag{14}$$

where $\mathbf{S}$ is a constraint on the sum of the maximum memory usage of scheduled UAVs. We first let $S = \sum_{j=1}^{M} d_j$ and $\mathbf{S} \leftarrow S$, we get the initial UAV scheduling set $U_{N'}$ through continuous recursive iterations.

Algorithm 1. UAV Scheduling Decision Algorithm

Input: The set of UAVs available for scheduling $\mathcal{U}$, the set of UAV initial coordinates $\mathcal{C}_N$, the sum of the task size S
Output: The set of scheduled UAVs $\mathcal{U}_{N'}$
1: $S^* \leftarrow S, \mathcal{F} \leftarrow \emptyset, j \leftarrow 1$
2: Assign the value of S^* to $\mathbf{S}$ in Eq. (14) to calculate the initial set of scheduled UAV $\mathcal{U}_{N_1}$
3: Record the minimum memory m_k^* of the UAV in the set $(\mathcal{U} - \mathcal{U}_{N_1})$
4: Sort all tasks in ascending order according to the task size: $t_1^*, t_2^*, ..., t_M^*$
5: **while** $S^* < S + m_k^*$ **do**
6:     $S^* \leftarrow S^* + t_j^*$
7:     Assign the value of S^* to $\mathbf{S}$ in Eq. (14) to update $\mathcal{U}_{N'}$
8:     **if** $\sum_{i=1}^{N'} \alpha m_i \geqslant S$ **then**
9:         Calculate the scheduling energy consumption by Eq. (3)
10:         Add the flight energy consumption to the set $\mathcal{F}$
11:     **end if**
12:     $j \leftarrow j + 1$
13: **end while**
14: Compare the value of the elements in $\mathcal{F}$ and select the set $U_{N'}$ corresponding to the smallest element
15: **return** $\mathcal{U}_{N'}$

However, the constraint $\mathbf{S}$ of the UAV scheduling problem is different from that of the knapsack problem. In the knapsack problem, the constraint needed to be satisfied is that the weight of all items loaded into the knapsack cannot exceed the capacity of the knapsack, which will however cause the knapsack to appear "not full". In the UAV scheduling problem, it is necessary to satisfy that the sum of the maximum memory usage of scheduled UAVs is not less than the sum of the users' task size, in order to guarantee that all tasks can be executed in parallel. To solve this problem, we propose a UAV scheduling decision algorithm to find the optimal UAV scheduling scheme, as shown in Algorithm 1. We analyze two situations that may occur in the line 2 of the algorithm to obtain the initially scheduled UAV set $U_{N'}$:

(1) $\sum_{i=1}^{N'} \alpha m_i < S$: At this time, the scheduled UAVs cannot execute all tasks in parallel. Therefore, it is necessary to appropriately increase the value of

the constraint **S**. We sort the tasks in an ascending order according to the task size, and add the size value into **S** in turn until it satisfies the condition of $\sum_{i=1}^{N'} \alpha m_i \geq S$.

(2) $\sum_{i=1}^{N'} \alpha m_i = S$: The scheduled UAVs can execute all tasks in parallel at this time, but the UAV deployment scheme may not be the one with the least scheduling energy consumption. For example, there may be some UAVs that have the lower flight energy consumption but are not scheduled due to their large memory. Therefore, we need to further increase the value of the constraint **S** to obtain the optimal solution. We change the value of **S** in the same way as (1) sequentially until the sum of the increased task size is no less than the minimum memory of the remaining UAVs after the first scheduling.

Algorithm 2. Task Allocation Algorithm

Input: The set of scheduled UAVs $\mathcal{U}_{N'}$, the set of users' tasks $\mathcal{T}$, the set of task clusters $\tilde{\mathcal{R}}$, the set of tasks coordinates $\mathcal{C}_M$, the set of centroid coordinates $\mathcal{C}_P$

Output: The set of scheduled UAVs coordinates $\mathcal{C}_{N'}$, the task allocation scheme $\mathcal{P}$

1: $\mathcal{T}_{M'} \leftarrow \emptyset, i \leftarrow 1, j \leftarrow 1$,
2: **while** $i < N'$ **do**
3:    **if** Meet any one of (1) or (2) or (3) or (4) **then**
4:        Select any element $(x_u, y_u, 0)$ from the set $\mathcal{C}_P$
5:        Arrange the tasks in the set $\mathcal{T}$ according to the distance from the centroid $(x_u, y_u, 0)$ from near to far: $t_1^\star, t_2^\star, ..., t_M^\star$
6:        **while** $\sum_{n=1}^{|\mathcal{T}_{M'}|} d_n < \alpha m_i$ **do**
7:            Add $t_j^\star$ to the set $\mathcal{T}_{M'}$
8:            $j \leftarrow j + 1$
9:        **end while**
10:       $\mathcal{T} \leftarrow \mathcal{T} \backslash \mathcal{T}_{M'}, \mathcal{T}_{M'} \leftarrow \emptyset$
11:       $\mathcal{C}_P \leftarrow \mathcal{C}_P \backslash \{(x_u, y_u, 0) \in \mathcal{C}_P\}, i \leftarrow i + 1$
12:       Get the task allocation scheme $\mathcal{P}$ in the above four cases
13:    **else**
14:       **for** each $v \in [1, N']$ **do**
15:           Calculate the energy consumption by Eq. (9) and Eq. (11) of UAV u_i processing task cluster $\mathcal{R}_v$
16:           **if** $\sum_{n=1}^{|\mathcal{R}_v|} d_n < \alpha m_i$ **then**
17:               Replace the corresponding position in the cost matrix $\mathcal{E}$ as $+\infty$
18:           **end if**
19:       **end for**
20:       Call the Hungarian algorithm to process the cost matrix $\mathcal{E}$
21:    **end if**
22:    Get the task allocation scheme $\mathcal{P}$ in this case
23: **end while**
24: **return** $\mathcal{P}$

4.2   Task Allocation Algorithm

After obtaining the scheduled UAV set $U_{N'}$, we need to further determine the hovering positions of these UAVs and the task allocation scheme. Due to the heterogeneity of the UAVs and the different distances between the UAVs and the tasks, the energy consumption of each UAV executing each task is different. From the perspective of the tasks geographical location, it is clear that the task executing energy consumption higher when the UAV executes a task with a longer distance. Therefore, it is necessary to guarantee that the tasks executed by each UAV are as close as possible geographically, which can not only avoid the extra flight energy consumption of the UAV, but also reduce the transmission energy consumption when the UAV executes long-distance tasks. In the task allocation problem, our goal is to minimize the sum of task execution energy consumption and hovering energy consumption, that is:

$$\min_{N', a_{ij}} \sum_{i=1}^{N'} E_i^H + \sum_{i=1}^{N'} \sum_{j=1}^{M} a_{ij} E_{ij}^C \tag{15}$$

$$s.t. \quad C4, \ C5 \ and \ C6$$

Aiming at solving the above problems, we propose to cluster all tasks, where the number of clusters is the same as the number of scheduled UAVs. Since the tasks have obvious geographical features, we adopt the binary K-means algorithm [11], which overcomes the problem of getting stuck in a local optimum compared with the K-means algorithm. To make sure of good intra-cluster similarity as well as poor inter-cluster similarity, the objective function of the clustering algorithm is defined as:

$$J = \sum_{j=1}^{M} \sum_{u=1}^{N'} w_{ju} \| (x_j, y_j, 0) - (x_u, y_u, 0) \|^2$$

$$= \sum_{j=1}^{M} \sum_{u=1}^{N'} w_{ju} [(x_j - x_u)^2 + (y_j - y_u)^2] \tag{16}$$

where $(x_u, y_u, 0)$ is the centroid coordinate of the task cluster R_u, w_{ju} means 1 when the task t_j belongs to task cluster R_u, and 0 otherwise:

$$w_{ju} = \begin{cases} 1, & w_{ju} \in R_u \\ 0, & w_{ju} \notin R_u \end{cases} \tag{17}$$

The set of the task clusters after clustering is denoted as $\widetilde{\mathcal{R}} = \{\mathcal{R}_1, \mathcal{R}_2, ..., \mathcal{R}_{N'}\}$ ($|\widetilde{\mathcal{R}}| = N'$), in which each element is the subset containing several tasks. The centroid coordinates of the task clusters obtained by clustering are the ground mapping coordinates of the hovering position of the scheduled UAVs, which denoted as $\mathcal{C}_P = \{(x_u, y_u, 0), \forall \mathcal{R}_u \in \widetilde{\mathcal{R}}\}$ ($|\mathcal{C}_P| = N'$). We can get the hovering position of the scheduled UAVs from above. The clustering results are shown in Fig. 3, taking the five scheduled UAVs as an example.

The task allocation algorithm is shown in Algorithm 2, which trades off the UAV memory capacity and users' task size. To guarantee that all tasks are executed in parallel, it is necessary to compare the memory capacity of each scheduled UAVs in $\mathcal{U}_{N'}$ with the total size of each task cluster in $\widetilde{\mathcal{R}}$, and then formulate different task allocation schemes according to different comparison results. The four cases described in the line 3 of the algorithm are as follows:

(1) There is at least one task cluster in $\widetilde{\mathcal{R}}$ that cannot be executed by any UAV in $\mathcal{U}_{N'}$ in parallel.
(2) There is at least one UAV in $\mathcal{U}_{N'}$ that cannot execute any task cluster $\widetilde{\mathcal{R}}$ in parallel.
(3) There are at least two task clusters in $\widetilde{\mathcal{R}}$ that can only be executed by the same UAV in $\mathcal{U}_{N'}$ in parallel.
(4) There at least two UAVs in $\mathcal{U}_{N'}$ that can only execute the same task cluster in $\widetilde{\mathcal{R}}$ in parallel.

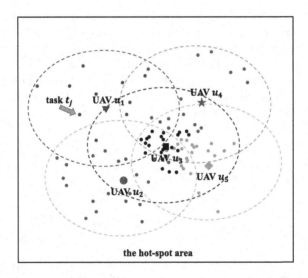

Fig. 3. The clustering result for tasks, taking the five scheduled UAVs as an example.

When the above four situations occur, we take all the centroid coordinates in $\mathcal{C}_P$ as the center, and re-cluster the tasks from near to far according to the ratio of the maximum memory usage of the scheduled UAVs in $\mathcal{U}_{N'}$. Then we allocate the task clusters to the UAVs for execution with corresponding memory ratios.

Except for the above four situations, we regard the task allocation problem as an assignment problem [27]. We sequentially calculate the energy consumption of each UAV in $\mathcal{U}_{N'}$ executing each task cluster in $\widetilde{\mathcal{R}}$, which is also the sum of the task execution energy consumption and hovering energy consumption, and obtain the cost matrix $\mathcal{E}$:

$$\mathcal{E} = \begin{bmatrix} e_{1,1} & e_{1,2} & \cdots & e_{1,N'} \\ e_{2,1} & e_{2,2} & \cdots & e_{2,N'} \\ \vdots & \vdots & \ddots & \vdots \\ e_{N',1} & e_{N',2} & \cdots & e_{N',N'} \end{bmatrix} \tag{18}$$

where $e_{i,u}$ represents the sum of the task execution energy consumption and hovering energy consumption when the UAV u_i executes the task cluster R_u. Since the number of scheduled UAVs is the same as the number of task clusters, the matrix $\mathcal{E}$ is ideally an N'-order square matrix. However, we cannot guarantee that each UAV in $\mathcal{U}_{N'}$ can execute each task cluster in $\widetilde{\mathcal{R}}$ in parallel, which is also the case as follows:

$$\sum_{t_j \in \mathcal{R}_u} d_j > \alpha m_i, \exists u_i \in \mathcal{U}_{N'}, \exists \mathcal{R}_u \in \widetilde{\mathcal{R}} \tag{19}$$

The task allocation problem in this case is an unbalanced assignment problem, and the cost matrix $\mathcal{E}$ is not a square matrix. In order to transform the unbalanced assignment problem into a balanced assignment problem, we set the value of the corresponding element satisfied to Eq. (19) in $\mathcal{E}$ to $+\infty$. This is because when a UAV cannot execute a task cluster in parallel, we will not allocate the task cluster to this UAV in the end. After setting the corresponding value in $\mathcal{E}$ to $+\infty$, on the one hand, it can facilitate a series of subsequent operations on the cost matrix $\mathcal{E}$, on the other hand, the changing of value will be ignored no matter what operations are operated. In order to minimize the sum of the task executing energy consumption and the hovering energy consumption, we apply the Hungarian algorithm [10] to operate on the cost matrix $\mathcal{E}$. Based on such processing results of $\mathcal{E}$, we get the task allocation scheme in this case.

5   Performance Evaluation

In this section, we simulate our proposed joint optimization method JoSA with the other two benchmarks in the same scenario and obtain different system energy consumption and time cost by changing the independent variable. What's more, we also analyze the experimental results in detail.

5.1   Experimental Settings

We employ numerical simulations to evaluate the performance of our proposed joint optimization method JoSA in the heterogeneous UAV-assisted MEC system. The default values of simulation settings are shown in the Table 2 reference

Table 2. Simulation parameters.

Parameter	Value	Parameter	Value
d_j	[50,150]MB	B_0	1MHz
c_j	[15,1500]MCycles	P^{TE}	10W
m_i	[6,12]GB	G_0	1.42×10^{-4}
q_i	[0.2,0.8]GHz	N_0	10^{-15}W/Hz
p_i^f	[60,90]W	C_1	2.2846
p_i^h	[45,70]W	ς	10^{-10}
s_i	[50,80]km/h	C_2	3
b_i	[20,60]Wh	α	0.7
r_i	[150,200]m	H	100m

to [12,31]. Specifically, we evaluate the system energy consumption and time cost respectively under the conditions of changing the number of tasks, the number of UAVs, the side length of the hot-spot area and the radius of the UAV area, as shown in Table 3. To evaluate the effectiveness of our proposed algorithm, the comparison algorithms are used as follow:

- HOLD [30]: All tasks are merged into several areas into several areas and UAVs iteratively find their appropriate hovering locations to serve the users.

- HPSO [37]: UAVs are scheduled based on the particle swarm optimization algorithm. The particles are eliminated in combination with the search space and the particle update scheme is designed according to the number of particles.

We adjust the above methods under our experimental scenario and make reasonable adjustments to relevant parameters.

Table 3. Experimental independent variables.

Experiment Num	The number of tasks	The side length of hot-spot area(m)	The number of UAVs	The radius of UAV area(km)
Experiment I	100, 200, 300, 400, 500	300	15	3
Experiment II	300	100, 200, 300, 400, 500	15	3
Experiment III	300	300	7, 9, 11, 13, 15	3
Experiment IV	300	300	15	2, 2.5, 3, 3.5, 4

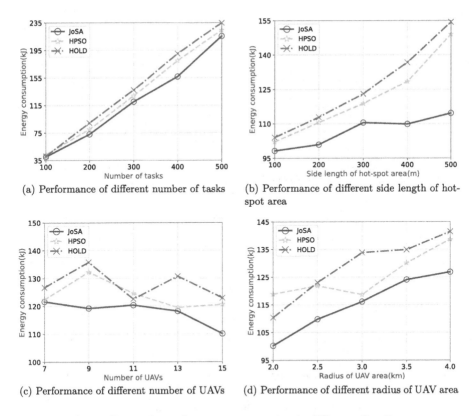

(a) Performance of different number of tasks

(b) Performance of different side length of hot-spot area

(c) Performance of different number of UAVs

(d) Performance of different radius of UAV area

Fig. 4. Comparison of energy consumption in different situations.

5.2   Experimental Results

Experiment I. Performance Under Different Number of Tasks: By increasing the number of tasks, we observe the variation trend of energy consumption as well as the time cost of our proposed system using different algorithms. To be specific, We set the radius of the UAV area as R=3km and randomly generate several UAVs as N=15 in this circular UAV area. We also set the side length of the hot-spot area as L=300m and randomly generate tasks as M= 100, 200, ..., 500 in this square the hot-spot area. As Fig. 4(a) and Fig. 5(a) shown, with the number of tasks increasing, more UAVs are needed to be scheduled for task executing, resulting in the high system energy consumption and time cost. What's more, our algorithm makes a reduction of 7% and 12% in system energy consumption, and 4% and 5% in system time cost, respectively, compared with the other two benchmarks, as it can flexibly deploy the number and location of UAVs based on the total number of tasks.

Experiment II. Performance Under Different Area of the Hot-Spot Area: By changing the side length of the hot-spot area, we observe the same parts as above using different algorithms. Same with experiment I, we set the

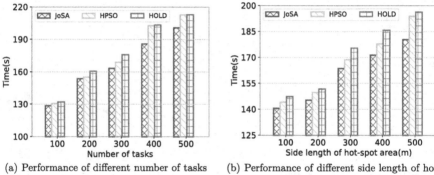

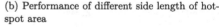
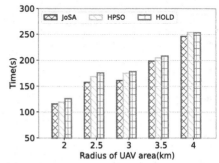

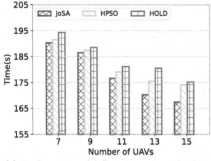

(a) Performance of different number of tasks

(b) Performance of different side length of hot-spot area

(c) Performance of different number of UAVs

(d) Performance of different radius of UAV area

Fig. 5. Comparison of time in different situations.

radius of the UAV area as R=3km and randomly generate several UAVs as N = 15 in this circular UAV area. The difference is that we change the side length of the hot-spot area as L = 100, 200, ..., 500 m and randomly generate tasks as M = 300 in these different square hot-spot areas separately. As Fig. 4(b) and Fig. 5(b) shown, as the side length of the hot-spot area increases, the distance between the UAV and the task becomes longer, resulting in the high system energy consumption and time cost. What's more, since HOLD processes tasks from near to far straightly, while HPSO leads the final hovering positions of all scheduled UAVs too close. Both of them result in more system energy consumption and time cost when processing long-distance tasks. However, our algorithm solves the problems the above algorithms faced through controlling the hovering distances among scheduled UAVs when the locations of tasks are relatively scattered. As a result of this, our algorithm performs much better. To be specific, our algorithm makes a reduction of 12% and 15% in system energy consumption, and 4% and 6% in system time cost, respectively, compared with the other two benchmarks.

Experiment III. Performance Under Different Number of UAVs: By increasing the number of UAVs, we observe the same parts as above using different algorithms. We set the side length of the hot-spot area as $L = 300$ m and randomly generate tasks as $M = 300$ in this square hot-spot area. We also set the radius of the UAV area as $R = 3$ km and randomly generate UAVs as $N = 7, 9, ..., 15$ in this circular UAV area. As Fig. 4(c) and Fig. 5(c) shown, with the number of UAVs available for scheduling increasing, we can formulate more candidate schemes to make full use of these UAVs, in order to descend the system energy consumption. Compared with the other two benchmarks, simulation results show that our algorithm makes a reduction of 4% and 7% in system energy consumption, and 2% and 3% in system time cost, respectively. Therefore, our algorithm can make the optimal decision. The more the number of UAVs, the more obvious this advantage will be.

Experiment IV. Performance Under Different Radius of the UAV Area: By changing the radius of the UAV area, we observe the same parts as above using different algorithms. Same with experiment III, we set the side length of the hot-spot area as $L = 300$ m and randomly generate tasks as $M=300$ in this square hot-spot area. The difference is that we change the radius of the UAV area as $R = 2, 2.5, ..., 4k$ m and randomly generate UAVs as $N=15$ in these different circle areas in order. As Fig. 4(d) and Fig. 5(d) shown, as the radius of the UAV area increases, the distance between the initial positions of scheduled UAVs and the hot-spot areas becomes longer, resulting in the high system energy consumption and time cost. What's more, since HOLD schedules UAVs from near to far, while HPSO has randomness in the choice of scheduled UAVs. Both of them result in more system energy consumption and time cost if the scheduled UAVs have a long distance to the hot-spot areas. However, our algorithm solves the problems the above algorithms faced through making flexible choices based on the distribution state of UAVs as a result of better performance. Specifically, our algorithm makes a reduction of 8% and 10% in system energy consumption, and 4% and 2% in system time cost, respectively, compared with the other two benchmarks.

6   Conclusion

In this paper, we propose a heterogeneous UAV-assisted MEC system, which improves the performance of traditional MEC systems by using multiple heterogeneous UAVs. What's more, we also propose a joint optimization scheme, named JoSA, for the UAV scheduling and task allocation in this system. As to the optimization of UAV scheduling, the system regards it as a knapsack problem. Based on this, a UAV scheduling decision algorithm is proposed by us to obtain the optimal solution so as to minimize energy consumption. As to the optimization of task allocation, we propose a parallel task allocation algorithm by trading off the UAV memory capacity and users' task size, so that it can realize network deployment on demand. This paper fully considers the correlation between UAV scheduling and task assignment and proves these two parts can promote each other. What's more, simulation results show that our method makes a reduction of 8% and 11% in system energy consumption, and 3% and 4% in system time cost, respectively, compared with the other two benchmarks.

In future, we can use the UAVs as relay nodes to interact with location-fixed edge servers, to make full use of the resources of the location-fixed edge servers. In addition, it is also necessary to consider the priority of users' tasks and the temporal sequence of tasks completed by UAVs. The UAVs that complete the tasks first can be scheduled continuously to execute other tasks to realize the co-ordinating conjunction of multiple UAVs and to further improve QoE.

Acknowledgments. This work was supported by the National Key Research and Development Program of China under Grant 2021ZD0112400, the National Natural Science Foundation of China under grant 62202080, the Science and Technology Project of Liaoning Province under grant 2021JH1/10400009, the CCF-Tencent Open Fund under grant IAGR20220114, the Liaoning Revitalization Talents Program under grant XLYC2008017, the Fundamental Research Funds for the Central Universities under grant DUT20RC(3)039.

References

1. Akter, S., Kim, D.Y., Yoon, S.: Task offloading in multi-access edge computing enabled UAV-aided emergency response operations. IEEE Access **11**, 23167–23188 (2023)
2. Alzenad, M., El-Keyi, A., Yanikomeroglu, H.: 3-D placement of an unmanned aerial vehicle base station for maximum coverage of users with different QoS requirements. IEEE Wirel. Commun. Lett. **7**(1), 38–41 (2017)
3. Asim, M., ELAffendi, M., El-Latif, A.A.A.: Multi-IRS and Multi-UAV-Assisted MEC system for 5G/6G networks: efficient joint trajectory optimization and passive beamforming framework. IEEE Trans. Intell. Transp. Syst. **24**(4), 4553–4564 (2023)
4. Cao, Y., Long, C., Jiang, T., Mao, S.: Share communication and computation resources on mobile devices: a social awareness perspective. IEEE Wirel. Commun. **23**(4), 52–59 (2016)
5. Chen, J., Xie, J.: Joint task scheduling, routing, and charging for multi-UAV based mobile edge computing. In: ICC 2022 - IEEE International Conference on Communications, pp. 1–6 (2022)
6. Devagiri, J.S., Paheding, S., Niyaz, Q., Yang, X., Smith, S.: Augmented Reality and Artificial Intelligence in industry: Trends, tools, and future challenges. Expert Systems with Applications, p. 118002 (2022)
7. Diao, X., Yang, W., Yang, L., Cai, Y.: UAV-relaying-assisted multi-access edge computing with multi-antenna base station: offloading and scheduling optimization. IEEE Trans. Veh. Technol. **70**(9), 9495–9509 (2021)
8. Dzardanova, E., Kasapakis, V.: Virtual reality: a journey from vision to commodity. IEEE Ann. Hist. Comput. **45**(1), 18–30 (2023)
9. Fan, J., Liu, Y., Sun, G., Pan, H., Wang, A., Liang, S., Liu, Z.: 3D position scheduling of UAV secure communications with multiple constraints. In: 2022 IEEE International Conference on Systems, Man, and Cybernetics (SMC), pp. 143–149 (2022)
10. Hamuda, E., Mc Ginley, B., Glavin, M., Jones, E.: Improved image processing-based crop detection using Kalman filtering and the Hungarian algorithm. Comput. Electron. Agric. **148**, 37–44 (2018)

11. He, K., Wen, F., Sun, J.: K-means hashing: an affinity-preserving quantization method for learning binary compact codes. In: Proceedings of the IEEE Conference on Computer Vision and Pattern Recognition, pp. 2938–2945 (2013)
12. Jeong, S., Simeone, O., Kang, J.: Mobile edge computing via a UAV-mounted cloudlet: Optimization of bit allocation and path planning. IEEE Trans. Veh. Technol. **67**(3), 2049–2063 (2017)
13. Jiang, X., Hou, P., Zhu, H., Li, B., Wang, Z., Ding, H.: Dynamic and intelligent edge server placement based on deep reinforcement learning in mobile edge computing. Ad Hoc Netw. **145**, 103172 (2023)
14. Keshavamurthy, B., Michelusi, N.: Multiscale adaptive scheduling and path-planning for power-constrained UAV-Relays via SMDPs. In: 2022 56th Asilomar Conference on Signals, Systems, and Computers, pp. 1091–1097 (2022)
15. Lai, C.C., Bhola, Tsai, A.H., Wang, L.C.: Adaptive and fair deployment approach to balance offload traffic in multi-UAV cellular networks. IEEE Trans. Vehicular Technol. **72**(3), 3724–3738 (2023)
16. Li, R., Lim, C.S., Rana, M.E., Zhou, X.: A trade-off task-offloading scheme in multi-user multi-task mobile edge computing. IEEE Access **10**, 129884–129898 (2022)
17. Liu, H., Li, X., Wu, G., Fan, M., Wang, R., Gao, L., Pedrycz, W.: An iterative two-phase optimization method based on divide and conquer framework for integrated scheduling of multiple UAVs. IEEE Trans. Intell. Transp. Syst. **22**(9), 5926–5938 (2021)
18. Liu, Y., Peng, M., Shou, G., Chen, Y., Chen, S.: Toward edge intelligence: multiaccess edge computing for 5G and internet of things. IEEE Internet Things J. **7**(8), 6722–6747 (2020)
19. Long, Y., Zhang, W., Gong, S., Luo, X., Niyato, D.: AoI-aware scheduling and trajectory optimization for multi-UAV-assisted wireless networks. In: GLOBECOM 2022–2022 IEEE Global Communications Conference, pp. 2163–2168 (2022)
20. Luna, M.A., Ale Isaac, M.S., Ragab, A.R., Campoy, P., Flores Peña, P., Molina, M.: Fast multi-UAV path planning for optimal area coverage in aerial sensing applications. Sensors **22**(6), 2297 (2022)
21. Luo, Q., Rao, Y.: Heuristic algorithms for the special knapsack packing problem with defects arising in aircraft arrangement. Expert Syst. Appl. **215**, 119392 (2023)
22. Mach, P., Becvar, Z.: Mobile edge computing: a survey on architecture and computation offloading. IEEE Commun. Surv. Tutorials **19**(3), 1628–1656 (2017)
23. Mahr, D., Heller, J., de Ruyter, K.: Augmented reality (AR): the blurring of reality in human-computer interaction. Comput. Hum. Behav. **145**, 107755 (2023)
24. Meng, K., Wu, Q., Ma, S., Chen, W., Quek, T.Q.S.: UAV trajectory and beam-forming optimization for integrated periodic sensing and communication. IEEE Wireless Commun. Lett. **11**(6), 1211–1215 (2022)
25. Qin, L., Lu, H., Wu, F.: When the user-centric network meets mobile edge computing: challenges and optimization. IEEE Commun. Mag. **61**(1), 114–120 (2023)
26. Salaht, F.A., Desprez, F., Lebre, A.: An overview of service placement problem in fog and edge computing. ACM Comput. Surv. (CSUR) **53**(3), 1–35 (2020)
27. Shi, Y., Mei, Y.: Efficient sequential UCB-based Hungarian algorithm for assignment problems. In: 2022 58th Annual Allerton Conference on Communication, Control, and Computing (Allerton), pp. 1–8 (2022)
28. Sigwele, T., Hu, Y.F., Susanto, M.: Energy-efficient 5G cloud RAN with virtual BBU server consolidation and base station sleeping. Comput. Netw. **177**, 107302 (2020)

29. Wang, C., Yu, X., Xu, L., Wang, W.: Energy-efficient task scheduling based on traffic mapping in heterogeneous mobile-edge computing: a green IoT perspective. IEEE Trans. Green Commun. Networking **7**(2), 972–982 (2023)

30. Wang, J., Liu, K., Pan, J.: Online UAV-mounted edge server dispatching for mobile-to-mobile edge computing. IEEE Internet Things J. **7**(2), 1375–1386 (2020)

31. Wang, Y., Ru, Z.Y., Wang, K., Huang, P.Q.: Joint deployment and task scheduling optimization for large-scale mobile users in multi-UAV-enabled mobile edge computing. IEEE Trans. Cybern. **50**(9), 3984–3997 (2020)

32. Wei, Z., Zhao, X.: Multi-UAVs cooperative reconnaissance task allocation under heterogeneous target values. IEEE Access **10**, 70955–70963 (2022)

33. Wu, X., Yin, Y., Xu, L., Wu, X., Meng, F., Zhen, R.: Multi-UAV task allocation based on improved genetic algorithm. IEEE Access **9**, 100369–100379 (2021)

34. Yang, J., Wang, Y., Li, Z.: Inverse order based optimization method for task offloading and resource allocation in mobile edge computing. Appl. Soft Comput. **116**, 108361 (2022)

35. Zhang, R., Feng, Y., Yang, Y., Li, X.: A deadlock-free hybrid estimation of distribution algorithm for cooperative multi-UAV task assignment with temporally coupled constraints. IEEE Trans. Aerospace Electron. Syst. (2022)

36. Zhang, T., Xu, Y., Loo, J., Yang, D., Xiao, L.: Joint computation and communication design for UAV-assisted mobile edge computing in IoT. IEEE Trans. Industr. Inf. **16**(8), 5505–5516 (2020)

37. Zhang, W., Zhang, W.: An efficient UAV localization technique based on particle swarm optimization. IEEE Trans. Veh. Technol. **71**(9), 9544–9557 (2022)

38. Zhang, X., Wang, Y.: DeepMECagent: multi-agent computing resource allocation for UAV-assisted mobile edge computing in distributed IoT system. Appl. Intell. **53**(1), 1180–1191 (2023)

39. Zhang, X., Chen, X.: UAV task allocation based on clone selection algorithm. Wirel. Commun. Mob. Comput. **2021**, 1–9 (2021)

40. Zhou, R., Wu, X., Tan, H., Zhang, R.: Two time-scale joint service caching and task offloading for UAV-assisted mobile edge computing. In: IEEE INFOCOM 2022 - IEEE Conference on Computer Communications. pp. 1189–1198 (2022)

Multi-UAV Collaborative Face Recognition for Goods Receiver in Edge-Based Smart Delivery Services

Yi Xu[1]([✉]), Fengguang Luan[1], Jonathan Kua[2], Haoyu Luo[3], Zhipeng Wang[1], and Xiao Liu[2]

[1] School of Computer Science and Technology, Anhui University, Hefei, China
08055@ahu.edu.cn, {e20301324,e22301221}@stu.ahu.edu
[2] School of Information Technology, Deakin University, Geelong, Australia
{jonathan.kua,xiao.liu}@deakin.edu.au
[3] College of Mathematics and Informatics, South China Agricultural University, Guangzhou, China
luohy@whu.edu.cn

Abstract. The accurate and efficient recognition of goods receivers is a crucial task for Unmanned Aerial Vehicle (UAV) based smart delivery services in edge computing. Meanwhile, face recognition has become one of the major recognition methods in smart delivery services given its high efficiency and reliability. However, unlike many other application scenarios, the face recognition algorithm's accuracy and efficiency are often limited due to the unique camera shooting angles by the UAVs flying in the sky. To address such an issue, in this paper, we propose a multi-UAV collaborative face recognition method to enhance both the accuracy and efficiency of face recognition for multiple goods receivers at the same time, named UAVs4FR. Specifically, instead of each UAV running its own face recognition process, multiple UAVs first use face detection models to collaboratively detect and capture pedestrian faces, then transmit these images to the edge server. Afterwards, the edge server uses a face recognition algorithm to perform preliminary recognition of faces. Meanwhile, the face fusion model is also deployed in the edge server to fuse the face images to improve the recognition accuracy. Finally, the edge server utilizes a face recognition algorithm to further recognize target goods receivers using the fused face images and send out the corresponding instructions to the group of UAVs through broadcasting. Comprehensive experiments based on a real-world edge-based smart UAV delivery service system successfully demonstrate the effectiveness of our proposed framework and its better performance compared with other representative methods.

Keywords: Edge Computing · UAV · UAV Smart Logistics · Multi-UAV Collaboration · Face Recognition

© The Author(s), under exclusive license to Springer Nature Singapore Pte Ltd. 2024
Z. Tari et al. (Eds.): ICA3PP 2023, LNCS 14490, pp. 217–235, 2024.
https://doi.org/10.1007/978-981-97-0859-8_13

1   Introduction

Smart logistics is a modern logistics model supported by modern information technology to achieve intelligent control and improve operational efficiency in all stages of cargo transportation and distribution [1,2]. In recent years, indepth integration of Internet of Things (IoT) and Artificial Intelligence (AI) in the logistics industry has further promoted the development of smart logistics, making it gradually become an effective solution for improving the quality and efficiency in the logistics industry [3–5]. As an important intelligent delivery device, Unmanned Aerial Vehicles (UAVs) have broad application prospects in many fields such as smart logistics and modern agriculture because of their flexibility, low costs, and ease of deployment [6,7]. Due to its important role in the "last mile", such as epidemic sealing and control and complex terrain areas, UAV delivery has gradually been widely used in various smart logistics scenarios [8,9]. For example, ecommerce companies such as JD and Amazon have started establishing a UAV delivery system that uses UAVs for goods delivery [10,11]. However, traditional cloud-based UAV smart logistics cannot guarantee real-time and efficient services due to factors such as network instability and large data transmission [12]. Edge computing-based UAV smart logistics has been proposed to address these issues by performing data computation closer to the edge nodes [13,14].

Despite the many advancements in edge computing, there are still many open technical challenges deploying UAVs for smart logistics. One of the key challenges is how we can accurately and efficiently identify goods receivers with face recognition technologies [15]. However, many of the current face recognition algorithms only have high recognition accuracy for frontal faces, and in actual scenarios, the recognition accuracy will be significantly reduced due to several factors such as variable face angles, which is a key challenge when applied to UAV smart logistics [16].

There are several studies on improving face recognition accuracy in the context of UAV scenarios. For example, Shao et al. [18] use facial detection algorithms to detect pedestrians' faces and then used face recognition networks to identify them. However, in practical applications, most of the faces captured by UAVs have random postures, and this method cannot guarantee high accuracy in face recognition. Amato et al. [19] designed a specialized network structure for face recognition in low-resolution conditions for UAVs. This method ensures face recognition in low-quality facial images. However, this method ignores the negative impact of facial poses. Xu et al. [20] used a generative adversarial network to train a face frontalization model and then used this model to frontalize faces. This method can eliminate the negative impact of facial angle on face recognition to some extent, and can improve face recognition accuracy. However, the effectiveness of face frontalization model varies for different faces, resulting in some instability in practical scenarios. Shen et al. [21] proposed a facial scoring mechanism and UAV flight control algorithm that adjusts the flight direction according to the captured facial pose, enabling the UAV to automatically capture frontal facial photos for face recognition. Because this method ensures capturing the front face of a pedestrian every time, this method significantly increases face recognition accuracy in smart logistics for UAVs because it ensures capturing the front face of the pedestrians. However, UAV trajectory adjustments require considerable time, hence reducing the efficiency of identifying the target recipient. Abate et al. [22] proposed a three-dimensional facial reconstruction method based on UAVs, where the pedestrian first attracts the UAV's attention using gestures. The UAV then hovers around the pedestrian to collect sufficient face information to construct a 3D facial model, which is used for face recognition. Although this method constructs a 3D facial model with good distinctive features, recognizing pedestrian gestures and constructing 3D faces is time-consuming, significantly impacting face recognition efficiency. Gao et al. [23] first used posture recognition to filter pedestrians and then used frontal facial capture to perform face recognition. This method also significantly improves face recognition accuracy. However, the posture recognition process requires high hardware requirements, which cannot be guaranteed without appropriate hardware capabilities, thus affecting the efficiency of identifying the target goods receiver. In this paper, we propose a face recognition method based on multi-UAV collaboration to improve the recognition accuracy of face recognition algorithms in UAV smart logistics scenarios. Firstly, the UAVs use face detection models to detect and capture pedestrian faces, then transmits them to the edge server. Secondly, the edge server uses a face recognition algorithm to perform preliminary recognition of faces. Thirdly, the face fusion model deployed in the edge server is used to fuse the faces of pedestrians, which results in fused faces. Finally, the edge server utilizes a face recognition algorithm to further recognize the fused face. If successful, confirmation instructions are sent to the UAVs through broadcast and the subsequent commodity delivery task is accomplished. Our experimental results using real-world scenarios and datasets demonstrate that the proposed framework further improves the efficiency of face recognition while ensuring high recognition accuracy of the face recognition algorithms.

2   Framework Design and Workflow

In this section, we provide an overview of our proposed framework and explore in detail the working details of each module in it.

2.1   Framework Overview

In this section, we will give an overall introduction to our proposed framework, named UAVs4FR. As shown in Fig. 1, the framework is divided into three main modules: (i) cloud server module, (ii) edge server module, and (iii) UAV group module as a terminal device. The cloud server module is responsible for storing information such as frontal photos and delivery addresses uploaded by users according to system prompts during registration, and continuously provide intelligent services for the UAV smart logistics system. The edge server has sufficient computing resources and is closer to the terminal device compared to the cloud

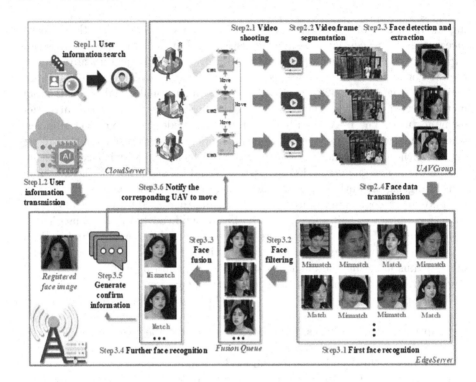

Fig. 1. The architecture of UAVs 4FR

server, which can effectively reduce the data transmission time during the service. Therefore, AI models and algorithms are required for the service, such as face fusion models and face recognition algorithms, which can be deployed in the edge server. As terminal devices in the delivery system, the UAVs use their equipped hardware devices (e.g., cameras, batteries) to perform flight and commodity delivery-related tasks. In addition, the UAVs can run some lightweight applications such as LFFD [24] or tiny-yolo [25] to perform data pre-processing tasks (such as face detection or face interception) by using limited computational resources.

2.2 Framework Execution Process

We now describe the workflow of our proposed framework in detail through a practical example. The relevant definitions are described in Table 1. The specific process can be divided into the following steps:

Step 1: CS processing user data.
Step 1.1: CS search P according to order information.
Step 1.2: CS transfer and temporarily store P into ES.

Step 2: U processing of captured pedestrian videos.
Step 2.1: U uses the equipped high-definition camera to take pictures of pedestrians.

Table 1. Relevant definitions

Symbol	Meaning
k	Number of users with the same receiving location
w	Number of pedestrian faces
CS	Cloud Server
ES	Edge Server
$R = \{\gamma_1, \gamma_2 \ldots \gamma_k\}$	The set of goods receivers
$P = \{p_1, p_2 \ldots p_k\}$	The set of registered faces corresponding to goods receivers
$U = \{u_1, u_2 \ldots u_k\}$	Group of UAVs distributed by the system for goods receivers
$F = \{f_1, f_2 \ldots f_w\}$	The set of pedestrian faces
α	The threshold for first face recognition
β	The threshold for further face recognition
$Q = \{q_1, q_2 \ldots q_k\}$	The threshold for further face recognition
$C = \{c_1, c_2 \ldots c_k\}$	The threshold for further face recognition
t	Time for face fusion queue
$L = \{l_1, l_2 \ldots l_k\}$	The set of confirmation instructions
v	Face feature value

Step 2.2: U uses the limited on-board resources to split the captured pedestrian video into video frames and retain some of the video frame data at a specific ratio.

Step 2.3: The lightweight face detection model deployed in U performs face detection on the retained video frames, and the detected faces are cropped out using an image segmentation algorithm, each cropped face is bound with pedestrian location information and the corresponding UAV information, these face images constitute F.

Step 2.4: U transmits F to ES.

Step 3: ES performs target goods receiver identification tasks according to F

Step 3.1: The face recognition model deployed in s matches the faces in F with the registered faces in P in turn, and the matching result is compared with α. If the matching result is less than or equal to α, the matching is successful, otherwise the matching fails.

Step 3.2: ES filter the faces in F, the specific steps are as follows: discard the faces that do not match with all registered faces in P, deposit f_j that matches successfully with p_i into the corresponding q_i, and set $c_i = $ t for q_i, which can continue to deposit other pedestrian faces before $c_i \leq 0$.

Step 3.3: If $c_i \leq 0$, the face fusion model deployed in ES fuses the faces in q_i in turn in order to obtain more discriminative faces.

Step 3.4: The face recognition model deployed in s matches the fused faces with p_i, and the matching result is compared with β. If it is less than or equal to β, the matching is successful, otherwise the matching fails.

Step 3.5: For the fused face successfully matched with p_i, ES generatesl_i, and at the same time obtains the position of u_i, calculates the landing direction of u_i based on the pedestrian positioning information bound in the face and the current position of u_i, and stores it in l_i Step 3.6: ES sends l_i to U with broadcast.

Step 4: u_i adjusts the direction of landing according to the information stored in l_i and completes subsequent delivery tasks.

3   Related Technologies and Algorithms

In this section, we describe the relevant technologies and the algorithms used in the framework.

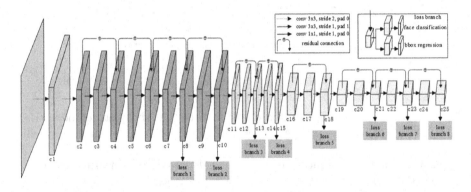

Fig. 2. The architecture of LFFD

3.1   Face Detection

Face detection technology is a critical aspect of automatic face recognition systems, which uses certain strategies and algorithms to determine whether a face is present in a given image. In our proposed architecture, the face detection task is performed by the lightweight face detection model LFFD deployed in the UAV, which is suitable for application to UAVs with limited on-board computational resources because of its small size and fast detection. The framework of LFFD is shown in Fig. 2. As shown in Fig. 2, LFFD has 25 convolution layers divided into four parts for various Receptive Field (RF) strategies. There are also eight loss branches which are used to detect faces with different scales. Each branch has two sub-branches for facial classification and bounding box regression. LFFD adopts softmax with cross-entropy loss over two classes for facial classification. For bounding box regression, it uses L2 loss directly. The regression ground truth is defined as follows:

$$\frac{RF_x - b_x^{tl}}{RF_s/2}, \frac{RF_y - b_y^{tl}}{RF_s/2}, \frac{RF_x - b_x^{br}}{RF_s/2}, \frac{RF_y - b_y^{br}}{RF_s/2} \tag{1}$$

where RF_x and RF_y are the center coordinates of the RF, RF_s is the RF size and $RF_s/2$ is the normalization constant, b_x^{tl} and b_y^{tl} are coordinates of the top-left corner of the bounding box, b_x^{br} and b_y^{br} are coordinates of the bottom-right corner of the bounding box. Based on this method, faces with different scales in video frames will be predicted by the RF of a certain convolution layer and then extracted accordingly.

3.2   Face Fusion

With the continuous improvement and development of AI technology, the image processing technology using face fusion is being applied to many different scenarios. Face fusion refers to the process of fusing two face pictures into one face picture, and the face obtained through face fusion will have the features of the faces in both pictures [26]. At this stage, face fusion technology has a wide range of applications in image processing, video editing, and other fields. Face fusion is mainly divided into the following steps: First, face key point extraction, through which 68 key points specific to the face will be obtained; Second, the triangulation of the face region based on the face key points; Third, an affine transformation of the face triangle region; Fourth, set the transparency for face fusion, with the fusion formula shown below:

Algorithm 1. Face Fusion Process

Input: f_a, f_b
Output: fused face f_c
1: Extract face feature points for f_a and get the set of face key points kp_a
2: Extract face feature points for f_b and get the set of face key points kp_b
3: Delaunay triangulation of f_a according to kp_a to obtain the corresponding set of triangles T_a
4: Delaunay triangulation of f_b according to kp_b to obtain the corresponding set of triangles T_b
5: **for** each $i \in [1, 68]$ **do**
6:    Calculate the mean value T_c^i of T_a^i and T_b^i
7:    Affine transform T_a^i and adjust its position to coincide with T_c^i
8: **end for**
9: Perform face fusion based on (2) to obtain f_c
10: **return** f_c

$$M(x, y) = \alpha \cdot f_a(x, y) + (1 - \alpha) \cdot f_b(x, y) \tag{2}$$

$$p_1 \qquad\qquad p_2 \qquad\qquad p_3$$

Fig. 3. Users' registered face images

where $M(x, y)$ denotes the pixel value of the face image after fusion, α is the fusion degree parameter, $f_a(x, y)$ denotes the pixel value of the face image f_a with coordinates (x, y) to be fused, and $f_b(x, y)$ denotes the pixel value of the face image f_b with coordinates (x, y) to be fused. The face fusion code is process is presented in Algorithm 1.

3.3   Face Recognition

Face recognition is a biometric identification technology based on human facial feature information, which provides the pre-requisites for identification due to the uniqueness of the human face and the difficulties in replication. In the face recognition process, the Euclidean distance (ED) between face feature values is usually used as the judgment criterion, and the function is defined as shown in (3).

$$ED(F_i, F_j) = \sum_{n=1}^{128} \sqrt{(x_{in} - x_{jn})^2} \tag{3}$$

The values of face features F_i and F_j comprise 128 feature vectors x_{in} and x_{in} respectively (where n = 1, 2, ..., 128). The Euclidean Distance (ED) is calculated between F_i and F_j. If the ED result is less than the pre-defined threshold value, then it is concluded that the two face features correspond to the same person, otherwise, they are not considered to be the same [27]. In the context of UAV smart logistics, the face recognition system based on deep learning can achieve satisfactory results under ideal acquisition conditions and user cooperation. However, the face photos captured by image devices are often in random poses, which adversely affects the accuracy face recognition algorithms. To address this challenge, we used face fusion technology that combines the faces captured by multiple UAVs to improve the accuracy of face recognition during the goods receiver identification process. The specific process is presented in Algorithms 2 and 3.

Algorithm 2 instructs the edge server to execute varying operations depending on the results of face recognition. In cases where pedestrian faces exhibit a high level of similarity, the edge server immediately produces related confirmation information. However, for pedestrian faces with lower similarity, the decision to

---

Algorithm 2. Face Recognition Process

Input: P, F, α, β and t
Output: Q,L
 1: Initialize $Q = \{q_1, q_2 \ldots q_k\}$
 2: Initialize $C = \{c_1, c_2 \ldots c_k\}$
 3: **for** each $i \in [1, w]$ **do**
 4:     Calculate v_i corresponding to f_i
 5:     **for** each $j \in [1, k]$ **do**
 6:       **if** $c_j > 0$ **then**
 7:         Calculate v_i corresponding to p_j
 8:         Calculate ED between v_i and v_j according to (3)
 9:         **if** $ED \leq \beta$ **then**
10:           $l_i \longleftarrow$ ture
11:           $c_j = 0$
12:         **else**
13:           **if** $\beta \leq ED$ and $ED \leq \alpha$ **then**
14:             **if** $q_j = null$ **then**
15:               $q_j \longleftarrow f_i$
16:               $c_j = t$ and start the countdown
17:             **else**
18:               **if** $c_j > 0$ **then**
19:                 $q_j \longleftarrow f_i$
20:               **end if**
21:             **end if**
22:           **end if**
23:         **end if**
24:       **end if**
25:     **end for**
26: **end for**
27: **return** Q, L

---

add them to the corresponding face fusion queue is made based on time. For pedestrians in the face fusion queue, we used fusion techniques to create more distinctive faces. Algorithm 3 elaborates on this process.

4 Experiments and Evaluation

4.1 Case Studies

This section provides a practical example of how the proposed framework operates in identifying a goods receiver. Three users have placed orders with delivery addresses. The framework queries them as $p = \{p_1, p_2, p_3\}$. The details are shown in Fig. 3. Subsequently, the system assigns one UAV to each goods receiver for efficient delivery of the goods. A UAV group, denoted by $U = \{u_1, u_2 \ldots u_k\}$. Additionally, ES located close to the delivery addresses is responsible for downloading P from CS and temporarily storing it. When U arrives at the destination, it captures footage of the surrounding pedestrians using the configured camera.

Algorithm 3. Face Fusion-Based Goods Receiver Recognition

Input: P, Q,β
Output: L
 1: **for** each $i \in [1, k]$ **do**
 2:     **if** $l_i = null$ **then**
 3:         Calculate the number of elements in q_j and note it as N
 4:         **if** $N \leq 1$ **then**
 5:             $l_i \longleftarrow flase$
 6:         **else**
 7:             $q_i^1 = f_a$
 8:             **for** each $j \in [2, N]$ **do**
 9:                 $q_i^j = f_b$
10:                 Fuse f_a, f_b into f_c according to Algorithm (1)
11:                 Calculate v_c of f_c
12:                 Calculate v_i corresponding to p_i
13:                 Calculate ED between v_c and v_i according to (3)
14:                 **if** $ED \leq \beta$ **then**
15:                     $l_i \longleftarrow ture$
16:                 **else**
17:                     Select the image more similar to p_i from f_a andf_b as f_a
18:                 **end if**
19:             **end for**
20:             **if** if all elements in q_i are fused and l_i=null **then**
21:                 $l_i \longleftarrow flase$
22:             **end if**
23:         **end if**
24:     **end if**
25: **end for**
26: **return** L

The recorded video is split into frames at a 30:1 ratio and the resulting partial video frames are displayed in Fig. 4. The face detection algorithm LFFD deployed in U is capable of detecting faces in the video frames and capturing the intercepted faces. A set of face images obtained from this interception is denoted as $F = \{f_1, f_2 \ldots f_w\}$. These images are subsequently transmitted by U to ES. A few of the captured face images are displayed in Fig. 5. ES calculates ED between the face in F and each registered face in P using the face recognition algorithm. If the ED calculated by f_7 and p_1 satisfies $ED \leq \beta$, then ES sets $c_1 = 1$. Additionally, ES calculates the landing direction of u_1 based on the positioning information bounded by f_7 and the location of u_1 at that moment. The landing direction is stored in l_1, and ES then sends l_1 to U in the form of a broadcast. Once u_1 receives l_1, it proceeds to perform the commodity delivery task. Subsequently, f_9 and p_2 calculate ED to satisfy $\beta \leq ED \leq \alpha$. ES then makes $c_2 = 30$ and begins counting down, storing f_9 in q_2. Similarly, f_{12} and q_3 calculate ED satisfy $\beta \leq ED \leq \alpha$. ES makes $c_3 = 30$ and starts counting down, storingf_{12} in q_3. Suppose q_2 and q_3 each deposit four faces before $c_2 = 0$ and $c_3 = 0$, the faces deposited by q_2 are f_9, f_{13}, f_{17}, andf_{20}, while the faces deposited by q_3

are f_{12}, f_{15}, f_{18}, and f_{27}. The faces stored in q_2 and q_3 are illustrated in Fig. 6. The fused facial features are shown in Fig. 7, where q_2^1 is the face obtained by fusing f_9 with f_{13}, q_2^2 is the face obtained by fusing f_{13} with f_{17}, q_2^3 is the face obtained by fusing f_{13} with f_{20}, q_3^1 is the face obtained by fusing f_{12} with f_{15}, and q_3^2 is the face obtained by fusing f_{15} with f_{18}. During further facial recognition, neither q_2^1 and q_2^2 were successfully matched, while q_2^3 was matched with p_2. Based on the positional information of f_{20} and the current location of u_2, ES calculated the landing direction of u_2 and stored it in l_2. ES then broadcasts l_2 to U. q_3^1 did not match with p_3, while q_3^2 matched successfully with p_3. In this case, ES terminates the face fusion and face recognition process in advance, then calculated the landing direction of u_3 based on the positional information of f_{18} and the current location of u_3, and stored it in l_3. ES subsequently broadcasts l_3 to U, and u_3 received l_3 to complete the delivery task.

Fig. 4. Video frames

f_1 f_2 f_3 f_w

Fig. 5. Pedestrians' faces

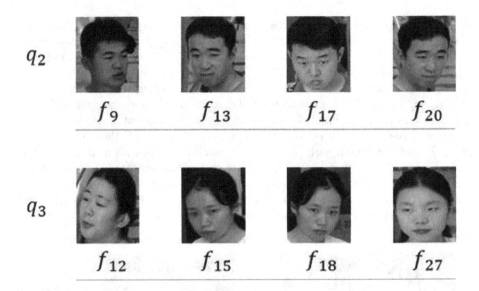

Fig. 6. Faces in q_2 and q_3

4.2   Evaluation

In this section, we compare the accuracy and efficiency of our proposed framework with several other representative device-edge collaboration frameworks to evaluate the effectiveness of the proposed architecture. The representative architectures are shown as follows:

- **Baseline:** After receiving the face image transmitted by the UAV, the edge server performs direct face recognition.
- **Edge4FR:** The edge server initially receives the face image uploaded from the UAV and then processes it using a face frontalization model to generate a corresponding frontal face image. It then conducts face recognition tasks on the frontal face image.
- **Edge4Sys:** The edge server first receives face images transmitted by the UAV and performs pose recognition to filter out pedestrians that meet recognition conditions. Eligible pedestrians' frontal faces are then captured and further processed through face recognition.
- **PC4FR:** The private cloud server first receives the face images uploaded by the UAVs and processes them through a face recognition algorithm to identify eligible face images. Eligible face images then undergo face fusion operations using the face fusion model deployed in the private cloud server. Finally, fused face images undergo further face recognition through the face recognition algorithm. The dataset used in the experiments was collected by a DJI Matrice 600 Pro UAV and consists of images of approximately 150 people. All experiments were conducted on computers running Windows 10

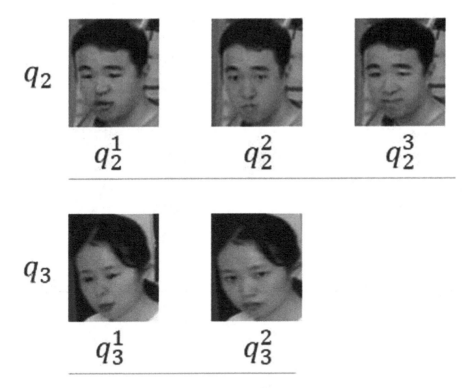

Fig. 7. Fused faces

(64-bit) operating system, using an 11th generation Intel® Core^{TM} i5-11400 processor and 16 GB of RAM.

Accuracy. The accuracy comparison results for the four architectures are shown in Fig. 8. The Baseline framework has the lowest face recognition accuracy since it cannot effectively address potential issues caused by random UAV captures such as random face poses. The Edge4FR framework improves face recognition accuracy to some extent. However, it should be noted that face frontalization through Generative Adversarial Network (GAN) training. Due to the training data or the network itself, the face frontalization model may not have a positive effect on every face image, which limits the improvement of face recognition accuracy. As demonstrated in our experimental results, the Edge4Sys framework obtains the highest recognition accuracy. However, the process of locating the goods receiver using pose recognition is time-consuming and unreliable. The UAVs4FR framework prescreens face images more similar to the goods receiver through preliminary face recognition before performing more accurate face recognition. It then performs face fusion operations on the screened face images to obtain face images with more distinctive features. Although its recognition accu-

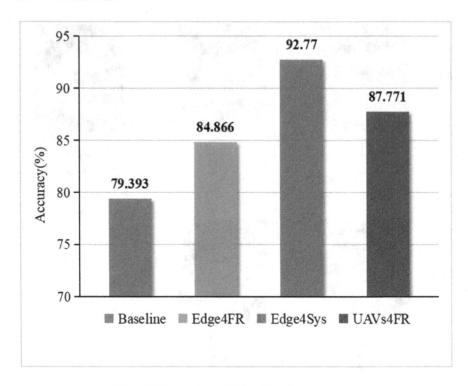

Fig. 8. Comparison of identification accuracy

racy is slightly lower than the Edge4Sys framework, it is significantly higher than the Baseline framework and Edge4FR framework.

Efficiency. We now compare our proposed framework with several other frameworks in terms of face recognition efficiency. The time taken to recognize the goods receiver under different frameworks is illustrated in Fig. 9. Specifically, blue color denotes the time used for face detection and face extraction in the Baseline, Edge4FR, and UAVs4FR frameworks (FP). Purple color denotes the time used for face recognition in the Baseline, Edge4FR, and Edge4Sys framework (FR). Yellow color denotes the time taken for face turnaround in the Edge4FR (FF), and orange is the time consumed for pedestrian detection and pedestrian extraction in the Edge4Sys framework (PP). Green color is the time used for pose recognition in the Edge4Sys framework (PR), and red color indicates the time used for face recognition and face fusion in the UAVs4FR framework (TFR). We can observe that the Edge4Sys framework spends a considerable amount of time in the pose recognition process, making its overall time the highest among the four frameworks. Furthermore, as the number of UAVs in the scene increases, the time taken by the other three frameworks to complete the goods receiver recognition task increases significantly. Conversely, the UAVs4FR framework has a slower over-all time growth and takes significantly less time than the other

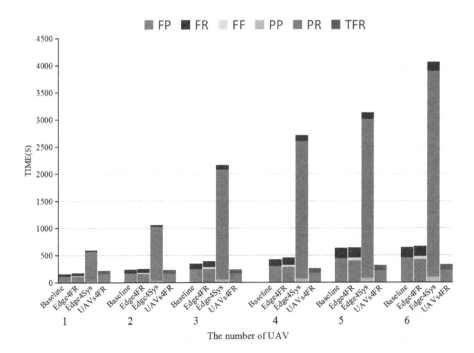

Fig. 9. Time comparison of identifying the goods receiver through different frameworks (Color figure online)

three frameworks, thanks to its multiUAV collaboration scheme. The results of the time comparison between UAVs using collaborative and non-collaborative approaches to complete goods receiver identification are presented in Fig. 10. We can observe that as the number of UAVs increases, the time taken to complete goods receiver identification in a noncollaborative manner increases significantly whereas the time taken in a collaborative approach by multiple UAVs increases slowly. This highlights the effectiveness of the collaborative UAV solution proposed in our architecture as it can significantly reduce the time required to complete goods receiver identification, thereby improving the overall efficiency of UAV-enabled smart logistics delivery. To further demonstrate the importance of the edge server in the proposed framework for face recognition, we compare the efficiency of our proposed framework with Private4FR. The experimental results are illustrated in Fig. 11, with different colors indicating different stages of the process. The orange color represents the time taken for face detection and extraction in the PC4FR framework ($PC_F P$). The yellow color represents the time for data transfer between the PC4FR framework and the end device ($PC_D T$). The green color represents the time for completing face fusion and recognition in the PC4FR framework ($PC_T FR$). The blue color represents the time taken for face detection and extraction in the UAVs4FR framework ($UAVs_F P$). The red color represents the time taken for face fusion and recognition in the UAVs4FR

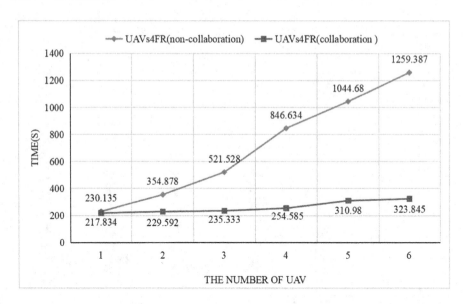

Fig. 10. Time comparison of identifying the goods receiver through collaborative and non-collaborative approach

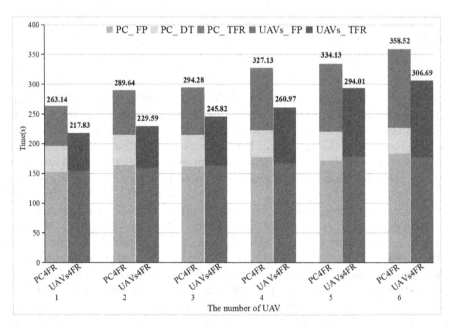

Fig. 11. Time comparison of identifying the goods receiver through Private4FR and UAVs4FR (Color figure online)

framework $(UAVs_TFR)$. The results clearly show that the time taken for face recognition in the UAVs4FR framework is significantly less compared to the private cloud server-based architecture. In the private cloud server-based architecture, the terminal devices upload all the captured and extracted face images to the private cloud server for recognition and fusion. This leads to a substantial amount of facial data, which can cause delays in data transmission between the cloud server and terminal devices.

5   Conclusions

In this paper, focusing on the goods receiver recognition task in the edge-based smart UAV delivery service system, we proposed a multi-UAV collaborative face recognition framework called UAVs4FR to address both the accuracy and efficiency issues in multiple goods receiver recognition. Our experimental results based on a real-world edge-based smart UAV delivery service scenario and datasets have successfully demonstrated the effectiveness of our proposed framework. There are still some potential future research directions in this area. For example, we only considered the involvement of one edge server in our proposed framework. The collaboration between multiple UAVs and multiple edge servers could be much more complex and will also require multiple levels of collaboration. We will explore this research problem as part of future work to further improve the performance of our proposed framework.

Acknowledgements. This work was supported by the National Natural Science Foundation of China (Nos. 62076002, 61402005, 61972001), and the Natural Science Foundation of Anhui Province of China (No. 2008085MF194).

References

1. Song, Y., Yu, F., Zhou, L., Yang, X., He, Z.: Applications of the Internet of things (IoT) in smart logistics: a comprehensive survey. IEEE Internet Things J. **8**(8), 4250–4274 (2020)
2. Ding, Y., Jin, M., Li, S., Feng, D.: Smart logistics based on the internet of things technology: an overview. Int. J. Logist. Res. Appl. **24**(4), 323–345 (2021)
3. Liu, W., Wang, S., Lin, Y., Xie, D., Zhang, J.: Effect of intelligent logistics policy on shareholder value: evidence from Chinese logistics companies. Transport. Res. Part E Logist. Transport. Rev. **137**, 1–24 (2020)
4. Tang, X.: Research on smart logistics model based on Internet of Things technology. IEEE Access **8**, 151150–151159 (2020)
5. Bag, S., Yadav, G., Wood, L., Dhamija, P., Joshi, S.: Industry 4.0 and the circular economy: resource melioration in logistics. Resour. Policy **68**, 1–16 (2020)
6. Arafat, M., Moh, S.: A survey on cluster-based routing protocols for unmanned aerial vehicle networks. IEEE Access **7**, 498–516 (2018)
7. Arafat, M., Moh, S.: Routing protocols for unmanned aerial vehicle networks: a survey. IEEE Access **7**, 99694–99720 (2019)

8. Engesser, V., Rombaut, E., Vanhaverbeke, L., Lebeau, P.: Autonomous delivery solutions for last-mile logistics operations: a literature review and research agenda. Sustainability **15**(3), 2774–2791 (2023)

9. Yoo, H., Chankov, S.: UAV-delivery using autonomous mobility: an innovative approach to future last-mile delivery problems. In: Proceedings of IEEE International Conference on Industrial Engineering and Engineering Management, pp. 1216–1220. IEEE Press (2018)

10. Shahzaad, B., Bouguettaya, A., Mistry, S.: Robust composition of UAV delivery services under uncertainty. In: Proceedings of International Conference on Web Services, pp. 675–680. IEEE Press (2021)

11. Song, B., Park, K., Kim, J.: Persistent UAV delivery logistics: MILP formulation and efficient heuristic. Comput. Ind. Eng. **120**, 418–428 (2018)

12. Mao, Y., You, C., Zhang, J., Huang, K.: A survey on mobile edge computing: the communication perspective. IEEE Commun. Surv. Tutor. **19**, 2322–2358 (2017)

13. Shi, W., Cao, J., Zhang, Q., Li, Y., Xu, L.: Edge computing: vision and challenges. IEEE Internet Things J. **3**, 637–646 (2016)

14. Jararweh, Y., Doulat, A., AlQudah, O., Ahmedet, E.: The future of mobile cloud computing: integrating cloudlets and mobile edge computing. In: Proceedings of 23rd International Conference on Telecommunications, pp. 1–5. IEEE Press (2016)

15. Qian, Y., Deng, W., Hu, J.: Unsupervised face normalization with extreme pose and expression in the wild. In: Proceedings of IEEE/CVF Conference on Computer Vision and Pattern Recognition, pp. 9851–9858. IEEE Press (2019)

16. Ali, W., Tian, W., Din, S., Iradukunda, D., Khan, A.: Classical and modern face recognition approaches: a complete review. Multimedia Tools Appl. **80**(3), 4825–4880 (2021)

17. Hsu, H., Chen, K.: Face recognition on UAVs: issues and limitations. In: Proceedings of the First Workshop on Micro Aerial Vehicle Networks, Systems, and Applications for Civilian Use, pp. 39–44. ACM Press (2015)

18. Shao, Y., Zhang, D., Chu, H., Zhang, X., Chang, Z.: Aerial photography pedestrian target recognition based on YOLO and Face Net. Manuf. Autom. **42**(11), 56–60 (2020)

19. Amato, G., Falchi, F., Gennaro, C., Massoli, F., Vairo, C.: Multi-resolution face recognition with UAVs. In: Proceedings of the 3rd International Conference on Sensors, Signal and Image Processing, pp. 13–18. ACM (2020)

20. Xu, Y., Luan, F., Liu, X., Li, X.: Edge4FR: a novel device-edge collaborative framework for face recognition in smart UAV delivery systems. In: Proceedings of the IEEE 8th International Conference on Cloud Computing and Intelligent Systems (CCIS), pp. 95–101. IEEE (2022)

21. Shen, Q., Jiang, L., Xiong, H.: Person tracking and frontal face capture with UAV. In: Proceedings of the IEEE 18th International Conference on Communication Technology, pp. 1412–1416. IEEE (2018)

22. Abate, A., De, M., Distasi, R., Narducci, F.: Remote 3D face reconstruction by means of autonomous unmanned aerial vehicles. Pattern Recogn. Lett. **147**, 48–54 (2021)

23. Gao, H., et al.: Edge4Sys: a device-edge collaborative framework for MEC based smart systems. In: Proceedings of the IEEE/ACM International Conference on Automated Software Engineering (ASE), pp. 1252–1254. IEEE (2020)

24. He, Y., Xu, D., Wu, L., Jian, M., Xiang, S., Pan, C.: LFFD: a light and fast face detector for edge devices. arXiv:1904.10633 (2019)

25. Nikouei, S., Chen, Y., Song, S., Xu, R., Choi, B., Faughnan, T.: Real-time human detection as an edge service enabled by a lightweight CNN. In: Proceedings of the IEEE International Conference on Edge Computing (EDGE), pp. 125–129. IEEE (2018)
26. The source code of Face recognition. https://github.com/age-itgey/face~recognition. Accessed 30 May 2023
27. The source code of FaceMorph. https://github.com/Largefreedom/Opencvpra/tree/master/Face20Morph. Accessed 30 May 2023

Distributed Generative Adversarial Networks for Fuzzy Portfolio Optimization

Xueying Yang[1,2]([✉]), Chen Li[1], Zidong Han[1,2], and Zhonghua Lu[1,2]

[1] Computer Network Information Center, Chinese Academy of Sciences,
Beijing 100190, China
{yangxueying,zdhan,zhlu}@cnic.cn, lichen@sccas.cn
[2] University of Chinese Academy of Sciences, Beijing 100049, China

Abstract. Financial time series is one of the most important data in the field of economics and finance, and it is important to forecast and simulate such data effectively based on historical patterns and trends. Existing forecasting models mainly forecasting one-step ahead, and cannot retain the complex characteristics of financial time series data such as serial correlation and the long-term time-dependent relationship. On the other hand, the large-scale data makes the training of the deep learning models a time-consuming process. Therefore, how to forecast financial time series multi-step ahead efficiently has become a key point to improve the asset management capability. At the same time, constructing a fuzzy portfolio optimization for different distributions is also an important direction to improve the robustness of a portfolio model. This paper proposes a distributed financial time series simulating model AssetGANs that simulating multi-step ahead based on GANs, and apply GANs as a parameter simulation method to fuzzy portfolio optimization to provide users with better strategy choices. The paper carries on numerical experiments on real market stock data, compares the results with LSTM and achieves a training speedup of over 573 with 8 GPUs compared to the CPU version.

Keywords: Generative Adversarial Networks · Fuzzy Simulation · Portfolio Optimization · Parallel computing

1 Introduction

Financial time series forecasting and simulation are vital for investment management decisions, and has become one of the hot spots of scientific research today. In the early research, scholars used Autoregressive Model (AR), Moving Average Model (MA) for financial time series forecasting. However, the above models only reflect linear patterns and are not capable of tackling nonlinear data. In addition, these methods mostly require that the time-series data are stable, which is difficult to achieve in real financial markets. With the increasing complexity, efficiency and volatility of financial markets, the pattern of correlations

Z. Tari et al. (Eds.): ICA3PP 2023, LNCS 14490, pp. 236–247, 2024.
https://doi.org/10.1007/978-981-97-0859-8_14

among data becomes more complex, and accurate simulation of financial time series becomes more difficult. The rapid development of deep learning techniques has provided new solutions for this problem and have become a popular research direction today.

In 2014, Goodfellow et al. [1] proposed Generative Adversarial Networks (GANs), which consist of a generator and a discriminator to generate samples that cannot be discriminated from the real data. GANs are proposed to meet the research and industrial needs of many fields [2–4]. Currently, the fields of image generation are one of the most widely studied and applied fields for GANs, while there are few applications in the field of financial time series analysis [5,6].

Ricardo and Carrillo predicted whether the price would increase one day ahead by using LSTM as the generator and Convolutional Neural Network as the discriminator [7]. In this case, the accuracy of GANs is slightly lower than LSTM. However, Zhang et al. forecasted the closing price of the stock one day ahead with the LSTM as the generator and Multi-Layer Perceptron as the discriminator [8]. The GANs they proposed performs better than LSTM. Lin et al. simulated future stock price three days ahead with Gated Recurrent Units as the generator and Convolutional Neural Network as the discriminator. It is proved that the GANs got a lower RSME than LSTM when apply to one single stock [9].

GANs can fully exploit the important information in financial time series data and improve the cognitive ability of financial markets [10,11], but their generators and discriminators can oscillate significantly during the iterative process, resulting in longer training time [12]. In order to meet the requirements of practical applications, distributed training is usually used to train models [13].

This paper proposes a financial time series simulating model that can fully capture the complex features of financial time series data, and can effectively reflect the nonlinear dynamic interactions among data, based on the WGAN-GP [14], and achieves the distributed training of the model based on the high-performance heterogeneous computing system. In addition, this paper conducts comparison experiments on stock data sets to verify the effectiveness of the model and the distributed efficiency. We construct a fuzzy investment portfolio model with the simulated sampling data, and optimizing and solving the portfolio model.

2   Generative Adversarial Networks Architecture

2.1   Theoretical Background

Traditional GANs consist of a generator and a discriminator. The generator captures the potential distribution of real data samples and generates new data samples; the discriminator discriminates whether the input is real data or generated samples. Traditional GANs are difficult for training and are prone to pattern collapse problem. This problem can be avoided by improving the loss function. Gulrajani et al. proposed a new loss function using the Wasserstein distance. The Wasserstein distance is used to calculate the distance for two data distributions $x \sim P_r$ and $y \sim P_g$, which is mathematically defined as the minimum cost to transform from the data distribution P_r to the data distribution P_g:

$$W(P_r, P_g) = \inf_{\gamma \in \prod(P_r, P_g)} E_{(x,y) \sim \gamma}[\|x - y\|]. \tag{1}$$

The objective of the generator is to minimize the Wasserstein distance. The output of the discriminator is to estimate the Wasserstein distance between the simulated distribution and the real distribution, and the objective function to be maximized is:

$$L(D) = -E_{x \sim P_r}[D(x)] + E_{x \sim P_g}[D(x)] \tag{2}$$

The WGAN loss function (the opposite of Eq. 2) can reflect the quality of the generated distribution. The smaller it is, the better the generator's ability to generate sequences. Therefore, it can be judged from the loss function curve whether the model training has converged. WGAN is particularly sensitive to parameters, and WGAN-GP uses a gradient penalty to avoid this problem. The objective function of the WGAN-GP discriminator is:

$$L(D) = -E_{x \sim P_r}[D(x)] + E_{x \sim P_g}[D(x)] + \lambda E_{x \sim P_x} \sim [\|\nabla D(x)\| - 1]^2 \tag{3}$$

2.2 Architecture

The generator G is constructed with convolutional neural networks. The input parameter of the convolutional neural network is the normalized M_k, which is used to learn the patten of asset returns over the past k days. The input of the simulator is the patten and the random noise vector, which is used to generate the asset return trend for the next l days. The training purpose is to minimize the Wasserstein distance between the synthetic data and the real data M_l based on historical data. The architecture is shown in Fig. 1.

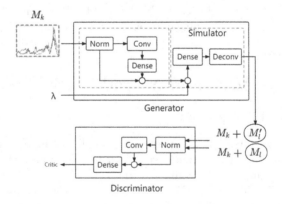

Fig. 1. AssetGANs architecture.

2.3   Generator

We set a 4-layer convolutional neural networks for Generator at first. The kernel size is 5, the stride is 2, the padding is 2, and the fifth layer is the dense layer. Then, we concatenate the noise vector to the output of the return feature. A dense layer and 2-layer deconvolutional neural networks (the kernel size is 4, the stride is 2, and the padding is 1) are used to generate the simulation returns. The details can be seen in Fig. 2, *bs* refers to batch size, while *sn* means number of stocks.

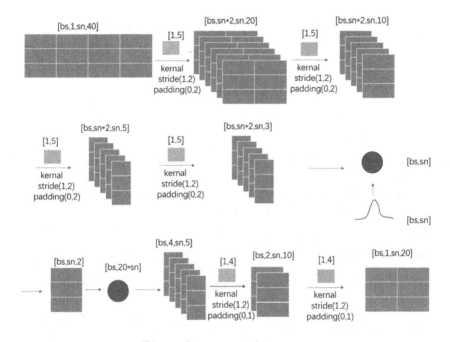

Fig. 2. Generator architecture.

2.4   Discriminator

Discriminator D is based on a 5-layer convolutional neural networks. The kernal size is 5, the stride is 2, the padding is 2. The input is the concatenation of the past real data to future real data or future synthetic data. The details can be seen in Fig. 3.

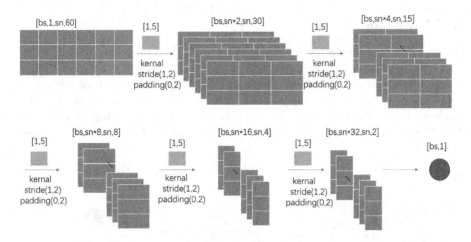

Fig. 3. Discriminator architecture.

2.5 Adversarial Training

As the network adversarial training continues, the difference between the real data and the generated data becomes smaller and smaller, and finally D cannot distinguish the authenticity of the input data. At this time, the network reaches a balanced state and the training is completed.

3 Generative Adversarial Networks Architecture

In deep learning, the weight matrix is updated by backpropagation, which is conducted by a gradient-based optimization algorithm. All workers have a copy of the model. The global data batch is divided into smaller batches and are processed by different workers. Each worker computes the corresponding loss and gradient for the data it owns. Our AssetGANs are implemented by Pytorch. With the help of the GPUs, we use DistributedDataParallel method to train the GANs. We use init_process_group() to initialize the process group. DistributedSampler() is used to partition data sets, make sure data in every batch is distributed equally to each process, and each process can obtain different data. We use DistributedDataParallel() to package the model, and perform all reduce for the gradients obtained on different GPUs. The losses and gradients are averaged across all GPUs by the Ring All-reduce algorithm before the parameters are updated in each period. Communication library Gloo is used to communicate among multiple nodes.

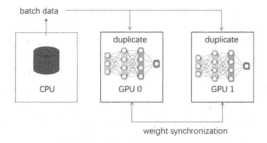

Fig. 4. Distributed training for AssetGANs.

4   Experimental Results

4.1   Experimental Setting

The dataset is the daily prices of stocks selected from Shenzhen Stock Exchange stocks from 2007-12-01 to 2021-12-30. Normalization method is Standard Scale. The parameters are estimated using the Adam optimizer, the learning rate adopts the LR attenuation method, and the number of epochs is set to 32. All experiments run on the computing platform with the following parameters shown in Table 1.

Table 1. The parameters.

Parameter	ConfRev
Each Node	32-core CPU, 4 GPUs
Operating System	Centos7.6
MPI	hpcx-2.4.1
Network	HDR Infiniband (200 Gb)

4.2   Experimental Results

The loss function curve using the AssetGANs is visualized as Fig. 5. According to the mathematics meaning of the loss function of the WGAN network introduced above, it proves that the simulated data generated by the generator are already getting closer in distribution as the loss function of the discriminator has been declining. The simulations for asset 000002.SZ are shown in Fig. 5.

Strong Scalability and Weak Scalability. We analyze Strong scalability and Weak scalability by training the GANs on 1 Node 1 GPU, 1 Node 4 GPUs, 2 Node 8 GPUs and compute their speedup over the computing time on CPU. We also test when the number increased, the speedup changed.

4.3   Experimental Evaluation

The simulation results of GANs are compared to LSTM(Long short-term memory network), the traditional network in Fig. 7. We set a 5-layer network for LSTM model. The first three layers are the LSTM layer (dimension: 64,64,32), the fourth layer is the dropout layer (dropout = 0.054, to prevent over fitting), and the fifth layer is the full connection layer (the number of neurons is 20, to predict the future 20 prices). The parameters are estimated using the Adam optimizer, the learning rate adopts the LR attenuation method, and the maximum number of generations is set to 70. Loss curves for the training set and the validation set are as follows. After 100 epochs, the convergence is achieved. The same data frame as the GANs experiment is shown in Fig. 8. We evaluate the performance for the models through Root Mean Square Error (RMSE), which is defined as:

$$RMSE = \sqrt{\frac{\sum_{i=1}^{N}(x_i - \hat{x})^2}{N}} \tag{4}$$

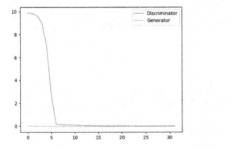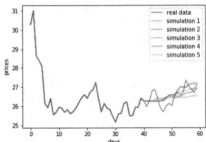

Fig. 5. AssetGANs loss and simulation.

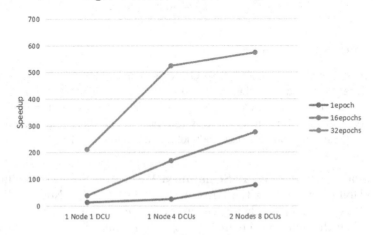

Fig. 6. Strong and weak scalability.

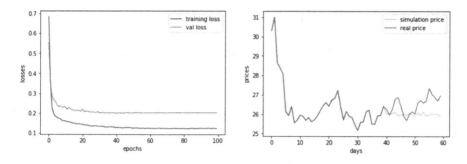

Fig. 7. LSTM loss and simulation.

Table 2. The parameters.

	LSTM	AssetGANs
RMSE	0.663826	0.461514
MSE	0.440666	0.212995

Table 2 shows the RMSE and MSE of LSTM and AssetGANs. It turns out that our AssetGANs has a lower RMSE than LSTM, and therefore performs better than LSTM in the terms of simulating multiple days in the future.

5   Parallel Fuzzy Portfolio Optimization

In recent years, fuzzy portfolio selection theory has been fully developed while the returns of assets were described as fuzzy variables [15]. This paper applies AssetGANs to predict asset returns for fuzzy portfolio optimization. Fuzzy simulation is used to obtain the training data and testing data for the Simulated Annealing Resilient Back Propagation (SARPROP) neural network [16]. Then, Genetic Algorithms are used for global optimization search to solve the optimal solution of the fuzzy Mean-CVaR model. To further improve the model solving efficiency and shorten the model solving time, the MPI-based algorithm is parallelized.

5.1   Fuzzy Mean-CVaR Portfolio Model

Take the fuzzy CVaR (Conditional value at risk) as the risk metric, the fuzzy Mean-CVaR portfolio model is:

$$
\begin{aligned}
&\min \xi_{CVaR}(\alpha) \\
&s.t. E[\sum_{i=1}^{n} x_i \xi_i] \geq r \\
&\sum_{i=1}^{n} x_i = 1 \\
&0 \leq x_i \leq 1, i = 1, 2, ..., n
\end{aligned}
\tag{5}
$$

x_i is the weight of asset i in the portfolio, ξ_i is the rate of return of asset i, which can be set as a triangular fuzzy variable $\xi_i \sim (a_i, b_i, c_i)$, then the expected return of the portfolio is $E[\sum_{i=1}^{n} x_i \xi_i]$. $\xi_{CVaR}(\alpha)$ is the CVaR risk metric

$$\xi_{CVaR}(\alpha) = (\int_{\alpha}^{1} \xi_{VaR}(\beta)d\beta)/(1-\alpha) \tag{6}$$

C_r is credibility measure, while $\xi_{VaR}(\alpha) = \inf\{x|Cr\{\xi \leq x\} \geq \alpha\}$.

5.2   Optimization Algorithms

Fuzzy Simulation. At the end of the training, the posterior probability distribution of future asset returns learned from the adversarial training process is sampled to generate simulations of future asset returns:

$$s_1 = (s_{1,1}, ..., s_{i,1}, ..., s_{i=n,1})... \\ s_m = (s_{1,m}, ..., s_{i,m}, ..., s_{i=n,m}) \tag{7}$$

The value interval of future return of asset i is $[b_i, c_i]$, then $b_i = \min\limits_{1 \leq j \leq m} s_{i,j}$, $c_i = \max\limits_{1 \leq j \leq m} s_{i,j}$. In addition, calculate the mean value of future return of asset i, $a_i = avaerage\limits_{1 \leq j \leq m} s_{i,j}$. Then, the triangular fuzzy return of asset i is $a_i = avaerage\limits_{1 \leq j \leq m} s_{i,j}$, the triangle-shape grade of membership function is:

$$u_{\xi_i} = \begin{cases} 1 - \frac{a_i - x}{a_i - b_i} & , b_i \leq x \leq a_i \\ 1 - \frac{x - a_i}{c_i - a_i} & , a_i \leq x \leq c_i \\ 0 \end{cases} \tag{8}$$

Fuzzy simulation is an application of Monte-Carlo methods, and the details was described in the book [17]. The minimum r that satisfies $Cr\{\sum_{i=1}^{n} x_i \xi_i \geq r\} \geq \alpha$ is fuzzy VaR risk metric. Fuzzy CVaR risk metric can be calculated as:

$$L(r) = \frac{1}{2}(\max_{0 \leq k \leq N}\{u_k| \sum_{i=1}^{n} x_i \xi_i \geq r\} + 1 - \max_{0 \leq k \leq N}\{u_k| \sum_{i=1}^{n} x_i \xi_i < r\}) \tag{9}$$

Fuzzy simulation is used to generate training data sets for the neural network.

SARPROP Neural Network. The trained SARPROP neural network is used to approximate the fuzzy return expectation and fuzzy CVaR of the portfolio to improve the model solving speed. The output side is the objective function of the model $E[\sum_{i=1}^{n} x_i \xi_i]$ and the fuzzy CVaR risk measure $\xi_{CVaR}(\alpha)$.

Genetic Algorithms. Genetic algorithms are heuristic algorithms that search for optimal solutions by simulating natural evolutionary processes and are suitable for solving global optimization problems. A general genetic algorithm usually consists of chromosome representation, constraint processing, population initialization, individual selection, crossover and variation.

The genetic algorithm is the final step to solve the fuzzy Mean-CVaR model. The next generation population is selected by a rotating roulette wheel method, and the population is changed with a certain probability (crossover and variation). The best individual is selected as the optimal solution of the model.

5.3   Parallelization and Results

The algorithm is parallelized by MPI, and the schematic diagram is shown in Fig. 8. The assets used to perform the portfolio are 000963.SZ, 300146.SZ, 600535.SH and 601318.SH. The experiment runs on the same machine and has the same settings as in Chapter 4. The weights for different assets are (0.2491255, 0.3388552, 0.2820555,0.1299638), and the return of the portfolio in 20 days under

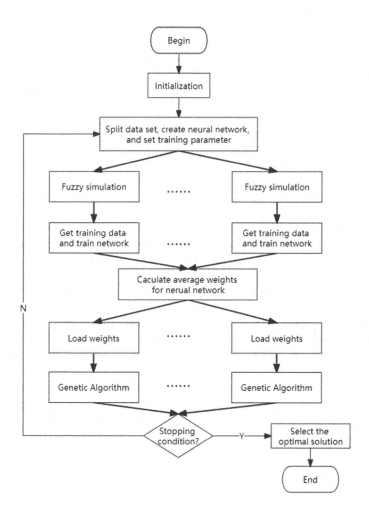

Fig. 8. The schematic diagram of parallel algorithm.

the model is 4.940837533, which is higher than the return of 4.785934489 when divided the capital equally.

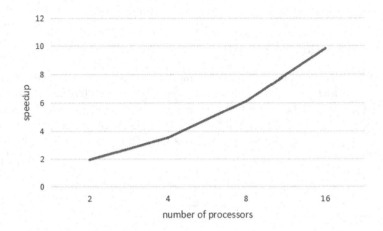

Fig. 9. Speedup for fuzzy simulation and portfolio optimization.

Figure 9 shows the speedup under different processor scores, and the maximum parallel efficiency is 96.3% when the number of processors is 2.

6   Conclusion

The development of the financial market provides fund managers with a wealth of investment choices. Simulating financial time series is crucial for constructing efficient investment portfolios. Traditional statistical simulations to capture the complexity of the market are not comprehensive enough. This paper implements distributed AssetGANs based on WGAN-GP to simulate the expected return in the future, and applies the multiple simulated distributions to the fuzzy portfolio. The structure and effect of the neural network and portfolio model are introduced and evaluated by empirical analysis, and the effectiveness of the model is proved by comparing it with the LSTM model. The correlations among time series stock data are preserved when generating data. We solve the problem of long training time for AssetGANs and long simulating time in the process of fuzzy investment portfolio optimization. Our future research plan is to implement to apply AssetGANs to high-frequency financial time series.

Acknowledgements. This work is partially supported by the China Postdoctoral Science Foundation (Grant No. 2021M693226) and Beijing Natural Science Foundation (Grant No. 4232039).

References

1. Goodfellow, I., et al.: Generative adversarial nets. In: Advances in Neural Information Processing Systems, vol. 27 (2014)
2. Polamuri, S.R., Srinivas, K., Mohan, A.K.: Multi-model generative adversarial network hybrid prediction algorithm (MMGAN-HPA) for stock market prices prediction. J. King Saud Univ.-Comput. Inf. Sci. **34**(9), 7433–7444 (2022)
3. Staffini, A.: Stock price forecasting by a deep convolutional generative adversarial network. Front. Artif. Intell. **5**, 837596 (2022)
4. Mariani, G., et al.: Pagan: portfolio analysis with generative adversarial networks. arXiv preprint arXiv:1909.10578 (2019)
5. Faraz, M., Khaloozadeh, H.: Multi-step-ahead stock market prediction based on least squares generative adversarial network. In: 2020 28th Iranian Conference on Electrical Engineering (ICEE), pp. 1–6. IEEE (2020)
6. Jiang, J.: Stock market prediction based on SF-GAN network. In: 6th International Symposium on Computer and Information Processing Technology (ISCIPT), pp. 97–101. IEEE (2021)
7. Romero, R.A.C.: Generative adversarial network for stock market price prediction. CD230: Deep Learning, p. 5. Stanford University (2018)
8. Zhang, K., Zhong, G., Dong, J., Wang, S., Wang, Y.: Stock market prediction based on generative adversarial network. Procedia Comput. Sci. **147**, 400–406 (2019)
9. Lin, H., Chen, C., Huang, G., Jafari, A.: Stock price prediction using generative adversarial networks. J. Comput. Sci. 17–188 (2021)
10. Zhou, X., Pan, Z., Hu, G., Tang, S., Zhao, C.: Stock market prediction on high-frequency data using generative adversarial nets. Math. Probl. Eng. (2018)
11. Sonkiya, P., Bajpai, V., Bansal, A.: Stock price prediction using BERT and GAN. arXiv preprint arXiv:2107.09055 (2021)
12. Kumar, D., Sarangi, P.K., Verma, R.: A systematic review of stock market prediction using machine learning and statistical techniques. Mater. Today Proc. **49**, 3187–3191 (2022)
13. Using the latest advancements in AI to predict stock market movements. https://github.com/borisbanushev/stockpredictionai. Accessed 4 Oct 2023
14. Gulrajani, I., Ahmed, F., Arjovsky, M., Dumoulin, V., Courville, A.C.: Improved training of Wasserstein GANs. In: Advances in Neural Information Processing Systems, vol. 30 (2017)
15. Zadeh, L.A.: Fuzzy sets. Inf. Control **8**(3), 338–353 (1965)
16. Treadgold, N., Gedeon, T.: The Sarprop algorithm, a simulated annealing enhancement to resilient back propagation. In: Proceedings International Panel Conference on Soft and Intelligent Computing, pp. 293–298 (1996)
17. Liu, B., Liu, Y.-K.: Expected value of fuzzy variable and fuzzy expected value models. IEEE Trans. Fuzzy Syst. **10**(4), 445–450 (2002)

A Novel Transaction Processing Model for Sharded Blockchain

Xiang Ying[1], Wei Luo[2], and Jianrong Wang[1(✉)]

[1] College of Intelligence and Computing, Tianjin University, Tianjin, China
{xiang.ying,wjr}@tju.edu.cn
[2] Tianjin International Engineering Institute, Tianjin University, Tianjin, China
luoweimss@tju.edu.cn

Abstract. The sharding-based protocols provide an efficient scaling solution for blockchain networks. However, the sharded blockchain suffers from transaction verification inefficiency while facing the threat of attacks from malicious peers. To solve aforementioned issues, this paper proposes a novel transaction processing model applicable to sharded blockchain. Specifically, a transaction admission control algorithm is established based on queuing model for single-shard flooding attacks to avoid transactions from being injected in a short period of time. And then, we present a transaction verification mechanism based on threshold signature, which is more efficient compare with PBFT protocol. The results of simulation experiments show that our transaction processing model can bring better performance compare to other advanced sharding protocols. In the case of 16 shards, the model achieves 3500 TPS throughput improvement and 4 s transaction latency reduction, and it exhibits good robustness in the case of peer misbehavior.

Keywords: Blockchain · Sharding · Flooding Attack · Transaction Verification

1 Introduction

Blockchain technology has advanced rapidly since the introduction of Bitcoin [1], because it allows for the preservation of data transparency and integrity between untrusted network peers. The potential of blockchain technology is enormous, with applications excepted in fields like digital finance and the industrial Internet of Things. However, scalability remains a significant bottleneck in the development of blockchain technology, which means that the ability of blockchain to processing transactions can not meet the needs of users. To address this issue, researchers have put out a number of solutions recently to scale out blockchain, and sharding is always a hot research topic. The main idea of sharding is to divide the entire blockchain into several small groups called *shards*, where the peers of each shard can process transactions in parallel. Compared with traditional blockchain system, sharded blockchain has achieved significant benefits in improving transaction throughput and reducing latency. For the peers in the network, they do not need to be involved in the processing of all transactions, which enables them to significantly reduce communication and storage overhead. Therefore, sharding is considered an important technology for building efficient and serviceable blockchains.

However, the security and efficiency of transaction processing have become major factors restricting the development of the sharded blockchain. Malicious peers may try to launch attacks on the blockchain system, causing the degradation of service quality. In particular, the single-shard flooding attack [2] is an easy to launch attack method. Malicious peers cause the congestion of transaction mempool by injecting a large number of transactions into the target shard, thus slowing down the verification process of transactions from other peers. We notice that there are few sharding protocols that propose countermeasures against single-shard flooding attack. Furthermore, sharding eliminates unnecessary computational work between shards but not within shards. As the most sharding models run PBFT protocol internally to reach consensus, all the peers in the shard need to go through a three-stage network communication to determine the validity of transactions. Actually, for the sharding network that can tolerate at most f malicious peers, a transaction can be considered valid if it can be accepted by $f + 1$ peers. The redundant transaction verification processes have severely limited the transaction throughput and latency of the sharded blockchains.

To tackle these aforementioned issues, we propose a novel transaction processing model for sharded blockchains, which makes the processing of transactions more robust and efficient. The main contributions of this work are as follows:

- To tackle the single-shard flooding attacks against sharded blockchains, we design a transaction admission control algorithm based on Lyapunov optimization theory, which avoids massive transactions injected while maximizing the peer satisfaction in placing transactions.
- To improve the efficiency of transaction verification, we propose a threshold signature-based transaction verification approach, which allows transactions to be verified by partial peers, and they can be verified in parallel.
- We implement the transaction processing model in simulation platform OMNeT++ and compare it with other advanced sharding models. Experiment results reveal that the model we proposed perform better in both throughput and latency. Moreover, it is capable of resisting the transaction flooding attack.

The rest of this paper is structured as follows. Section 2 summarizes related work. Section 3 presents the proposed model including the model overview and threat model. Section 4 and Sect. 5 describe the design of the transaction admission control algorithm and transaction verification method, respectively. Section 6 is about the simulation experiment results. Finally, conclusions are drawn in Sect. 7.

2   Related Work

2.1   Blockchain Sharding Protocols

Elastico [3] is considered the first sharding protocol for public blockchain. In this protocol, all nodes are required to solve a PoW puzzle to determine their corresponding consensus committee. The sharding committee runs the PBFT protocol internally to achieve consensus and the result is committed to final committee. And directory committee helps blockchain identities find their committee peers. OmniLedger [4] builds on the work of Elastico and makes some improvements. It utilises a bias-resistant randomness generation protocol to assign blockchain nodes to different shards. The authors propose a client-driven mechanism with fund lock/unlock operations to assure the atomicity of cross-shard transactions.

Considering the issue of uneven transaction allocation caused by sharding, OptChain [5] proposes a novel transaction placement method inspired by PageRank algorithm. Compared with the random placement method used in Elastico, this method reduces the number of cross-shard transactions while improving the throughput of blockchain. Monoxide [6] is another representative work on blockchain sharding. The authors describe the Asynchronous Consensus Zone, which allows blockchain to scale linearly while maintaining decentralization and security. Furthermore, *Chu-ko-nu mining* mechanism is designed to ensure the effective mining power in each zone is at the same level.

2.2   Transaction Processing

In the blockchain sharding network, each transaction only need to be verified in one shard to be accepted by all the peers. As a result, sharding technology brings a huge efficiency improvement to blockchain. In the sharded blockchain, each shard maintains a set of unconfirmed transactions called *transaction pool*. After generating a transaction, it is transmitted to transaction pool, at the same time transactions are continuously taken out and broadcasted to the intra-shard peers to obtain consensus on the validity. Valid transactions are added to the blockchain, otherwise they are discarded.

Several recent works have investigated transaction processing utilizing queueing theory. Kawase et al. [7] investigate the stochastic behavior of the transaction confirmation process using a queueing model with batch service and broad input. Easley et al. [8] highlight the fact that delays are rarely employed to prioritize transactions in the Bitcoin pool of unconfirmed transactions. The authors then argue in favor of new prioritization mechanisms. Misic et al. [9] apply the Jackson network model to Bitcoin network, with individual peers acting as priority $M/G/1$ queueing system. This model's utility is proved by effectively computing the bitcoin forking probability. Li et al. [10] adopt a queueing theory methodology to study transaction delays. Based on their work, Ricci et al. [11] present a basic queueing model to demonstrate the latency incurred by Bitcoin transactions. Huang et al. [12] study how to elastically allocate limited network resources to permissioned sharded blockchain, and the authors define the problem based on the Lyapunov optimization theory.

In most of the sharding models, after a transaction is taken out from the transaction pool, shard runs PBFT protocol [13] internally to reach consensus on the validity of the transaction. Several studies have proposed scalable BFT protocols to improve the efficiency of transaction verification. Mir-BFT [14] enhances PBFT protocol and proposes signature verification sharding to allow transactions to be validated in batch. But due to possible single faulty, verification sharding may revert to full verification. In Red Belly [15], the authors propose sharded verification to verify transactions in a more efficient and robust way. However, these works do not have mechanism to track transaction dependency, therefore the usage scenarios are very limited.

3 Proposed Model

3.1 Overview of Model

We assume the blockchain adopts account/balance model, peers need to maintain and update data in their own shard, including the state of other peers and local ledger. The proposed transaction processing model is illustrated in Fig. 1. After a new transaction is generated, it is sent to the transaction pool to be verified. However, transactions may not admitted by transaction pool, as there is a quantity limit to the transaction that can be submitted by peers, and the excess ones need to be resubmitted in the next period. On the other hand, transactions in the pool are constantly taken out and verified. We propose a threshold signature-based transaction verification method, which allows transactions to be verified in parallel by different sets of peers.

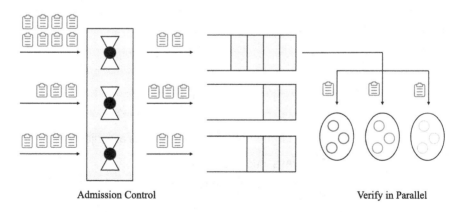

Fig. 1. Overview of proposed transaction processing model.

3.2   Threat Model

There are two kinds of peers in the sharded blockchain: *honest* and *malicious*. Like the most protocols, we consider the Byzantine fault tolerant scenario, that is the fraction of malicious peers in each shard is no more than $1/3$. Our model can be combined with other blockchain security solutions to cope with attacks like sybil attack [16]. There we focus on two ways malicious peers doing evil in the transaction processing phase: single-shard flooding attack and tampering with transaction verification results.

Single-Shard Flooding Attack. It refers to the malicious peers injecting massive transactions into the specific shard, causing a backlog of transactions in that shard and preventing timely processing of transactions generated by other peers, as shown in Fig. 2. Single-shard flooding attacks are launched in different ways in different sharding models. For the transaction sharding model like Elastico, transactions are assigned to specific committees for processing according to the last s bits of transaction hash. Malicious peers can inject transactions into the specific shard by constructing the transaction hashes. As for the state sharding model like BrokerChain [17], attack is even simpler, as it only requires a large number of transactions generated by malicious peers in the target shard.

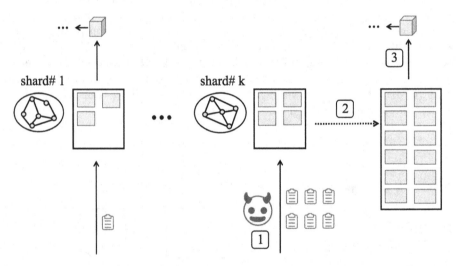

1 : A large number of transactions are injected.

2 : Transaction congestion.

3 : Transactions are taken out for verification.

Fig. 2. Single-shard flooding attack in sharded blockchain.

Tampering with Transaction Validation Results. It indicates that malicious peers collude with each other to provide incorrect transaction verification results, resulting in invalid transactions being recorded into blockchain.

4 Transaction Admission Control Algorithm

4.1 Problem Formulation

In this paper, we consider the system runs in discrete time with timeslot indexed by $t \in \{0, 1, 2, ...\}$. Table 1 summarizes some important notations we define. Suppose the blockchain is divided into S shards, and the transaction queue in each shard is indexed by $i \in I$, where $I = \{1, 2, 3, ..., S\}$. For convenience, we assume that each shard has the same number of peers, which is denoted by N.

Table 1. Main Symbols

Notation	Description
I	set of queues, $|I| = S$ denotes the number of shards
T	set of all timeslots, $t \in T$
N	number of peers in one shard
$G_i^n(t)$	number of TXs generated by peer n in shard i
$A_i^n(t)$	number of TXs admitted by peer n in shard i
$A_i(t)$	number of TXs admitted in shard i
$B_i(t)$	number of TXs processed in shard i
α	parameter for adjusting $A_i^n(t)$
$\kappa_i^n(t)$	satisfaction of peer n in shard i
$Q_i(t)$	queue size of shard i
$\boldsymbol{Q}(t)$	vector of queue size
V	parameter reflecting the emphasis on peer satisfaction
$L(\boldsymbol{Q}(t))$	Lyapunov function of $\boldsymbol{Q}(t)$
$\Delta(\boldsymbol{Q}(t))$	Lyapunov drift of $\boldsymbol{Q}(t)$
$\boldsymbol{Z}(t)$	vector of virtual queues
$z_i(t)$	increment of virtual queue in shard i
$\boldsymbol{\Theta}(t)$	concatenated vector

We employ multi-queue model to describe the process of transactions waiting to be verified in transaction pool. We use $G_i^n(t)$ to denote the number of transactions generated by peer n in shard i at timeslot t. In most sharded blockchains, any transactions generated by peers can be submitted to the transaction pool, which provides conditions for malicious peers to launch single-shard flooding attack. In the transaction processing model we proposed, there is a quantity restriction for transactions that peer can submit, which is denoted by $A_i^n(t)$. If $G_i^n(t) > A_i^n(t)$, it means that the peer has consumed its transaction quota at timeslot t and the excess transactions need to be resubmitted in the next timeslot. The number of transaction admitted in shard i can be calculated as:

$$A_i(t) = \sum_{n=1}^{N} A_i^n(t) \tag{1}$$

And we use vector $\boldsymbol{A}_i(t) = [A_i^1(t), A_i^2(t), A_i^3(t), ..., A_i^N(t)]$ to represent the number of admitted transactions for different peers in shard i.

For each peer, the more transactions that can be admitted, the more satisfied it is. We use $\kappa_i^n(t)$ to represent the peer satisfaction, which is defined based on the quantity of transactions generated in previous timeslot and be expressed as:

$$\kappa_i^n(t) = \frac{A_i^n(t)}{\alpha \cdot G_i^n(t-1)} \tag{2}$$

where $\alpha(\geq 1)$ is the amplification parameter for the transactions quantity in the previous timeslot, and $A_i^n(t)$ satisfies:

$$A_i^n(t) \leq \alpha \cdot G_i^n(t-1) \tag{3}$$

To measure the congestion of queues, we define $Q_i(t)$ to denote the queue size of shard i at timeslot t, and $Q_i(t)$ evolves in accordance with the following expression:

$$Q_i(t+1) = \max\{Q_i(t) - B_i(t) + A_i(t), 0\} \tag{4}$$

Our transaction admission control algorithm amis to maximize the overall satisfaction of peers while maintaining the transaction queue stable. According to [18], the conditions for multi-queue stability can be expressed as:

$$\lim_{t \to \infty} \frac{1}{t} \frac{1}{S} \sum_{\tau=0}^{t-1} \sum_{i=1}^{S} \mathbb{E}\{Q_i(\tau)\} = 0 \tag{5}$$

Therefore, the transaction admission control problem can be expressed as follows:

$$\max \sum_{i=1}^{S} \sum_{n=1}^{N} \kappa_i^n(t)$$

$$\text{s.t.} \begin{cases} A_i^n(t) \leq \alpha \cdot G_i^n(t-1), \forall i \in I, n \in [N], t \in T \\ \lim_{t \to \infty} \frac{1}{t} \frac{1}{S} \sum_{\tau=0}^{t-1} \sum_{i=1}^{S} \mathbb{E}\{Q_i(\tau)\} = 0 \end{cases}$$

$$\text{var}: \ \boldsymbol{A}_i(t), \forall i \in I, t \in T \tag{6}$$

4.2   Algorithm Design

To address the proposed transaction admission control problem (6), we design a control algorithm based on the stochastic Lyapunov optimization technique [18].

Since the blockchain is regarded as a multi-queue model, we use vector $\boldsymbol{Q}(t) = [Q_i(t), Q_2(t), Q_3(t), ..., Q_S(t)]$ to denote queue size of all shards. A quadratic Lyapunov function $L(\boldsymbol{Q}(t))$ is defined to measure the size of $\boldsymbol{Q}(t)$.

$$L(\boldsymbol{Q}(t)) = \frac{1}{2} \sum_{i=1}^{S} Q_i^2(t) \tag{7}$$

Correspondingly, the Lyapunov drift can be expressed as:

$$\Delta(\boldsymbol{Q}(t)) = L(\boldsymbol{Q}(t+1)) - L(\boldsymbol{Q}(t)) \tag{8}$$

$\Delta(\boldsymbol{Q}(t))$ is in fact the change of $\boldsymbol{Q}(t)$ over one timeslot, which can represent the variation of queue size. According to the proof of [18], the condition for queue stability is that $\Delta(\boldsymbol{Q}(t))$ remains a sufficiently small value. Thus, the second constraint of problem (6) can be transferred to minimizing the Lyapunov drift of $\boldsymbol{Q}(t)$.

Next, we must address the first constraint of problem (6) to ensure that the number of admitted transaction is limited by the budget. We convert this inequality constraint into another queue stability problem by defining a virtual queue $Z_i(t)$, with $Z(0) = 0$ and update equation:

$$Z_i(t+1) = \max\{Z_i(t) + z_i(t), 0\} \tag{9}$$

where $z_i(t) = \sum_{n=1}^{N} [A_i^n(t) - \alpha \cdot G_i^n(t-1)]$.

Theorem 1. *If virtual queue $Z_i(t)$ is mean rate stable, the first constraint of problem (6) holds automatically.*

Proof. We can get $Z_i(t+1) - Z_i(t) \geq z_i(t)$ from Eq. (9). Summing the inequality over timeslots $t = 0, 1, ..., T-1$, and dividing both sides by T, we have:

$$\frac{Z_i(T) - Z_i(0)}{T} \geq \frac{1}{T} \sum_{t=0}^{T-1} z_i(t) \tag{10}$$

As $Z_i(0) = 0$, consider the expectation on both sides and let $T \to \infty$, we have:

$$\limsup_{T\to\infty} \frac{\mathbb{E}\{Z_i(T)\}}{T} \geq \limsup_{T\to\infty} \bar{z}_i(t) \tag{11}$$

If $Z_i(t)$ is mean rate stable, there is $\limsup_{T\to\infty} \bar{z}_i(t) \leq 0$, which indicates the constraint of admitted transaction quantity in problem (6) is met.

To unify $\boldsymbol{Q}(t)$ and $\boldsymbol{Z}(t)$, we devise a concatenated vector $\boldsymbol{\Theta}(t) = [\boldsymbol{Q}(t), \boldsymbol{Z}(t)]$. The Lyapunov function and Lyapunov drift of $\boldsymbol{\Theta}(t)$ is defined as follows:

$$L(\boldsymbol{\Theta}(t)) = \frac{1}{2} \sum_{i=1}^{S} Q_i^2(t) + \frac{1}{2} \sum_{i=1}^{S} Z_i^2(t)$$
$$\Delta(\boldsymbol{\Theta}(t)) = L(\boldsymbol{\Theta}(t+1)) - L(\boldsymbol{\Theta}(t)) \tag{12}$$

Since we aim to maximize the peer satisfaction while ensuring the queue stability, we focus on the drift minus utility as follows:

$$\Delta(\boldsymbol{\Theta}(t)) - V\sum_{i=1}^{S}\sum_{n=1}^{N}\kappa_i^n(t) \tag{13}$$

where V is a tunable weight, denoting how important maximizing peer satisfaction is compare with maintaining queue stability. Equation (13) can be further bounded by:

$$\Delta(\boldsymbol{\Theta}(t)) - V\sum_{i=1}^{S}\sum_{n=1}^{N}\kappa_i^n(t)$$

$$\leq \frac{1}{2}\sum_{i=1}^{S}[A_i^2(t)+B_i^2(t)]+\sum_{i=1}^{S}Q_i(t)[A_i(t)-B_i(t)]-\sum_{i=1}^{S}A_i(t)B_i(t)$$

$$+\frac{1}{2}\sum_{i=1}^{S}z_i^2(t)+\sum_{i=1}^{S}Z_i(t)z_i(t)-V\sum_{i=1}^{S}\sum_{n=1}^{N}\kappa_i^n(t) \tag{14}$$

$$\leq B+\frac{1}{2}\sum_{i=1}^{S}A_i^2(t)+\sum_{i=1}^{S}Q_i(t)[A_i(t)-B_i(t)]-\sum_{i=1}^{S}A_i(t)B_i(t)$$

$$+\sum_{i=1}^{S}Z_i(t)z_i(t)-V\sum_{i=1}^{S}\sum_{n=1}^{N}\kappa_i^n(t)$$

where B is a positive constant which satisfies:

$$B \geq \frac{1}{2}\sum_{i=1}^{S}\mathbb{E}\{B_i^2(t)|\boldsymbol{\Theta}(t)\}+\frac{1}{2}\sum_{i=1}^{S}\mathbb{E}\{z_i^2(t)|\boldsymbol{\Theta}(t)\} \tag{15}$$

Thus, the objective of problem (6) can be transformed into minimizing the right-hand-side of Eq. (14). We rearrange the objective as:

$$obj(\boldsymbol{A}_i(t)) = \frac{1}{2}\sum_{i=1}^{S}(\sum_{n=1}^{N}A_i^n(t))^2+\sum_{i=1}^{S}Z_i(t)\cdot\sum_{n=1}^{N}[A_i^n(t)-\alpha\cdot G_i^n(t-1)]$$

$$+\sum_{i=1}^{S}Q_i(t)[\sum_{n=1}^{N}A_i^n(t)-B_i(t)]-\sum_{i=1}^{S}\sum_{n=1}^{N}[A_i^n(t)\cdot B_i(t)] \tag{16}$$

$$-V\sum_{i=1}^{S}\sum_{n=1}^{N}\frac{A_i^n(t)}{\alpha\cdot G_i^n(t-1)}$$

Obviously, $obj(\boldsymbol{A}_i(t))$ is a convex function, and we get:

$$\frac{\partial obj(\boldsymbol{A}_i(t))}{\partial A_i^n(t)} = A_i^n(t)+\sum_{i=1}^{S}[Q_i(t)-B_i(t)+Z_i(t)]-\frac{V}{\alpha\cdot G_i^n(t-1)} \tag{17}$$

We can find a valley point $A_i^n(t) = \frac{V}{\alpha \cdot G_i^n(t-1)} - \sum_{i=1}^{S} [Q_i(t) - B_i(t) + Z_i(t)]$, which is the optimal transaction admission solution to problem (6). Given Eq. (17), the transaction admission control algorithm is presented as Algorithm 1.

Algorithm 1: Transaction Admission Control Algorithm

Input: T, V, α

1 $Q(0) \leftarrow \varnothing,\ Z(0) \leftarrow \varnothing,\ G(0) \leftarrow \varnothing,\ B(0) \leftarrow \varnothing$;

2 **foreach** $t \in T$ **do**

3   **for** $\forall i \in I, n \in [N]$ **do**

4          $A_i^n(t) \leftarrow \frac{V}{\alpha \cdot G_i^n(t-1)} - \sum_{i=1}^{S} [Q_i(t) - B_i(t) + Z_i(t)]$

5   **end**

6   Update $Q(t)$ and $Z(t)$ according to Eq. (4) and Eq. (9), respectively.

7 **end**

---

5   Transaction Verification Design

Threshold signature scheme (TSS) [19,20] is a cryptographic digital signature protocol based on multi-party computation. Unlike traditional digital signature techniques, TSS allocates the signature capability to multiple participants, as any number of participants that reach the threshold can sign the network message. This section proposes a TSS based transaction verification mechanism. While the sharding technique eliminates duplicate work between shards, the duplicate work within shard does not eliminated, as all the intra-shard peers need to involve in the transaction verification. Actually in the Byzantine fault tolerant scenario, valid transactions only required to be verified by $f + 1$ honest peers. Based on TSS, We can achieve this goal in a more direct way, instead of conducting multi-stage network communication like the PBFT protocol, thus improving the efficiency of transaction verification.

5.1   Verification Based on TSS

After a transaction is taken out from the transaction pool, it is first verified by *verification group*(shorten as VG) composed of $f + 1$ peers. The VG members check whether the balance of the sender account is sufficient to complete it. If the balance is enough, the transaction is digitally signed; otherwise, it is rejected. To restrict malicious peers from deliberately delaying the verification process, the verification time limit for VG members is set to T_{inf}. If the transaction verification process goes well that all the required $f + 1$ signatures are collected in VG, the transaction is validated and broadcasted to other peers by the VG members. Conversely, if the transaction fails to meet the threshold signature requirement within T_{inf}, it needs to be verified by another f secondary peers one by one.

Secondary peers are randomly assigned a waiting time, with the wait ending no earlier than T_{inf} and no later than T_{sup}, which is the maximum verification time. The complete transaction validation timeline is shown in Fig. 3(a). When this transaction is verified by some secondary peer and meets the signature threshold, this peer should broadcast the transaction to other peers within the shard. With the public key of TSS, peers can easily check whether a broadcasted transaction has finished verification. For an invalid transaction that cannot be verified within the given time T_{sup}, it will eventually be discarded.

Figure 3(b) illustrates the transaction validation process for a shard composed of 4 peers. For each transaction, two peers are randomly selected to form a VG, we use $(VG)_i$ to denote the VG of transaction TXi. For example, peer P1 and peer P3 are in $(VG)_1$. Since $(VG)_1$ and $(VG)_3$ do not intersect, $TX1$ and $TX3$ can be validated in parallel.

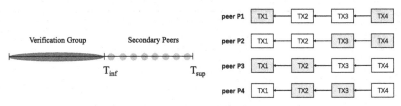

(a) Timeline for transaction verification. (b) Partial peers form VGs to verify transactions.

Fig. 3. Transaction Verification Based on TSS.

5.2 Verification Peers Selection

Each transaction requires the selection of $f + 1$ peers to form a verification group along with f secondary peers. The VG members must be unpredictable, otherwise malicious peers will find ways to join VGs, resulting in transactions need to be verified by more secondary peers, causing the efficiency of transaction verification to decline. Moreover, given a transaction, each peer must be able to determine the verification group members and secondary peers independently, so that all peers can reach consensus on the verification qualifications without exchanging messages. Therefore, we design the following approach to select the verification peers:

1. Each peer makes the hash operation as follow:

$$\mathcal{H} = hash(ID, hash(TX)) \tag{18}$$

where ID is the peer identity, $hash(TX)$ represents the transaction hash. After that, peer broadcast hash result $\mathcal{H}$ to other peers in the shard.
2. Peers sort all the hash results.

3. The $f + 1$ peers with smallest hash result form the verification group, while the second smallest f peers are the secondary peers. Other peers ignore the verification of this transaction.

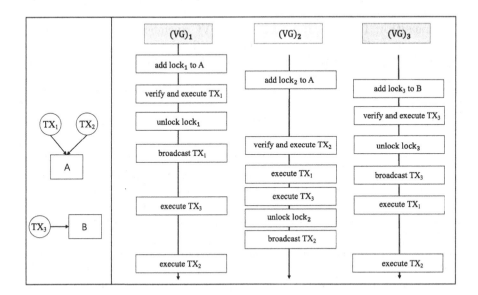

Fig. 4. Example of transaction verification in parallel.

5.3 Transaction Dependency Handling

Parallel verification of transactions must take transaction dependency into account. For example, suppose the $TX1$ and $TX3$ in Fig. 3(b) are "TX1:A→ B, 4" and "TX3:A→ C, 10", respectively. The verification results of these two transactions are jointly dependent on the account balance of A. It is dangerous to verify $TX1$ and $TX3$ in parallel in this case. Because if the account balance of A happens to be 10, both transactions may be considered to have sufficient balance. Therefore, there should be mechanism to avoid verifying such transactions with dependency simultaneously.

We address the transaction dependency issue through the account lock mechanism, and the process is shown in Fig. 4. For a transaction to be verified, VG first locks the sender account, indicating that this account is pending one transaction verification. Next, it checks whether there are other locks on the sender account. If not, it starts to verify the transaction, otherwise it needs to wait until this account has no other lock, that is all the dependent transactions have been verified. For the VG members, there are three possible outcomes for the transaction:

- **Transaction is validated by VG.** The VG members execute the transaction and unlock the sender account, and finally broadcast the transaction to other peers.
- **Transaction is validated by secondary peer.** After receiving the signed transaction from secondary peer, the VG members execute the transaction and unlock the sender account.
- **Transaction fails Verification.** If the VG members do not receive the signed transaction within the maximum verification time T_{sup}, this transaction is considered invalid and does not need to be executed, and the sender account is unlocked as well.

6   Experiment and Evaluation

6.1   Implementation

We develop a sharded blockchain simulation and run our proposed transaction processing model. The simulation is conducted on the OMNeT++ 6.0 [21], which is a discrete event-based network simulator. We measure the different performance of the model with and without single-shard flooding attack and compared it with other models.

In the experiments simulating flooding attack, we divide the blockchain network into 16 shards, each containing 70 peers. We set the block size to 1MB following the setting of Bitcoin, and each block contains 1000 transactions, and the network latency between peers is set to 100 ms. All the peers randomly generate transactions at a certain rate during the simulation, and the parameter α is set to 2. We simulate the operation of 10 timeslots, and malicious peers inject massive transactions in the second timeslot. The typical PBFT protocol is compared with the suggested method. As for the simulation experiments without attack, we measure the blockchain performance with different number of shards, and compare it with Monoxide [6] and Pyramid [22].

6.2   Performance Evaluation

Effect of Parameter V. In our proposed admission control algorithm, the parameter V reflects how much the system emphasize on maximizing the peer satisfaction. We set V to 3000, 3500, 4000 and the effect of different values of V is shown in Fig. 5. In the case without single-shard flooding attack, the larger the parameter V is set, the longer transaction queue is, because the restriction on transactions admission is lenient. In addition, the peer satisfaction is close to 1, which indicates that our admission control algorithm does not prevent peers from submitting transactions at normal rate. In the case of single-shard flooding attack, the transaction queue with a larger parameter V is more significantly affected, but the peer satisfaction decreases less, which is consistent with the meaning of the parameter V.

Overall, the setting of the parameter V depends on the system's trade-off between shard transaction congestion and peer satisfaction. It should be noted that the value of parameter V in the case without attack has little impact on transactions, which indicates that our admission control algorithm does not sacrifice transaction efficiency for robustness.

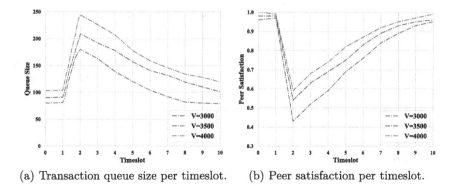

(a) Transaction queue size per timeslot.   (b) Peer satisfaction per timeslot.

Fig. 5. Effect of parameter V on admission control performance.

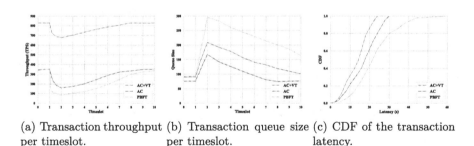

(a) Transaction throughput per timeslot.   (b) Transaction queue size per timeslot.   (c) CDF of the transaction latency.

Fig. 6. Robustness performance under single-shard flooding attack.

Robustness Performance. Robustness is an important aspect of model performance. The robustness of our proposed model is measured from three perspectives: transaction throughput, transaction queue size and transaction latency. In dealing with single-shard flooding attack, our model employs two approaches. On the one hand, it restricts the number of transactions injected by malicious peers through admission control. On the other hand, it enhances transaction verification speed by utilizing threshold signature. Experiments compare the robustness performance under three transaction processing methods:

– admission control and transaction verification based on TSS (AC+VT)
– admission control only (AC)
– PBFT protocol

Figure 6 shows the robustness performance comparison under single-shard flooding attack. From Fig. 6(a), we observe that before the single-shard flooding is launched, the transaction throughput with only admission control is comparable to the throughput with PBFT consensus, because the proposed admission control do not speed up the verification of transactions. In contrast, the TPS of attacked shard with verification based on TSS significantly improves from 350 to 820. After the attack is launched, we can notice that the TPS with PBFT consensus protocol drops dramatically from 350 to less than 100 due to the lack of effective countermeasure strategies. Combined with Fig. 6(b), we can observe that the size of transaction queue is controlled to some extent, that is most of the transactions generated by malicious peers are blocked from the transaction pool, indicating that our admission control is working.

Figure 6(c) depicts the latency distribution of each transaction during the experiment. It can be seen that both admission control and verification based on TSS have positive significance in reducing latency. The longest transaction latency with PBFT protocol reaches nearly 60 s, while the latency with admission control only and admission control plus verification based on TSS is 30 s and 24 s, respectively. The overall transaction latency also decreases in the same order.

Throughput and Latency Performance Without Attack. In this experiment, we fix the number of peers and set the number of shards to 4, 8, 12 and 16. Figure 7 shows the transaction throughput and latency performance. As the number of shards rises, we can notice that transaction throughput and latency increase together. This is because the rise of shard quantity implies enhancement in transaction parallel processing and increasement of cross-shard transaction quantity. Regardless of the number of shards, our model can achieves higher throughput and lower latency compare with baselines. In the case of 16 shards, it improves throughput by 3500 TPS and reduce latency by 4 s. The results demonstrate that the model can also outperform in the normal scenario without attack.

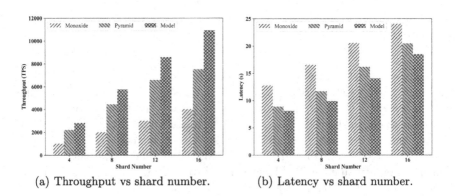

(a) Throughput vs shard number.   (b) Latency vs shard number.

Fig. 7. Performance under different number of shards.

7   Conclusion

In this paper, we proposed a novel transaction processing model for sharded blockchain, aiming to improve the robustness and efficiency of transaction processing. We noticed the damage caused by single-shard flooding attack to the service quality of sharded blockchain. To tackle this issue, we proposed a transaction admission control algorithm to avoid massive transactions being injected. To improve the efficiency of transaction verification, we introduced a threshold signature-based transaction verification mechanism. We implemented the model on simulation platform and experiment results show that the transaction processing model we proposed has better performance and it is resistant to single-shard flooding attack.

References

1. Nakamoto, S.: Bitcoin: a peer-to-peer electronic cash system. Decentralized Bus. Rev. 21260 (2008)
2. Nguyen, T., Thai, M.T.: Denial-of-service vulnerability of hash-based transaction sharding: attack and countermeasure. IEEE Trans. Comput. (2022)
3. Luu, L., Narayanan, V., Zheng, C., Baweja, K., Gilbert, S., Saxena, P.: A secure sharding protocol for open blockchains. In: Proceedings of the 2016 ACM SIGSAC Conference on Computer and Communications Security, pp. 17–30 (2016)
4. Kokoris-Kogias, E., Jovanovic, P., Gasser, L., Gailly, N., Syta, E., Ford, B.: Omniledger: a secure, scale-out, decentralized ledger via sharding. In: 2018 IEEE Symposium on Security and Privacy (SP), pp. 583–598. IEEE (2018)
5. Nguyen, L.N., Nguyen, T.D., Dinh, T.N., Thai, M.T.: Optchain: optimal transactions placement for scalable blockchain sharding. In: 2019 IEEE 39th International Conference on Distributed Computing Systems (ICDCS), pp. 525–535. IEEE (2019)
6. Wang, J., Wang, H.: Monoxide: scale out blockchains with asynchronous consensus zones. In: NSDI, vol. 2019, pp. 95–112 (2019)
7. Kawase, Y., Kasahara, S.: A batch-service queueing system with general input and its application to analysis of mining process for bitcoin blockchain. In: 2018 IEEE International Conference on Internet of Things (iThings) and IEEE Green Computing and Communications (GreenCom) and IEEE Cyber, Physical and Social Computing (CPSCom) and IEEE Smart Data (SmartData), pp. 1440–1447. IEEE (2018)
8. Easley, D., O'Hara, M., Basu, S.: From mining to markets: the evolution of bitcoin transaction fees. J. Financ. Econ. **134**(1), 91–109 (2019)
9. Misic, J., Misic, V.B., Chang, X., Motlagh, S.G., Ali, M.Z.: Block delivery time in bitcoin distribution network. In: ICC 2019–2019 IEEE International Conference on Communications (ICC), pp. 1–7. IEEE (2019)
10. Li, Q.-L., Ma, J.-Y., Chang, Y.-X.: Blockchain queue theory. In: Chen, X., Sen, A., Li, W.W., Thai, M.T. (eds.) CSoNet 2018. LNCS, vol. 11280, pp. 25–40. Springer, Cham (2018). https://doi.org/10.1007/978-3-030-04648-4_3
11. Ricci, S., Ferreira, E., Menasche, D.S., Ziviani, A., Souza, J.E., Vieira, A.B.: Learning blockchain delays: a queueing theory approach. ACM SIGMETRICS Perform. Eval. Rev. **46**(3), 122–125 (2019)

12. Huang, H., et al.: Elastic resource allocation against imbalanced transaction assignments in sharding-based permissioned blockchains. IEEE Trans. Parallel Distrib. Syst. **33**(10), 2372–2385 (2022)
13. Castro, M., Liskov, B., et al.: Practical byzantine fault tolerance. In: OsDI, vol. 99, pp. 173–186 (1999)
14. Stathakopoulou, C., David, T., Pavlovic, M., Vukolić, M.: MIR-BFT: high-throughput robust BFT for decentralized networks. arXiv preprint arXiv:1906.05552 (2019)
15. Crain, T., Natoli, C., Gramoli, V.: Red belly: a secure, fair and scalable open blockchain. In: 2021 IEEE Symposium on Security and Privacy (SP), pp. 466–483. IEEE (2021)
16. Douceur, J.R.: The sybil attack. In: Druschel, P., Kaashoek, F., Rowstron, A. (eds.) IPTPS 2002. LNCS, vol. 2429, pp. 251–260. Springer, Heidelberg (2002). https://doi.org/10.1007/3-540-45748-8_24
17. Huang, H., et al.: Brokerchain: a cross-shard blockchain protocol for account/balance-based state sharding. In: IEEE INFOCOM 2022-IEEE Conference on Computer Communications, pp. 1968–1977. IEEE (2022)
18. Neely, M.J.: Stochastic network optimization with application to communication and queueing systems. Synth. Lect. Commun. Netw. **3**(1), 1–211 (2010)
19. Boldyreva, A.: Threshold signatures, multisignatures and blind signatures based on the gap-Diffie-Hellman-group signature scheme. In: Desmedt, Y.G. (ed.) PKC 2003. LNCS, vol. 2567, pp. 31–46. Springer, Heidelberg (2003). https://doi.org/10.1007/3-540-36288-6_3
20. Shoup, V.: Practical threshold signatures. In: Preneel, B. (ed.) EUROCRYPT 2000. LNCS, vol. 1807, pp. 207–220. Springer, Heidelberg (2000). https://doi.org/10.1007/3-540-45539-6_15
21. Varga, A., Hornig, R.: An overview of the Omnet++ simulation environment. In: 1st International ICST Conference on Simulation Tools and Techniques for Communications, Networks and Systems (2010)
22. Wang, J., Shou, L., Chen, K., Chen, G.: Pyramid: a layered model for nested named entity recognition. In: Proceedings of the 58th Annual Meeting of the Association for Computational Linguistics, pp. 5918–5928 (2020)

Computation Offloading Based on Deep Reinforcement Learning for UAV-MEC Network

Zheng Wan, Yuxuan Luo, and Xiaogang Dong[⊠]

Jiangxi University of Finance and Economics, Nanchang, Jiangxi, China
dxg110@aliyun.com

Abstract. The existing MEC technology is not suitable for the situations where the number of Mobile Users (MUs) increases explosively or network facilities are sparsely distributed, and general MEC solutions cannot fully address the MU emergency communication and task offloading requirements in post disaster emergency communication networks. Unmanned Aerial Vehicle (UAV) can play an important role in wireless systems because it can be flexibly deployed to quickly help disaster areas improve signal coverage and restore communication quality, enabling MUs in disaster areas to unload tasks normally. This article investigates the deployment of UAV-MEC systems in emergency situations of network communication interruption in the region after a disaster, and studies the optimization strategy for task offloading in edge environments. Based on the relevant advantages of UAV and MEC technology, we have constructed a UAV-MEC system model for emergency disaster relief scenarios. Under the constraints of UAV flight trajectory and MU calculation mode, the problem of minimizing the total system delay is formulated. Due to the non convexity of this problem, there are high-dimensional state spaces and continuous action spaces, making it difficult to find the optimal solution. This article proposes a Split DQN (SDQN) algorithm based on Reinforcement Learning (RL) and uses it to solve optimization problems. The results of simulation experiments show that our proposed SDQN algorithm can approach the global optimal solution of the optimization problem more closely than the benchmark algorithm, and thus reduce the total system delay. Thus, it is an effective optimization decision-making scheme.

Keywords: Mobile edge computing · Unmanned Aerial Vehicle · Reinforcement learning

Supported by the National Natural Science Foundation of China (Grant No. 61961021), the Science and Technology Project of Jiangxi Education Department (Grant No. GJJ180251, GJJ2201915).

Z. Tari et al. (Eds.): ICA3PP 2023, LNCS 14490, pp. 265–276, 2024.
https://doi.org/10.1007/978-981-97-0859-8_16

1   Introduction

In recent years, with the rapid growth of the number of mobile users (MUs), the issue of coverage has become more serious. Although traditional cloud computing is attractive, latency and rigid architecture are unbearable for new services. At the same time, connectivity and spectral efficiency are challenging traditional computing net-works. Future applications, such as the industrial internet of things (IoT), image recognition and virtual reality (VR), need a delay of less than milliseconds. Mobile edge computing (MEC) can transfer computing and storage to edge nodes close to MU devices [1,2]. When a cloud server serves as a computing and information ex-change center, edge nodes can handle secondary tasks on the MU side and reduce the burden on the cloud server.

Unmanned Aerial Vehicle (UAV) is considered a promising technology in future networks. Compared with traditional cellular networks, UAV communication has many advantages, especially flexibility and economy. MEC can add storage, computing, and processing functions at the edge of communication networks. In UAV assisted wireless communication systems utilizing MEC, ground MU and IoT devices may have insufficient computing resources and need to offload computing tasks to surrounding UAVs for computation. In addition, due to the introduction of MEC technology, task offloading strategies need to be studied while planning UAV trajectories [3].

The task transmission delay during the task offloading process of MU has a significant impact on the service quality of MU. In response to this issue, Xu et al. [4] proposed the task computation delay problem of multiple UAVs assisted in MEC systems. In order to improve the system's service quality, they jointly optimized the time slot size, computing resource allocation, UAV flight trajectory, and MU device scheduling during the cycle, and decided on two task offloading modes to minimize the total system delay. Yu et al. [5] proposed a low latency and efficient task offloading scheme to jointly optimize the optimal location of UAVs, allocation of computing and communication resources, and task segmentation decisions in partial task offloading, in order to minimize the weighted sum of UAV energy consumption in the system and the total delay of all MU tasks offloading services in the system. Hu et al. [6] studied a UAV assisted MEC system by jointly optimizing task offloading ratio, UAV flight trajectory, and MU scheduling to minimize the sum of maximum delays for all MUs in the system within each time slot. To solve this non convex optimization problem with discrete binary variables and coupling constraints, an algorithm based on penalty dual decomposition is proposed to solve the optimization problem, minimizing the average delay of all MUs in the system within one cycle. Yu et al. [7] designed a UAV assisted MEC system based on millimeter wave mobile communication backhaul. In this system, MU devices can pass through ground based Ad_HOC network or UAV as a continuation node, offloading computing tasks to nearby ground BS through Ad_HOC. We designed a novel iterative algorithm framework to address the problem of minimizing the weighted delay sum of the two task offloading methods within the system, namely, hoc routing allocation and UAV flight trajectory.

In summary, most researchers currently use UAVs as auxiliary tools to assist MEC and improve its service quality, with differences mainly reflected in application scenarios and optimization objectives. Researchers have proposed different optimization algorithms and strategies for solving optimization problems, but the application scenarios are relatively single. Based on this, the purpose of this article is to envision a universal emergency rescue scenario for UAV assisted MEC in edge environments, reasonably plan UAV flight trajectories and task offloading decisions.

2   System Model and Problem Description

2.1   System Model

The UAV assisted MEC system model is shown in Fig. 1. Consider a UAV assisted MEC system, deploying an ECs (Edge Computing Server) carrying UAVs to provide computing services for K randomly distributed MUs on the ground. In the system, each MU carries a fixed computing task, and the way the MU offloads the task is binary offloading, that is, the task can be offloaded to the UAV and assisted by its ECs for computation; You can also perform calculation tasks separately locally. For the convenience of analysis, this chapter assumes K MUs, $K = \{1, 2, ..., K\}$. UAV can only serve one MU at a time, and when serving MUs, UAVs need to fly directly above them until the calculation task of that MU is unloaded and completed before flying to the next MU to provide services. Repeat this process until all MUs in the system are serviced.

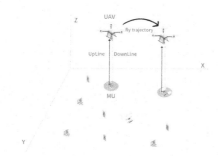

Fig. 1. UAV-assisted MEC system model

2.2   Communication Model

This chapter defines coordinates using a three-dimensional Cartesian coordinate system, where the coordinates of MU within the system are represented as $W_k = (X_k, Y_k, 0)$. In this chapter, we can assume that the fixed flight altitude of the

UAV is H and the unit is m, and the entire communication cycle T is divided into I time slots. If the initial coordinates of the UAV are $W_{uav} = (X_{uav}, Y_{uav}, H)$, then the coordinates of the UAV at time i are $W_{uav} = (X_{uav}(i), Y_{uav}(i), H)$.

The multiple access technology used in this chapter is time division multiple access (TDMA), which implements multiple access communication with different time slots. Each time slot UAV can only serve one MU. In each time slot, the UAV hovers in a fixed position and establishes communication with one of the MUs. After unloading its computing task and completing the calculation, the service ends. The unloading decision is represented by the binary variable a_k, where $a_k = 0, 1$ represents whether the k-th MU will be unloaded using UAV. $a_k = 1$ indicates that the MU completely offloads the task to the ECs carried by the UAV for calculation, and $a_k = 0$ indicates that the MU executes the calculation task locally. Therefore, the channel Power gain $h_{k,u}$ from MU to UAV at time i can be described as a free space path loss model, as shown in Eq. (1).

$$h_{k,u}(i) = \frac{\beta_0}{\|W_k W_{uav}(i)\|^2} \tag{1}$$

where, β_0 represents the channel gain at a reference distance of $1m$. So, according to the Shannon formula, the transmission rate R_k between the UAV and the k-th MU at time i is shown in Eq. (2).

$$R_{k,u}(i) = B_k Log_2(1 + \frac{P_k h_{k,u}(i)}{\sigma^2}) \tag{2}$$

where, B_k is the channel bandwidth of the k-th MU; P_k is the transmission power of the k-th MU; σ^2 represents the Gaussian noise power at the receiver.

2.3   Task Calculation Model

In this chapter, we assume that each MU has a specific computing task $Task_k = L_k, C_k$, where L_k represents the size of the task in bits and C_k represents the number of CPU cycles required to compute per bit task. The UAV first flies to the MU that needs to provide services, and then hovers over it to provide task of-floading services. The MU offloads the task to the UAV through the uplink UpLine or performs local calculations. After the UAV offloads and calculates the task, it sends the task calculation results back to the MU through the downlink DownLine.

Before the mission is unloaded, the UAV needs to fly over the target MU. In this chapter, we assume that the UAV is flying at a constant speed V, and the flight time t_k^{fly} required for the UAV to fly towards the k-th MU at time i is shown in Eq. (3).

$$t_k^{fly} = \frac{\|W_k - W_{uav}(i)\|^2}{V} \tag{3}$$

where, W_k represents the position of the target MU, and $W_{uav}(i)$ represents the position of the UAV at time i.

In the UAV-MEC system, the time t_k^{up} for the k-th MU to unload tasks to the UAV through UpLink at time i is shown in Eq. (4).

$$t_k^{up} = \frac{L_k}{R_{k,u}(i)} \qquad (4)$$

where, $R_{k,u}(i)$ represent the transmission rate between the UAV and the k-th MU at time i.

The calculation time in the UAV-MEC system can be divided into two categories: calculation on UAV ECs and local calculation. When MU chooses to offload to UAV calculation, the calculation time t_k^{ECs} on UAV is shown in Eq. (5).

$$t_k^{ECs} = \frac{L_k C_k}{f_k^{ECs}} \qquad (5)$$

where, L_k represents the amount of tasks that need to be unloaded by the k-th MU, and C_k represents the number of CPU cycles required for the k-th MU to calculate per one task. f_k^{ECS} represent the computing resources provided by the UAV to the k-th MU.

When MU leans towards executing tasks locally, the time required for local computation, t_k^{local} is shown in Eq. (6).

$$t_k^{local} = \frac{L_k C_k}{f_k^{mu}} \qquad (6)$$

where, f_k^{mu} represents the local CPU computing resources of the k-th MU.

In summary, we can obtain that in the UAV-MEC system, the total time consumed by the unloading calculation task of a single MU is four parts. UAV flight time from initial position to k-th MU; The k-th MU calculates the task offloading time to the UAV through UpLink offloading, and the UAV uses the ECs carried to calculate the time for the k-th MU offloading task; When the k-th MU selects the local calculation method, the local calculation time of the MU. In summary, the total delay required for UAV to provide services to the k-th MU is shown in Eq. (7).

$$t_k = t_k^{fly} + t_k^{up} + a_k t_k^{ECs} + (1 - a_k) t_k^{local} \qquad (7)$$

2.4   Problem Formulation

This Based on the model introduced above, we have developed optimization problems in the UAV network assisted MEC system. In order to ensure the effective utilization of limited computing resources, we aim to minimize the sum of maximum delays for all MUs by jointly optimizing the MU task calculation method and UAV flight trajectory in the system. Specifically, according to Eq. (7), the optimization problem can be derived as shown in Eq. (8).

$$\min_{\{W_{uav}(i)\},\{a_k\}} \sum_{k=1}^{K} t_k \tag{8}$$

Subject to :

$$\forall k \in \{1, 2, ..., k\}, \ \forall i \in \{1, 2, ..., I\}, \ a_k = \{1, 0\}$$

$$W_{uav} \in \{(x(i), y(i)) \mid x(i) \in \{0, L\}, y(i) \in \{0, W\}\}, \ \forall i$$

$$W_k \in \{(x_k, y_k) \mid x_k \in \{0, L\}, y_k \in \{0, W\}\}, \forall k$$

Constraint (8a) restricts the MU of the service to only MU in the system; Constraint (8b) constrains the time slot i of one service cycle; Constraint (8c) constrains the UAV's flight range; Constraint (8d) constrains that MUs that require task offloading services can only be distributed within the specified range of the system; Con-straint (8e) represents the task offloading mode constraint selected by each MU in the system, where $a_k = 0$ represents the local computing mode and $a_k = 1$ represents the complete offloading mode.

3   Proposed Scheme

In this chapter, the problem is represented as MDP and deep RL will be used to solve problem (8). This chapter will propose an improved SDQN algorithm based on the DQN algorithm framework.

3.1   SDQN Algorithm

Considering the existence of the UAV's service order decision in the UAV-MEC system and the large number of composite pattern that will exist in the task unloading mode of each MU, as well as the existence of the UAV's flight distance and other time-varying variables, it is difficult to solve the problem (8) with the traditional convex optimization method. This article proposes an RL based SDQN algorithm to reasonably solve optimization problem (8).

We have improved the traditional DQN algorithm by completely splitting the Q function into the state value function V and the action advantage function Q, as shown in Fig. 2. The SDQN algorithm is similar to the DQN algorithm and also uses two completely independent Q networks during the algorithm training process. One is used to estimate the Q value in the current state, and the other is used to calculate the target Q value. The parameters of the target Q network are not frequently updated, which can ensure the stability of the algorithm and improve the convergence speed and effectiveness.

The state transition function of the SDQN algorithm considers the state value function and the action advantage function separately, thereby making learning more efficient and stable.

3.2   Reinforcement Learning Modeling

In the UAV-MEC system, we use the UAV as the intelligent agent of the algorithm, and assume that the UAV has already stored information of all nearby

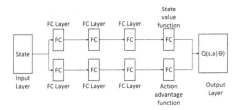

Fig. 2. UAV-assisted MEC system model

MUs before an emergency event occurs, and can make quick responses when an emergency event occurs, thereby quickly making system offloading decisions. The environmental details are described below.

In the UAV-MEC system, the state space is jointly determined by a UAV carrying ECs, MUs, and their accompanying Tasks. The system state s_i at time slot i is shown in Eq. (9).

$$s_i = (W_{uav}(i), W_1, ..., W_k, \ Task_1(i), ..., Task_1(i)) \tag{9}$$

where, the position of MU in our hypothetical system is fixed and invariant, so W_k does not change over time in the state space. UAV will change with its time slot i, so $W_{uav}(i)$ may change at each time slot i. $Task_k(i)$ represents the task completion status of the k-th MU in time slot i.

In order to facilitate the training of the RL algorithm, we normalized the state space of the system and converted the original s_i to s_i'. Define vectors $T = \{T_1, ..., T_k\}, T_k \in \{-1, 0, 1\}$. Among them, T_k represents whether the task of the k-th MU, $Task_i$, has been completed. $T_k = 1$ indicates that the UAV has per-formed task uninstallation service on the k-th MU and is a complete uninstallation service. $T_k = 0$ indicates that the UAV has performed task uninstallation service on the k-th MU and is a local computing service. $T_k = -1$ indicates that the UAV has not yet performed task uninstallation service on the k-th MU. So the system state s_i' at slot i after standardization is shown in Eq. (10).

$$s_i' = (W_{uav(i)}, T(i)) \tag{10}$$

Based on the current state of the system and the observed environment, the agent should jointly consider which MU to choose for service and whether the task is completely unloaded to the UAV or calculated locally in the MU. Based on system requirements, the action decision variable a_i at time slot i is shown in Eq. (11).

$$a_i = (k(i), a_k(i)) \tag{11}$$

where, $k(i)$ represents the k-th MU in the system selected by ECs at time slot i for service, and $a_k(i)$ represents the task calculation method selected by MU at time slot i.

Our goal is to maximize the reward r_i by minimizing the processing delay defined in Problem 3.8. If a MU that has already been served (i.e. a MU whose task has been completed) is selected, $r_i = 100$, otherwise as shown in Eq. (12).

$$r_i = r(s_i', a_i) = -\tau_{delay}(i) \tag{12}$$

If all MUs in the system are completed by UAV services within a cycle T, then $r_i = 100$ and the training is completed. If all MUs in the system are not fully serviced by the UAV within one cycle T, then $r_i = -100$, and the training is completed.

Due to TDMA, ECs can only serve one MU within a time slot. So the processing delay at time slot i is equal to the delay of serving the k-th MU at time slot i, as shown in Eq. (13).

$$\tau_{delay}(i) = t_k \tag{13}$$

Using the SDQN algorithm, you can find actions to maximize the Q value. The long-term average return of the system can be calculated using the Bellman equation as shown in Eq. (14).

$$Q_u\left(s_i', a_i\right) = \mathbb{E}_u\left[r\left(s_i', a_i\right) + \gamma Q_u\left(s_{i+1}', a_{i+1}\right)\right] \tag{14}$$

3.3   Training and Testing

In order to learn and evaluate the performance of the SDQN-based task offloading algorithm in reducing and minimizing the total delay in the UAV-MEC system, it is divided into two stages of training and testing.

The task offloading training algorithm based on SDQN is shown in Algorithm 1. During training, minimizing the loss function $L(\theta)$ is similar to reducing the mean square error in the Bellman equation, and updating the parameter θ by utilizing the Adam algorithm.

The task offloading test algorithm based on SDQN in the UAV-MEC system is shown in Algorithm 2. Algorithm 2 describes the SDQN-based task offloading testing process in the UAV-MEC system, which uses the Q-network trained in Algorithm 1.

4   Simulation Results

4.1   Simulation Setting

In the UAV-MEC system, we consider a 3D square area. Assume that $K = 10$ MUs are randomly distributed in a three-dimensional area of $100 \times 100 \times 100 m^3$, and the UAV is initially located in the center of the square area. Assume that the task data size generated by MU is $L_k \in [1.6 \times 10^6, 2.0 \times 10^6]$ bit. Assume a complete cycle is $T = 200s$. Assume that the flying speed of the UAV is $V = 3m/s$. Assume that the channel power gain at a reference distance of 1m

---

Algorithm 1: Task offloading training algorithm based on SDQN

Input: Initialize prediction network $Q(s, a|\theta)$, target network $Q(s, a|\theta')$ with weights θ and θ'; Initialize replay memory D with capacity $|D|$;

1   **for** *each episode e in* $[1, 2, ..., E]$ **do**

2   Reset simulation parameters in the UAV-MEC system and obtain initial observation state s_i;

3   **for** *each i in* $[1, 2, ..., I]$ **do**

4   Normalize state $s_i \rightarrow s_i'$;

5   When $0 \leqslant p < \varepsilon$, randomly chosen from $A(s)$ as a_i. When $\varepsilon \leqslant p < 1$, $a_i = max_a Q(s, a|\theta)$;

6   Takes an action a_i and receive the current step reward r_i, the state of the next step s_{i+1}';

7   Save $\left\{ s_i', a_i, r_i, s_{i+1}' \right\}$ to the replay memory D;

8   Random samples a batch of experiences $\widehat{D}$ from D;

9       $R_i = R_i + r_i, s = s_{i+1}$;

10   Calculates loss function $L(\theta)$;

11   Number of parameter updates completed every N rounds $\theta' \leftarrow \theta$;

Algorithm 2: Task offloading testing algorithm based on SDQN

Input: Testing episode length E, testing steps sample length I;
       Trained Q network $Q(s, a)$;
       Current states: $W_{uav}(i), W_1, \ldots, W_K, Task_1(i), \ldots, Task_k(i)$;

Output: Current episode reward R_i and total system delay $delay$;

1   **for** *each episode e in* $[1, 2, ..., E]$ **do**

2   Reset simulation parameters in the UAV-MEC system and obtain initial observation state s_i;

3   **for** *each i in* $[1, 2, ..., I]$ **do**

4   Normalize state $s_i \rightarrow s_i'$;

5   Get the action with the trained Q network: $a_i = Q(s_i', a)$;

6   Perform action a_i, and obtain reward $r_i = r(s_i', a_i)$, the state of the next step s_{i+1}';

7       $R_i = R_i + r_i, s = s_{i+1}'$;

8   **if** *All tasks in the system are completed* **then**

9   Return R_i, s;

---

is set to $\beta_0 = -50dB$. It is assumed that the noise power of both MU and ECs receivers is $\sigma^2 = -100dBm$, and there is no signal blocking. Assume that the bandwidth transmitted in the Ad_hoc route is $B_k = 1MHZ$. Assume that the computing resource allocated by ECs to the k-th MU is $f_k^E Cs = 1.2GHz$. Assume that the computing resource of MU is $f_k^m u \in [0.1, 0.6]GHz$. Assume that the number of CPU cycles $C_k = 1000cycles/bit$ required to calculate each

1bit of data. Assume that the transmission power of the k-th MU is $P_k = 0.1W$. All simulation processes, experiments, and algorithm implementations in this chapter were conducted on the Python platform. The simulation experiment in this chapter is based on the assumed values of the parameters mentioned above. In order to evaluate the performance of our proposed SDQN-based task offloading algorithm for optimizing the total delay of UAV-MEC systems, we also compare it with other RL algorithms.

4.2   Simulation Results and Discussions

To verify the convergence of the SDQN algorithm, we first trained the four RL based algorithms 1000 times in the same UAV-MEC system learning environment to take the average value. Each training randomly reset the MU coordinates and task information, and fixed the iteration step of 1000 for each training. The training process is shown in Fig. 3. From the graph, it can be observed that SDQN algorithm has a 11.1% faster convergence speed compared to the Dueling DQN algorithm.

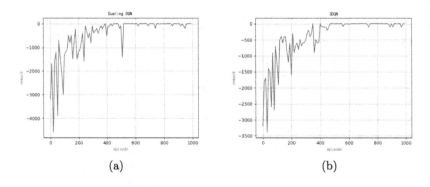

(a)   (b)

Fig. 3. Relationship Convergence process of two RL algorithms

In order to provide a more intuitive explanation of the optimized unloading algorithm used in the UAV-MEC system, through simulation experiments, we have obtained a typical example of the UAV flight trajectory corresponding to the SDQN algorithm, as shown in Fig. 4. In this scenario of the UAV-MEC system, there is a UAV carrying ECs and 10 MUs that require emergency services from the UAV, distributed within the characteristic area of $100 \times 100m^2$.

We evaluate the final performance of the SDQN algorithm by studying the changes in the total delay of the UAV-MEC system under different ECs calculation frequencies, with other conditions remaining unchanged. In Fig. 5, it can be observed that as the calculation frequency of ECs increases, the total delay of the system initially decreases. However, as the UAV calculation frequency continues to increase, the total delay of the system begins to decrease and gradually

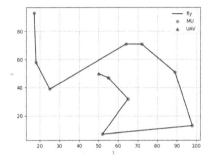

Fig. 4. Example of UAV trajectory corresponding to optimal delay

approaches a stable straight line. The main reason for the above changes is that at lower UAV calculation frequencies (such as $f_k^{ECs} = 0.2GHz$), the delay associated with task offloading performed by UAVs is an important factor limiting the overall system delay (mainly the calculation delay of UAVs). However, with the improvement of UAV computing power, the main limiting factor of overall system delay has become the UAV's own flight delay. At this point, the task offloading calculation delay of each MU is much smaller than the flight delay of the UAV. Continuing to observe, it is not difficult to find that our proposed SDQN algorithm significantly reduces the total delay of the system compared to the other four algorithms. At $f_k^{ECs} = 1.2GHz$, compared to the benchmark algorithm, the total delay under the SDQN algorithm was reduced by 37.6%, the total delay under the Dueling DQN algorithm was reduced by 35.4%, the total delay under the DQN algorithm was reduced by 30.3%, and the total delay under the Q-learning algorithm was reduced by 20.8%.

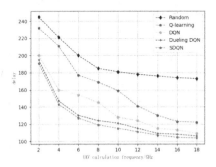

Fig. 5. Total delay at different UAV calculation frequencies

5   Conclusion

This chapter considers a network architecture for a UAV assisted MEC system, where there is a UAV carrying ECs and providing task offloading computing services for all MUs in the emergency system. The MU in the system can choose to unload the task to the MU or directly execute the task locally. We assume that the UAV, as an intelligent agent, can collect relevant information (such as coordinate information and task information) from all MUs around the system before a disaster or emergency occurs. Based on the above assumptions, we propose an optimization problem with the goal of minimizing the total delay of the UAV-MEC system by jointly optimizing the flight trajectory of the UAV during service and the task offloading decision of the MU in the system. To optimize this problem, we propose an RL based SDQN algorithm, which is an improvement on the DQN algorithm and can dynamically determine the flight trajectory of UAVs and the unloading mode of MUs. Our simulation experimental results show that our proposed SDQN algorithm significantly reduces the total delay of the UAV-MEC system compared to other RL algorithms used in this chapter, and shortens the number of iterations required for RL algorithm convergence. It can approach the global optimal solution of the optimization problem with a small gap in the UAV-MEC system in an environment with large action space.

References

1. Mao, Y., You, C., Zhang, J., Huang, K., Letaief, K.B.: A survey on mobile edge computing: the communication perspective. IEEE Commun. Surv. Tutorials **19**(4), 2322–2358 (2017)
2. Wang, F., Xu, J., Wang, X., Cui, S.: Joint offloading and computing optimization in wireless powered mobile-edge computing systems. IEEE Trans. Wireless Commun. **17**(3), 1784–1797 (2018)
3. Kim, K., Park, Y.M., Hong, C.S.: Machine learning based edge-assisted UAV computation offloading for data analyzing. In: 2020 International Conference on Information Networking, ICOIN 2020, Barcelona, Spain, January 7–10, 2020, pp. 117–120. IEEE (2020)
4. Xu, Y., Zhang, T., Loo, J., Yang, D., Xiao, L.: Completion time minimization for UAV-assisted mobile-edge computing systems. IEEE Trans. Veh. Technol. **70**(11), 12253–12259 (2021)
5. Yu, Z., Gong, Y., Gong, S., Guo, Y.: Joint task offloading and resource allocation in UAV-enabled mobile edge computing. IEEE Internet Things J. **7**(4), 3147–3159 (2020)
6. Hu, Q., Cai, Y., Yu, G., Qin, Z., Zhao, M., Li, G.Y.: Joint offloading and trajectory design for UAV-enabled mobile edge computing systems. IEEE Internet Things J. **6**(2), 1879–1892 (2019)
7. Yu, Y., Bu, X., Yang, K., Yang, H., Han, Z.: UAV-aided low latency mobile edge computing with mmWave backhaul. In: 2019 IEEE International Conference on Communications, ICC 2019, Shanghai, China, May 20–24, 2019, pp. 1–7. IEEE (2019)

Privacy-Aware Scheduling Heuristic Based on Priority in Edge Environment

Yue Hong, Caie Wang, and Wei Zheng[(✉)]

Department of Computer Science, Xiamen University, Xiamen 361000, China
zhengw@xmu.edu.cn

Abstract. The need for edge computing has been increasing recently with the rise of the Internet of Things (IoT). This leads to an urgent demand for suitable scheduling strategies in the edge computing environment. However, edge nodes are more vulnerable to privacy breaches. Thereby, workflow scheduling algorithms in edge computing systems are required to fully consider privacy issues. Additionally, edge applications usually desire real-time responsiveness. So makespan is also an important quality of service (QoS) metric considered in edge environments. This paper proposes a privacy-aware and priority-based algorithm (PAPBS) based on the dynamic priority-based heuristic (PB) to address privacy issues and real-time scheduling problems in edge environments. The proposed approach aims at minimizing application completion time while satisfying privacy task scheduling requirements. Extensive simulation experiments have been conducted to compare our approach with other related scheduling algorithms. The results showed that our proposed algorithm outperforms its competitors on makespan while satisfying privacy scheduling requirements.

Keywords: Workflow scheduling · Privacy Awareness · Edge computing

1 Introduction

Edge computing is a distributed computing paradigm that has become a critical technology for various applications driven by the rise of Internet of Things (IoT) devices and demands for real-time data processing. As an increasingly popular form of computing, Edge computing distributes computing resources and data storage to edge devices and edge nodes that are located closely to the data source. It enables faster, more secure, and more reliable data processing and computation [1]. Various research works related to edge computing have received wide attention, including workflow scheduling which refers to the process of assigning tasks in a workflow to available resources [2–4]. The typical goals of workflow scheduling are to optimize the execution efficiency of workflow, to improve resource utilization, and to meet various QoS requirements.

However, in the context of edge computing, real-time responsiveness is often desired. For example, in the scenarios such as autonomous driving, industrial

© The Author(s), under exclusive license to Springer Nature Singapore Pte Ltd. 2024
Z. Tari et al. (Eds.): ICA3PP 2023, LNCS 14490, pp. 277–294, 2024.
https://doi.org/10.1007/978-981-97-0859-8_17

automation, and smart homes, data needs to be processed and analyzed in real-time to achieve control and management of IoT devices and sensors [5,6]. This requires edge devices to complete data processing and computation in a short amount of time to ensure system reliability and efficiency. Besides, edge nodes are exposed to a higher risk of privacy breaches compared to cloud nodes since they are usually deployed in open environments such as public places and industrial control systems [7,8]. In workflow scheduling algorithms, workflows are typically modeled as directed acyclic graphs (DAGs). Determining the order of task execution in the DAG and mapping tasks to resources has been proven and known as a typical NP-hard problem [9], which is difficult to obtain optimal solutions in polynomial time. Therefore, designing a scheduling algorithm that schedules applications with a DAG structure to satisfy the requirements of real-time, reliability, and privacy in edge computing environments has become an urgent issue with research significance.

To simultaneously meet the real-time and security requirements of scheduling in edge environments, we aim to find an efficient scheduling algorithm and modify it to have the capability of privacy-aware scheduling. A just-in-time (JIT) scheduling approach, which postpones the assignment of tasks until they are prepared for execution, allowing task allocation decisions to be made on-the-fly, is an ideal approach for highly heterogeneous environments [10]. Among JIT scheduling algorithms, Zheng et al. [11] proposed a Priority-Based (PB) scheduling algorithm, which has been indicated to be more effective than its competitors and is suitable for any topology structure. In this work, the PB algorithm has been extended to be applied to workflow scheduling scenarios in edge environments and addressed the issue of scheduling privacy tasks. As a result, we achieved better scheduling performance than other dynamic scheduling algorithms. This paper has three main contributions as follows:

- We have improved the calculation of task priority of the PB algorithm by incorporating critical path analysis, with the goal of minimizing application completion time and making it more suitable for the QoS of makespan in edge environments, resulting in enhanced scheduling performance.
- In the context of workflow scheduling in edge environments, we have taken into account the scheduling demands for privacy tasks within applications and optimized the processor selection strategy of the PB algorithm series. By ensuring the privacy requirements of tasks while further reducing application completion time, we have achieved significant improvements in scheduling performance.
- Extensive simulation experiments have been conducted for comparing the improved algorithm with the PB series algorithms in edge environments, and the experimental results verified their performance differences, proving the superiority of our algorithm.

The paper is structured as follows: Sect. 2 reviews related work, Sect. 3 discusses problem modeling, Sect. 4 presents the novel PAPBS heuristic algorithm, Sect. 5 evaluates the algorithm's performance through simulations, and Sect. 6 concludes the paper.

2   Related Work

Efficient scheduling algorithms are required to flexibly schedule DAG applications in highly diverse and dynamic distributed infrastructures such as edge computing, grid computing, and cloud computing.

Workflow scheduling algorithms can be categorized as either static or dynamic, depending on whether the scheduling mapping of tasks is before or during workflow execution. In static scheduling algorithms, HEFT [12] divides task priorities into two parts, considering the computation time and communication time of tasks, and sorts them by priority to select the most suitable task and assign it to the server that can complete the task as early as possible. Static scheduling algorithms are more usually adopted in a situation where tasks have relatively fixed numbers and predetermined execution time. In dynamic algorithms, ICO algorithm [13] aimed at achieving the IC optimal target, which is to select the task with the largest number of ready-to-execute subtasks after the selected task is completed. Cordasco et al. [14] improved the ICO algorithm and proposed the AO algorithm. The AO algorithm first transforms the task DAG into an SP-DAG and then uses the idea of region-oriented scheduling, which is to make the average number of ready-to-execute tasks achieve the maximum value, to generate a scheduling sequence. Based on the above two works, Zheng et al. [11] proposed the PB algorithm, which is a priority-based heuristic algorithm. The algorithm considers the total impacts potentially acting on descendant tasks after the current task is executed, and calculates three indicator values for each task: direct indicators, hierarchical indicators, and potential indicators to comprehensively measure task priorities. The experimental results demonstrate PB's superior scheduling performance compared to ICO and AO, and PB is easy to implement without complex DAG decomposition steps, suitable for DAGs of any topology type. Based on the PB algorithm, there were a series of algorithms that further improve its scheduling performance. The APB algorithm [15] improved the calculation of priority indicators by incorporating estimated task execution time. The EPB algorithm [16] was proposed to comprehensively consider estimated task execution time and estimated communication time between tasks, and rediscover the sequence of PB algorithm's three indicators. Due to the superiority of the PB algorithm itself and the continuously improved scheduling performance, we choose to propose our scheduling algorithm for edge environment based on the PB algorithm.

The algorithms mentioned above primarily focus on minimizing completion time as the optimization objective. However, with the increasing emphasis on data security, privacy issues have become critical in distributed environments. [17] built a security-enhanced scheduling system for a hybrid cloud environment, consisting of three modules. The security enhancement module was designed to improve the security of scheduling services while reducing scheduling costs without improving application completion time. In another work. [18] proposed a security-aware scheduling method that satisfies certain security requirements and reduces the secure task placement threat of task interactions in a heterogeneous cloud environment. Literature [19] first proposed a multi-objective scheduling algorithm that simultaneously optimizes task execution time and monetary cost under the premise of satisfying privacy constraints in the cloud. Paper [20] designed for a four-tier architecture, which includes IoT, mist, fog, and cloud layers and presents a scheduling heuristic for real-time workflow jobs that handle IoT data with different security needs. However, these studies have not directly placed the computing of the workflow on the edge nodes which are closer to the user side where data originates, but still rely more on cloud or fog nodes. Additionally, there is still a lack of research on balancing real-time optimization and security demands in dynamic scheduling. Our main objective is to propose a dynamic scheduling algorithm that handles security issues in the edge environment while meeting real-time requirements.

3   Problem Modeling

3.1   Application Model

The privacy workflow applications we discuss can be represented as a DAG graph model described as follow: a DAG $G =< V_G, E_G, D_G, P >$. Specifically, the set of nodes in the DAG graph V_G is defined as follows: $V_G = \{v_1, v_2, v_3, ..., v_n\}$, $v_i \in V_G$ is the ith task of DAG graph, and the information of each node is represented as $v_i = \{i, name_i, in, out\}$, where in represents the in-degree of the current node, out represents the out-degree of the current node, and a node with an in-degree of 0 is the entrance node, a node with an out-degree of 0 is the exit node. The set of directed edges in the DAG graph E_G is defined as follows: $E_G = \{e_1, e_2, e_3, ..., e_m\}$, $e_m \in E_G$ represents the number of directed edges, each element in E_G is an ordered triple $e_{i,j} =< v_i, v_j, c_{i,j} >$, where $e_{i,j}$ represents the edge pointing from node i to node j, and $c_{i,j}$ indicates the volume of data transmitted between node i and node j. If there exists a path from node i to node j, then node i is considered the parent node, while node j is classified as the child node. The set of computational loads D_G is defined as follows: $D_G = \{d_1, d_2, d_3, ..., d_n\}$, where n denotes the total number of nodes and the computational load of node v_1 is represented by d_1. Since privacy data is only placed in trusted nodes, these tasks are designated to be executed in trusted nodes and are therefore called tasks with privacy requirements. The set of privacy tasks P is defined as follows: $P = \{p_1, p_2, ..., p_j\}$, where j is less than or equal to n. Privacy tasks represent tasks that include privacy data to process. For a specified node v, its earliest start time is represented as EST_v, where the earliest start time for the starting point is 0. LST_v represents the latest start time of node v. The EST_v and LST_v is calculated as

$$EST_v = \begin{cases} 0 , v \in S_{entry} \\ \max_{u \in Pred(v)} \{EST_u + DT_{u,v} + ET_u\} , v \in otherwise \end{cases} \quad (1)$$

$$LST_v = \begin{cases} 0 , v \in S_{exit} \\ \min_{u \in Succ(v)} \{LST_u - DT_{u,v} - ET_u\} , v \in otherwise \end{cases} \quad (2)$$

where $DT_{u,v}$ indicates the estimated communication time between node u and node v, ET_u indicates the estimated execution time of node u. If $EST_v = LST_v$, the directed edge pointing from node u to node v is part of the critical path, and nodes u and node v are critical nodes. Critical nodes are important elements that determine the overall completion time of the DAG.

3.2 Resource and Processing Time Model

Our proposed trusted edge system consists of a network architecture that includes communication links between edge nodes, wireless access points, intermediate switching devices, and device nodes. Edge nodes refer to heterogeneous servers and base station resources, where DAG tasks are offloaded from wireless access points for execution [21]. To ensure privacy, specific edge nodes are equipped with privacy modules and marked as trusted nodes. Only these nodes can execute tasks with privacy requirements.

In the context of a trusted edge environment, the server and base station are distributed in different edge devices to form edge nodes. These nodes are interconnected through wired links with high-speed bandwidth, providing high-speed and stable data transmission. Nodes can collaborate with each other to complete data processing and storage. Edge nodes communicate with each other through a metropolitan area network, and there is no communication competition between them. Therefore, the average network transmission rate can be estimated using existing mechanisms [22]. Hence, this paper reasonably assumes a static network environment, where the average network bandwidth between edge node m and node n is represented as $b_{m,n}$.

Two scenarios are considered when calculating the transmission time between tasks. Taking tasks u and v as an example, if both tasks are executed on the same edge node, then the transmission time is 0, ignoring the communication overhead between tasks within the edge node. If tasks u and v are executed on different nodes m and n, we denote the data transmission volume between them as $c_{u,v}$. The transmission time is $\frac{c_{(u,v)}}{b_{(m,n)}}$. So the transmission time can be formulated as

$$T_{U,V} = \begin{cases} \frac{c_{u,v}}{b_{m,n}} &, if N_m \neq N_n \\ 0 &, if N_m = N_n \end{cases} \tag{3}$$

where N_m is node m and N_n is node n, $c_{u,v}$ is the size of data that needed to be transferred.

The processing speed of each edge node is represented as a collection: $F = \{f_1, f_2, f_3, ..., f_m\}$, where m represents the total number of edge nodes. Let the mth edge node be a designated trusted node, which is granted permission to execute privacy tasks. The designated trusted node can not only execute privacy tasks but also non-privacy tasks, while the rest of the edge nodes can only execute non-privacy tasks. f_k represents the processing speed set of the kth edge node, where each node k has a processing speed for different tasks, denoted as $f_k = \{f_{k,1}, f_{k,2}, f_{k,3}, ..., f_{k,n}\}$. Here, n represents the total number of tasks, and each task v_i has a computation amount defined as d_i, so its processing time on node k is $AET_{k,i} = \frac{d_i}{f_{k,i}}$.

4   Method

4.1   Task Selection Phase

In PB series algorithms, the priority of a task depends on three indicators: the direct quotient (DQ), the level quotient (LQ), and the potential quotient (PQ). The DQ is defined as the number of subsequent ready tasks for a given task after it is executed. The LQ is defined as the maximum length from a task to the exit task, which has an LQ of 0. The PQ is defined as a task's contribution to subsequent ready child nodes. If a task can make more tasks become ready faster after it is completed, then the task has more priority. PB series algorithms determine a task's priority by combining these three indicators and using a decision tree for comparison.

However, this method ignores the critical path's influence on task completion time. In the workflow, tasks have varying completion time and dependencies exists between tasks. If there exist certain activities whose completion time can affect the overall efficiency of the workflow, either making it faster or slower, then these activities can be referred to as critical activities. The sequence of critical activities in the workflow constitutes the critical path, which determines the shortest possible time required to complete the entire workflow. Considering that extending the execution time of critical activities may result in a delay in completing the entire DAG application, PAPBS categorizes tasks as critical or non-critical. Critical tasks are located on the critical path, while non-critical tasks are not. PAPBS uses the PB algorithm's original priority calculation method but assigns higher priorities to critical task nodes.

We use an example with eight task nodes to explain the superiority of the PAPBS algorithm, as shown in Fig. 1.

Figure 1 shows 8 nodes. The edges between the nodes represent the estimated transmission time, and the values on the nodes represent the estimated execution time. Based on the graph, the earliest start time, latest start time, and corresponding LQ for each node are calculated, and the numerical results are presented in Table 1.

Table 1. Table of Execution Time and LQ Value for Each Node.

Node	V0	V1	V2	V3	V4	V5	V6	V7
EST	0	3	2	2	6	4	8	10
LST	0	3	3	3	6	5	8	10
LQ	11	8	7	8	5	6	3	1

According to critical path theory, the critical path for the DAG depicted in Fig. 1 is $v0 - v1 - v4 - v6 - v7$. However, the APB and EPB algorithms

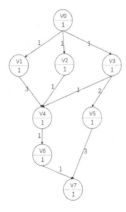

Fig. 1. A DAG example with 8 nodes

prioritize the LQ value of each node, and the LQ value remains constant during scheduling. Additionally, for nodes with the same LQ, such as $v1$ and $v3$, $v3$ has higher priority due to its larger RQ value. In both APB and EPB algorithms, RQ is defined the same as DQ mentioned earlier. Therefore, when scheduling the above example, the APB and EPB algorithms generate a scheduling sequence S_1 of $v0 - v3 - v1 - v2 - v5 - v4 - v6 - v7$. If, upon meeting the ready conditions, tasks $v1$ and $v4$ on the critical path can be given priority in scheduling, then the resulting scheduling sequence S_2 would be $v0 - v1 - v3 - v2 - v4 - v5 - v6 - v7$.

We assume that resources arrive in batches, with 1, 1, 2, 2, 1, 1, 1 resources in each batch and a two-time unit arrival interval between batches. Table 2 displays the execution details for S_1 and S_2. It is evident that prioritizing critical path tasks in the scheduling sequence can lead to a shorter makespan in this case.

Table 2. Execution process for schedule S_1 and S_2.

	Number of Res	Execution for S_1	Execution for S_2
batch0	1	t_0	t_0
batch1	1	t_3	t_1
batch2	2	t_1, t_2	t_3, t_2
batch3	2	t_5	t_4, t_5
batch4	1	t_4	t_6
batch5	1	t_6	t_7
batch6	1	t_7	

We can see that the APB and EPB algorithms delayed critical tasks on the critical path, including $v1$ and $v4$, resulting in a longer overall completion time for the application.

Next, we will describe the three steps of the algorithm for generating the scheduling sequence. Firstly, critical nodes are determined. Let ET_v denote the estimated execution time of task v, and $DT_{u,v}$ denote the estimated communication time between task u and task v. Both ET_v and $DT_{u,v}$ are assumed to follow a normal distribution. Based on ET_v, $DT_{u,v}$, and the dependency relationship between nodes in the DAG, the total slack time TF of each task is calculated as $TF = LST_u - EST_u$. Nodes with $TF = 0$ are identified as critical path nodes, marking them as critical nodes. The procedure of determining the critical nodes is given in Algorithm 1. The next step involves calculating three indicator values for each node in the DAG graph. We follow PB's initialization method but with slight modifications during the update process, due to the improved processor selection algorithm described in the next section. After computing the critical path and initializing the indicators, nodes with no predecessors are added to the ready task pool. Task scheduling is the third step, where critical tasks in the pool are given priority. If there are multiple critical tasks, their indicators are compared using PB's decision tree. Otherwise, indicators of each ready task are compared based on the same decision tree to select the next task for scheduling. The system will wait for resources, if necessary. The pseudocode for this step is provided in Algorithm 2.

Algorithm 1: Determine The Critical Nodes

Input: A DAG application G
Output: All critical nodes in DAG
1 Perform a topological sort on the DAG graph;
2 Initialize the earliest start time (EST) of each task node to 0 and the latest completion time (LCT) to infinity;
3 Create a queue Q to store nodes with an in-degree of 0.;
4 **while** *Q is not empty* **do**
5 　Compute the earliest start time(EST) of node v in Q;
6 　**for** *each child node x of v* **do**
7 　　add new ready task x into Q;
8 　　remove v from Q;
9 　**end**
10 **end**
11 Create a queue H to store nodes with out-degree 0, and traverse the original DAG graph.;
12 **while** *H is not empty* **do**
13 　Compute the latest start time(LST) of node v in H;
14 　**for** *each parent node x of v* **do**
15 　　add new ready task x into H;
16 　　remove v from H;
17 　**end**
18 **end**
19 Traverse the DAG graph;
20 Mark the nodes where TF=0, with their k value set to 1;

Algorithm 2: Task Selection

Input: A DAG application G
Output: A task for next execution
1 Create a ready queue Q and a critical ready queue KQ;
2 Calculate the initial DQ, LQ, EQ, and IQ for each node in G;
3 Add the entry task to Q;
4 **while** *Q or KQ is not empty* **do**
5   **if** *KQ is not empty* **then**
6   Schedule task v with the highest priority P in KQ;
7   Remove v from KQ;
8   Remove v from Q;
9   **else**
10   Schedule the task v in Q with the highest priority P;
11   Remove v from Q;
12   **end**
13   **for** *each new ready task k* **do**
14   Add k into Q;
15   **if** *k is critical task* **then**
16   Add k into KQ;
17   **end**
18   **end**
19 **end**

Compared with the PB series algorithm, the PAPBS algorithm based on the critical path has two significant advantages: 1) it narrows the range of selectable ready tasks, improves the execution speed and efficiency of the entire dynamic scheduling algorithm; 2) it incorporates the influence of the critical path in the DAG graph, leading to better scheduling solutions.

4.2 Processor Selection Phase

After establishing task priorities, selecting the appropriate processor is crucial to fully utilize resources and complete tasks ahead of schedule. Due to the high heterogeneity of edge nodes and specific processing requirements of privacy tasks processors., we must consider the impact of edge node attributes on processing and transmission time when selecting.

Before scheduling, the processor selection algorithm estimates the execution time of tasks on different nodes using a processing time model in Subsect. 3.2. In addition, to identify and validate trusted nodes, each node has a unique identity identifier, and the scheduler maintains a set of trusted nodes, where the set contains the identifiers of trusted nodes. Next, PAPBS uses the insertion method to select the edge node that can complete the task first after task selection. If the task is a privacy task, only the trusted node's gaps are traversed. Otherwise, the scheduler would traverse the resource gaps of each node and obtain the processing time of each node for the current task. Let ACT_j denote the actual

---

Algorithm 3: Processor Selection

Input: DAG G; task execution time set S, communication amount between tasks, communication bandwidth between nodes

Output: Processor assigned to the task

1 Create a list to store the set of unscheduled nodes;
2 Find the node with the shortest execution time for each task across all nodes, update the task-node mapping, and update the communication time between tasks.;
3 **while** L *is not empty* **do**
4   Select tasks based on Algorithm 3 to obtain the pending tasks v_i;
5   **if** *tasks v_i is privacy task* **then**
6   Traverse gaps of designated processor p and calculate ACT_i using the insertion method;
7   **else**
8   Traverse all processor gaps and calculate ACT_i using the insertion method;
9   **end**
10   Schedule the task to the selected processor and update the processor's time slot;
11   Update the execution time, selected node, and completion time of task vi;
12   Compute and update the LQ and PQ of task v_i;
13   remove v_i from L;
14   update G;
15 **end**

---

completion time of task v_j. Let MCT_i denote the maximum time required for all predecessor tasks to transfer data to the current node i. MCT_i is calculated as

$$MCT = \max_{j \in pred(i)} \{DT_{i,j} + ACT_j\} \tag{4}$$

Let bt denote the earliest starting time for node resource gaps that satisfy the current task's predecessor dependency. Let $FT_{k,i}$ denote the completion time of task v_i at the current node k. FT_k is calculated as

$$FT_{k,i} = \max\{bt, MCT_i\} + AET_{k,i} \tag{5}$$

where MCT is derived by Eq. (4).

During the search for gaps in node resources, if a gap does not meet the duration required for the current task, the search continues downward until a suitable gap is found for insertion. If no gaps are found, the task is inserted at the end of the queue of scheduled tasks, provided that its dependencies have been met.

The above process calculates the processing time of a given task on each edge node. The node with the shortest completion time is then selected as the execution node for the task.

The formula for calculating the completion time of task i is as follows:

$$DT_{U,V} = \begin{cases} \min_{k \in m} \{FT_{k,i}\} & , i \in otherwise \\ FT_{m,i} & , i \in P \end{cases} \quad (6)$$

After obtaining the execution time of the currently scheduled task, it is necessary to update the priority indicators of the unscheduled tasks, which will affect the selection of tasks for the next scheduling. Since the initial stage uses the ideal shortest execution time of tasks, the actual task execution time may change, which will affect the LQ and PQ indicators of tasks. Therefore, the LQ and PQ of the scheduled tasks are updated first, followed by the LQ and PQ of the unscheduled tasks.

Then, continue updating the DAG graph and scheduling the new tasks until all tasks are completed. When scheduling a task that has privacy requirements, there is no need to traverse other edge nodes. Instead, it can be directly inserted at the end of the scheduling queue of a trusted node, provided that pre-dependencies are met. The pseudocode is as Algorithm 3.

In comparison to the PB series algorithm, we have made innovative enhancements to the resource selection approach of PB, taking into account the privacy demands of the edge environment. After obtaining the task priority and the task to be processed, we schedule it to the edge node with the earliest potential to finish it and consider the resource availability of the edge node. This ensures the optimal makespan of the entire DAG application while maximizing the utilization of node resources.

5   Evaluation

Due to the improvements made to the original PB algorithm in both stages, this section will validate the superiority of the algorithm through simulation experiments in two different application scenarios and with different resource settings. Privacy applications refer to workflow applications that include privacy tasks for processing sensitive data, while general applications refer to workflow applications that do not include any privacy tasks. The algorithm not only improves the performance of privacy application scheduling in a specific scenario but also achieves better performance in general application scheduling scenarios.

5.1   Settings

This paper empirically evaluates the proposed algorithm's performance and demonstrates its applicability using three common DAG configurations which are CBBBC-DAG, SP-DAG, and LRD-DAG. This paper applies the algorithm from paper [23] to randomly combine different CBBBs structures into a CBBBC-DAG graph. Figure 2 depicts common CBBBs building blocks. SP-DAG is also generated based on the approach proposed in paper [23], as depicted in Fig. 3. The left portion of the figure comprises two DAGs connected in series and then merged, while the right portion comprises two DAGs connected in parallel and then merged. The LRD-DAG (Levelized Random DAG), illustrated in Fig. 4, has a clear hierarchical structure. This paper generates it using the algorithm proposed by Zheng et al. [15].

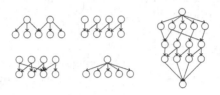

Fig. 2. This CBBBC-DAG is created using four types of CBBBs, the left image shows these four building blocks.

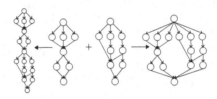

Fig. 3. An example of SP-DAG serial merge and parallel merge process.

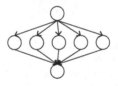

Fig. 4. An example of an LRD-DAG with 7 nodes.

For the experiments on privacy applications, we will use the CBBBC-DAG application type. Compared to the other two types of DAGs, the CBBBC-DAG has a higher degree of node parallelism, which can better differentiate algorithm performance. We will design DAGs of sizes 50, 100, 200, and 300 and privacy task numbers 5, 10, 15, and 20, resulting in 16 different DAG application configurations. The total number of edge nodes is set to 3, 4, and 5 with one trusted node. These 16 application configurations will be applied to edge nodes, and each experiment will be conducted 100 times, with the average value taken for analysis.

For the experiments on general applications, in order to verify their generality, another two types of DAGs were used: SP-DAG and LRD-DAG. Regarding the DAG size, instances ranging from 100 to 1000 nodes were generated with a step size of 100, resulting in a total of 30 instances. Considering that the APB algorithm, EPB algorithm, and PAPBS algorithm require additional information, the estimated execution time (ET) and estimated communication time (DT) of the task nodes were randomly assigned. Two types of resource configurations were generated: low heterogeneity resources and high heterogeneity resources. For the low heterogeneity resources, ET and DT followed a normal distribution on [1,9], while for the high heterogeneity resources, ET and DT followed a normal distribution on [1,99].

The way resources arrive can affect the execution time of the entire DAG application, and the experiment is conducted in batch mode for comparison. Considering the unpredictability of edge node resources, the number of available resources r_i for each batch is modeled using an exponential distribution. The batch size, which represents the available resource quantity, is also selected from an exponential distribution with different rate parameters: $\lambda = 1/2, 1/4, 1/8, 1/16, 1/32, 1/64$, and $1/128$, respectively. This results in an average batch size (μ) of 2, 4, 8, 16, 32, 64, and 128, respectively. For each batch of resource arrival time, this paper selects the time interval between adjacent batches is selected from an exponential distribution with a rate parameter of $\gamma = 1/10$. It implies that, on average, resources arrive once every 10-time units for each batch.

5.2 The Performance Comparison Experiment for Privacy Application

In this section, the PB algorithm is used as the baseline for the experiments. Since the HEFT algorithm is also a rank-based method based on critical paths, it is chosen as the comparative algorithm for this section, along with the heuristic algorithm PAPBS we proposed. The performance of the PAPBS algorithm is evaluated through four sets of experiments that quantify task completion time. These experiments examine the algorithm's behavior under different conditions, such as variations in DAG size, number of edge nodes, and number of privacy tasks.

To better illustrate the differences between the algorithms, the application completion time ratio is defined as follows:

$$CTR = \frac{makespan}{T(PB)} \tag{7}$$

We conducted three comparative experiments to evaluate the performance of the PABPS algorithm and its competing algorithms under different scheduling configurations, and investigated the reasons for the differences in algorithm performance under different configuration conditions, considering three primary influencing factors: DAG size, edge node number, and privacy task number. The experimental results are presented in Figs. 5 and 6.

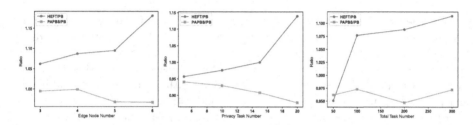

Fig. 5. The completion time ratio of a workflow changing with different variables. (Left: edge node number, Mid: privacy task number, Right: task number)

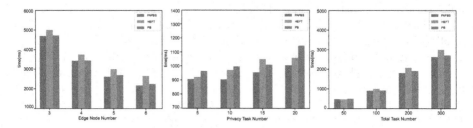

Fig. 6. The completion time of a workflow changing with different variables. (Left: edge node number, Mid: privacy task number, Right: task number)

Based on the experimental results, we derived the following two observations: The PABPS algorithm exhibited superior scheduling performance under different DAG sizes and resource quantities due to its consideration of resource heterogeneity. In edge environments, resource heterogeneity affects the actual execution time of tasks, which is overlooked by PBS and HEFT algorithms when creating scheduling plans. However, PABPS considers this point and uses an insertion-based dynamic scheduling and processor allocation method, allowing it to make reasonable scheduling decisions for each task even when the number of tasks increases. Moreover, with an increase in resources, PABPS has an advantage due to its ability to make full use of resources, resulting in a more significant decrease in completion time.

The PABPS algorithm has the potential to schedule high-privacy data volume DAG applications. Privacy tasks must be executed on trusted nodes, reducing parallelism and increasing serialization. However, even with an increase in the number of privacy tasks, PABPS can still maintain a short completion time due to its ability to dynamically update each task's metric based on its actual execution time and consideration of privacy tasks' impact on the critical path. In contrast, the HEFT algorithm does not consider resource heterogeneity, and the PBS algorithm does not consider the impact of task execution time on priority, resulting in increased time as parallelism is weakened due to an increase in the number of privacy tasks.

Overall, compared to the PBS algorithm, the PABPS algorithm improves performance by 1%-13%, and compared to the HEFT algorithm, it improves performance by 2%-18%, with the greatest advantage observed in scheduling scenarios with abundant node resources and a high volume of privacy tasks.

5.3 The Performance Comparison Experiment for General Application

To visually compare the scheduling performance of algorithms, this section uses the ratio defined in Eq. (8) as the evaluation metric, where the batched_makespan average of the PB algorithm is used as the denominator, and the numerator is the average batched_makespan of each comparative algorithm. If R is less than 1, it indicates that the algorithm performs better than the PB algorithm, and if R is greater than 1, it indicates that the algorithm performs worse than the PB algorithm.

$$R_x = \frac{T(X)}{T(PB)} \tag{8}$$

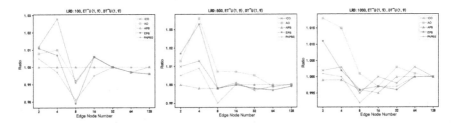

Fig. 7. Comparative experiment using LRD-DAG with different sizes under low heterogeneity.

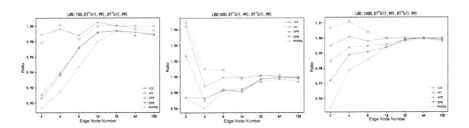

Fig. 8. Comparative experiment using LRD-DAG with different sizes under high heterogeneity.

Figures 7, 8 and 9 illustrates the performance of seven algorithms on LRD-DAG and SP-DAG applications, from which we can draw the following two conclusions:

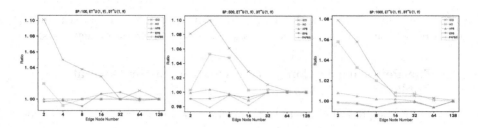

Fig. 9. Comparative experiment using SP-DAG with different sizes under low heterogeneity.

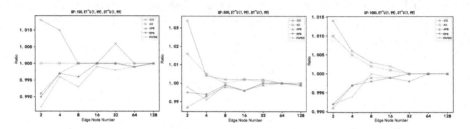

Fig. 10. Comparative experiment using SP-DAG with different sizes under high heterogeneity.

For general-purpose applications, the PAPBS algorithm demonstrates a significant advantage in scheduling highly heterogeneous DAG applications. In high-heterogeneity scenarios, compared to the ICO, AO, and PB algorithms that are insensitive to the heterogeneity of DAGs, and the APB and EPB algorithms that are sensitive to it, the PAPBS algorithm exhibits a clear superiority in the completion time metric, especially for LRD-type DAGs, when the resource quantity is limited. In low-heterogeneity scheduling, the PAPBS algorithm performs similarly to the PB algorithm but generally outperforms the APB, EPB, AO, and ICO algorithms (Fig. 10).

For general-purpose applications, the PAPBS algorithm exhibits superior scheduling performance compared to other algorithms when resources are insufficient. In fact, the performance of each algorithm is best reflected when the resource quantity is limited, especially for resource providers in edge environments, where resources are often limited due to higher monetary costs. As resources become more abundant, the performance of the various algorithms gradually becomes similar, because as the number of resources per batch increases, the resource saturation required for task execution is gradually reached, and all ready tasks can be maximally scheduled, reducing the importance of priority allocation. Therefore, the performance differences among all algorithms are not significant when the resource quantity is maximal. The significant advantage of the PAPBS algorithm over its competitive algorithms in low-resource scenarios indicates its potential for practical edge scheduling scenarios.

6   Conclusion

In this paper, we propose the PAPBS algorithm, which addresses the privacy-aware scheduling problem in edge environments. Our algorithm enhances task priority calculation of the PB algorithm by incorporating critical path analysis, which minimizes application completion time and makes it more suitable for the QoS of makespan in edge environments. We also optimize the processor selection strategy of the PB series algorithm while considering the scheduling requirements of privacy tasks. Our extensive simulation experiments confirm the superiority of the PAPBS algorithm compared to the PB series algorithms and the HEFT algorithms in edge environments, as it further reduces application completion time while satisfying the privacy scheduling requirements.

References

1. Alameddine, H.A., Sharafeddine, S., Sebbah, S., Ayoubi, S., Assi, C.: Dynamic task offloading and scheduling for low-latency IoT services in multi-access edge computing. IEEE J. Sel. Areas Commun. **37**(3), 668–682 (2019)
2. Lin, K., Lin, B., Chen, X., Lu, Y., Huang, Z., Mo, Y.: A time-driven workflow scheduling strategy for reasoning tasks of autonomous driving in edge environment. In: 2019 IEEE International Conference on Parallel & Distributed Processing with Applications, Big Data & Cloud Computing, Sustainable Computing & Communications, Social Computing & Networking (ISPA/BDCloud/SocialCom/SustainCom), pp. 124–131. IEEE (2019)
3. Alsurdeh, R., Calheiros, R.N., Matawie, K.M., Javadi, B.: Hybrid workflow provisioning and scheduling on cooperative edge cloud computing. In: 2021 IEEE/ACM 21st International Symposium on Cluster, Cloud and Internet Computing (CCGrid), pp. 445–454. IEEE (2021)
4. Xu, X., Cao, H., Geng, Q., Liu, X., Dai, F., Wang, C.: Dynamic resource provisioning for workflow scheduling under uncertainty in edge computing environment. Concurrency Comput. Pract. Experience **34**(14), e5674 (2022)
5. Li, H., Ota, K., Dong, M.: Learning IoT in edge: deep learning for the internet of things with edge computing. IEEE Network **32**(1), 96–101 (2018)
6. Stojmenovic, I., Wen, S., Huang, X., Luan, H.: An overview of fog computing and its security issues. Concurrency Comput. Pract. Experience **28**(10), 2991–3005 (2016)
7. Aljumah, A., Ahanger, T.A.: Fog computing and security issues: a review. In: 2018 7th International Conference on Computers Communications and Control (ICCCC), pp. 237–239. IEEE (2018)
8. Ferrag, M.A., Derhab, A., Maglaras, L., Mukherjee, M., Janicke, H.: Privacy-preserving schemes for fog-based IoT applications: Threat models, solutions, and challenges. In: 2018 International Conference on Smart Communications in Network Technologies (SaCoNeT), pp. 37–42. IEEE (2018)
9. Wangsom, P., Lavangnananda, K., Bouvry, P.: Multi-objective scientific-workflow scheduling with data movement awareness in cloud. IEEE Access **7**, 177063–177081 (2019)
10. Liu, J., Pacitti, E., Valduriez, P., Mattoso, M.: A survey of data-intensive scientific workflow management. J. Grid Comput. **13**, 457–493 (2015)

11. Zheng, W., Tang, L., Sakellariou, R.: A priority-based scheduling heuristic to maximize parallelism of ready tasks for DAG applications. In: 2015 15th IEEE/ACM International Symposium on Cluster, Cloud and Grid Computing, pp. 596–605. IEEE (2015)

12. Topcuoglu, H., Hariri, S., Wu, M.Y.: Performance-effective and low-complexity task scheduling for heterogeneous computing. IEEE Trans. Parallel Distrib. Syst. **13**(3), 260–274 (2002)

13. Malewicz, G., Foster, I., Rosenberg, A.L., Wilde, M.: A tool for prioritizing DAG-Man jobs and its evaluation. J. Grid Comput. **5**(2), 197–212 (2007)

14. Cordasco, G., Rosenberg, A.L., D'Ambra, P., Guarracino, M., Talia, D.: Area-maximizing schedules for series-parallel DAGs. In: Euro-Par (2), pp. 380–392 (2010)

15. Zheng, W., Zhang, X., Tang, L., Zhang, D., Chen, J.: An adaptive priority-based heuristic approach for scheduling DAG applications with uncertainties. In: 2017 IEEE International Symposium on Parallel and Distributed Processing with Applications and 2017 IEEE International Conference on Ubiquitous Computing and Communications (ISPA/IUCC), pp. 72–79. IEEE (2017)

16. Zhang, X., Zhang, D., Zheng, W., Chen, J.: An enhanced priority-based scheduling heuristic for DAG applications with temporal unpredictability in task execution and data transmission. Futur. Gener. Comput. Syst. **100**, 428–439 (2019)

17. Hammouti, S., Yagoubi, B., Makhlouf, S.A.: Workflow security scheduling strategy in cloud computing. In: Chikhi, S., Amine, A., Chaoui, A., Saidouni, D.E., Kholladi, M.K. (eds.) MISC 2020. LNNS, vol. 156, pp. 48–61. Springer, Cham (2020). https://doi.org/10.1007/978-3-030-58861-8_4

18. Abazari, F., Analoui, M., Takabi, H., Fu, S.: Mows: multi-objective workflow scheduling in cloud computing based on heuristic algorithm. Simul. Model. Pract. Theory **93**, 119–132 (2019)

19. Wen, Y., Liu, J., Dou, W., Xu, X., Cao, B., Chen, J.: Scheduling workflows with privacy protection constraints for big data applications on cloud. Futur. Gener. Comput. Syst. **108**, 1084–1091 (2020)

20. Stavrinides, G.L., Karatza, H.D.: Security and cost aware scheduling of real-time IoT workflows in a mist computing environment. In: 2021 8th International Conference on Future Internet of Things and Cloud (FiCloud), pp. 34–41. IEEE (2021)

21. Lin, L., Liao, X., Jin, H., Li, P.: Computation offloading toward edge computing. Proc. IEEE **107**(8), 1584–1607 (2019). https://doi.org/10.1109/JPROC.2019.2922285

22. Guo, S., Liu, J., Yang, Y., Xiao, B., Li, Z.: Energy-efficient dynamic computation offloading and cooperative task scheduling in mobile cloud computing. IEEE Trans. Mob. Comput. **18**(2), 319–333 (2018)

23. Cordasco, G., De Chiara, R., Rosenberg, A.L.: Assessing the computational benefits of area-oriented DAG-scheduling. In: Euro-Par (1), pp. 180–192 (2011)

24. Kashyap, R., Vidyarthi, D.P.: Security driven scheduling model for computational grid using NSGA-II. J. Grid Comput. **11**, 721–734 (2013)

A Novel Multi-objective Evolutionary Algorithm Hybrid Simulated Annealing Concept for Recommendation Systems

Yu Du, Haijia Bao, and Ya Li[(✉)]

College of Computer and Information Science, Southwest University,
Chongqing 400715, China
`swu_yali@163.com`

Abstract. Nowadays, recommendation systems have been widely used in information systems and internet applications. In order to solve the problem that most traditional recommendation algorithms mainly focus on accuracy and neglect other requirements such as diversity and novelty, multi-objective evolutionary algorithm has been introduced into recommendation system. But there is still much room for improvement in their performance. In this paper, we design a novel multi-objective evolutionary algorithm for recommendation system with accuracy, diversity, and novelty as its objective functions. In the proposed model, an efficient uniform distribution initialization and a mutation operator that integrates the concept in simulated annealing algorithm has been designed to improve the probability of producing more high-quality offspring. And an adaptive hybrid selection strategy is designed to select more valuable or promising individuals from offspring. Experimental results demonstrate the effectiveness of the proposed algorithm in terms of accuracy, diversity, novelty, and hypervolume compared to some state-of-the-art recommendation algorithms.

Keywords: Multi-objective evolutionary algorithm · Recommendation system · Genetic operator

1 Introduction

With increasing use of internet, data in people's daily life is growing explosively. Therefore, it has become crucial to know how to use data reasonably and how to find the content users need in massive data. As an information system, recommendation systems(RSs) can explore user interests and recommend suitable products, information, and services in e-commerce, social media, medical, and many other fields.

Traditional recommendation algorithms, such as collaborative filtering based recommendation systems, analyze users' historical behavior data to find other users or items with similar interests and recommend items that users may be interested in. However, these algorithms mainly focus on providing accurate recommendations and ignore diversity and novelty of recommendation system.

Z. Tari et al. (Eds.): ICA3PP 2023, LNCS 14490, pp. 295–306, 2024.
https://doi.org/10.1007/978-981-97-0859-8_18

This can lead to a single and biased recommendation of popular items. This means there is an imbalance in performance of diversity, novelty, and other performance indicators. Therefore, balancing accuracy, diversity, novelty, and other performance indicators is a challenge for researchers.

In order to solve the problem mentioned above, multi-objective evolutionary algorithm has been introduced to recommendation system. First, traditional recommendation algorithms are utilized to generate user-item rating matrix, such as collaborative filtering [1]. Then, accuracy, diversity and other performance indicators are set as objective functions for multi-objective evolutionary algorithm such as non-dominated sorting genetic algorithm II (NSGA-II) [2]. Finally, a set of recommendation lists with different performances are obtained to achieve a balance between different indicators.

In this paper, we propose a novel multi-objective evolutionary algorithm (HSA-MOEA) to improve the performance of multi-objective recommendation systems. The main contributions of this paper are as follows:

1. An efficient uniform distribution initialization strategy is proposed to make initial recommended item distribution in population more comprehensive and different extreme points are added to guide population evolve towards Pareto-optimal front.
2. A mutation operator that integrates the simulated annealing concept is proposed, which assigns different mutation probabilities to different individuals in different period of iteration.
3. An adaptive hybrid selection strategy is proposed, which combines the crowding distance selection strategy in NSGA-II and the hypervolume contribution selection strategy in FV-MOEA and adaptively switches the two strategies.
4. We conduct experiments on the Movielens dataset to evaluate the proposed HSA-MOEA, and compare it with some well-known recommendation systems. The experimental results indicate HSA-MOEA can effectively improve performance in accuracy, diversity, novelty, and hypervolume.

2   Related Work

Most traditional recommendation algorithms focus only on accuracy and ignore other objectives such as diversity and novelty, which are equally important for both system and users. High accuracy often leads to lower diversity and novelty. To address the conflict between accuracy and non-accuracy indicators [3], multi-objective evolutionary algorithms have been introduced into recommendation systems.

In [4], a multi-objective recommendation algorithm named MOEA-ProbS was proposed. MOEA-Probs achieves a good balance between accuracy and diversity. However, the performance of MOEA-ProbS is not good enough on account of it mainly depending on the framework of NSGA-II and it only considers accuracy and diversity metrics. Cui et al. proposed a probabilistic multi-objective recommendation algorithm (PMOEA) [5], which uses the frequency of gene occurrence in different parents to determine whether the gene should be retained by offspring. This method has some improvements on recommendation, but the calculation of the genetic probability of each gene dramatically increases the required running time. MOEA-EPG [6] introduced extreme points of accuracy, diversity, and novelty in the population initialization stage, i.e., individuals with the best corresponding objective function values, the multi-objective recommendation algorithm is guided to search toward the true Pareto front. This method dramatically improves the performance of the recommendation system with almost no increase in computational complexity. Milojkovic et al. [7] devised a multi-objective RS based on stochastic multigradient descent, in which semantic relevance, quality, and price are considered. Furthermore, Jain and colleagues introduced a multi-objective approach for item evaluation recommendations. Their study developed a novel similarity calculation model to enhance the prediction accuracy of existing methods in rating predictions, while also taking into account the gene order in multi-objective optimization. Ultimately, the proposed multi-objective evolutionary algorithm optimized both the accuracy and topic diversity of the recommendation list. A hybrid probabilistic multiobjective evolutionary algorithm (HP-MOEA) [8] is proposed for commercial recommendation systems. It adaptively combines both NSGA-II and SMS-EMOA to have their complementary advantages so that hypervolume can rise effectively at each stage and some specifically designed genetic operators are proposed. It has better multi-objective recommendation performances in terms of hypervolume and stability. [9] proposed a community-based evolutionary algorithm named ComEA, for large-scale multi-objective recommendations. This approach significantly reduces the search space of MOEA from both row and column perspectives by utilizing community partitioning.

3 Problem Definition

3.1 Multi-objective Evolutionary Algorithm

Multi-objective optimization algorithm is designed to solve the problem consists of multiple conflicting objectives. They can be stated as follows:

$$max \ F(x) = (f_1(x), f_2(x), \ldots, f_m(x))^T, s.t. \ x \in \Omega \quad (1)$$

where Ω is the decision space and $F : \Omega \to R_m$ represents m real-valued objective functions.

On account of objectives always contradict with each other, and it is rare to find a solution with the best fitness value for all objectives. Thus defining Pareto dominance that a solution u dominates another solution v if :

$$\forall i \in \{1, \ldots, m\} \ f_i(u) \geq f_i(v), \exists j \in \{1, \ldots, m\} \ f_j(u) > f_j(v) \quad (2)$$

If there is no solution that dominates the solution u, the solution u could be named Pareto optimal. Furthermore, a set of Pareto optimal is called Pareto front(PF). Finding all of Pareto optimal solutions is hard, while using the multi-objective algorithm to find its approximate set is more feasible.

NSGA-II is the most popular dominance-based MOEA. It applies a non-dominated sorting for the population, and every individual is allocated a non-dominated rank. When selecting individuals from the population into the next generation, the individuals with the first non-dominated rank are selected firstly, and the second non-dominated rank is selected later...until the individuals with nth non-dominated rank can not be selected on account of the limit of population size. And then, the crowding distance of an individual is defined as the perimeter of the cuboid formed by using the neighboring individuals on its PF, and it is used to select individuals with the nth non-dominated rank into the next generation. The individual with a lower crowding distance means better diversity for the population and will be chosen firstly.

FV-MOEA is an indicator-based MOEA with an efficient hypervolume contribution update method applicable to any dimension. Different from NSGA-II, FV-MOEA uses hypervolume contribution as the secondary selection criterion which makes the hypervolume of the population grow steadily.

4   The Proposed Model

The proposed HAS-MOEA is introduced in this section. A flowchart of the proposed algorithm is shown in Fig. 1, which includes two integral components: Item rating evaluation and Multi-objective optimization. In the former part, the ProbS method is used to form the user-item rating matrix. The optimization process is mainly based on the NSGA-II and improved with three essential strategies: efficient uniform distribution initialization, mutation operator integrates simulated annealing concept, and adaptive hybrid selection strategy.

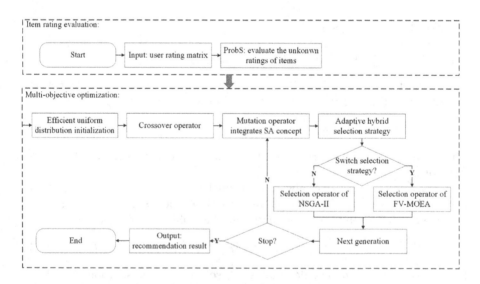

Fig. 1. The flowchart of HSA-MOEA.

4.1 Objective Functions

Objective functions are the core of multi-objective optimization recommendation algorithms, they lead the population to evolve in objective space. In this work, we set accuracy function, diversity function, and novelty function as objective functions. In fact, it is impossible to compute the true preferences of users in the training stage. Therefore, the sum of predicted ratings for users' recommendation lists presents the recommendation system's accuracy. According to the objective functions in [6], the objection function in this work is described in Eq. 3–Eq. 6.

$$PR = \frac{\sum_{u \in U} \sum_{i \in L} r_{u,i}}{|U| \cdot |L|} \tag{3}$$

where $|U|$ is the number of users to recommend, $|L|$ is the length of a recommendation list for every user, and $r_{u,i}$ is the predicted rating of item i given by user u.

The diversity of recommendations reflects the difference between items in a recommendation list. Here, coverage is chosen to evaluate the diversity of recommendations, which is computed as follows:

$$COV = \frac{Nu}{I} \tag{4}$$

where I is the number of item in the system.

Novelty reflects the ability of the system to recommend unpopular items, which can be evaluated by the average self-information of items.

$$Nov = \frac{1}{MN} \sum_{u=1}^{M} \sum_{i \in O_N^u} SI_i \tag{5}$$

$$SI_i = \log_2 \left(\frac{M}{k_i} \right) \tag{6}$$

where M is the number of user, O_N^u is the recommended list with N items generated by the recommendation system for user u. And SI_i is the self-information of the item i.

4.2 Individual Encoding and Initialization

In the proposed algorithm, an individual is encoded as a two-dimensional matrix, where each entry represents an item ID, and the i-th row is a recommendation list for user i generated by the recommendation system. An individual contains all users' recommendation lists.

Given a number M of users in a cluster and a length L of the recommendation list, the individual is encoded by a matrix with size $M \times L$. Generally speaking, the recommendation system will not recommend the same item to a user more than once.

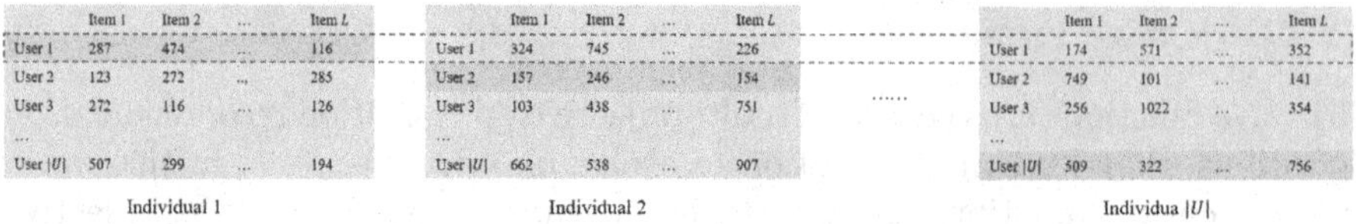

Fig. 2. An example of efficient uniform distribution initialization.

We design a new initialization method named efficient uniform distribution initialization shown in Fig. 2.

Firstly, we need the population to have all items for every user in order to ensure every item can be chosen as a recommended item. For instance, as the dashed box shows in Fig. 2, recommendation items for every individual are selected randomly, but each item needs to be recommended to each user at least once. Secondly, [6] indicates the extreme points with the highest objective fitness value having a positive guiding significance for the evolution direction. Inspired by this, every individual is designed to have one user's recommendation list that consists of the highest predicted rating items to guide the population to evolve. As shown in Fig. 2, the deeper background color means the recommended items are the items with the highest predicted rating. At last, we add two extreme points with the highest accuracy and coverage into the population.

4.3   Genetic Operators

Genetic operator is one of the most critical components, which is helpful to break the local optimality and find the optimal solution. In this work, we design a mutation operator mixing simulated annealing ideology to increase the probability of generating excellent individuals. The uniform crossover [4] is adopted in our approach on account of its effectiveness.

In the traditional evolutionary algorithm, the mutation probability for every individual is the same for the whole period. Inspired by the simulated annealing(SA) algorithm [10], when the solution performs better, it is accepted and when it performs worse there is still a probability p to retain this solution. Probability p is changed depending on two factors: the temperature T and energy level difference ΔE, shown below:

$$p = exp(-\Delta E/T) \tag{7}$$

Inspired by SA, we design the mutation probability for individual i as follows:

$$pm_i = p_0 * exp\left(-\alpha * T * (1/\Delta E_i)\right) + p_{min} \tag{8}$$

$$\Delta E_i = f'_{max} - f'_i \tag{9}$$

$$f'_i = \sum_{x \in \Theta} \frac{f_{i,x} - min(f_x)}{max(f_x) - min(f_x)} \tag{10}$$

Figure 3 indicates the trend of mutation probability changes during the iteration. In Fig. 3, the point is the mutation probability for an individual, and the upper bound and lower bound are the worst and best individual's mutation probability, respectively.

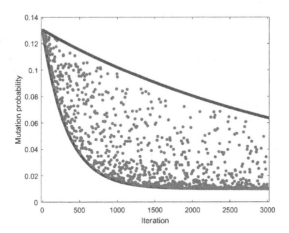

Fig. 3. The changing trend of mutation probability during the iteration process.

4.4 Adaptive Hybrid Selection Strategy

On account of NSGA-II is not effective enough to deal with the situation when the whole population is non-dominated solutions and may result in stagnation for the growth of hypervolume. Thus, we design an adaptive hybrid selection strategy that combines the selection strategy in NSGA-II and FV-MOEA. In the early stage of iteration, we use the selection operator of NSGA-II to quickly improve the hypervolume of the population. When hypervolume growth stagnates for multiple times, the selection operator automatically switches to FV-MOEA to steadily improve hypervolume at the cost of increased time complexity. Specifically, the selection strategy is switched automatically when the hypervolume of the population is less than the maximum hypervolume retained so far for more than 1000 times.

5   Experiment

5.1   Datasets and Metrics

For our experiments, we utilize the classical Movielens dataset, which can be obtained from the GroupLens Research website (http://www.grouplens.org/), to evaluate the performance and effectiveness of HAS-MOEA. This dataset comprises 943 users and 1682 movies, with original ratings ranging from 1 to 5. However, we employed a binary rating system, where a user's rating is classified as either "like" or "dislike". Specifically, a user's rating is considered as "like" if their original rating is greater than 3. We randomly selected 80% of the data as the training set, which contains known information for recommendations, while the remaining data was used as the test set. To ensure a fair comparison, we divided users into four clusters using the k-means algorithm [11], and each cluster is used as the test set. Table 1 provides an overview of the properties of these datasets. Sparsity refers to the total number of rated items in a matrix divided by the product of its rows and columns.

Besides, hypervolume, accuracy, diversity, and novelty are adopted as the performance metrics, which are explained next.

1) HV metric [12] is utilized to assess the convergence and diversity of solutions, which is a commonly employed method for evaluating MOEAs. Formally, it is computed as follows:

$$HV(S) = VOL \left(\bigcup_{x \in S} [f_1(x), z_1^r] \times \cdots \times [f_m(x), z_m^r] \right) \tag{11}$$

where $VOL(\cdot)$ denotes the Lebesgue measure and $(z_1^r, z_2^r, \ldots, z_m^r)$ is the reference point in objective space, which is the point dominated by all solutions. The reference point is set to $(0,0,0)$ in this experiment.

2) In our experiments, accuracy is defined as the ratio of recommended items that are accepted to the total number of items in the recommendation list. In this paper, the Eq. 12 is used as a performance metric to measure the accuracy of RSs.

$$P_u = \frac{NR_u}{L} \tag{12}$$

3) The coverage metric of RSs illustrates the proportion of different items in the recommendation list over all items. In this paper, Eq. 4 is used as a performance metric to measure the diversity of RSs.

4) The novelty exhibited by RSs indicates their ability to suggest unfamiliar items to users. RSs that possess a high level of novelty typically broaden the user's scope of interest. In this paper, Eq. 5 is used as a performance metric to measure the novelty of RSs.

Table 1. Properties of the movielens data sets.

Data sets	Users	Items	Sparsity
Movielens 1	190	1682	0.0747
Movielens 2	254	1682	0.0450
Movielens 3	245	1628	0.0661
Movielens 4	254	1682	0.0694

5.2 Comparison Algorithms and Parameter Settings

HAS-MOEA is compared with three multi-objective optimization recommendation algorithms, these are MOEA-ProbS [4], PMOEA [5], and MOEA-EPG [6] and they are used to consider three objectives mentioned above as accuracy, diversity, and novelty. Besides, two traditional approaches only aiming for accuracy are selected to be compared with, these are Item-based-CF [1] and NBI [13]. ProbS is used for every comparison algorithm to generate the user-item rating matrix. Parameters are set as suggested in their literature listed in Table 2, where L is the number of item recommended to users, T is the number of iteration, p_m is mutation probability, p_c is crossover probability, k is the number of the most similar items for item-based collaborative filtering. Parameter p_0, p_{min}, α is only used in HAS-MOEA, and p_0, p_{min} present initial probability of mutation and minimum probability of mutation, α is the scale factor, λ is threshold value controls whether to switch the selection strategy. Parameter p_n is only used in PMOEA and it represents the number of parents in a crossover operation.

Table 2. The related parameters in comparison algorithms.

Algorithm	Parameter Setting
HSA-MOEA	$L = 10, T = 3000, p_0 = 0.12, p_{min} = 0.01, \alpha = 0.0015, \lambda = 1000, p_c = 0.8$
MOEA-ProbS	$L = 10, T = 3000, p_m = 0.1, p_c = 0.8$
PMOEA	$L = 10, T = 3000, p_m = 0.1, p_c = 0.8, k = 20$
MOEA-EPG	$L = 10, T = 3000, p_m = 0.1, p_c = 0.8$
Item-based-CF	$L = 10$
NBI	$L = 10$

5.3 Experimental Results

HSA-MOEA is compared with five state-of-the-art algorithms (i.e., MOEA-ProbS, PMOEA, MOEA-EPG, Item-based-CF, and NBI) on Movielens data sets. Table 3 lists hypervolume for different multi-objective recommendation algorithms. The best result for each case is highlighted in bold. Due to the fact that item-based CF and NBI can only obtain one optimal value in a single run, HV cannot be used as a comparison metric. We can observe from Table 3 that HSA-MOEA achieves the highest hypervolume value, indicating that it outperforms the other MOEA-based algorithms.

PFs obtained by the HAS-MOEA and other comparison algorithms are shown in Figs. 4 and 5. In our experiments, PF performance comparison was conducted in different objective spaces, namely accuracy-coverage objective space and accuracy-novelty objective space. We can see that the single objective approaches, Item-based-CF and NBI, perform worse than other multi-objective approaches as they only consider one objective. Based on Figs. 4 and 5, it is evident that the PFs of the other three multi-objective recommendation algorithms are nearly enclosed by the PFs of HSA-MOEA in all spaces. As a result, it can be inferred that HSA-MOEA is capable of providing superior recommendations compared to other algorithms in the majority of cases.

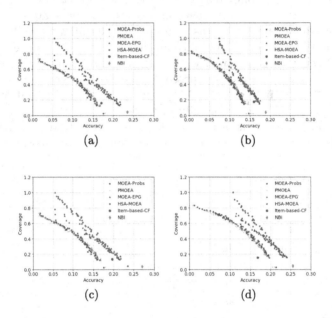

Fig. 4. PFs of comparison algorithms under the Accuracy-Coverage space for (a) Movielens 1, (b) Movielens 2, (c) Movielens 3, (d) Movielens 4.

5.4   Ablation Experiments of HAS-MOEA

In this section, we conduct ablation studies to verify the effectiveness of HAS-MOEA. We compare HSA-MOEA with its three variants. HSA-MOEA-I is the variant without efficient uniform distribution initialization, HSA-MOEA-II is the variant without mutation operator that integrates simulated annealing concept and HSA-MOEA-III is the variant without adaptive hybrid selection strategy.

Table 4 presents the comparative results of the algorithms on HV indicator. It can be seen that HSA-MOEA outperforms other variants of HSA-MOEA in terms of accuracy-diversity-novelty objective space, accuracy-novelty objective space, and accuracy-diversity objective space. In particular, the efficient uniform distribution initialization strategy and mutation operator that integrates the simulated annealing concept significantly impact the algorithm's performance. They promote the effectiveness of evolutionary search in a better direction, proving the proposed HSA-MOEA is effective.

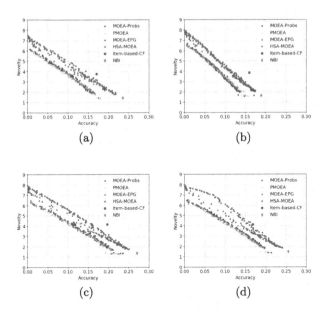

(a)   (b)

(c)   (d)

Fig. 5. PFs of comparison algorithms under the Accuracy-Novelty space for (a) Movielens 1, (b) Movielens 2, (c) Movielens 3, (d) Movielens 4.

Table 3. HV results for different comparison algorithms.

Data sets	Test Instance	MOEA-ProbS	PMOEA	MOEA-EPG	HSA-MOEA
Movielens 1	Accuracy-coverage	0.084	0.085	0.132	**0.145**
	Accuracy-novelty	0.711	0.705	0.964	**1.026**
	Accuracy-coverage-novelty	0.390	0.380	0.622	**0.725**
Movielens 2	Accuracy-coverage	0.081	0.083	0.122	**0.127**
	Accuracy-novelty	0.640	0.664	0.856	**0.883**
	Accuracy-coverage-novelty	0.405	0.408	0.616	**0.672**
Movielens 3	Accuracy-coverage	0.122	0.118	0.172	**0.182**
	Accuracy-novelty	0.885	0.903	1.148	**1.234**
	Accuracy-coverage-novelty	0.570	0.551	0.826	**0.941**
Movielens 4	Accuracy-coverage	0.119	0.115	0.169	**0.178**
	Accuracy-novelty	0.885	0.863	1.124	**1.266**
	Accuracy-coverage-novelty	0.567	0.542	0.827	**0.957**

Table 4. HV result for different HSA-MOEA variants.

Data sets	HSA-MOEA	HSA-MOEA-I	HSA-MOEA-II	HSA-MOEA-III
Movielens 1	**0.725**	0.463	0.658	0.665
Movielens 2	**0.672**	0.475	0.637	0.648
Movielens 3	**0.941**	0.651	0.873	0.874
Movielens 4	**0.957**	0.676	0.863	0.867

6 Conclusions

In order to improve the recommendation performance of multi-objective recommendation algorithms, we have designed a novel multi-objective evolutionary algorithm called HSA-MOEA. In this work, three enhancement strategies are designed. Through comparison with some well-known recommendation algorithms, and experimental results indicate HSA-MOEA has excellent recommendation performance, especially in terms of diversity and novelty. And HSA-MOEA achieves the highest HV in different solution spaces, which presents the PFs obtained by HSA-MOEA has best diversity and are closer to the true PFs.

References

1. Ekstrand, M.D., Riedl, J.T., Konstan, J.A., et al.: Collaborative filtering recommender systems. Found. Trends® Hum. Comput. Interact. 4(2), 81–173 (2011)
2. Deb, K., Pratap, A., Agarwal, S., Meyarivan, T.: A fast and elitist multiobjective genetic algorithm: NSGA-II. IEEE Trans. Evol. Comput. 6(2), 182–197 (2002)
3. Zheng, Y., Wang, D.X.: A survey of recommender systems with multi-objective optimization. Neurocomputing 474, 141–153 (2022)
4. Zuo, Y., Gong, M., Zeng, J., Ma, L., Jiao, L.: Personalized recommendation based on evolutionary multi-objective optimization [research frontier]. IEEE Comput. Intell. Mag. 10(1), 52–62 (2015)
5. Cui, L., Ou, P., Fu, X., Wen, Z., Lu, N.: A novel multi-objective evolutionary algorithm for recommendation systems. J. Parallel Distrib. Comput. 103, 53–63 (2017)
6. Lin, Q., et al.: Multiobjective personalized recommendation algorithm using extreme point guided evolutionary computation. Complexity 2018, 1–18 (2018)
7. Milojkovic, N., Antognini, D., Bergamin, G., Faltings, B., Musat, C.: Multi-gradient descent for multi-objective recommender systems. arXiv preprint arXiv:2001.00846 (2019)
8. Wei, G., Wu, Q., Zhou, M.: A hybrid probabilistic multiobjective evolutionary algorithm for commercial recommendation systems. IEEE Trans. Comput. Soc. Syst. 8(3), 589–598 (2021)
9. Zhang, L., Zhang, H., Liu, S., Wang, C., Zhao, H.: A community division-based evolutionary algorithm for large-scale multi-objective recommendations. IEEE Trans. Emerg. Top. Comput. Intell. 7(5), 1470–1485 (2022)
10. Kirkpatrick, S., Gelatt Jr, C.D., Vecchi, M.P.: Optimization by simulated annealing. Science 220(4598), 671–680 (1983)
11. Yu-xia, L., Jian-ping, L., Guo-xing, Y.: K-means clustering analysis based on genetic algorithm. Comput. Eng. 34(20), 200–202 (2008)
12. Jiang, S., Zhang, J., Ong, Y.-S., Zhang, A.N., Tan, P.S.: A simple and fast hypervolume indicator-based multiobjective evolutionary algorithm. IEEE Trans. Cybern. 45(10), 2202–2213 (2014)
13. Zhou, T., Ren, J., Medo, M., Zhang, Y.-C.: Bipartite network projection and personal recommendation. Phys. Rev. E 76(4), 046115 (2007)

Eco-SLAM: Resource-Efficient Edge-Assisted Collaborative Visual SLAM System

Wenzhong Ou[✉], Daipeng Feng, Ke Luo, and Xu Chen

School of Computer Science and Engineering, Sun Yat-sen University,
Guangzhou, China
{ouwzh3,fengdp3,luok7}@mail2.sysu.edu.cn, chenxu35@mail.sysu.edu.cn

Abstract. Collaboration among multiple smart agents such as robots and UAVs is critical for the key tasks of simultaneous localization and mapping (SLAM), which are essential for many robotics applications. Visual SLAM maps by multiple collaborative agents can be promptly generated with the assistance of edge servers in proximity as the computing infrastructures. However, as the number of agents connected to an edge server continues growing, the pressure on bandwidth consumption and resource utilization also climbs dramatically, which may trigger the failure of map generation. To tackle these challenges, this article proposes Eco-SLAM, a resource-efficient edge-assisted collaborative multi-agent visual SLAM system. Eco-SLAM has been designed to enable large-scale parallelism in SLAM framework and optimized for use in both edge servers and intelligent agents. The unique Core-tr and Co-Map library design ensures efficient utilization of resources and consistent data. Additionally, Eco-SLAM incorporates the ORB-based image compression algorithm, which optimizes data transmission with constrained networking resources. We implement and evaluate the Eco-SLAM system in a real environment and demonstrate its effectiveness through extensive experiments ranging from the public SLAM datasets to realistic deployment scenarios. Thorough evaluations show that Eco-SLAM can reduce the memory consumption of other multi-agent SLAM frameworks by up to 20.1% during runtime on the edge server, and save up to 25.1% wireless bandwidth consumption without compromising the accuracy of the map generation.

Keywords: Edge Computing · Collaborative SLAM · Robotics Applications

1 Introduction

Visual simultaneous localization and mapping (vSLAM) serve as a core technology that can fuse image data collected by visual sensors on smart agents (e.g., robots, UAVs) into a map [23]. In many scenarios, multiple agents are necessary to perceive the surrounding environment, enabling intelligent situation-aware planning and informed operational decisions. Thus, multi-agent collaborative

Z. Tari et al. (Eds.): ICA3PP 2023, LNCS 14490, pp. 307–324, 2024.
https://doi.org/10.1007/978-981-97-0859-8_19

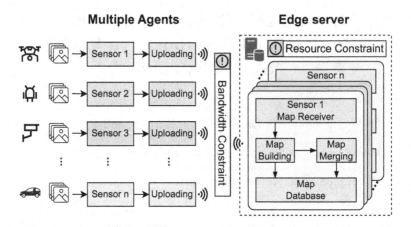

Fig. 1. Overview of challenges in edge-assisted multi-agent collaborative SLAM system. Each agent uploads its captured data to the edge server, then the edge server builds, merges the submaps and finally generates the global map. The wireless bandwidth constraint and the edge server's resource constraint must be carefully considered as the number of participating agents increases.

SLAM has diverse potential applications in various fields, such as smart transportation, home services, manufacturing site inspections, and outdoor rescue operations. However, in practice, many smart agents such as robots and drones typically hold considerably limited power and battery capacity. In order to save their resources as much as possible, the emerging edge intelligence paradigm [26] allows the agents only run light-weight and time-sensitive tasks such as data acquisition and task execution, and offload the data to an edge server (e.g., 5G MEC server) in proximity to accomplish the compute-intensive tasks [25] such as SLAM map generation.

While acknowledging the prominent advantages that edge intelligence brought to multi-agent collaborative SLAM, it is critical to keep in mind that the memory and computational resources of edge server is limited in many application scenarios, and communication bandwidth between the edge server and agents can also be the bottleneck when the network fluctuation is obvious. Specifically, as illustrated in Fig. 1, the growing number of involved agents exacerbates those issues that challenge the serving capacity of edge-assisted collaborative multi-agent visual SLAM systems:

On the one hand, network capacity constrains the number of services provided by a SLAM framework. In a SLAM task, wireless networks are typically used for communication between intelligent agents and edge servers, where network capacity is relatively limited. When a large amount of data converge towards the edge server, network bandwidth will be rapidly consumed, causing problems such as data loss and delays. Data loss can result in a decline in mapping quality, while delayed data can cause tracking and localization failures. As the number of agents serviced simultaneously increases, this situation becomes more severe.

Therefore, we need a data compression method optimized for SLAM tasks that reduces the relative amount of data transmitted by each agent, thereby increasing the number of services the network can support without affecting SLAM mapping performance.

On the other hand, limited edge server resources restrict the number of services provided by a SLAM framework. Unlike cloud services, edge servers usually have relatively fixed amount of CPU and memory resources, meaning that we cannot increase resource provisioning easily. The server needs to establish a submap for each agent and provide real-time tracking, localization, and fast merging services. As the number of agents increases, various operations will use the servers more frequently, causing significant resource costs. When the edge server is greatly overloaded, various data cannot be processed effectively and rapidly, resulting in a decline in mapping quality and tracking failures. We also note that agents run dynamically in the environment, and the server maintains an active submap for each agent. When there are overlapping areas between multiple submaps, they need to be merged to obtain the global map. In case of agent anomalies that lead to loss of tracking, the server needs to quickly recognize and recover from the anomaly. Therefore, we need to run a robust multi-agent collaborative SLAM framework on the edge server that provides lightweight thread support to reduce server resource usage. We also need to correctly operate the map database to ensure quick responses and high consistency between different agents.

To address the above challenges in multi-agent collaborative SLAM system, in this work, we propose Eco-SLAM: a resource efficient edge-assisted collaborative multi-agent visual SLAM system. Eco-SLAM supports large-scale multi-agent parallel execution and includes a map database system to ensure data consistency. Moreover, we propose an efficient data transmission for SLAM under limited network resources. Our main contributions are highlighted as follows:

1. We optimize the SLAM framework to facilitate collaborative mapping among multi-agent. The framework employs Core-tr (the core processing thread) and Co-Map (collaborative map library) to optimize for the edge server. Core-tr is a lightweight thread that optimizes data processing by integrating tracking and localization functions, reducing redundant resource consumption and memory overhead for the edge server, and improving the efficiency of the edge server in receiving and utilizing agents' information. Based on the workflow of cross-thread mapping, we develop Co-Map and propose an operator-level abstraction for map control which is called the ESDI model. ESDI is a specially designed four-state machine including *Exclusive*, *Shared*, *Dormant*, and *Invalid* states. Based on Co-Map, we effectively implement the need for multi-threaded map switching and collaborative operations, enabling agents to work in independent operational environments and perform cross-thread analysis of data from other maps. This framework effectively reduces edge server's memory consumption during initialization and runtime, as well as addresses issues with map fragmentation.

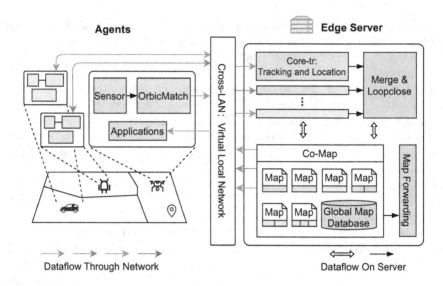

Fig. 2. System architecture of Eco-SLAM. Each agent individually collects its sensory data, compresses the data by the OrbicMatch algorithm, and transfers them to the edge server for merging and processing. The edge server leverages fine-grained parallelism (Core-tr) and the consistency-oriented Co-Map to accommodate more agents for collaborative map fusion.

2. We present OrbicMatch, an ORB-based [18] front and rear frame-matching image compression technique. In order to increase concurrently serving capacity on the edge server under limited bandwidth and reduce the mapping effect losses brought on by traditional data transmission compression, we fully utilize the characteristics of ORB algorithms and perform image compression based on feature point matching data obtained from the front and rear frames. By doing so, communication overhead is reduced while detection performance is guaranteed.
3. We implement and evaluate Eco-SLAM in both simulation and realistic deployment. The extensive simulation experiments demonstrate the effectiveness of Eco-SLAM and its components, showing up to 25.1% bandwidth saving and 20.1% runtime memory saving upon other approaches. The realistic prototype of multiple agents in a real-life experimental scene verifies its feasibility and validity in rendering efficient SLAM services for edge robotics applications.

The rest of this paper is organized as follows. Section 2 presents the design of the proposed system. Section 3 presents the implementation of the proposed system and evaluates it in terms of both simulation and real environment. Section 4 reviews related works. Section 5 discusses and Sect. 6 concludes.

2 Eco-SLAM System Design

2.1 System Overview

In order to boost the system serving capacity, Eco-SLAM has been optimized in terms of storage resources and bandwidth consumption, leveraging collaboration between the multiple agents and the edge server for real-time mapping. Figure 2 depicts the workflow of Eco-SLAM.

On the agent side, the high-resolution input images are continuously captured by the onboard visual sensors (e.g., RGB-D cameras). It is impractical to transmit the original image directly to the edge server due to its high data volume. The transmission of numerous uncompressed images can cause significant network pressure for the edge server. Therefore, efficient compression is crucial before sending the image data. Since the general SLAM systems leverage ORB-based [18] feature point extraction, the mapping performance has a strong correlation with the ORB feature points of the images. Common image compression techniques such as JPEG, results in significant detail loss. Thus, we propose OrbicMatch (ORB-based Image Compression with Matching) algorithm to avoid the negative effect of standard image compression on SLAM systems without sacrificing the SLAM mapping performance of agents. The algorithm can perform differential compression on various regions of the image based on ORB feature point matching information, reducing the data volume while retaining the information required for SLAM as much as possible.

On the edge server side, in order to increase the service capacity of the server, we propose a parallel SLAM framework that supports multiple agents. The main responsibility of the edge server is to obtain data in real time and perform low-latency mapping, positioning, and fusion operations based on the data to offer the agents with the generated local maps. In order to achieve this goal, Eco-SLAM system maintains only one minimum tracking and positioning thread for each agent, which we can call Core-tr. The Core-tr synchronously analyzes one data stream and generates a sub-map of the corresponding agent. These data will be passed to independent fusion and loop detection threads for real-time checking of whether there are co-visual (i.e., overlapping) regions between sub-maps of different agents. At the same time, the edge server runs a parallelized multi-map library: Co-Map, which is used to manage all agents' local map data globally. It employs the ESDI model which can support cross-thread map data fusion, reduce map fragmentation, and ensure the consistency of global map data.

2.2 Memory-Efficient SLAM Framework

Traditional unoptimized multi-agent collaborative mapping usually leads to overly complicated frameworks due to excessive threading. Additionally, the memory consumption caused by redundant scanning and map fragments cannot be ignored. Excessive memory usage reduces service capacity in the edge server with constrained resources. In general, the memory consumption of an edge server can be divided into two parts: the SLAM framework itself and map data, so our proposed Core-tr and Co-Map mainly optimize these two parts.

Core-tr: Minimum Core Support Threads

In an end-to-end SLAM framework, there are usually several components including tracking, localization, merging, loop detection, and map management, which run as threads and occupy memory space at the same time. Since agents are located in different geographical locations and transmit information continuously, we need to maintain tracking and localization threads for each agent, and we refer to this part as Core-tr, which is the minimum supporting thread of Eco-SLAM. Typically, the SLAM framework also separately maintains map management and merging threads for agents, but in Eco-SLAM, these parts will be managed together. As shown in Fig. 2, each Core-tr in Eco-SLAM runs in parallel, ensuring the reliability of each agent's map. As the number of agents grows, Eco-SLAM just needs to create new Core-trs, and all other operations are run uniformly.

Core-tr streamlines tracking and localization services, extracting ORB feature points from data frames and triggering area similarity functions. When sufficient differences exist, the frame is identified as a keyframe, and the keyframe's feature point vector is appended to DBoW2 (a BoW base dictionary). DBoW2 dictionary is a widely used data model in ORB-SLAM that can immediately find frame similarity in $O(1)$ time complexity [7]. Because Core-tr only operates on adjacent keyframes, it occupies only a small amount of system resources, ensuring real-time processing.

To guarantee the freshness of the map for the sake of accuracy, Eco-SLAM defines the longest time interval T for data reception. For each agent, if Core-tr has not continued to receive new data after processing the previous frame for more than T, it will disconnect from the agent and enter the tracking loss or localization failure mechanism. If the data uploaded by the agent is obtained after an interval greater than T, Core-tr will create a new map. This step easily generates map fragments, leading to unnecessary memory overhead. Therefore, Core-tr will notify Co-Map for cleaning, which will be discussed after.

Meanwhile, we set up a detection buffer between Core-tr and fusion and loop threads. The data in the buffer is composed of keyframes filtered by all Core-tr during the data detection period. The data in the buffer may not come from the same agent, and we need to detect them frame by frame based on BoW. However, not all keyframes can successfully merge with other maps, so Core-tr will not significantly affect CPU resources.

Through the above methods, Eco-SLAM utilizes Core-tr to provide fine-grained parallelism in supporting multi-agent tasks, and many core functions can be reused, greatly reducing system memory consumption.

Co-Map: Collaborative Map Library Based on ESDI Model

Multi-map database often wastes memory due to map redundancy and poor management. To better support map operations for multiple threads such as Core-tr, map merging, and loopclosing detection, we propose the ESDI model

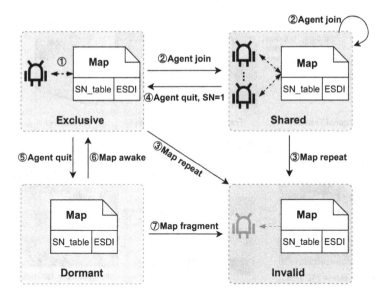

Fig. 3. ESDI model control flow. Co-Map maintains an ESDI state for each map. The map will transition between these four states depending on the circumstances during the lifecycle.

and design the multi-map library called Co-Map by adopting the idea of smart pointers in C++. The ESDI model is a state machine consisting of four states, representing four types of situations: *Exclusive*, *Shared*, *Dormant*, and *Invalid* state. Via Co-Map, we manage map creation, updating, merging, and saving operations in a unified manner.

In our system, Co-Map maintains keyframe data, SN_tables, and ESDI states for the map files, utilizing them to ensure data consistency and reduce map fragmentation. The constituent and state transition mechanisms of Co-Map can be seen in Fig. 3. At system initialization, a new map is created for each agent, with only one operator for each map recorded in the SN_table, as shown in ①. Therefore, the ESDI state for the map is *exclusive*, and data consistency is not an issue in this state.

If an agent matches an area of another existing map, the map merging operation is triggered, in which the agent's current map data is copied into the matching map, as shown in ②, and the transition information for the old map is saved. As the map merging operation results in data duplication, the old map will be invalidated and transformed to the state shown in ③. At this point, the merged map will have multiple operators, and the SN_table for it will record information about all of the agents, with the ESDI state transitioning to *shared* state.

Every operation on maps will check their status information in real time. If an agent discovers that its map is being changed into *invalid* state and has not been triggered by itself for a merging operation, it indicates that the map has

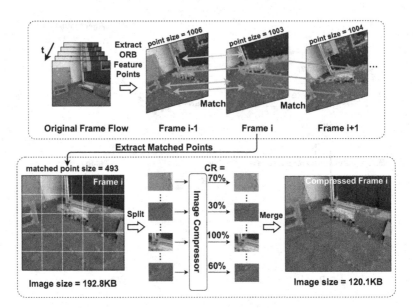

Fig. 4. The workflow of OrbicMatch. OrbicMatch consists of two steps. First, it extracts ORB feature points from each frame and matches them with neighboring frames. Then, it splits the frame into blocks and compresses each block according to the number of matching points.

already been transferred by other agents, and its map data is not the latest. The agent will not be able to write to the *invalid* state map and will automatically change its operations to the new map based on the transition information. When no members exist in the SN_table for an invalid map, the map will be deleted.

Data fluctuations and discontinuity issues are usually unavoidable. When an agent loses track, it is unable to accurately position itself in the original operating map, and it will be removed from the SN_table of the original map. When there is only one operator left in the SN_table or none at all, as shown in ④ or ⑤, the ESDI state will transition to *exclusive* state or *dormant* state.

Maps in *dormant* state are usually important because, as shown in ⑥, they can be reactivated at any time by the merging operation. However, when an agent continuously creates maps while experiencing continuous loss, the maps often do not preserve enough useful information. Therefore, we need to change maps with less than 5 frames of images (a threshold which depends on the operation scenario, you can change it to other empirical values) into *invalid* state and, as shown in ⑦, which will be finally deleted. This way can help to reduce memory usage generated by fragmented map data.

Algorithm 1 OrbicMatch

Require: Current Image: Ic, xBlockNum, yBlockNum
Ensure: Compressed Image Iz
 1:    $Pp \leftarrow Pc$ ▷ Previous ORB points set
 2:    $Pc \leftarrow$ ExtractORB(Ic)   ▷ Current ORB points set
 3: **for point** p **in** Pc **do**
 4:    $block_x \leftarrow p.x \times$ xBlockNum/imgWidth
 5:    $block_y \leftarrow p.y \times$ yBlockNum/imgHeight
 6:    matchNum$[block_x + block_y] \times$ xBlockNum $+= 1$
 7: **end for**
 8: $i \leftarrow 0$
 9: $j \leftarrow 0$
10: **while** $i \neq block_x$ **do**
11:    **while** $j \neq block_y$ **do**
12:       Coord$\leftarrow (i \times (\text{imgWidth}/\text{xBlockNum}), j \times (\text{imgHeight}/\text{yBlockNum}),$
                          imgWidth/xBlockNum,imgHeight/yBlockNum)
13:       $img_{block} \leftarrow$ Ic.ROI(Coord)
14:       $q \leftarrow$ stepDiff$(N, \text{matchNum}[i + j * \text{xBlockNum}])$
15:       $img_{compressed}(img_{block}, q)$
16:       $Iz_{ij} \leftarrow img_{compressed}$
17:       $j \leftarrow j + 1$
18:    **end while**
19:    $i \leftarrow i + 1$
20: **end while**

2.3 OrbicMatch: ORB-Based Image Compression with Matching

The wireless communication bandwidth is a limited and congested resource when a rising number of agents are involved in edge-assisted SLAM. Trivial image compression methods discard the pragmatic property of images, leading to a suboptimal compression result. The proposal of the OrbicMatch algorithm stems from the observation that the distribution of feature points on the image is unbalanced. Some vulnerable feature points may be wiped by uniform compression methods, which results in a decline in matching performance. OrbicMatch algorithm borrows the wisdom from the image processing community [5] and maximally preserves SLAM feature points while effectively reducing the offloading data by block-wise compression.

    Figure 4 illustrates how the OrbicMatch works. The agent first extracts the ORB feature of the current frame once obtaining it from the sensor and retrieves the feature point information of the previous frame from the historical information. Then the agent filters out the feature point data that can be matched between the two consecutive frames by running the ORB matching algorithm and records the number of matches as N. Afterward, the current frame is partitioned into blocks and each block is given a specific compression ratio (CR) based on the ratio of the feature points extracted in this block to feature points extracted in the entire picture.

In Algorithm 1, we provide a detailed explanation of our OrbicMatch method. To compress an image, the agent must input the video or image stream and the required number of blocks, after which the algorithm automatically compresses all images. We use Ic to represent the image to be processed, while $xBlockNum$ and $yBlockNum$ represent the block division in the width and height directions of the image. In the algorithm, $matchNum$ stores the number of matching feature points in each block, Pc represents the current image's ORB point set and Pp represents the previous ORB point set.

Lines 1–2 of the algorithm are used to save the previous frame's point set and generate the current frame's point set. Lines 3–7 traverse the point set, distinguishing the location of each feature point in the block, and using the $matchNum$ array to record the amount. Lines 8–20 compress the image based on the number of feature points in each block. In lines 12–13, we extract the coordinates of each block and segment image Ic using the OpenCV [2] ROI tool. Line 14 computes the compression ratio mapping of the block based on the ratio between the number of feature points in the block and the uniform distribution. This mapping is typically pre-set. In our experiments, we obtained compression ratio (CR) in stair functions with a step by 10%, and the compression ratio increases as the number of the block's feature points increases. Line 15 compresses the image according to the compression ratio obtained earlier. We can use various mixed compression methods [15, 24] in this step, but for fairness, we used JPEG encoding in our experiments. Line 16 utilizes OpenCV to integrate the segmented image blocks into Iz. Using the above algorithm, we can obtain an image result compressed based on the ORB information.

3   Evaluation

3.1   Experimental Setup

In this section, we will introduce the experimental environment and basic settings. The experimental equipment is shown in Fig. 5.

Prototype. The edge server is equipped with a 16-core, 32 GB, Intel(R) Xeon(R) W-2145 CPU running Ubuntu 20.04LTS, which requires over 200W of power when operating at maximum speed. On the agent side, to ensure basic data collection functionality, we employ the low-power NVIDIA(R) Jetson NX development board [13] with 6 ARM cores and 8 GB of shared memory, which consumes no more than 15 W of power at full speed. We also use the Intel(R) RealSense(R) D435i [8] mini camera, which can capture RGB-D image data and has a volume of only 55 cm^3. We use the resource manager built into the Linux system to monitor Eco-SLAM's memory usage and CPU utilization and use iperf3 to measure bandwidth performance between the agents and server.

Implementation. We compile Eco-SLAM on an open-source platform. Like most SLAM algorithms, Eco-SLAM's framework is built on ORBSLAM3 [4], a widely used and top-ranked SLAM framework. Our framework is also compatible with dense point cloud display and can provide real-time rendering results. To

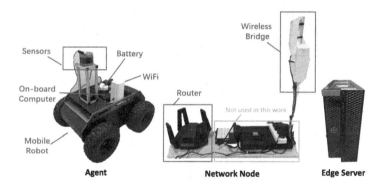

Fig. 5. Eco-SLAM experimental equipment.

enhance compatibility, our framework is built upon the ROS platform [17], which is one of the dominating OS in the robotics field.

Scenario. To ensure that our results can be compared with similar studies, we initially evaluated our method using publicly available SLAM datasets (TUM [21] and EuRoC [3]). Additionally, we deployed Eco-SLAM on SYSU campus, as depicted in Fig. 6, to obtain real-life network information. Furthermore, we conducted further testing in the same environment using a 16 laser-beam LiDAR to perform an accurate scan of the campus and gather the ground truth trajectory data for each agent. This provided us with data for evaluating the accuracy of vSLAM methods. Within an area of approximately $5000\,\mathrm{m}^2$, we deploy four network nodes, each of which is responsible for WiFi signal transmission in its respective area, and the nodes are connected using wireless base stations. By using iperf3 [11] for speed measurement, the network bandwidth between any two nodes can be stable at 700–800 mbps (1 mbps = 0.125 MBps). At the software level, we use cross-LAN transmission technology called Tailscale [22], which is a cross-platform VPN network based on WireGuard [6], to connect the four wireless router nodes' LANs into a virtual LAN, assign unique LAN IP addresses to devices, and enable direct communication between devices connected to different nodes.

Metrics. We use Absolute Trajectory Error (ATE, in cm) to evaluate the SLAM accuracy on two public datasets. ATE is a popular metric for evaluating SLAM tracking performance. Since ATE generates traces calibrated with ground truth trajectories before measuring the absolute error, it achieves less error than actual positional error. To evaluate the effects of image compression, we compare the bandwidth requirements (in $mbps$) of different compression schemes (defined as the sum of the average data volume transmitted per second by all agents). At the same time, we also evaluate the reduction in data volume brought by different compression methods in public datasets and real environments, using the number of matching feature points (MPN) to verify the compression effect.

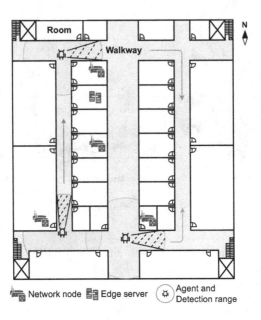

Fig. 6. Experimental environment on SYSU campus.

3.2   Overall Performance Comparison

We compare Eco-SLAM with CCM-SLAM [20] and Multi-UAV [19], two SOTA edge-assisted multi-agent SLAM systems, to evaluate the overall performance. Our evaluation is based on common benchmarks of public datasets [3,21], and we split these datasets and deploy them into agents. The data is transmitted by running the ROS node on the agent side, setting the ROS message to publish the topic, and using the server to obtain the messages. Due to the message transmission characteristics of ROS, channel delay and packet loss can cause not all images to be successfully transmitted, so the mapping accuracy depends on the SLAM framework and algorithm running on the server. Figure 7 depicts the ATE on the public datasets as the number of agents changes. When the number of agents increases, merging operations become more frequent, and data loss becomes more common, so in this experiment, the mapping error gradually increases. However, with slightly better ATE accuracy than CCM-SLAM and significantly better than Multi-UAV, Eco-SLAM demonstrates its superior end-to-end performance.

Figure 8 shows the data transmission requirements for multiple agents, with Eco-SLAM requiring only about 82% and 70% of the bandwidth compared with CCM-SLAM and Multi-UAV, respectively, and the overall requirements for multiple agents are also better than the other two solutions. The considerable performance advantage comes because Eco-SLAM uses Core-tr to enhance real-time support for agents' data, which can process the information received from agents promptly and effectively. Also, the system employs Co-Map to manage the map

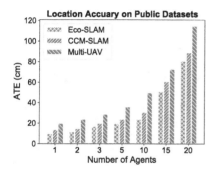

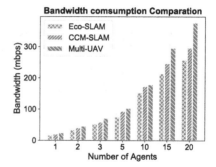

Fig. 7. ATE results on public datasets.

Fig. 8. Bandwidth usage between agents and edge server.

library effectively, broaden the effectiveness of cross-thread support and enable agents to utilize map data more extensively. Additionally, the use of OrbicMatch reduces bandwidth consumption, lowers data transmission loss, and enables the server to receive information more completely, thereby improving the accuracy of the map.

Given the fact that the Eco-SLAM shows apparent priority over the Multi-UAV in terms of the location accuracy on the public dataset, we further compare the Eco-SLAM with CCM-SLAM in a real-life scenario. Figure 9 depicts the trajectory discrepancies of those two multi-agent SLAM methods and the ground truth. Specifically, we leverage the results of 16 laser-beam LiDAR as the ground truth. The trajectory of Eco-SLAM matches the ground truth trajectory better than the CCM-SLAM does, which stems from that the OrbiMatch remains more feature points. Noteworthily, the Eco-SLAM obviously outperforms CCM-SLAM at the conjunction area (the circle). That's because the Co-Map mechanism has the virtue of tracking the map smoothly.

3.3 Core-tr and Co-Map Performance Analysis

In this experiment, we deployed Eco-SLAM, CCM-SLAM, and the original ORB-SLAM3 on a real edge server. They have similar architectures and support various sensors. The total resources of the edge server limit the scale of agents that can be supported simultaneously. Therefore, if the number of agents is increased, the SLAM framework needs to save memory on the edge server. We used the original ORBSLAM3 framework as a baseline to compare their initialization memory usage on the edge server to verify the performance of Eco-SLAM.

We link the edge server to the network node by wired connection to ensure it can receive data from all agents. Figure 10 compares the RAM consumption of an ORB-based SLAM framework during initialization. In the ORBSLAM3 framework, a significant amount of memory is wasted due to the need to allocate new memory space for each runtime and maintain inter-process communication. In comparison to CCM-SLAM, Eco-SLAM does not require a separate

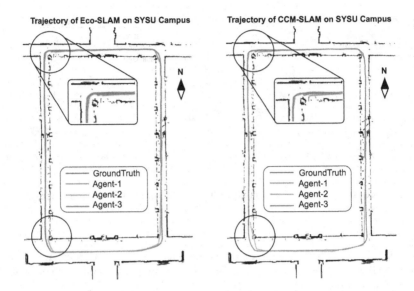

Fig. 9. Trajectory comparison of Eco-SLAM and CCM-SLAM in the real-life setting. The ATE of Eco-SLAM is 30.8 cm while CCM-SLAM is 38.9 cm, indicating a 20.8% improvement in accuracy.

communication thread for each agent and uses a lighter tracking and localization thread Core-tr, it reduces the average memory overhead per maintained agent by 13.6 MB.

Figure 11 shows the memory growth when 3 agents collaborate to build a map in the same environment. With the operation of Eco-SLAM, the rapid growth trend of memory is reduced due to the synchronization and fragmentation map management of Co-Map based on the ESDI model. Figure 12 shows the CPU utilization on the edge server under the same scenario. We found that the usage of this resource did not change significantly, indicating that the memory optimization of Eco-SLAM did not come at the expense of CPU resources. Eco-SLAM achieves up to 6.5% memory savings at the initialization phase and up to 20.1% during the runtime phase. Therefore, the Eco-SLAM framework can save memory resources for edge servers, improve resource utilization efficiency, and increase the number of agents that resource-limited edge servers can simultaneously serve.

3.4   OrbicMatch Performance Analysis

In this experiment, we test the compression performance of OrbicMatch. The test data is divided into two parts: one part is from public datasets and the other is collected in the SYSU campus environment. JPEG is a commonly used image compression format in ROS system. Therefore, in this experiment, we use it as a counterpart.

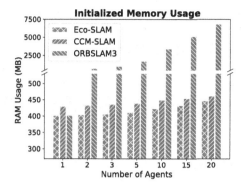

Fig. 10. ORB-based SLAM initialized memory usage.

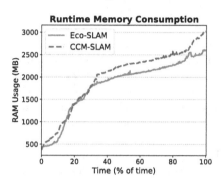

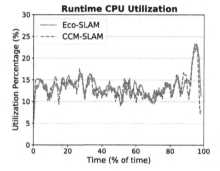

Fig. 11. Edge server runtime RAM usage consumption. (3 Agents)

Fig. 12. Edge server CPU utilization. (3 Agents)

Our proposed OrbicMatch helps optimize data transmission capacity. Under the same compression standard parameters as JPEG compression, OrbicMatch can ensure that sensitive areas are not affected by lossy compression, and uses a higher compression ratio in non-sensitive areas. Therefore, images can be compressed smaller while saving more ORB feature matching points. Table 1 compares the performance of OrbicMatch in different datasets. In terms of bandwidth, the average transmission bandwidth required by OrbicMatch-compressed images is only 74.9% of the original image, saving 25.1% of the data, and reducing 8.7% compared to JPEG compression. In terms of image quality, OrbicMatch can retain an average of 95.1% of the matching feature points of the original image, which is a 3.1% improvement over JPEG compression.

The utilization of OrbicMatch compression significantly diminishes the bandwidth consumption of agents, mitigates the reception load on edge server, and retains substantial feature point information. Consequently, it enhances the capacity of edge server to accommodate a greater number of agents concurrently in wireless network environments with constrained bandwidth, without compromising image quality.

Table 1. Comparision of Bandwidth consumption (BW, *mbps*) and Matched Points Number (MPN) result on different datasets. The two kinds of data items are obtained by taking the average of all image results.

Method	TUM (Difficult) - 3100 images	TUM (Medium) - 1670 images	EuRoC (Medium) - 1000 images	SYSU Campus - 800 images
	BW/MPN	BW/MPN	BW/MPN	BW/MPN
Original	13.15/395.3	13.97/471.7	16.02/399.0	48.06/211.4
JPEG	11.59/351.8	12.53/452.7	14.02/372.6	43.80/198.0
OrbicMatch	9.43/365.8	10.47/460.5	12.59/384.4	33.62/205.6

4 Related Work

Visual SLAM. In the past decade, visual SLAM has been one of the areas of great interest for mobile and robotic systems. This work mainly uses RGB, depth, or stereo cameras to construct real-time maps of the surrounding environment based on sensor data without any prior knowledge. Some well-known works in this field include ORBSLAM [12], RTAB-MAP [10], and VINS-mono [14], which focus on the algorithms themselves and aim to find a feasible SLAM solution. Our work is based on ORBSLAM3 [4], so Eco-SLAM can demonstrate strong sensor support features and apply many optimizations to further enhance performance. In the field of edge computing, Edge-SLAM [1], Multi-UAV [19], and CCM-SLAM [20] are well-known visual SLAM works. They split the ORB-SLAM integrated framework and offload some of the work to the edge server, using a central service with stronger computing power to construct maps.

Edge Collaborative SLAM. Using edge computing for multi-agent SLAM collaboration is one of the areas of recent interest, such as CVI-SLAM [9] based on VI-ORB, C2TAM [16] based on PTAM, and CCM-SLAM based on ORB-SLAM. They use edge computing to assist in mapping based on existing frameworks. However, as the number of agents increases, there are also problems such as rapid consumption of edge server resources and tight transmission bandwidth, which severely limits the service scale of the overall system. In contrast, Eco-SLAM optimizes these two points, and due to the scalability of the Eco-SLAM framework, we have also laid the foundation for larger-scale SLAM services in the future agent-edge-cloud collaboration.

5 Discussion

We briefly discuss limitations and future work in this section. In this paper, we introduce Eco-SLAM, which concentrates on enhancing the data uploading method from the agents and optimizing the framework for multi-agent SLAM on the edge server to decrease resource consumption for the collaborative multi-agent visual SLAM system. In the next stage, it's urgent to concentrate on exploring edge-to-edge collaboration methods and mechanisms, facilitating multi-edge servers collaboration to the whole system, which encounters more intricate network topology and resource allocation schemes. Moreover, we need to investigate the communication mechanisms between applications performing on a considerable quantity of agents and large-scale edge servers, to develop a more comprehensive and unified SLAM system.

6 Conclusion

Our work proposes Eco-SLAM, an Edge-Assisted Collaborative Multi-Agent Visual SLAM System that enhances the service capacity of nowadays edge-assisted SLAM systems. We present a Core-tr parallel framework and Co-Map library based on the ESDI model to reduce the resource consumption of a single agent service thread on the edge server. Additionally, we optimize network capacity by compressing data transmission using the OrbicMatch algorithm, thus improving visual SLAM performance. Eco-SLAM is built using open-source frameworks and adapt to the ROS platform. The implementation and enhancements of Eco-SLAM have laid a foundation for future collaborative edge-assisted SLAM systems.

Acknowledgements. This work was supported in part by the National Science Foundation of China (No. U20A20159, No. 61972432); Guangdong Basic and Applied Basic Research Foundation (No. 2021B151520008).

References

1. Ali, A.J.B., Kouroshli, M., Semenova, S., Hashemifar, Z.S., Ko, S.Y., Dantu, K.: Edge-slam: edge-assisted visual simultaneous localization and mapping. ACM Trans. Embed. Comput. Syst. **22**(1), 1–31 (2022)
2. Bradski, G.: The opencv library. Dr. Dobb's J. Softw. Tools Prof. Program.r **25**(11), 120–123 (2000)
3. Burri, M., et al.: The Euroc micro aerial vehicle datasets. Int. J. Robot. Res. **35**(10), 1157–1163 (2016)
4. Campos, C., Elvira, R., Rodríguez, J.J.G., Montiel, J.M., Tardós, J.D.: Orb-slam3: an accurate open-source library for visual, visual-inertial, and multimap slam. IEEE Trans. Rob. **37**(6), 1874–1890 (2021)
5. Chen, B., Yan, Z., Nahrstedt, K.: Context-aware image compression optimization for visual analytics offloading. In: MMSys '22: 13th ACM Multimedia Systems Conference, Athlone, Ireland, June 14–17, 2022. pp. 27–38. ACM (2022). https://doi.org/10.1145/3524273.3528178
6. Donenfeld, J.A.: Wireguard: next generation kernel network tunnel. In: NDSS, pp. 1–12 (2017)
7. Gálvez-López, D., Tardós, J.D.: Bags of binary words for fast place recognition in image sequences. IEEE Trans. Rob. **28**(5), 1188–1197 (2012). https://doi.org/10.1109/TRO.2012.2197158
8. Intel: Realsense depth camera d435i. https://www.intelrealsense.com/depth-camera-d435i

9. Karrer, M., Schmuck, P., Chli, M.: CVI-slam-collaborative visual-inertial slam. IEEE Robot. Autom. Lett. **3**(4), 2762–2769 (2018)

10. Labbé, M., Michaud, F.: RTAB-map as an open-source lidar and visual simultaneous localization and mapping library for large-scale and long-term online operation. J. Field Robot. **36**(2), 416–446 (2019)

11. Laboratory, L.B.N.: iperf3: A TCP, UDP, and SCTP network bandwidth measurement tool. https://github.com/esnet/iperf

12. Mur-Artal, R., Montiel, J.M.M., Tardos, J.D.: Orb-slam: a versatile and accurate monocular slam system. IEEE Trans. Rob. **31**(5), 1147–1163 (2015)

13. Nvidia corporation: Nvidia embedded systems with jetson. https://www.nvidia.com/en-us/autonomous-machines/embedded-systems

14. Qin, T., Li, P., Shen, S.: VINS-Mono: a robust and versatile monocular visual-inertial state estimator. IEEE Trans. Rob. **34**(4), 1004–1020 (2018)

15. Rabbani, M., Joshi, R.: An overview of the jpeg 2000 still image compression standard. Signal Process. Image Commun. **17**(1), 3–48 (2002)

16. Riazuelo, L., Civera, J., Montiel, J.M.: C2tam: a cloud framework for cooperative tracking and mapping. Robot. Auton. Syst. **62**(4), 401–413 (2014)

17. Robot operating system. http://wiki.ros.org

18. Rublee, E., Rabaud, V., Konolige, K., Bradski, G.: Orb: an efficient alternative to sift or surf. In: 2011 International Conference on Computer Vision. pp. 2564–2571. IEEE (2011)

19. Schmuck, P., Chli, M.: Multi-UAV collaborative monocular slam. In: 2017 IEEE International Conference on Robotics and Automation (ICRA), pp. 3863–3870. IEEE (2017)

20. Schmuck, P., Chli, M.: CCM-SLAM: robust and efficient centralized collaborative monocular simultaneous localization and mapping for robotic teams. J. Field Robot. **36**(4), 763–781 (2019)

21. Schubert, D., Goll, T., Demmel, N., Usenko, V., Stückler, J., Cremers, D.: The tum vi benchmark for evaluating visual-inertial odometry. In: 2018 IEEE/RSJ International Conference on Intelligent Robots and Systems (IROS), pp. 1680–1687. IEEE (2018)

22. Tailscale Inc.: The easiest, most secure way to use wireguard and 2fa. https://github.com/tailscale/tailscale

23. Tourani, A., Bavle, H., Sanchez-Lopez, J.L., Voos, H.: Visual slam: what are the current trends and what to expect? Sensors **22**(23), 9297 (2022)

24. Vijayvargiya, G., Silakari, S., Pandey, R.: A survey: various techniques of image compression. arXiv preprint arXiv:1311.6877 (2013)

25. Ye, S., Zeng, L., Wu, Q., Luo, K., Fang, Q., Chen, X.: ECO-FL: adaptive federated learning with efficient edge collaborative pipeline training. In: Proceedings of the 51st International Conference on Parallel Processing, pp. 1–11 (2022)

26. Zhou, Z., Chen, X., Li, E., Zeng, L., Luo, K., Zhang, J.: Edge intelligence: paving the last mile of artificial intelligence with edge computing. Proc. IEEE **107**(8), 1738–1762 (2019). https://doi.org/10.1109/JPROC.2019.2918951

POWERDIS: Fine-Grained Power Monitoring Through Power Disaggregation Model

Xinxin Qi, Juan Chen$^{(\boxtimes)}$, Rongyu Deng, Zekai Li, Lin Deng, Yuan Yuan, and Yonggang Che

National University of Defense Technology, Changsha, China
{qixinxin19,juanchen,dengrongyu21,zekaili,lindeng,yuanyuan,
ygche}@nudt.edu.cn

Abstract. In an era where power and energy are the first-class constraints of computing systems, precise power information is crucial for energy efficiency optimization in parallel computing systems. Current power monitoring techniques rely on either software-centric power models that suffer from poor accuracy or hardware measurement schemes that have coarse granularity. This paper introduces POWERDIS, a new technique for accurately measuring and forecasting the power consumption of computing components, such as CPUs and memory, within the compute node on parallel computing systems. POWERDIS combines coarse-grained node power readings collected through hardware measurement and a software power disaggregation model to improve monitoring granularity while achieving high accuracy. Additionally, it can be used to predict future power usage by mining spatiotemporal patterns with minimal modifications to dataset construction building on the existing POWERDIS framework. We evaluate POWERDIS on both ARM-based and X86-based platforms. Extensive results show that POWERDIS has great accuracy for power estimation and forecasting, reducing the mean absolute percentage error (MAPE) by up to 20% compared to other power modeling methods.

Keywords: power monitoring · power disaggregation model · parallel computing

1 Introduction

The increased importance of energy efficiency in parallel computing systems highlights the need for more stringent power monitoring requirements. Accurate power readings, comprehensive coverage that extends from system-wide to computing components like CPUs and DRAMs, and quick capture of power consumption changes are all essential for parallel computing systems to respond promptly to changes in workload demand and behavior. Efficient job scheduling, reduced energy consumption, prevention of overheating, and maintenance of system stability all benefit from these capabilities [1–4].

Unfortunately, none of the existing power monitoring techniques achieves a good balance between accuracy, coverage, latency, scalability, and deployment cost—all of which are important for the wide adoption of power monitoring [5]. Existing power monitoring solutions can be generally categorized into

Z. Tari et al. (Eds.): ICA3PP 2023, LNCS 14490, pp. 325–346, 2024.
https://doi.org/10.1007/978-981-97-0859-8_20

two groups: hardware measurement and software-centric power modeling techniques. Hardware measurement obtains power readings either through the use of external power instruments or by way of power sensors, baseboard management controllers (BMC), or FPGA components. Although these techniques can provide accurate readings, power sensors are not consistently available in computing components, rendering it difficult to obtain power consumption data for computing components by providing only coarse-grained node power readings. Software-centric power modeling techniques, such as Intel® RAPL [6], estimate the power consumption by establishing a function between the hardware performance counters (PMCs) and the power consumption. While they have low deployment costs and can offer timely power readings, they are limited by their less-than-reliable accuracy [7].

The combination of hardware measurement and software power modeling can present an ideal solution for power monitoring, as it harnesses the strengths of each while leveraging the other to offset their respective weaknesses. In situations where power sensors or BMC may be absent, hardware measurement can circumvent these limitations using power models to break down the coarse-grained node power consumption into individual component power consumption. Furthermore, because node power consumption is strongly linked to component power consumption, incorporating hardware measurement (i.e., coarse-grained node power readings) into power models can notably optimize model accuracy. This bi-directional interplay between hardware measurements and software power modeling allows for the integration of individual solutions to construct a unified framework that transcends the limitations of an isolated technology component.

While the combination of hardware measurement and software power modeling presents a promising approach to power monitoring, translating these high-level ideas into a practical system is a complex task. One of the major challenges confronting is power disaggregation, namely the separation of component power consumption from node power consumption in power modeling, as this process requires exploration of the complex and non-linear relationship between PMCs and component power consumption. This relationship is also prone to change over time, further complicating the modeling process. Although some power modeling studies have attempted to address this challenge by exploring the hidden relationship between PMCs and power consumption over time [8] (temporal dependence), they have focused primarily on past values within the same feature (PMC), neglecting the temporal dependence across different features, which has contributed to low accuracy. Furthermore, the readings obtained from PMCs may be imprecise or noisy, which further underscores the need for robust modeling techniques that can accurately simulate the complex relationship between PMCs and power consumption in order to distribute node-level power readings to low-level hardware components.

We present POWERDIS, a power monitoring framework that operates on a component level and utilizes a power disaggregation model to break down node power readings gathered through hardware measurement, into individual component power consumption. To enhance the accuracy of the disaggregation, we consider the crucial observation of spatiotemporal dependencies across different features and propose three distinct modules within the power disaggregation model: the graph structure learning (GL) module, the spatial convolution (SC) module, and the temporal-pattern convey (TC) module. The GL module and the SC module explicitly represent spatial dependencies among PMCs and reveal spatial patterns. Concurrently, the TC module works together with the SC module to capture temporal patterns that represent the influence of historical values on current values. The integration of spatial and temporal patterns in POWERDIS allows for the distribution of node power readings to lower-level computing components. Besides real-time power estimation on the component level, POWERDIS can also function as a power predictor, capable of forecasting the future power consumption of components. This can be accomplished by modifying the dataset construction and employing the same time-space dependence modeling approach.

We implement and evaluate POWERDIS on an ARM-based computing system, which integrates hardware BMC with software interfaces to provide node-level power readings. We test POWERDIS on 96 benchmarks and compare it with 12 prior methods. Extensive evaluation results show that POWERDIS can effectively reduce MAPEs by up to 20% compared to prior component-level power modeling methods.

This paper makes the following contributions.

- It formally describes the node power disaggregation problem and points out that the complex relationship between PMCs and component power consumption is influenced by the historical values of different features.
- It proposes a new approach to power disaggregation modeling. POWERDIS decouples the spatial-temporal dependence and identifies their variation patterns over time. More importantly, POWERDIS enables not only power estimation but also power forecasting.
- Extensive evaluation results show that POWERDIS is highly accurate for both CPU power consumption and DRAM power consumption on both ARM-based and X86-based platforms, with a MAPE reduction of up to 20% compared to other methods.

2   Motivation

Our study provides observations on the spatiotemporal interdependence of PMCs and power consumption, and their interactions, which have significant implications for exploring the distribution of node power consumption among its components. This section will first discuss both the spatial and temporal dependences of power consumption and PMCs and then illustrate their importance to power modeling using an example.

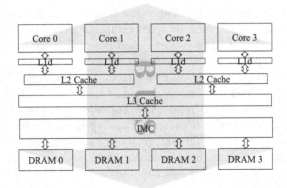

Fig. 1. Spatial dependence of components within a compute node.

2.1 Spatio-Temporal Dependence

1. Behind the change in power consumption are system activities caused by program execution, which interact with each other due to their direct physical connection. → **Spatial dependence**
2. Power consumption data is a set of dynamic time series in which each data point is related to the previous data point. → **Temporal dependence**

Spatial Dependence. The structural connection of computing components like CPU and DRAM induces spatial dependence on the hardware events that flow through them. As a result, power modeling based on PMCs must take into account the interaction between different computing components, and the efficacy of the power model is contingent on the hardware event selection across various computing components. For instance, cache misses generally lead to stalled instruction cycles, highlighting the spatial dependence between PMCs. Figure 1 portrays a schematic diagram of diverse compute node components, including CPUs, L1, L2, L3 cache, and so on. It is evident from the diagram that these components are structurally interconnected, and hardware events originating from the upper-level computing components manifest changes in the lower-level hardware events.

Temporal Dependence. The PMCs change dynamically over time and exhibit trend, cyclic, and randomness characteristics. The *trend characteristics* are that the PMCs change with time, displaying a homogeneous trend, but the variation range can vary. The *cyclic characteristics* are that power consumption and some PMCs fluctuate with irregular periods. *The randomness characteristics* are most visible in irregular changes in power consumption, such as abrupt peak and valley values. In general, power consumption fluctuates on a regular basis, with some unexpected spikes.

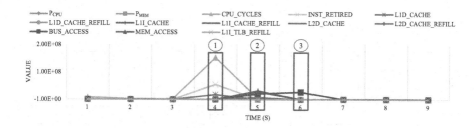

Fig. 2. An example illustrating temporal dependence of PMC events.

In addition to the above three characteristics of the power consumption data, Fig. 2 reveals an important observation that the current PMC value is influenced by the same and other PMCs of the previous moment or longer duration. The red boxes in Fig. 2 respectively represent the three PMC groups that have reached their peaks successively. At time 3, *CPU_CYCLES*, *INST_RETIRED*, *L1D_CACHE* (red box ①) begin to escalate and attain their peak at time 4, while *MEM_ACCESS*, and *BUS_ACCESS* (red box ②) start increasing at time 4. At time 5, *MEM_ACCESS* reaches its highest, followed by *BUS_ACCESS* and at time 6 (red box ③). It is apparent that PMCs situated within the same red box display similar patterns, and the current group of PMCs is impacted by the preceding grouping of PMCs. This signifies that PMCs within the same grouping are interdependent and affect PMCs from other groupings in the temporal domain.

2.2 A Motivation Example

Our work aims to improve the accuracy of the power model that is integrated with the hardware measurement. Upon analyzing the spatiotemporal dependence of PMCs, we observe a previously overlooked point that the distinct spatial dependence among PMCs is not constant across time, but is influenced by other PMCs in the spatial dimension. In this section, we illustrate this finding with an example showing how spatial and temporal dependence within the same feature, as well as temporal dependence across different features, impact the accuracy of the power consumption model.

We employ convolutional neural networks to investigate the influence of spatial and temporal dependence within the same feature, as well as temporal dependence across different features, on the accuracy of the power model. In our experiment, we utilize a 1D convolutional neural network to explore the significance of space and (same dimension) temporal dependence. The feature extraction direction is illustrated in Fig. 3. The spatial dependence model slides from left to right to capture relationships between features, while the time-dependent model slides from top to bottom to capture changes over time. Additionally, we use a 2D convolutional neural network to examine the effect of temporal dependence across different dimensions on the accuracy of the power model, which involves feature extraction in both time and space dimensions.

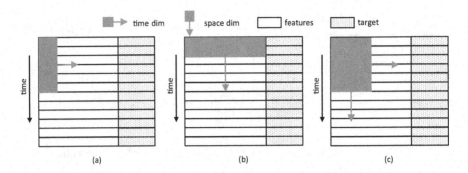

Fig. 3. Feature extract directions in convolutional neural networks.

We utilize the same training and test sets across all three models. The experimental results demonstrate that the mean absolute percentage error (MAPE) of the spatial-dependent model is 20.56%, the MAPE of the temporal-dependent (same feature) model is 17.80%, and the MAPE of the space-dependent (across different features) model is 10.25%. The findings demonstrate that in accurately modeling power consumption, the hierarchy of significance is as follows: temporal dependence of distinct features > temporal dependence of identical features > spatial dependence. Further investigations in Sect. 5 will validate this point. When constructing dependable power models, it is critical to take into account spatial dependence as well as both types of temporal dependence (within and between features).

3   Our Approach

3.1   Problem Formulation

Node power disaggregation is a methodology that involves breaking down the node power consumption into individual components, such as CPU, memory, and peripherals, using hardware performance counters, CPU utilization rates, and other similar metrics. Traditional power modeling for computing components typically focuses on the hardware components of interest, using microbenchmarks to highlight each component and collecting system states to build component-level power models [9]. In contrast, node power disaggregation does not require extensive domain knowledge of architecture and is accessible to non-experts. Rather than delving into the intricacies of each computing component, this approach treats them as a "black box" and employs three steps to accomplish power breakdown: feature selection, modeling, and power distribution.

The objective of node power disaggregation is separating the power consumption of the different components, which is presented by Eq. (1),

$$y = x^T \mathbf{WP} \tag{1}$$

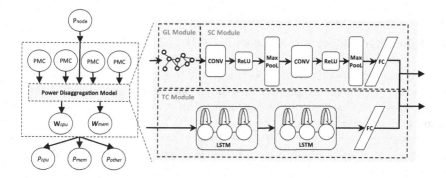

Fig. 4. A concept map of POWERDIS.

where the vector $x^T(\delta T \times 1)$ represents the node power consumption, $\mathbf{W}(\delta T \times n)$ represents the weight matrix of features, $\mathbf{P}$ is the power disaggregation model $(n \times k)$, δT represents the number of node power readings in the past, typically $\delta T = 1$. n represents the number of features (s) and k is the number of computing components. The vector $y(k \times 1)$ represents the component power consumption. As demonstrated by Eq. (1), the development of a reliable power disaggregation model hinges on the ability to identify the appropriate $\mathbf{P}$. This task necessitates an extensive examination of the connection between features and the power consumption of individual components. Building upon the observation that spatial-temporal dependence highly impacts the accuracy of power models, we propose a novel approach to power disaggregation that involves identifying patterns of spatial and temporal dependence across different features.

3.2 Overview

Figure 4 presents the proposed POWERDIS framework, which is integrated with hardware measurement deployed on the computing system. The core component of POWERDIS is its power disaggregation model that comprises three modules: the graph structure learning (GL) module, the spatial convolution (SC) module and the temporal-pattern convey (TC) module. The GL module extracts a sparse graph adjacency matrix based on node power consumption and PMCs to explicitly represent the feature dependencies. Leveraging this adjacency matrix, the SC module mines spatial feature patterns using an enhanced two-dimensional convolutional neural network. Meanwhile, the TC module captures temporal patterns with a Long Short-Term Memory (LSTM) neural network. Finally, the outputs from the SC and TC modules are fused to generate the final results. It is notable that POWERDIS operates on each node and is fine-tuned during program execution, which implicitly accounts for power variations between nodes.

3.3 Power Disaggregation Model

Module 1: Graph Structure Learning (GL) Module. The objective of the GL module is to acquire the spatial dependencies among features by utilizing a graph consisting of said features. As illustrated in Fig. 4, each feature is depicted as a vertex within the graph and their interrelationships are conveyed through weighted edges. Consequently, when using PMCs as input, a weighted feature graph may be formed through the GL module.

To begin with, we select PMCs (see Table 1) that exhibit strong correlations with the components' power consumption. Following this, we gather the PMCs and node power readings and combine them to create the training data (See Fig. 5 (a)). The training data, denoted by X, is represented as a matrix comprised of n rows and $m + 3$ columns. Each row corresponds to a sample $s^{(i)}$, which consists of $m + 3$ elements including m features and two labels, P_{node}, P_{cpu} and P_{mem}. Specifically, at the i-th moment, we amalgamate the collected PMCs and node power consumption to obtain the i-th sample $s_{(i)}$, as shown in Eq. (2),

$$s_{(i)} = << PMC_1^{(i)}, ...PMC_m^{(i)}, P_{node}^{(i)} >, < P_{cpu}, P_{mem} >> \qquad (2)$$

where $PMC_1^{(i)}...PMC_2^{(m)}$, $P_{node}^{(i)}$ are features and P_{cpu} and P_{mem} are labels.

Then, the adjacent matrix $\mathbf{A}$ ($\mathbf{A} \in R^{(N+1)\times(N+1)}$) is defined to describe the connectivity of features. Note that one additional dimension in $\mathbf{A}$ stands for the CPU power or the memory power. The degree matrix $\mathbf{D}$ contains a lot of A_{ij}, which stores the correlation coefficient between x_i and x_j. Finally, we construct the weighted graph in terms of the adjacency matrix.

Module 2: Spatial Convolution (SC) Module. Given an adjacency matrix $\mathbf{A}$ and the feature matrix $\mathbf{H}$, the SC module can be expressed as Eq. (3),

$$\mathbf{H}^{l+1} = \sigma(\mathbf{D}^{-\frac{1}{2}}\mathbf{A}\mathbf{D}^{\frac{1}{2}}\mathbf{H}^{(l)}\mathbf{W}^{(l)}) \qquad (3)$$

where $\mathbf{H}^{(l)}$ is the model's output at the lth layer, $\sigma(\cdot)$ is the nonlinear transformation.

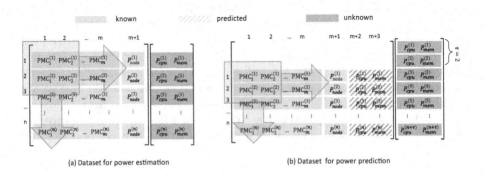

(a) Dataset for power estimation   (b) Dataset for power prediction

Fig. 5. Training data used in POWERDIS.

As depicted in Fig. 4, the SC module comprises a standard two-layer convolutional neural network. Each layer consists of a Convolutional Layer (CONV), a Rectified Linear Unit (ReLU) Layer for applying non-linear transformations, a MaxPooling Layer (MaxPool), and a Fully Connected Layer (FC). The SC module's output is then merged with that of the Temporal Convolutional (TC) module to achieve accurate power predictions. To extract spatiotemporal dependencies among features, POWERDIS employs two feature extractors (illustrated by the orange arrows in Fig. 5). While one extractor extracts features by sliding horizontally across features, the other slides vertically across features, capturing changes over time.

Module 3: Temporal-Pattern Convey (TC) Module. As exemplified in Fig. 4, the TC module executes in a parallel manner to the SC module and houses a recurrent neural network endowed with Long Short-Term Memory (LSTM) units, tailored to capture temporal patterns. Specifically, the TC module includes two consecutive LSTM units followed by a fully connected layer (FC), which adapts the output shape to maintain conformity with the SC module's output shape. The same shape was used to make it easier to connect the two and transform them into a predicted target through the FC layer.

3.4   Future Power Forecasting

In addition to estimating power consumption at the component level, POWERDIS can also be used for forecasting instantaneous power consumption in the subsequent time step (such as the instantaneous power consumption after $10\,\mathrm{s}$). The POWERDIS framework necessitates only minor adjustments to the dataset construction process in power forecasting while maintaining a consistent modeling approach as power estimation. Specifically, apart from capturing the existing power consumption of nodes, the forecasting technique involves incorporating CPU and memory power consumption of preceding time steps as supplementary features. As shown in Fig. 5 (b), we displace the labels by a user-specified period τ (which is equal to 2 in Fig. 5), to rebuild the training set $\tilde{X}$. The training data is represented as an n-by-$(m+5)$ matrix, including $m+3$ significant characteristics such as m PMCs, P_{node}, P_{cpu}, and P_{mem} at the i-th moment and 2 labels, encompassing P_{cpu} and P_{mem} at the $(i+\tau)$-th moment. Each row pertains to a sample $s'^{(i)}$, embodying $m+5$ constituents. To be specific, at the i-th moment, we amalgamate the collected PMCs and node power consumption to obtain the i-th sample $s'_{(i)}$, as shown in Eq. (4),

$$s'_{(i)} =<< PMC_1^{(i)}, ...PMC_m^{(i)}, P_{node}^{(i)}, P_{cpu}^{(i)}, P_{mem}^{(i)} >, < P_{cpu}^{(i+\tau)}, P_{mem}^{(i+\tau)} >> \quad (4)$$

where $PMC_1^{(i)}...PMC_2^{(m)}$, $P_{node}^{(i)}$, $P_{cpu}^{(i)}$ and $P_{mem}^{(i)}$ are features and $P_{cpu}^{(i+\tau)}$ and $P_{mem}^{(i+\tau)}$ the are labels. This dataset design approach leverages the trending characteristics of power consumption data and POWERDIS's ability to handle long time series.

4   Experimental Setup

4.1   Platform

To test POWERDIS's effectiveness, we implement and evaluate it on an ARM-based system where a BMC is deployed on a plug-in consisting of multiple nodes. Each compute node is equipped with 64-core ARMv8 processors, and 128 GB DDR4. While this system provides only node power consumption, we can capture CPU and memory power consumption using a special direct measurement method (See Fig. 6). Although this direct measurement method is accurate, it is not suitable for large-scale deployments, as the hardware cost will become unacceptable as the system size increases. We emphasize that the modeling approach described in this paper is generic, making POWERDIS compatible with any system. To validate this point, we also perform POWERDIS on an X86-based platform to determine the extent of its generalizability, as detailed in Sect. 5.3. The processor operates from 2.0 to 2.3 GHz, and the results given in Sect. 5 were obtained using a frequency of 2.0 GHz. Additionally, to test the sensitivity of to different frequencies, we set three frequency levels: min (1.4 GHz), mid (1.8 GHz), and max (2.2 GHz). More details are given in Sect. 5.4.

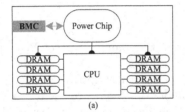 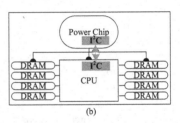

Fig. 6. Power measurement of ARM-based platform.

4.2   Measurement

We use two methods to collect real power consumption. First, we use the BMC to read the power consumption from the power chip using an integrated measurement method based on IPMI. Second, we cascade the I^2C interfaces of the power supply and CPU, and the operating system directly accesses the registers $0x8b$ and $0x8c$ to read the real-time processor voltage and current. This allows us to obtain the real-time power consumption of the processor (P_{cpu}) and memory (P_{mem}) at a sampling rate of 1 Sa/s. The power reading error is 0.1 W, which is lower than the error range of 1 W in the power consumption reading tool provided by the board supplier. We consider the power readings through direct measurement as the most accurate. In addition to P_{cpu} and P_{mem}, we also account for the peripheral's power consumption (P_{other}). During power modeling, P_{other} is assumed to be constant at 25 W, which is measured when no workload is running on the compute node. Our experiments have shown that P_{other} varies very little, within a range of under 1 W. We keep it constant to simplify measurements and ensure sufficient model accuracy.

Table 1. PMCs used in POWERDIS.

Unit		PMC Events
Core	Cycles	CPU_CYCLES
	Instructions	INST_RETIRED
	Branch instructions	BR_PRED
Lx Cache	Load instructions	LxI_Cache_LD
	Store instructions	LxD_CACHE_ST
Main Memory	Bus access	BUS_ACCES
	Memory access	MEM_ACCESS

Aside from collecting power consumption data, performance data is also required for modeling purposes. We collect PMC events (as shown in Table 1) utilizing a Linux loadable kernel module at a sampling rate of 1 Sa/s. The readings from various per-core counters are aggregated to achieve the desired performance data. All measurements are obtained by running experiments in standalone mode to minimize disturbances from fluctuations in system performance.

4.3   Dataset

We need empirical data to train our power model. For that purpose, a total of 96 benchmarks are used, including 43 benchmarks from SPEC CPU 2017 [10], 36 from PARSEC [11], 12 from HPCC [12], 2 from Graph500 [13], as well as HPL-AI [14], SMG2000, and HPCG [15]. The configurations of benchmarks are listed in Table 2. These benchmarks cover a wide range of application domains from scientific applications to consumer workloads. These benchmarks are either compute-intensive or memory-intensive and intended to stress the CPU and DRAM components.

To fully evaluate POWERDIS, we use 5-fold cross-validation and design two methods to construct a training set. The first approach involves assessing seen programs, whereby the training set encompasses samples that are associated with the target program. Alternatively, the second method evaluates unseen programs, in which none of the samples within the training set pertain to the target program. Specifically, we have grouped the programs belonging to the same benchmark suite that comprises different kinds of programs, resulting in a total of seven sets. From these, we randomly select one set as the target program(s) and use the remaining sets as the corresponding training set. We collect

Table 2. Configurations of benchmarks.

Benchmark suite	Parameter	Range	Step	# Nodes
Graph500	SCALE	[8–26]	2	[1, 2, 4, 8, 16, 32]
	Edgefactor	[8, 16, 32, 64]	NAN	
SMG2000	nx/ny/nz	[100–250]	50	[1, 2, 4, 8, 16, 32]
	Px/Py/Pz	[4, 8, 16, 64]	NAN	
	cx/cy/cz	[0.1–1.0]	0.1	
HPCC	N	[80000–160000]	20000	[1, 2, 4, 8, 16, 32]
	P/Q	[1]		
SPEC	scale	test, train, ref	NAN	[1, 2, 4, 8, 16, 32]
PARSEC	scale	test, simsmall, simmedium, simlarge	NAN	
HPCG	nx/ny/nz	[52, 104, 208]	NAN	[1, 2, 4, 8, 16, 32]
HPL-AI	N	[80000–160000]	20000	[1, 2, 4, 8, 16, 32]

1000 samples from each set in chronological order to form a training set containing 6000 samples (unseen set) and 6300 samples (seen set) and a test set containing 1000 (unseen) and 700 (seen) samples. Our selection of combinations is presented in Table 3. Due to page limitations, we only provide averaged results for both the seen and unseen sets in Sect. 5.

4.4   Models

We employ 12 baseline models to evaluate the accuracy of PowerDis, including ten spatial-dependent methods(four linear and six nonlinear), and two temporal-dependent methods. It should be noted that all baseline models adopt the same features as PowerDis and undergo fine-tuning. All features listed in Table 1 are used when building the power models for CPU and DRAM. The linear and nonlinear models are constructed using algorithms from the *scikit-learn* package, while the two RNN models are built based on the structure of PowerDis,

Table 3. Combination of seen (unseen) sets used for evaluation. The number of programs included in the benchmark suite is denoted in parentheses.

Training set	Test set
SPEC(43), PARSEC(36), HPCC(12), Graph500(2), HPL-AI(1), SMG2000(1)	HPCG(1)
SPEC(43), PARSEC(36), HPCC(12), Graph500(2), HPL-AI(1), HPCG(1)	SMG2000(1)
SPEC(43), PARSEC(36), HPCC(12), Graph500(2), SMG2000(1), HPCG(1)	HPL-AI(1)
SPEC(43), PARSEC(36), HPCC(12), SMG2000(1), HPL-AI(1), HPCG(1)	Graph500(2)
SPEC(43), PARSEC(36), SMG2000, HPL-AI(1), HPCG(1), Graph500	HPCC(12)
SPEC(43), SMG2000(1), HPL-AI(1), HPCG(1), Graph500(2), HPCC(12)	PARSEC(36)
SMG2000(1), HPL-AI(1), HPCG(1), Graph500(2), HPCC(12), PARSEC(36)	SPEC(43)

with GridSearch used to tune the hyperparameters in each cross-validation. The hyperparameters of each model are provided in Table 4. The experimental settings for all methods are consistent.

Table 4. Hyperparameters of baseline models for comparisons.

Type	Model	Abbreviation	Hyperparameters
Spatial	Linear Regression [16]	LR	automatic options
	Lasso Regression [17]	LaR	automatic options
	Ridge Regression [18]	RR	solver=auto
	SGD Regression [18]	SGD	criterion=squared_error, max_iter=10000
	Decision Tree [19]	DT	criterion=squared_error
	Random Forest [18]	RF	#trees=10
	Gradient Boosting Tree [18]	GB	#trees=10
	K Nearest Neighbor	KNN	#neighbors=3, algo.=auto
	Support Vector Machine [17]	SVM	automatic options
	Neural Network [20]	NN	#hidden_size=30, solver=adam, learning_rate=adaptive, max_iter=10000
Temporal	Gated Recurrent Unit	GRU	#units=2
	Long Short-Term Memory [8]	LSTM	#units=2

4.5   Metric

We use the Mean Absolute Percentage Error (MAPE), Root Mean Square Error (RMSE), and Mean Absolute Error (MAE) between the observed power consumption and the predicted power consumption to evaluate the accuracy of POWERDIS.

5   Evaluation

In this section, we first discuss the evaluation results of power estimation, followed by the evaluation of power forecasting. We further investigate its sensitivity to different frequencies, hyperparameters, and its potential for extending to other peripherals, as well as its overhead. Finally, we investigate its sensitivity to different frequencies, hyperparameters, and its potential for extension to other peripherals, as well as overhead.

Table 5. Comparisons of power estimation of POWERDIS and other models on the ARM-based system.

Model	Power estimation						Power forecasting					
	P_{cpu}			P_{mem}			P_{cpu}			P_{mem}		
	MAPE	RMSE	MAE	MAPE	RMSE	MAE	MAPE	RMSE	MAE	MAPE	RMSE	MAE
LR	23.04%	8.59	7.18	9.57%	1.37	0.94	23.05%	12.56	10.49	20.61%	3.14	2.66
LaR	23.62%	8.82	7.50	10.72%	1.59	1.08	21.23%	10.54	9.12	21.67%	3.33	2.83
RR	23.04%	8.59	7.21	9.53%	1.37	0.93	22.21%	11.63	9.87	20.18%	3.08	2.59
SGD	23.13%	8.59	7.21	9.53%	1.37	0.93	21.38%	10.81	9.26	19.76%	3.03	2.52
DT	24.13%	11.91	8.60	11.97%	1.96	1.20	27.82%	14.53	12.21	30.61%	8.66	3.99
RF	18.94%	9.59	6.81	10.52%	1.67	1.08	17.59%	9.28	7.81	27.71%	6.34	3.66
GB	25.43%	11.13	8.67	9.96%	1.54	1.01	20.84%	10.03	8.81	19.65%	2.95	2.52
KNN	20.37%	10.58	7.40	10.99%	1.84	1.13	21.36%	12.79	9.98	21.83%	3.42	2.87
SVM	21.40%	8.45	7.27	8.40%	1.55	1.01	25.40%	15.45	12.27	23.40%	3.65	3.09
NN	22.84%	8.52	7.24	9.18%	1.34	0.91	26.33%	17.44	12.78	19.82%	3.01	2.52
GRU	26.48%	8.86	7.79	14.17%	1.52	1.33	23.48%	10.01	9.01	19.19%	2.57	2.26
LSTM	17.80%	10.23	6.51	10.29%	1.48	1.06	20.95%	10.00	8.85	19.26%	2.84	2.45
POWERDIS	10.31%	2.90	1.75	10.24%	0.98	0.95	8.27%	4.90	3.75	14.68%	2.38	1.95

5.1 Power Estimation

POWERDIS has been demonstrated as an effective method for modeling both spatial dependence and temporal dependence across different features of power consumption behavior, exhibiting clear advantages over other models. Table 5 presents the evaluation results of power estimation using POWERDIS. As evident from the table, POWERDIS achieved significant improvement for the seen application, with metrics of 10.31% (MAPE), 2.90 (RMSE), and 1.75 (MAE) for P_{cpu} and 10.24% (MAPE), 0.98 (RMSE), and 0.99 (MAE) for P_{mem}.

For Unseen Application. The model accuracy of POWERDIS for unseen applications is commendable. Table 5 demonstrates that the metrics for estimating P_{cpu} are 8.27% (MAPE), 4.90 (RMSE), and 3.75(MAE), respectively, while those for estimating P_{mem} are 14.68% (MAPE), 2.38 (RMSE), and 1.95 (MAE). This is attributed to the 2D convolution in POWERDIS, which precisely models power consumption trends even for unseen applications. Therefore, based on the assessment of unseen applications, resetting POWERDIS through re-execution of the initialization step seems unnecessary if the workload is non-repeated and exhibits power characteristics that undergo transient fluctuations.

Comparisons with Baseline Models. We have conducted a comparative study of spatial-dependent and temporal-dependent (from the same feature) methods to estimate power consumption. Specifically, we have evaluated four common linear methods and six machine learning-based methods, and two temporal-dependent methods that are based on deep learning (GRU and

LSTM [8]). Our results indicate that POWERDIS outperforms other models in terms of MAPE, with a reduction of 5%–16%, and MAE, with only a reduction of $2w$ for P_{cpu} and $1w$ for P_{mem} when compared to other models in power estimation. In terms of unseen applications, POWERDIS achieves a reduction of 9%–19% in MAPE and a reduction of $4w$ for P_{cpu} and $2w$ for P_{mem} in MAE compared to other models. This improvement can be attributed to two key factors. Firstly, POWERDIS can detect the spatiotemporal dependence of power measurement counters (PMCs), which makes it particularly well-suited for environments with complex spatial structures. Secondly, it considers node power consumption, leading to improved accuracy for P_{mem}. P_{node}, which is the cumulative power consumption of all components within the compute node, exhibits a strong correlation between component power consumption.

Table 6. Comparisons of power forecasting of POWERDIS and other models on the ARM-based system.

Model	Power estimation						Power forecasting					
	P_{cpu}			P_{mem}			P_{cpu}			P_{mem}		
	MAPE	RMSE	MAE	MAPE	RMSE	MAE	MAPE	RMSE	MAE	MAPE	RMSE	MAE
LR	23.30%	8.79	7.32	9.53%	1.36	0.93	24.89%	14.30	11.41	20.20%	3.10	2.58
LaR	23.86%	9.01	7.62	10.74%	1.60	1.08	22.28%	11.52	9.60	21.69%	3.34	2.84
RR	23.30%	8.79	7.32	9.53%	1.36	0.93	23.81%	13.04	10.68	20.28%	3.10	2.60
SGD	23.57%	8.83	7.39	9.58%	1.36	0.93	22.42%	11.72	9.70	20.02%	3.06	2.56
DT	23.97%	12.09	8.60	12.08%	1.81	1.21	46.53%	24.86	20.5	33.02%	10.00	4.38
RF	19.03%	10.03	6.94	10.60%	1.62	1.08	19.25%	10.77	8.79	31.43%	7.39	4.23
GB	25.11%	10.93	8.54	10.09%	1.54	1.02	21.44%	10.82	9.29	20.61%	3.13	2.66
KNN	20.33%	10.62	7.40	10.88%	1.68	1.11	22.18%	13.03	10.24	21.40%	3.46	2.82
SVM	22.46%	12.76	8.43	10.20%	1.61	1.06	25.61%	15.79	12.43	23.20%	3.61	3.06
NN	23.19%	8.77	7.40	9.19%	1.35	0.91	30.06%	22.85	14.96	20.15%	3.08	2.56
GRU	26.53%	9.13	7.60	23.57%	2.31	2.13	25.04%	13.46	11.06	21.38%	3.24	2.71
LSTM	18.15%	10.17	6.56	10.32%	1.48	1.06	23.94%	13.04	10.90	19.92%	2.98	2.56
POWERDIS	12.30%	6.83	3.84	11.24%	1.38	0.99	14.23%	9.83	6.84	15.47%	2.40	1.99

5.2 Power Forecasting

The evaluation results of power forecasting using POWERDIS are presented in Table 6. As can be observed from the table, POWERDIS has demonstrated a significant improvement over known applications with respect to the performance metrics of 12.30% (MAPE), 6.83 (RMSE), and 3.84 (MAE) for P_{cpu}, and 11.24% (MAPE), 1.38 (RMSE), and 0.99 (MAE) for P_{mem}.

For Unseen Applications. We assess the efficacy of POWERDIS and other approaches for power consumption forecasting in previously unseen applications.

The prediction targets comprise P_{cpu} and P_{mem} after a 10-s interval. The comparison methods employ the same dataset construction approach, whereas POW-ERDIS utilizes the methodology outlined in Sect. 3. The forecast results are presented in the rightmost column of Table 6, with a MAPE of 14.23% and an MAE of 6.84 W. It is noteworthy that power forecasting presents a greater challenge than power estimation, resulting in marginally lower prediction accuracy.

Compariosns with Baseline Models. Similarly, we compare POWERDIS with spatial- and temporal-dependent methods. Compared with others, the MAPE of POWERDIS is reduced at least by 5%–30%, the RMSE is reduced by 5–9, and the MAE is reduced by 4–9 W. Although POWERDIS's power forecasting is slightly inferior to power estimation, it still outperforms other baseline models by exploring temporal dependencies, especially those from different features, thus greatly improving model accuracy.

5.3 On X86-Based Platform

To validate the applicability of POWERDIS as a general-purpose approach for X86-based platforms, we conducted an implementation and evaluation on our X86-based computing cluster that emulates the architecture of Tianhe-1A [21]. The cluster is comprised of 64 compute nodes, each of which is equipped with Intel® Xeon® E5-2660 v3 processors [22]. In order to leverage the high-precision capabilities of RAPL that the processors support, we utilized the *perf* tool to retrieve performance monitoring events */power/energy-pkg/* and */power/energy-ram/* at 1-second intervals to obtain the CPU and memory power consumption. As demonstrated in [23], PMC sampling does not introduce unacceptable inaccuracies. To verify the accuracy of POWERDIS, we compared RAPL readings with predicted values generated by our system.

Table 7 shows the evaluation results of POWERDIS for unseen applications on the X86-based platform. Only a comparison of unseen applications is given here since the baseline model can easily overfit applications seen. As seen from Table 7, POWERDIS, achieves the lowest error rate no matter in power estimation or power forecasting. When forecasting the future power consumption, the MAPEs are reduced by up to 25% for P_{cpu} and up to 30% or more for P_{mem}. POWERDIS has a considerable advantage over other models no matter in estimating or predicting P_{mem} since it incorporates node power information, despite the fact that it estimates P_{cpu} similarly to them.

5.4 Discussion

Frequency. We selected three different CPU frequency levels, namely *min* (1.4 GHz), *mid* (1.8 GHz), and *max* (2.2 GHz), to investigate POWERDIS's sensitivity to variations in CPU frequency. Figure 7 illustrates the Mean Absolute Percentage Errors (MAPEs) of POWERDIS for P_{CPU} and P_{MEM} when the Graph500 program is executed at different frequency levels. As depicted in the figure, our

Table 7. Comparisons of POWERDIS and other models on the X86-based system.

Model	Power estimation						Power forecasting					
	P_{cpu}			P_{mem}			P_{cpu}			P_{mem}		
	MAPE	RMSE	MAE	MAPE	RMSE	MAE	MAPE	RMSE	MAE	MAPE	RMSE	MAE
LR	16.54%	25.22	15.76	38.61%	21.37	18.73	30.75%	48.21	24.88	42.21%	23.52	20.82
LaR	17.22%	26.27	15.88	37.89%	23.72	20.52	31.26%	49.31	24.98	43.93%	24.30	21.97
RR	16.37%	25.07	15.50	38.19%	21.36	18.68	30.80%	48.31	24.88	42.40%	23.59	20.96
SGD	16.40%	25.50	15.76	37.62%	21.36	18.66	32.19%	49.68	27.01	43.19%	23.92	21.41
DT	10.42%	19.44	13.73	42.94%	23.53	21.28	37.01%	62.05	37.42	47.72%	26.51	23.54
RF	8.40%	15.06	10.40	34.96%	21.03	18.79	33.49%	49.97	32.67	43.72%	25.36	22.77
GB	17.16%	26.17	15.52	40.8%	23.65	21.41	31.35%	48.63	26.00	45.57%	24.52	22.81
KNN	9.21%	16.99	10.58	28.06%	19.35	16.38	34.55%	52.74	31.63	39.65%	22.79	20.25
SVM	19.93%	30.79	17.36	33.03%	20.96	18.39	31.92%	50.78	24.95	41.77%	22.54	20.54
NN	14.93%	23.00	14.04	29.36%	18.83	16.38	31.15%	48.87	25.22	39.07%	23.13	20.18
GRU	5.53%	10.66	7.85	36.74%	26.45	23.11	30.55%	48.31	24.29	48.02%	26.99	24.91
LSTM	5.50%	11.43	8.48	39.27%	26.43	23.56	31.40%	49.80	24.75	49.24%	27.15	25.21
POWERDIS	4.27%	8.90	5.75	10.64%	5.54	3.44	10.23%	11.83	6.84	15.32%	18.45	7.32

proposed method demonstrates high accuracy in predicting instantaneous CPU and memory power consumption across all frequency levels. However, the accuracy of predicting P_{CPU} and P_{MEM} decreases with the increase in frequency, mainly because of the escalating CPU activity, which makes modeling more challenging. Despite this, even at the highest frequency (10% for P_{CPU} and 14% for P_{MEM}), our approach surpasses other existing modeling techniques.

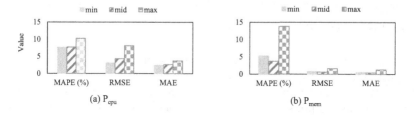

(a) P_{cpu} (b) P_{mem}

Fig. 7. Sensitivity to different frequency levels of the POWERDIS framework.

Extension to Peripheral Devices. While the scope of this research paper mainly centers on the CPU and memory components, the present discourse endeavors to explore the feasibility of extending the applicability of POWERDIS to encompass other peripheral devices that have hardware performance counters. We believe that this technology could be adapted to accelerator and network devices, given that power modeling for these devices also relies on PMCs, albeit utilizing different monitoring units. For example, the use of POWERDIS may be advantageous in GPU power modeling, owing to the similarities shared with CPU power modeling. However, more comprehensive architecture-oriented considerations are typically required when assessing the former [24]. As a part of our future research, we aim to delve deeper into resolution recovery for other peripheral devices.

Hyperparametric Analysis. We have conducted hyperparameter analysis on POWERDIS, where the number of layers of the LSTM was varied from 1 to 100 and the number of convolution layers was varied from 1 to 10 to determine the optimal network structure for utilizing node information. Our findings indicate that the accuracy first increases and then decreases as the number of layers increases, with the two-layer model yielding the best performance. Using this approach, we established the optimal model structure, which is further detailed in Sect. 3.

Overhead. As a software-based model, POWERDIS does not bear any associated hardware expenses. Moreover, it capitalizes on power consumption and performance data to fabricate a streamlined process on each node, thereby facilitating its operation in both single-node and multi-node scenarios. Additionally, a slight memory overhead is incurred, however, this does not impede program performance, and the runtime overhead is negligible, with time costs being dependent on the velocity of PMCs collected and clocking at less than 1 ms. It took less than 10 min to train our model offline and less than 2 s to fine-tune the trained model. For example, for an application with a runtime of 10 min, active learning fine-tuning led to a performance loss of approximately 3% after 10 adjustments, with a fine-tuning time of around 2 s.

6 Related Work

In large-scale parallel computing systems, power monitoring is a critical aspect that must satisfy several requirements such as accuracy, coverage, latency, scalability, and deployment cost [5]. Component-level power monitoring involves the real-time capture of power consumption for each component within a compute node. It facilitates the development of optimization strategies that target specific components, for instance, CPU/DRAM dynamic voltage frequency scaling (DVFS) and uncore frequency scaling, rather than treating the entire node as a black box. The existing component-level power monitoring techniques are categorized as hardware-centric measurement or software-centric power modeling techniques, each having its limitations. A detailed overview of these methods is provided in the rest of this section.

6.1 Hardware-Centric Measurement

Hardware-centric measurement schemes for component-level power monitoring rely on power sensors deployed within the compute node or external devices directly connected to the compute node in an intrusive manner. Currently, there are some tools, such as PowerPack [25], PowerMon, and PowerMon2 [26] are typical devices built into commodity servers. However, both employ autonomous measuring boards, resulting in significant hardware expenditures and measurement equipment complexity. Ilsche et al. [5] propose another hardware measurement device, HAEC, whose sampling rate is up to 500 kSa/s. However, it is only suitable for a single node rather than a cluster. Libri et al. [27] utilize a dedicated embedded computer to collect power measurements, inevitably introducing additional hardware overhead. The intelligent platform management interface (IPMI) [28] is a popular integrated measurement solution used in most supercomputers worldwide. However, despite its widespread use, IPMI-based solutions suffer from long readout delays and limitations by the hardware-exposed sensor interfaces and cannot provide detailed information on computing components like CPUs and memory. Overall, Hardware-centric measurement schemes offer several advantages such as high accuracy, good temporal resolution, and scalability. However, the major drawback is its high hardware cost, which increases as the system scales and the monitoring granularity becomes finer. POWERDIS differs from hardware-centric methods in that we rely on established hardware measurement techniques for production systems and leverage power models, rather than incurring extra hardware expenses, to gather component power consumption data.

6.2   Software-Centric Power Modeling

Utilizing models that are generally formulated with PMCs located in the CPU for power monitoring is an indirect, non-intrusive solution. Due to its advantages including high spatial and temporal resolution, low cost, and ease of scalability, power modeling has received a great deal of research attention, with improving the accuracy of power consumption models being a top priority. Earlier studies have attempted to explore the linear (nonlinear) relationship between PMC and power consumption. For example, Singh et al. [29] employ a power model based on multiple linear regression using PMCs. Powell et al. [16] use a linear regression model based on PMCs to estimate the activity factor and further estimate the power consumption. In addition to linear methods, machine learning (ML)-based methods are also used to mine non-linear relationships between variables. For example, Song et al. [20] propose a power and energy model using a configurable back-propagation artificial neural network (BP-ANN) as the modeling method. Sagi et al. [30] use negative feedback neural networks to model the power consumption of multi-core processors. ML-based modeling approaches exhibit superior accuracy compared to linear approaches; however, they come with high model implementation complexity, repeated experimental testing, a large amount of training data, and careful selection of kernel programs [31]. These issues are demanding and require attention when using ML-based power modeling methods. Later, some studies explored the implicit relationship between PMCs and power consumption over time. For example, Sagi et al. [8] forecast the future power consumption using an LSTM neural network. However, these studies have solely concentrated on the dependencies within the same dimension while disregarding the influence of previous values from other dimensions. POWERDIS considers both spatial dependence and temporal dependence across different features, improving the model accuracy significantly.

Besides academia, the industry also cares about component-level power consumption. Intel® introduced RAPL [6] in its sandy bridge architecture to provide power readings of CPU and memory. In essence, RAPL uses a power model that was derived from the correlation between hardware counters, temperature, and leakage power consumption. Based on their generality, we classify power models into two categories: vendor-specific power models and general-purpose power models. The same goes for APM [32] for AMD processors. However, its customization is a barrier to generality, especially since RAPL only supports parts of Intel® processors (related to the CPU model), let alone ARM-based platforms. Furthermore, relying solely on it does not enable system-wide power monitoring because RAPL is limited to a single processor. In contrast, POWERDIS not only covers the entire system but also has greater generality compared to other power monitoring schemes.

7 Conclusions

This paper presents a novel technique, named POWERDIS, which precisely measures and predicts the power consumption of computing components within a compute node. To achieve this, POWERDIS leverages a power disaggregation model consisting of three modules that analyze spatiotemporal patterns to accurately distribute power readings to low-level components. Extensive evaluations demonstrate that POWERDIS achieves remarkable accuracy in power estimation and forecasting on ARM-based and X86-based platforms. Due to the complexity of GPU power modeling, we postpone investigating fine-grained power monitoring for GPUs until future research endeavors.

References

1. Wu, F., et al.: A holistic energy-efficient approach for a processor-memory system. Tsinghua Sci. Technol. **24**(4), 468–483 (2019)
2. Gholkar, N., et al.: Uncore power scavenger: a runtime for uncore power conservation on HPC systems. In: Proceedings of the International Conference for High Performance Computing, Networking, Storage and Analysis, SC 2019, New York, NY, USA (2019)
3. Zhou, L., et al.: Gemini: learning to manage CPU power for latency-critical search engines. In: 2020 53rd Annual IEEE/ACM International Symposium on Microarchitecture (MICRO), pp. 637–349 (2020)
4. Juan, C., et al.: More bang for your buck: boosting performance with capped power consumption. Tsinghua Sci. Technol. **26**(3), 370 (2021)
5. Ilsche, T., et al.: Power measurement techniques for energy-efficient computing: reconciling scalability, resolution, and accuracy. Comput. Sci. Res. Dev. **34**, 03 (2019)

6. Rotem, E., et al.: Power-management architecture of the intel microarchitecture code-named sandy bridge. IEEE Micro **32**(2), 20–27 (2012)

7. Sagi, M., et al.: A lightweight nonlinear methodology to accurately model multicore processor power. IEEE Trans. Comput. Aided Des. Integr. Circuits Syst. **39**(11), 3152–3164 (2020)

8. Sagi, M., et al.: Long short-term memory neural network-based power forecasting of multi-core processors. In: 2021 Design, Automation & Test in Europe Conference & Exhibition (DATE), pp. 1685–1690 (2021)

9. Haj-Yihia, J., Yasin, A., Asher, Y.B., Mendelson, A.: Fine-grain power breakdown of modern out-of-order cores and its implications on skylake-based systems. ACM Trans. Archit. Code Optim. **13**(4) (2016)

10. Spec cpu 2017 (2017). http://www.spec.org/cpu2017/

11. The princeton application repository for shared-memory computers (2020). https://parsec.oden.utexas.edu/

12. Luszczek, P., et al.: S12-the HPC challenge (HPCC) benchmark suite. In: Proceedings of the ACM/IEEE SC2006 Conference on High Performance Networking and Computing, November 11–17, 2006, Tampa, FL, USA (2006)

13. Graph500 (2017). https://graph500.org/

14. Jack, D., et al.: HPL-AI mixed-precision benchmark (2019). https://icl.bitbucket.io/hpl-ai/

15. High performance conjugate gradients (2020). https://www.hpcg-benchmark.org/

16. Powell, M.D., et al.: Camp: a technique to estimate per-structure power at run-time using a few simple parameters. In: 15th International Conference on High-Performance Computer Architecture (HPCA-15 2009), 14–18 February 2009, Raleigh, North Carolina, USA (2009)

17. McCullough, J.C., et al.: Evaluating the effectiveness of model-based power characterization. In: Proceedings of the 2011 USENIX Conference on USENIX Annual Technical Conference, USENIX ATC'11, page 12, USA, 2011. USENIX Association (2011)

18. Wu, X., Taylor, V., Lan, Z.: Performance and power modeling and prediction using mummi and 10 machine learning methods. Concurr. Comput. Pract. Exp. **35**, e7254 (2022)

19. Borghesi, A. Bartolini,, A., et al.: Predictive modeling for job power consumption in HPC systems, pp. 181–199 (2016)

20. Song, S., et al.: A simplified and accurate model of power-performance efficiency on emergent GPU architectures. In: 2013 IEEE 27th International Symposium on Parallel and Distributed Processing, pp. 673–686 (2013)

21. Yang, X.J., Liao, X.-K., et al.: The tianhe-1a supercomputer: Its hardware and software. J. Comput. Sci. Technol. **26**(3), 344–351

22. Intel Xeon Processor E5–2660 v. 3, 2020. https://ark.intel.com/content/www/us/en/ark/products/81706/intel-xeon-processor-e5-2660-v3-25m-cache-2-60-ghz.html

23. Isci, C.: Runtime power monitoring in high-end processors: methodology and empirical data. In: Proceedings of International Symposium on Microarchitecture (2003)

24. Bridges, R.A., et al.: Understanding GPU power: a survey of profiling, modeling, and simulation methods. ACM Comput. Surv. **49**(3) (2016)

25. Ge, R., et al.: Powerpack: Energy profiling and analysis of high-performance systems and applications. IEEE Trans. Parall. Distrib. Syst. **21**(5) (2010)

26. Bedard, D., et al.: Powermon: fine-grained and integrated power monitoring for commodity computer systems. In: Proceedings of the IEEE SoutheastCon 2010 (SoutheastCon), pp. 479–484 (2010)

27. Libri, A., et al.: Dig: enabling out-of-band scalable high-resolution monitoring for data-center analytics, automation and control (extended). Clust. Comput. 24(4), 2723–2734 (2021)

28. Intelligent platform management interface (2020). https://www.intel.com/content/www/us/en/products/docs/servers/ipmi/ipmi-second-gen-interface-spec-v2-rev1-1.html

29. Singh, K., et al.: Real time power estimation and thread scheduling via performance counters. SIGARCH Comput. Archit. News 37(2), 46–55 (2009)

30. Sagi, M., et al.: Fine-grained power modeling of multicore processors using FFNNs. In: Orailoglu, A., Jung, M., Reichenbach, M. (eds.) Embedded Computer Systems: Architectures, Modeling, and Simulation, pp. 186–199. Springer, Cham (2020)

31. O'Brien, K., et al.: A survey of power and energy predictive models in HPC systems and applications. Comput. Surv. 50(3) (2017)

32. AMP. "amd opteron 6200series processors linux tuning guid (2012)

WBGT Index Forecast Using Time Series Models in Smart Cities

Kai Ding[1,2], Yidu Huang[1], Ming Tao[1(✉)], Renping Xie[1], Xueqiang Li[1],
and Xuefeng Zhong[1]

[1] School of Computer Science and Technology, Dongguan University of Technology,
Dongguan 523808, China
taom@dgut.edu.cn
[2] Guangdong Laboratory of Artificial Intelligence and Digital Economy (SZ),
Shenzhen 518107, China

Abstract. WBGT (Wet Bulb Globe Temperature) was originally championed by the United States military and has been widely implemented in daily training to prevent casualties or injuries among soldiers due to unfavorable high temperature and humidity conditions during the summer, and it has been widely used in various fields, such as marathons, military training, and travel. Starting from the daily periodicity of WBGT, this paper uses the Holt-Winters 24 h, and discusses the feasibility of its autocorrelation prediction. The prediction results were evaluated using the time series cross-validation method and RMSE. Two experiments were conducted with the dataset acquired by NCSCO and the self-collected Dongguan University of Technology dataset(DGUT-D). First, a preliminary experiment was conducted using NCSCO data to explore the feasibility of WBGT autocorrelation pre-diction, and then the DGUT-D was used to predict the 24-h WBGT of the Songshan Lake Campus of the Dongguan University of Technology (DGUT).

Keywords: WBGT · Holt-Winters · SARIMA · Prophet

1 Introduction

The climate change is closely related to human activities. In the field of sports science, both professional athletes and amateur people face relevant risks. In a hot

Supported by the Basic and Applied Basic Research Funding Program of Guangdong Province of China (No.2019A1515110303, No.2019A1515110800, No. 2021A1515010656 and 2022B1515120059); the Open Research Fund from Guangdong Laboratory of Artificial Intelligence and Digital Economy (SZ) (No. GMLKF-22-02); the National Natural Science Foundation of China (No. 62001113); the Guangdong University Key Project (No. 2019KZDXM012); the Guangdong Key Construction Discipline Research Ability Enhancement Project (No. 2021ZDJS086); the Guangdong University Key Project (No. 2019KZDXM012); the Dongguan Science and Technology of Social Development Program (No. 20221800902472 and No. 20211800904712); the Research Team Project of Dongguan University of Technology (No. TDY-B2019009).

Z. Tari et al. (Eds.): ICA3PP 2023, LNCS 14490, pp. 347–358, 2024.
https://doi.org/10.1007/978-981-97-0859-8_21

environment, environmental factors affecting the body of athletes are constantly present during outdoor sports. An increase in outdoor temperature will increase the pressure on the cardiovascular, respiratory, and metabolic systems. In addition to temperature, factors such as average radiation, humidity, air pressure, and air quality will also impact the human body [21]. Elderly people, patients with respiratory diseases, and individuals with high body mass index are more susceptible to the negative effects of high temperatures and air pollution [2]. On the other hand, with the advancement and application of Internet of Things (IoT) technology, the rapid increase in sensing devices [1, 16, 23] has significantly reduced the cost of real-time monitoring and predicting environmental changes. In highly developed regions, the federal deployment flexibility of IoT and urban prediction services can greatly improve the living standards of urban residents [22]. By establishing small-scale weather stations based on IoT architecture, it is possible to reduce costs while providing citizens with reliable and precise monitoring and forecasting services within smaller regions, thereby contributing to the construction of smart cities [3, 9, 24].

Taking into account the construction cost and general applicability, in this paper, we have chosen the Wet Bulb Globe Temperature (WBGT) [19, 27] as the index for designing some forecasting services for preventing heatstroke in the face of hot summer conditions. The WBGT was originally developed by the U.S. military. It has been widely used in various fields including sport and High-temperature work, as a screening tool for evaluating environmental heat stress, and has obtained certification from the International Organization for Standardization (ISO 7243). Although some studies suggest that WBGT may need to be replaced by indices that are more in line with human physiological characteristics in the future, its widespread application in sports cannot be denied at present [28]. In terms of long-distance running, research has shown that as the WBGT increases between 5 °C and 25 °C, the marathon runners' performance will decrease, and slower runners are more negatively affected [11]. Kashimura O et al. used a road bike and WBGT measuring devices to perform WBGT index testing on the marathon track in order to evaluate the risk of physical burden on the marathon runners due to high temperature for the Tokyo Summer Olympics in 2020 [15]. In China, almost all middle schools and universities conduct military training for freshmen every year to improve students' self-discipline in China. The month for freshmen enrollment is often August and September, during which the temperature is relatively high every year. Due to the students' physical and weather reasons, there are often cases of heat stroke.

The main work of this article is to use three commonly used time series forecasting methods, Holt-Winters, SARIMA, and Prophet, to predict the WBGT values for the next 24 h. The accuracy of the predictions is evaluated by comparing the errors between the predicted values and the actual values using RMSE. Provide a reference for the summer military training of college freshmen, reducing the risk of heatstroke.

2   WBGT and Forecasting Models

2.1   WBGT

At present, the WBGT (Wet Bulb Globe Temperature) has been widely used in sports training and high-temperature work [4,17,18].

$$Outdoor: WBGT = 0.7T_{nw} + 0.2T_g + 0.1T_a \tag{1}$$

$$Indoor: WBGT = 0.7T_{nw} + 0.3T_a \tag{2}$$

According to the ISO 7243 standard, the WBGT index value is calculated from the natural wet bulb temperature (T_{nw}), black bulb temperature (radiation temperature, T_g), and dry bulb temperature (air temperature, T_a) [26].

2.2   Holt-Winters

In this paper, the linear trend addition mode in the Holt-Winters seasonal method [5,6,12,14] is adopted, and the framework is:

$$ES_{t+T} = S_t^* + R_t T - P_{t+T-N} \quad T \in [1, N] \tag{3}$$

The Eq. (3) is jointly obtained from Eq. (4), Eq. (5), and Eq. (6), and its detailed derivation steps can be found in the original paper [12].

$$S_t^* = A(S_t + P_t) + (1 - A)(S_{t-1}^* + R_t) \tag{4}$$

$$P_t = B(S_t^* - S_t) + (1 - B)P_{t-N} \tag{5}$$

$$R_t = C(S_t^* - S_{t-1}^*) + (1 - C)R_{t-1} \tag{6}$$

Where t represents the time sequence, S is the real observation value, S* is the estimate of the expected value of the distribution, P is the periodic adjustment increment, and R is the trend adjustment increment. A, B, and C are the exponential weighting coefficients.

2.3   SARIMA

The predecessor of SARIMA [7,10,20] is ARIMA, which is a classic and commonly used time series forecasting model and integrates the autoregressive model (AR model), integral (I), and moving average model (MA). The ARIMA model is an extension of the ARMA. When the parameter d (order difference) of ARIMA is zero, the ARMA model is equivalent to ARIMA. As an extension of ARIMA, SARIMA takes into account the seasonal cycle factor and is written as SARIMA (p, d, q) (P, D, Q)S. Where p is the number of lag observations in AR, q determines the number of error terms included in the MA window, d is the differencing factors, and S is the cycle time interval. P, D, and Q values for seasonal parts of SARIMA.

$$\phi_p(\sigma)\varphi_P(\sigma_s)(1 - \sigma)^d(1 - \sigma_s)^D y_t = \theta_q(\sigma)\beta_Q(\sigma_s)\varepsilon_t \tag{7}$$

$(1 - \sigma)^d$ is the differentiating operator, $\phi_p(\sigma)$ is the $AR(p)$ polynomial, $\theta_q(\sigma)$ is the $MA(q)$ polynomial, σ is the backshift operator. $\varphi_P(\sigma_s)$ and $\beta_Q(\sigma_s)$ are for seasonal part.

2.4 Prophet

The Prophet model [8,25] is developed and open-sourced by Facebook, the framework is:

$$f(t) = g(t) + s(t) + h(t) + \epsilon_t \tag{8}$$

where $g(t)$ is the non-periodic changes trend function, $s(t)$ represents the periodic change, $h(t)$ is the special variation item, and ϵ is the noise. Among them, $g(t)$ is divided into a saturating growth model and a piecewise linear model. The short-term WBGT change trend is not fixed and is sensitive to weather changes, so piecewise linear is applied.

$$g(t) = (k + a(t)^T \delta)t + (m + a(t)^T \gamma) \tag{9}$$

k is the growth rate, δ is the adjustment of k, m is the offset parameter, $a(t)$ is the indicator function, and γ makes $g(t)$ continuous.

$$s(t) = \alpha_0 + \sum_{n=1}^{\infty} (a_n cos\frac{n\pi t}{e} + b_n sin\frac{n\pi t}{e}) \tag{10}$$

The periodic variation is approximated by a Fourier series, where e is the period

$$h(t) = \sum_{i=1}^{L} \kappa_i \cdot 1 \quad t \in D_i \tag{11}$$

D_i represents the range of the mutation, κ_i is the impact caused by the mutation, and L is the number of mutations.

2.5 Evaluation Criteria

In this paper, the root mean squared error (RMSE) is used as the model prediction evaluation criteria. N is set to 24 which means 24 h. The x_i and x_i^* represent actual measured and predicted values.

$$RMSE = \sqrt{\frac{\sum_{i=1}^{N}(x_i - x_i^*)^2}{N}} \tag{12}$$

3 Experiment and Results

3.1 Dataset

Data for the North Carolina State Climate Office (NCSCO) is provided by a range of weather web providers including government agencies, educational institutions, and private companies. The public can obtain the data for free, For details, please refer to the website address: https://api.climate.ncsu.edu/usage. The collection time of the North Stanley Middle School dataset (NSMS-D) used in this article is from March 7, 2023, to March 20, 2023, a total of 14 days.

DGUT-D was collected by the AZ87786 thermometer on the top floor of the School of Computer, Songshan Lake Campus, Dongguan University of Technology, from April 7, 2023, to April 21, 2023, for a total of 14 days. Collect once every five minutes, the data length is 4032, the value is the average amount per hour, and the length is 336.

3.2 WBGT of North Stanley Middle School

As shown in Fig. 1 and Fig. 2, from March 7, 2023, to March 20, 2023, the WBGT of North Stanley Middle School showed a daily periodicity but no obvious trend and the daily peaks were mainly concentrated at 12:00 Between 18:00 and 18:00, it is feasible to try to use the Holt-Winters seasonality method to predict WBGT in the next 24 hours using its daily cycle. In the prediction experiment in this paper, time series cross-validation [13] is used which means that independent training and testing sets are not divided, and the WBGT measurement values from the previous days are used to infer the index for the next day and no future variables are used as reference variables. In Fig. 3, the deep blue points represent past time, and the light blue are days to be predicted.

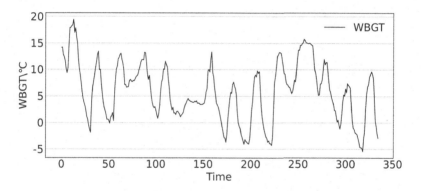

Fig. 1. WBGT of North Stanley Middle School.

The dataset contains 14 days of data. In the experiment using the NSMS-D, the last 5 days will be predicted. Therefore, the reference samples used for the first prediction is 9 days, the second is 10 days, and the third is 11 days, and so on. Prediction accuracy is evaluated through RMSE.

The Holt-Winters seasonal model is used for forecasting, the number of periods is 24, that is, one period per day, and all real values before the forecast time are used as input values to continuously forecast for five days, as shown in Fig. 4.

The prediction of SARIMA is more complicated than that of Holt-Winters. It needs to obtain p, q, d, P, Q, and D. The main steps are shown in Fig. 5.

The ADF test is used to judge whether there is a unit root in the WBGT original sequence, and if the sequence is not stable, there is a unit root (H0 null

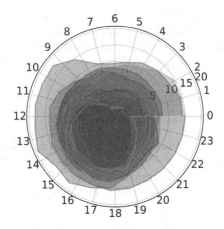

Fig. 2. Polar graph of NSMS-D.

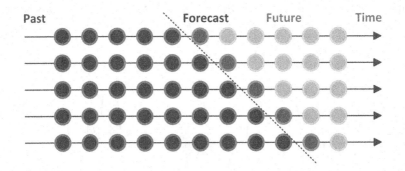

Fig. 3. Cross-validation.

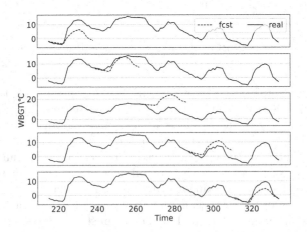

Fig. 4. Forecast by Holt-Winters used NSMS-D.

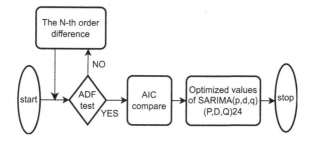

Fig. 5. Determine model parameters based on sequence.

hypothesis). The closer the P-value of the hypothesis test result is to zero, the better. When the P-value is greater than 0.05, the original assumption cannot be rejected, and the time series is unstable (Fig. 6).

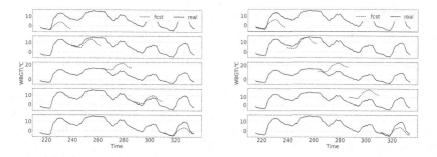

Fig. 6. Forecast by SARIMA(left) and Prophet(right) used NSMS-D

Table 1. The P-value of the differential sequence.

	no difference	1st difference
original	0.088611	2.177824e−12
seasonal difference	0.001354	8.144327e−11

It can be seen from Table 1 that the original WBGT sequence of NSMS-D is not stable. After 1st difference of WBGT, the P-value is 2.177824e−12 and less than 0.05, the null hypothesis can be rejected and the time series is stable, so $d = 1$. When the P-value after seasonal difference is 0.001354 and less than 0.05, the null hypothesis can be rejected and the time series is stable, so $D = 0$. The optimal model is selected through AIC, and the lower the AIC, the better (Table 2). NSMS-D uses SARIMA (2, 1, 0) (1, 0, 1)24 for prediction.

Table 2. The AIC values of SARIMA.

p	q	P	Q	AIC
2.0	0.0	1.0	1.0	347.935
1.0	1.0	1.0	1.0	348.475
2.0	3.0	1.0	1.0	348.666
1.0	2.0	1.0	1.0	349.781
3.0	0.0	1.0	1.0	349.927
2.0	1.0	1.0	1.0	349.929
1.0	0.0	1.0	1.0	349.980
4.0	1.0	1.0	1.0	350.380
0.0	4.0	1.0	1.0	350.829
3.0	3.0	1.0	1.0	350.871

The implementation of Prophet is also based on periods and trends, but from data loading to result in output, there is no need for too much manual intervention. The RMSE of all prediction results is shown in Table 3. Comparing the results of the five days comprehensively, the RMSE of Holt-Winters and SARIMA are the lowest on D5 in five days. From the image, the fitting situation is better than that of the other four days. From the prediction results of Holt-Winters and Prophet, the RMSE and fitting of D2, D4, and D5 are better than SARIMA. Overall, the five-day RMSE of the three models ranged from 2.88 °C to 10.73 °C, which fluctuated too much. This represents the instability of the data.Comparing the three models, Holt-Winters has the best Avg (RMSE) of 5.08 °C. It can be noticed that the three models all produced high RMSE in the prediction of the third day, which were 10.29 °C, 9.06 °C, and 9.82 °C respectively.

Table 3. The prediction results of the three models.

Model	3/16 (D1)	3/17 (D2)	3/18 (D3)	3/19 (D4)	3/20 (D5)	Avg (RMSE)
Holt-Winters	5.76	3.09	10.29	3.38	2.88	5.08
SARIMA	7.36	3.85	9.06	2.68	3.40	5.27
Prophet	6.11	3.26	9.82	10.73	3.33	7.19

Although the internal composition of the three models is different, their basic starting point is to use the periodicity and trend of the sequence to build a model. When the actual value of the forecast target does not meet the trend calculated by the model, it will cause a high RMSE. The WBGT in the first two days showed an upward trend but turned to a decline on the third day. The trend calculated by the model is calculated through the past time, so the forecast result is also an upward trend, which causes a large error in the prediction of D3.

3.3 WBGT of Dongguan University of Technology

In subsequent experiments using the DGUT-D, the weather conditions will be taken into account. The first day's prediction is based on 6 days prior to the first day, followed by a sample of 7 days prior to the second day, and so on (Fig. 7).

During the 14 days, in the Songshan Lake area, the weather was rainy and cloudy for 4 consecutive days starting on April 18, and the weather was fine on the rest of the days.

Because of the rainy days in the last four days, this gives the experiment the opportunity to compare the forecasting effects of rainy and sunny days, so the follow-up experiment will predict WBGT for eight days. April 14th to April 17th is sunny, and April 18th to April 21st is rainy. In Fig. 8, the left side is the sunny day forecast effect, and the right side is the rainy day forecast.

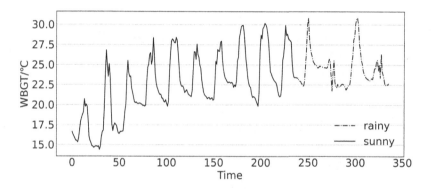

Fig. 7. WBGT of Dongguan University of Technology.

4   Discussion and Conclusion

Among the prediction results of the three models, RMSE is used as the evaluation criterion, and its specific performance is shown in Table 4. Among them, the prediction results of cloudy and rainy days are the worst. The average RMSE of the three models all exceeds 2 °C, and the RMSE of April 19 exceeds 3 °C.

Comparing the prediction results of sunny and rainy weather, combined with Fig. 8, and Table 5, it can be found that the prediction fit and RMSE value of sunny days are significantly better than those of cloudy and rainy days. Holt-Winters' average sunny day average RMSE is 1.328 °C, SARIMA's is 1.574 °C, and Prophet's is 1.412 °C. Obviously, Holt-Winters' forecast is the best.

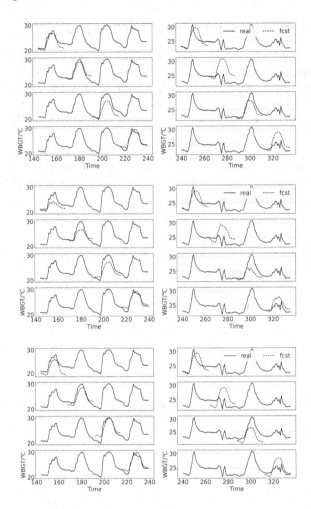

Fig. 8. Forecast by Holt-Winters (top), SARIMA(mid) and Prophet(bottom) used DGUT-D

Table 4. The prediction results of the three models tables.

Date	Holt-Winters	SARIMA	Prophet
2023-4-14	1.068	1.972	2.060
2023-4-15	1.104	1.670	1.233
2023-4-16	2.412	1.862	1.489
2023-4-17	0.727	0.793	0.869
2023-4-18	1.619	1.363	1.559
2023-4-19	4.433	3.340	3.446
2023-4-20	1.342	2.544	2.662
2023-4-21	2.383	1.857	1.865

During rainy weather, the prediction performance of the three models is worse than that of sunny days, no matter the fitting situation or the value of RMSE. Obviously, without the assistance of additional related variables, the autocorrelation of WBGT alone cannot ensure prediction accuracy when the weather changes or is unstable.

Table 5. Average RMSE.

RMSE	Holt-Winters	SARIMA	Prophet
SNY-Avg (RMSE)	1.328	1.574	1.412
RNY-Avg (RMSE)	2.444	2.276	2.383
TOTAL-Avg (RMSE)	1.886	1.925	1.898

The experiment in this paper only uses the autocorrelation of WBGT to carry out prediction experiments, and there are still many shortcomings. For the prediction of WBGT, it is necessary to take into account the local climate characteristics and future weather changes. Its autocorrelation alone cannot guarantee accurate predictions. Nevertheless, when the weather performance is relatively stable, such as on long-term sunny days, it is feasible to use univariate forecasting. This experiment provides a basis for the next step to use future variables (weather forecast temperature and humidity, etc.).

References

1. Alrashdi, I., Alqazzaz, A., Aloufi, E., Alharthi, R., Zohdy, M., Ming, H.: Ad-IoT: anomaly detection of IoT cyberattacks in smart city using machine learning. In: 2019 IEEE 9th Annual Computing and Communication Workshop and Conference (CCWC), pp. 0305–0310. IEEE (2019)
2. Bernard, P., et al.: Climate change: the next game changer for sport and exercise psychology. German J. Exercise Sport Res. 1–6 (2022)
3. Brito, R.C., Favarim, F., Calin, G., Todt, E.: Development of a low cost weather station using free hardware and software. In: 2017 Latin American Robotics Symposium (LARS) and 2017 Brazilian Symposium on Robotics (SBR), pp. 1–6. IEEE (2017)
4. Budd, G.M.: Wet-bulb globe temperature (wbgt)-its history and its limitations. J. Sci. Med. Sport **11**(1), 20–32 (2008)
5. Chatfield, C.: The holt-winters forecasting procedure. J. Roy. Stat. Soc.: Ser. C (Appl. Stat.) **27**(3), 264–279 (1978)
6. Chatfield, C., Yar, M.: Holt-winters forecasting: some practical issues. J. Roy. Stat. Soc. Ser. D: Stat. **37**(2), 129–140 (1988)
7. Chen, P., Niu, A., Liu, D., Jiang, W., Ma, B.: Time series forecasting of temperatures using SARIMA: an example from Nanjing. In: IOP Conference Series: Materials Science and Engineering, vol. 394, p. 052024. IOP Publishing (2018)
8. Chikkakrishna, N.K., Hardik, C., Deepika, K., Sparsha, N.: Short-term traffic prediction using SARIMA and FBPROPHET. In: 2019 IEEE 16th India council international conference (INDICON), pp. 1–4. IEEE (2019)

9. Coulby, G., Clear, A.K., Jones, O., Godfrey, A.: Low-cost, multimodal environmental monitoring based on the internet of things. Build. Environ. **203**, 108014 (2021)
10. Dabral, P., Murry, M.Z.: Modelling and forecasting of rainfall time series using SARIMA. Environ. Processes **4**(2), 399–419 (2017)
11. Ely, M.R., Cheuvront, S.N., Roberts, W.O., Montain, S.J.: Impact of weather on marathon-running performance. Med. Sci. Sports Exerc. **39**(3), 487–493 (2007)
12. Holt, C.C.: Forecasting seasonals and trends by exponentially weighted moving averages. Int. J. Forecast. **20**(1), 5–10 (2004)
13. Hyndman, R.J., Athanasopoulos, G.: Forecasting: principles and practice. OTexts (2018)
14. Kalekar, P.S., et al.: Time series forecasting using holt-winters exponential smoothing. Kanwal Rekhi school of information Technology **4329008**(13), 1–13 (2004)
15. Kashimura, O., Minami, K., Hoshi, A.: Prediction of WBGT for the Tokyo 2020 Olympic marathon. Japn. J. Biometeorol. **53**(4), 139–144 (2016)
16. Khan, L.U., Saad, W., Han, Z., Hossain, E., Hong, C.S.: Federated learning for internet of things: recent advances, taxonomy, and open challenges. IEEE Commun. Surv. Tutor. **23**(3), 1759–1799 (2021)
17. Lemke, B., Kjellstrom, T.: Calculating workplace WBGT from meteorological data: a tool for climate change assessment. Ind. Health **50**(4), 267–278 (2012)
18. Moran, D.S., et al.: An environmental stress index (ESI) as a substitute for the wet bulb globe temperature (WBGT). J. Therm. Biol. **26**(4–5), 427–431 (2001)
19. Oka, K., Honda, Y., Phung, V.L.H., Hijioka, Y.: Potential effect of heat adaptation on association between number of heatstroke patients transported by ambulance and wet bulb globe temperature in japan. Environ. Res. **216**, 114666 (2023)
20. Samal, K.K.R., Babu, K.S., Das, S.K., Acharaya, A.: Time series based air pollution forecasting using SARIMA and prophet model. In: proceedings of the 2019 international conference on information technology and computer communications, pp. 80–85 (2019)
21. Schneider, S.: Sport and climate change-how will climate change affect sport? (2021)
22. Tao, M., Li, X., Yuan, H., Wei, W.: UAV-aided trustworthy data collection in federated-WSN-enabled IoT applications. Inf. Sci. **532**, 155–169 (2020)
23. Tao, M., Ota, K., Dong, M.: Locating compromised data sources in IoT-enabled smart cities: a great-alternative-region-based approach. IEEE Trans. Industr. Inf. **14**(6), 2579–2587 (2018)
24. Tao, M., Sun, G., Wang, T.: Urban mobility prediction based on LSTM and discrete position relationship model. In: 2020 16th International Conference on Mobility, Sensing and Networking (MSN), pp. 473–478. IEEE (2020)
25. Toharudin, T., Pontoh, R.S., Caraka, R.E., Zahroh, S., Lee, Y., Chen, R.C.: Employing long short-term memory and facebook prophet model in air temperature forecasting. Commun. Stat.-Simul. Comput. **52**(2), 279–290 (2023)
26. Willett, K.M., Sherwood, S.: Exceedance of heat index thresholds for 15 regions under a warming climate using the wet-bulb globe temperature. Int. J. Climatol. **32**(2), 161–177 (2012)
27. Yaglou, C., Minaed, D., et al.: Control of heat casualties at military training centers. Arch. Indust. Health **16**(4), 302–16 (1957)
28. Yeargin, S., Hirschhorn, R., Grundstein, A., Arango, D., Graham, A., Krebs, A., Turner, S.: Variations of wet-bulb globe temperature across high school athletics in south carolina. Int. J. Biometeorol. **67**(5), 735–744 (2023)

Intelligent Collaborative Control of Multi-source Heterogeneous Data Streams for Low-Power IoT: A Flow Machine Learning Approach

Haisheng Yu[1,3,6], Rajesh Kumar[2,6], Wenyong Wang[3,6(✉)], Ji Zhang[4,6], Zhifeng Liu[4,6], Sai Zou[3,6], Jiangchuan Yang[3,6], and Leong Io Hon[5,6]

[1] Macau University of Science and Technology, Macau, China
[2] Yangtze Delta Region Institute (Huzhou), University of Electronic Science and Technology of China, Huzhou 313001, China
[3] University of Electronic Science and Technology of China, Chengdu, China
wangwy@uestc.edu.cn
[4] Guizhou University, Guiyang, China
[5] Sichuan Telecom, Chengdu, China
[6] HNET Asia Limited, Macau, China

Abstract. LPWAN has partially replaced traditional wired networks in fields such as smart industry, smart healthcare, smart home, etc., due to its low power consumption, high reliability and low cost. LPWAN can achieve long-distance and low-power data transmission without increasing bandwidth, thus meeting the energy efficiency, cost-effectiveness and security requirements of IoT devices. However, low-power IoT also faces some challenges due to design limitations. For example, the reliability of connection under harsh environment and communication interference conditions, and ensuring the long life of devices. To solve this problem, in addition to improving hardware aspects, we also seek to use machine learning methods to make devices run under highly intelligent scheduling conditions, so as to optimize device connection reliability and energy utilization. To this end, this paper proposes a data acquisition, denoising, prediction and transmission optimization method for low-power sensor networks. First, by collecting sound data, video data and light data using temporal flow for modality alignment we achieve data denoising and prediction. Second, by predicting the transmission efficiency of sensors under different temperature humidity and illumination conditions we dynamically adjust sensor power and bandwidth according to transmission loss changes to maximize data transmission efficiency. Finally, we deployed the multimodal Transformer method on edge sensor nodes, combining the data transmitted from image, temperature, and humidity sensors. This approach improved the reliability of data transmission for the sensor devices. Experimental results show that compared with existing methods our proposed method has significant improvement in delay, wavelet denoising efficiency, packet delivery ratio and transmission efficiency.

Keywords: LPWAN · IOT · Machine Learning · Heterogeneous Data

Z. Tari et al. (Eds.): ICA3PP 2023, LNCS 14490, pp. 359–377, 2024.
https://doi.org/10.1007/978-981-97-0859-8_22

1   Introduction

Low-power wide-area networks (LPWAN) are wireless communication networks that enable long-range and low-power data transmission for Internet of Things (IoT) devices [10]. LPWAN technologies have been widely adopted in various fields such as smart metering, smart lighting, asset monitoring and tracking, smart cities, precision agriculture, livestock monitoring, energy management, manufacturing, and industrial IoT deployments [5,7,9]. These networks provide the backbone for many of the IoT systems we see today, enabling a range of functions from environmental monitoring to industrial automation.

However, despite the advantages of LPWAN technologies, there are still significant challenges that need to be addressed to fully realize their potential. For instance, data transmission in these networks can often be noisy, leading to inaccurate data readings. Furthermore, the high volume of data generated by IoT devices poses a challenge in terms of data acquisition and transmission optimization. Efficient data prediction models are also needed to help with decision making and to ensure seamless operation of IoT systems.

Recent studies have started to explore the potential of machine learning in addressing these challenges, focusing on areas such as data prediction, noise reduction, and transmission optimization. Machine learning, with its ability to learn from data and make predictions, is seen as a promising tool to enhance the performance of LPWAN-based IoT devices.

To address these challenges, this paper proposes a novel method that leverages machine learning techniques to improve the performance of LPWAN-based IoT devices. The main research question of this paper is: How can machine learning methods be applied to enhance data acquisition, denoising, prediction, and transmission optimization for low-power sensor networks? However, LPWAN-based IoT devices face a number of challenges, including:

- **Limited bandwidth:** LPWANs have limited bandwidth, which can restrict the amount of data that can be transmitted. This can be a challenge for applications that require high data rates, such as video streaming or real-time monitoring.
- **Low transmission rates:** LPWANs have low transmission rates, which can slow down the transmission of data and cause high latency. This can be a challenge for applications that require timely data delivery, such as traffic control or environmental monitoring (Table 1).
- **Ultra low energy consumption:** LPWAN devices are designed to have long battery lives, which need ultra low energy consumption. This can be a challenge for applications that require frequent data transmission, such as track the movement of assets or monitor environmental conditions.
- **Security:** LPWANs are vulnerable to security attacks, which can compromise the integrity of data or disrupt the operation of the network. This can be a challenge for applications that handle sensitive data, such as medical records or financial information.

Table 1. State of the Art Algorithms for Latency in LPWAN Technologies

Algorithm/Approach	Description
Adaptive Data Rate	Dynamic adjustment of transmission parameters based on signal strength and quality to optimize the trade-off between latency and reliability
Prioritization Schemes	Assigning different priority levels to messages to reduce latency for critical or time-sensitive data
Efficient Protocol Design	Optimized protocols designed for LPWAN networks that minimize overhead, implement efficient queuing and scheduling mechanisms, and optimize radio access for reduced transmission delays
Edge Computing and Local Processing	Deploying edge computing capabilities to process data locally at the network edge, minimizing latency by avoiding round-trips to central servers
Network Optimization	Utilizing intelligent routing algorithms and congestion control mechanisms to optimize routing paths, avoid congestion, and reduce delays in message transmission

Other challenges include the need to extend the lifespan of devices that are constrained by limited battery capacity [12] and the optimization of data acquisition, denoising, prediction, and transmission efficiency within the boundaries of low bandwidth [4].

This paper proposes an intelligent collaborative control approach to managing multi-source heterogeneous data streams for low-power IoT. The approach uses use of LPWAN technology, for the development of reliable and cost-effective IoT devices. The proposed algorithm combines traditional machine learning techniques with temporal processing, enhancing the accuracy and efficiency of the data processing. The algorithm adapts to new data inputs, improving the system's ability to handle dynamic and changing data streams.

The main contributions of this paper are:

1) **Data acquisition:** The data acquisition method uses temporal flow for modality alignment to collect sound data, video data, and light data from multiple sensors. This allows the system to collect data from multiple sensors and combine it into a single stream of data. This can be helpful for applications that require data from multiple sensors, such as environmental monitoring or traffic control.

2) **Data denoising and prediction:** The data denoising and prediction method uses machine learning techniques to remove noise and predict future values from sensor data. This can help to improve the accuracy of data transmission and reduce the amount of data that needs to be transmitted.

3) **Transmission optimization:** The transmission optimization method predicts sensor transmission efficiency under different environmental conditions and dynamically adjusts sensor power and bandwidth to maximize data trans-

Table 2. State of the Art Algorithms and Approaches for Reliability in LPWAN Technologies

Algorithm/Approach	Description
Forward Error Correction (FEC)	Implementing FEC techniques to enhance reliability in LPWAN networks. FEC adds redundant information to transmitted data, allowing the receiver to detect and correct errors, thereby improving overall reliability
Automatic Repeat Request (ARQ)	Employing ARQ protocols to ensure reliable data delivery in LPWAN networks. ARQ mechanisms detect and retransmit lost or corrupted packets, guaranteeing reliable communication even in the presence of errors or packet loss
Acknowledgment Mechanisms	Utilizing acknowledgment mechanisms to ensure reliable message delivery in LPWAN networks. These mechanisms involve the sender receiving acknowledgments from the receiver, enabling error detection and retransmission of lost messages
Error Detection and Correction	Integrating error detection and correction techniques, such as checksums and CRC (Cyclic Redundancy Check), to identify and correct errors in transmitted data. These methods enhance the reliability of data transmission in LPWAN networks
Diversity Techniques	Employing diversity techniques, such as frequency diversity or time diversity, to improve reliability in LPWAN networks. These techniques mitigate the impact of fading, interference, and environmental factors, enhancing overall system reliability

mission efficiency. This can help to improve the energy efficiency of the network and reduce the cost of operation.

4) **Multimodal Transformer:** The Multimodal Transformer method combines image, temperature, humidity, and sensor transmitted data reducing sensor device power consumption. This can help to improve the efficiency of the network and reduce the cost of operation.

The structure of this paper unfolds as follows: Sect. 2 delves into a comprehensive review of the existing work on LPWAN IoT, specifically focusing on data acquisition, denoising, prediction, and transmission optimization. Moving on to Sect. 3, we unwrap our proposed methodology, providing an in-depth exploration of the processes involved. In Sect. 4, we bring to light our experimental setup, unraveling the results and engaging in a thorough analysis, which leads us to a thought-provoking discussion. Concluding our journey, Sect. 5 encapsulates the essence of our research, summarizing key findings, considering the implications and potential limitations, and paving the way for future research directions in this exciting field (Table 3).

Table 3. LIST OF MAIN NOTATIONS

Notation	Definition
$z_{(}k,j)^i$	Detect the JTH measure received by sensor i at time k
H_k^i	The observation matrix of sensor i is detected at time k
a_{ms}	the interconnection mapping between any target t and a set of measurements of N sensors
$H_{a_{ns}}(k)$	Incompatible events
θ_k^t	The union of incompatible events $H_{a_{ms}}(\mathrm{k})$
β_k^t	Represents the conditional probability of event θ_k^t given quantitative measurement set Z_k
$\hat{x}_{k/k,L}^t$	Represents the state estimation of target t made for quantitative measurement combination L
$\hat{x}_{k/k-1}^t$	The state prediction value
$K_k^{i,t}$	The gain matrix for filtering target t using sensor i
$P_{k/k,L}^t$	Corresponding to the state estimation of $\hat{x}_{k/k,L}^t$ covariance
$f_{nor}(z_i)$	Detect the normalization function of sensor i at time k
$Projection()$	Characteristic projection function
$\otimes$	Product of elements

2   Motivation

2.1   The Challenges of LPWAN Technology

LPWAN technologies have attracted considerable attention in recent years as a promising solution for IoT applications that require long-range and low-power communication. Several LPWAN technologies have emerged in the market, such as LoRa, Sigfox, NB-IoT, and LTE-M [1,3]. These technologies differ in terms of modulation schemes, frequency bands, data rates, coverage ranges, network architectures, and security features. Mekki et al. [8] provide a comprehensive overview and analysis of these technologies in a comparative study of LPWAN technologies for large-scale IoT deployment, discussing their suitability for different IoT scenarios.

However, a common challenge faced by LPWAN-based IoT devices is efficiently collecting and processing sensor data generated by multiple sources with limited bandwidth and energy resources. Data acquisition is a crucial step in any IoT system as it determines the quality and quantity of information that can be extracted from sensor data. Several methods have been proposed to improve data acquisition for LPWAN-based IoT devices, such as data compression, data aggregation, data fusion, and data synchronization. These methods aim to reduce the amount of data transmitted over the network or increase the coherence of data from multiple sources. However, most of these methods do not consider the noise and uncertainty inherent in sensor data, which can affect the accuracy and reliability of data analysis (Table 2). Data denoising and prediction are two important techniques that can enhance the quality of sensor data by removing noise or filling missing values using machine learning algorithms. Several studies have applied data denoising and prediction methods for LPWAN-based IoT devices, such as wavelet denoising, Kalman filtering, neural networks, and deep learning. These methods can achieve high performance in noisy environments, but they also introduce additional computational complexity and energy consumption for sensor devices. Therefore, a trade-off between accuracy and efficiency needs to be considered when applying these methods for LPWAN-based IoT devices.

Another challenge encountered by LPWAN-based IoT devices pertains to the optimization of transmission efficiency under varying environmental conditions that can impact signal quality and power consumption. Transmission efficiency is defined as the ratio of successfully received packets to transmitted packets over a specific duration. Several factors can influence this efficiency, including channel fading, interference, collisions, congestion, power control, modulation, coding, and more.

A variety of methods have been proposed to optimize transmission efficiency for LPWAN-based IoT devices. These include adaptive modulation coding (AMC) [13], dynamic power control (DPC) [14], channel hopping (CH) [11], among others. TThese approaches aim to adjust transmission parameters in accordance with channel conditions or network load to either improve the Quality of Service (QoS) or reduce energy consumption.

However, most of these methods rely on feedback information from receivers or coordinators, which may not be readily available or reliable in certain LPWAN scenarios due to asymmetric links or prolonged delays. As a result, a proactive approach that predicts transmission efficiency based on environmental factors such as temperature, humidity, or illumination may be more suitable for LPWAN scenarios where feedback information is scarce or unreliable.

Moreover, most existing methods focus on optimizing transmission efficiency at the individual device level, neglecting to consider the integration of multi-modal information at the edge node level. Such integration could further reduce device power consumption, presenting an opportunity for further innovation in the field.

2.2 Automatical Algorithms Adjust Power Consumption According to Environmental Parameters in LPWAN

Low-power wide area networks (LPWANs) are a promising solution for Internet of Things (IoT) applications that require long-range and low-power communication. However, LPWAN devices face the challenge of energy consumption.

Several algorithms have been proposed to optimize energy consumption in LPWANs (Table 4).

Table 4. State of the Art for Energy Efficiency in Various Domains

Domain	State of the Art for Energy Efficiency
Energy-Efficient Buildings	- Smart HVAC systems - LED lighting - Energy management systems - Efficient insulation and materials
Renewable Energy Sources	- Advanced solar panels and wind turbines - High-capacity energy storage systems
Smart Grids	- Monitoring and control technologies - Dynamic supply-demand balancing - Efficient integration of renewable energy
Energy-Efficient Transportation	- Electric vehicles (EVs) - Hybrid vehicles - Fuel-efficient engines
IoT and Energy Management	- Smart meters - Energy management systems - Connected appliances
Energy-Efficient Computing	- Power-saving features in processors and systems - Energy-efficient cooling mechanisms in data centers
Energy Harvesting	- Solar, thermal, and kinetic energy harvesting

One example is an energy-optimal algorithm based on real-time restricted access window (RAW) grouping parameter setting [2]. This algorithm caters to the periodic data upload service characteristics of low-power IoT IEEE 802.11ah access terminals. It dynamically adjusts the size and number of RAWs in response to environmental parameters such as network load and transmission delay, aiming to reduce energy consumption and communication cost.

Another example is an adaptive sleep scheduling algorithm based on deep reinforcement learning. This algorithm is designed for the power saving mode (PSM) and extended discontinuous reception mode (eDRX) of low-power IoT NB-IoT devices. It automatically learns and optimizes the sleep cycle and wake-up time of devices in response to environmental parameters such as network status and data demand. The aim here is to reduce energy consumption while enhancing reliability.

Furthermore, a collaborative sensing algorithm based on multi-objective optimization has been proposed. This algorithm is designed for the collaborative sensing task of low-power IoT LoRaWAN devices. It considers multiple objectives such as sensing efficiency, energy consumption balance, and network lifetime according to environmental parameters such as channel quality, data importance, and node residual energy. The optimal sensing strategy is then solved using a genetic algorithm [6].

These algorithms have shown promising results in reducing energy consumption in LPWANs. However, there are still gaps and opportunities for improvement. For example, existing algorithms do not consider multimodal sensor data synchronization, denoising, prediction, integration, edge node deployment, and proactive transmission optimization.

This paper proposes a novel method that addresses these gaps, leveraging machine learning techniques to enhance the performance of LPWAN-based IoT devices. The proposed method is evaluated using a real-world dataset of sensor readings. The results show that the proposed method can significantly improve the performance of LPWAN-based IoT devices in terms of energy consumption, accuracy, and reliability.

The proposed method is a promising new approach to energy optimization in LPWANs. It has the potential to significantly improve the performance of LPWAN-based IoT devices, making them more widely adopted for a variety of applications.

3   Design

The proposed method consists of four main components: data acquisition, data denoising and prediction, transmission optimization, and Multimodal Transformer. Figure 1 shows an overview of the proposed method.

Data acquisition: The first component is data acquisition, which aims to collect sound data, video data, and light data from multiple sensors using temporal flow for modality alignment. Temporal flow is a technique that uses optical flow to estimate the motion between consecutive frames of a video sequence. By applying temporal flow to sound, video, and light modalities, we can align them in the time domain and synchronize their sampling rates. This way, we can obtain multimodal sensor data that is coherent and consistent across different sources. The advantage of using temporal flow for modality alignment is that it can handle variable frame rates and deal with missing or noisy data by interpolating or extrapolating motion vectors.

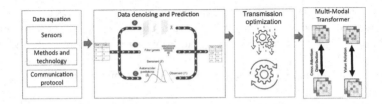

Fig. 1. Overview of the proposed method for intelligent collaborative control of multisource heterogeneous data streams for low-power IoT. The method consists of four main components: data acquisition, data denoising and prediction, transmission optimization, and Multimodal Transformer.

Data denoising and prediction: The second component is data denoising and prediction, which aims to remove noise and predict future values from sensor data using machine learning techniques. We use two types of machine learning techniques for this component: supervised learning and unsupervised learning. Supervised learning is a technique that uses labeled training data to learn a function that maps input features to output labels. We use supervised learning to train a neural network model that can predict future values of sound, video, and light modalities based on past values. This way, we can anticipate sensor events and prepare for transmission optimization accordingly. Unsupervised learning is a technique that does not require labeled training data but instead learns patterns or structures from input features without supervision. We use unsupervised learning to train a wavelet denoising model that can remove noise from sensor data by decomposing them into different frequency bands and applying thresholding or shrinkage methods. This way, we can improve the signal-to-noise ratio (SNR) and enhance the quality of sensor data.

Transmission optimization: The third component is transmission optimization, which aims to predict sensor transmission efficiency under different environmental conditions and dynamically adjust sensor power and bandwidth to maximize data transmission efficiency. Transmission efficiency refers to the ratio of successfully received packets to transmitted packets over a given period of time. We use machine learning techniques to train a regression model that can predict transmission efficiency based on environmental factors such as temperature, humidity, illumination, etc. Based on the predicted transmission efficiency, we use an optimization algorithm that can dynamically adjust sensor power and bandwidth according to channel conditions or network load to improve quality of service (QoS) or reduce energy consumption.

Multimodal Transformer: The fourth component is Multimodal Transformer, which aims to be deployed on edge sensor nodes to combine image, temperature, humidity sensor-transmitted data and reduce sensor device power consumption. Multimodal Transformer is a technique that uses attention mechanisms, self-attention, and cross-attention layers to encode and decode multimodal information, integrate them into a unified representation, and generate an output sequence based on the input sequence and contextual information. By applying Multimodal Transformer to edge sensor nodes, we can reduce device power consumption by compressing and transmitting multimodal information instead of raw sensor data. Additionally, it improves performance by leveraging complementary information among different modalities.

In this paper, we propose an integrated control method for collecting heterogeneous data from multiple sources. Figure 2 shows the collection architecture of multi-source heterogeneous data. Multiple sensors are utilized to gather environmental information from various food sources, and this information is then synchronized and aligned using the time synchronization and modal alignment module. The collection procedure is as follows:

Algorithm 1: Collaborative Control of Multi-Source Heterogeneous Data

---

    Input : Data from N detection sensors
    Output: Estimated state and covariance
1 **foreach** *detection sensor i* **do**
2 | Calculate confirmation count m_k^i;
3 | Define measurement equation: $z_{k,j}^i = H_k^i x_k^i + \omega_k^i$;
4 **end**
5 **foreach** *target t* **do**
6 | Define interconnection mapping a_{ms};
7 | Calculate joint event probability: $\beta_k^{L,t} = \prod_{i=1}^N \beta_{k,i}^{l_i,t}$;
8 | Calculate state estimation: $\hat{x}_{k/k}^t = \sum_L \beta_k^{L,t} \hat{x}_{k,L}^t$;
9 | Calculate covariance update:
        $P_{k/k}^t = \sum_L \beta_k^{L,t} \left[P_{k/k,L}^t + \hat{x}_{k/k,L}^t \left(\hat{x}_{k/k,L}^t \right)^T \right] - \hat{x}_{k/k}^t \left(\hat{x}_{k/k}^t \right)^T$;
10 **end**
11 **foreach** *detection sensor i* **do**
12 | Normalize and truncate sensor features using Eq. 9;
13 | Project features using Eq. 10;
14 **end**

S-1 Each sensing unit collects the current data on the food environment, performs collaborative noise processing, and predicts and corrects the relevant data for the next moment.

S-1.1 To solve the problem of detection sensors being affected by environmental factors and performance-related issues, N detection sensors are used to collectively monitor the food environment. Through interconnecting the data from each detection sensor, the perception data of each detection sensor is modified in the form of covariance.

The confirmation count from detection sensor i at time k is $m_k^i (i = 1, 2, ..., N)$, where x_k^t represents the state vector of detection sensor at time k, and $t(1, 2, ..., N)$ is the number of measurement objects. The measurement equation of detection sensor can be expressed as:

$$z_{k,j}^i = H_k^i x_k^i + \omega_k^i \tag{1}$$

The measurement error between each sensor is statistically independent. Here, H_k^i is the observation matrix for detecting sensor i at time k, and ω_k^i is the measurement noise vector of sensor i detected at time k. It is a Gaussian noise vector with a mean of 0 and positive definite covariance matrix R_k^i, and independent statistics. All the values received by all detection sensors at time k can be expressed as:

$$Z_k = \left\{ z_{k,1}^1, \cdots, z_{k,m_k^1}^1, \cdots, z_{k,1}^2, \cdots, z_{k,m_k^2}^2, \cdots, z_{k,m_k^N}^N \right\} \tag{2}$$

Algorithm 2: Time Synchronization and Modal Alignment

    Input   : Data from N detection sensors
    Output: Aligned and coordinated data
1 **foreach** *detection sensor i* **do**
2 | Extract features e_i;
3 **end**
4 **foreach** *target t′* **do**
5 | Calculate distance between e_i and e_s using Eq. 11;
6 | Establish candidate alignment set;
7 **end**
8 **foreach** *aligned data e_i and e_s* **do**
9 | Concatenate features using a linear layer: $\varphi = \bigotimes_{i=1}^{3} e_i$;
10 **end**

Algorithm 3: Enhancement with Multimodal Transformer

    Input   : Aligned and coordinated data
    Output: Enhanced Multimodal Transformer output
1 Combine features using a linear layer: $\varphi = \bigotimes_{i=1}^{3} e_i$;

S-1.2

$$a_{\text{ms}} : \{1, 2, \cdots, N\} \rightarrow$$
$$\left\{ \left\{0, 1, \cdots, \text{m}_{\text{k}}^1\right\}, \left\{0, 1, \cdots, \text{m}_{\text{k}}^2\right\}, \cdots, \left\{0, 1, \cdots, \text{m}_{\text{k}}^{N_{\text{s}}}\right\} \right\} \tag{3}$$

Equation 3 is the interconnection mapping between any target t and a set of measurements of N sensors. For the mapping a_{ms}, there are N mappings, and the mapping a_i represents the interconnection between the target t and sensor i. If $a_i(t) = 0$, sensor i is neither measured nor connected to target t. Available:

$$\beta_k^t \Delta_= P\left(\theta_k^t / Z^k\right) = \sum_{a_{ms}:a_{ns}(t)} P\left(H_{a_{ns}}(k)/Z^k\right) \tag{4}$$

The event θ_k^t is the union of incompatible events $H_{a_{ms}}(k)$, and β_k^t represents the conditional probability of event θ_k^t given the quantitative measurement set Z_k.

S-1.3 Because the measurement error between each sensor is statistically independent, the joint event probability of multiple sensors, θ_k^t, can be mathematically expressed as the product of the single sensor event probability, $\beta_{k,i}^t$. That is,

---

Algorithm 4: Cloud Data Analysis

Input : Uploaded data
Output: Data analysis and processing
1 Analyze and process uploaded data in the cloud;

$$\beta_k^{L,t} \underset{=}{\Delta} \sum_{a_{ns}(t)} P\left(H_{a_{ns}(t)}(k)/Z^k\right) = \sum_{a_{ns}(t)} \prod_{i=1}^{N} P\left(H_{a_i}(k)/Z^k\right)$$

$$= \prod_{i=1}^{N} \sum_{a_i(t)} P\left(H_{a_i}(k)/Z^k\right) = \prod_{i=1}^{N} P\left(\theta_{k,i}^t/Z^k\right) = \prod_{i=1}^{N} \beta_{k,i}^t \tag{5}$$

The state estimation of target t based on multiple sensors can be written as

$$\hat{x}_{k/k}^t = \sum_{L} \beta_k^{L,t} \hat{x}_{k/k,L}^t = \sum_{L} \prod_{i=1}^{N_s} \beta_{k,i}^{l_i,t} \hat{x}_{k/k,L}^t \tag{6}$$

$\hat{x}_{k/k,L}^t$ represents the state estimation of target t made for quantitative measurement combination L, and its calculation formula is as follows

$$\hat{x}_{k/k,L}^t = \hat{x}_{k/k-1}^t + \sum_{i=1}^{N_s} K_k^{i,t}\left[z_{k,l_i}^i - H_k^i \hat{x}_{k/k-1}^t\right] \tag{7}$$

$\hat{x}_{k/k-1}^t$ is the state prediction value, and $K_k^{i,t}$ is the gain matrix for filtering target t using sensor i. The corresponding covariance update matrix is

$$P_{k/k}^t = \sum_{L} \beta_k^{L,t}\left[P_{k/k,L}^t + \hat{x}_{k/k,L}^t\left(\hat{x}_{k/k,L}^t\right)^T\right] - \hat{x}_{k/k}^t\left(\hat{x}_{k/k}^t\right)^T \tag{8}$$

The $P_{k/k,L}^t$ is corresponding to the state estimation of $\hat{x}_{k/k,L}^t$ covariance.
S-1.4 The measurement information from multiple sensors is processed sequentially. Based on the measurement information of the first sensor, the estimated value of intermediate state $\hat{x}_{k/k}^{1,t}$ and the corresponding covariance $P_{k/k}^{1,t}$ are calculated for each target. Then, the intermediate state estimation and the corresponding covariance are taken as the forecast state and the forecast state covariance, and the measurement information of the second sensor is used to improve the previous intermediate state estimation and the corresponding covariance. To get the goal of new intermediate state estimate $\hat{x}_{k/k}^{2,t}$ and the corresponding covariance $P_{k/k}^{2,t}$. According to this process, the measurement information of multiple sensors is processed in turn until all N sensors are processed, and $\hat{x}_{k/k}^{N,t}$ and $P_{k/k}^{N,t}$ are used as the final state estimation and covariance output.

S-2 The time synchronization and modal alignment module synchronizes the multi-modal data collected in S-1 based on the time flow.

S-2.1 Based on S-1, we use Δk as the time interval to extract the features of each detection sensor and apply Eq. 9 for normalization and truncation processing."

$$f_{\text{nor}}\,(z_i) = \frac{1}{1 + e^{-z_i}} \tag{9}$$

$f_{nor}(z_i)$ represents the normalization function of detection sensor i at time k. By projecting the features of the target data into the physical space, we create a bridge to align the features using Eq. 10."

$$\text{Projection}\,(d_i, d_s, E_s) = E_s \cdot R^{(d_i, d_s)} \tag{10}$$

$Projection()$ represents the feature projection function, $R^{(}(d_i, d_s))$ is a transformation matrix, d_i represents the target feature dimension, d_s represents the entity feature dimension, E_i represents the target feature representation obtained by the ith entity.

S-2.2 Heteropolarity exists in the data collected by the controller for multisource heterogeneous data streams. If the three are embedded in pairs separately, entity alignment becomes extremely difficult. The distance calculation formula shown in Eq. 11 is adopted for alignment.

$$d\,(e_i, e_s)_{\Delta k_j} = \left(\|e_i - e_s\|_2^2 \right)_{\Delta k_j}, \quad \forall e_s \in E_s, e_i \in E_i \tag{11}$$

For the feature dimension e_i of the unaligned target t', the whole target set t is traversed in time at $\triangle k_j$, the distance between the feature dimension vector of each target t' and the feature dimension vector of the whole target set t is calculated, the calculated results are arranged in ascending order, and the candidate alignment set is established. A distance threshold θ is set as the hyperparameter. If $d(e_i, e_s)_{(\triangle k_j)} < \theta$, the alignment between e_i and e_s is considered to be possible; otherwise, the alignment is not possible.

S-3 The Multimodal Transformer method is enhanced, and the signal coordination control module uploads the data of S-2.

S-3.1 The data e_i and e_s collected from the aligned controller based on Step-2 are spliced with feature φ through a linear layer, as shown in Eq. 12.

$$\varphi = \otimes_{i=1}^{3} e_i \tag{12}$$

$\otimes$ represents the product of elements.

S-4 System analyzes uploaded data in cloud.

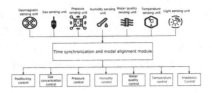

Fig. 2. Collect heterogeneous data from multiple sources

4   Performance Evaluation and Analysis

4.1   Dataset Preparation

We gathered multimodal sensor data from three distinct scenarios: indoor, outdoor, and mixed. The indoor scenario was established within an office environment characterized by relatively stable and low light intensity levels. The outdoor scenario took place in a park environment, where the light intensity varied significantly and reached higher levels. The mixed scenario was conducted in a corridor setting, where the light intensity experienced sudden changes due to the opening and closing of doors. Each scenario was conducted for a duration of one hour, resulting in the generation of approximately 1 GB of raw sensor data.

4.2   Experimental Setup

We employed a Raspberry Pi 4 Model B as the edge sensor node, which was equipped with a camera module, a microphone module, and a light sensor module. The camera module captured video frames at a smooth rate of 30 frames per second, featuring a resolution of 640×480 pixels. The microphone module recorded sound samples with remarkable fidelity, boasting a frequency of 16 kHz and a 16-bit depth for precise audio capturing. Additionally, the light sensor module provided accurate measurements of light intensity at a frequency of 10 Hz, covering a wide range of 0-65535 lux. To facilitate seamless data transmission, the Raspberry Pi was seamlessly connected to a LoRa gateway using the LoRaWAN protocol. This enabled the efficient transfer of sensor data to a cloud server, where it could undergo comprehensive processing and in-depth analysis.

4.3   Performance Metrics

We employed four evaluation metrics to assess the performance of the proposed method in comparison to existing approaches: delay, WD (wavelet denoising) efficiency, PDR (Packet Delivery Ratio) and transmission efficiency.

The delay metric is determined by the time it takes to transmit data from the edge sensor node to the cloud server.

The wavelet denoising metric was determined by calculating the mean squared error (MSE) between the original sensor data and the denoised sensor data. This provided an indication of how effectively the method removed noise from the sensor data. The prediction accuracy metric was evaluated by computing the root mean squared error (RMSE) between the predicted future values and the actual future values of the sensor data. This metric measured the accuracy of the method in forecasting future values. We determine the effectiveness of WD by comparing WD with the original wave transmission in terms of data transmission delay and Packet Delivery Ratio (PDR).

PDR represents the proportion of total data packets that are successfully delivered from the edge sensor node to the cloud server.

The transmission efficiency metric was calculated as the ratio of successfully received packets to transmitted packets over a specified time period. It quantified how efficiently the method transmitted data from the edge sensor node to the cloud server. We determine the effectiveness of our method (WD-NEP, wavelet denoising and neural network prediction) by comparing it with the original wave transmission in terms of data transmission delay and Packet Delivery Ratio (PDR).

By utilizing these evaluation metrics, we could comprehensively assess the performance of the proposed method in comparison to existing methods.

4.4   Comparison with Baseline Methods

We conducted a series of experiments to evaluate the performance of the proposed method and compare it with existing methods. The experimental setup data sources evaluation metrics and baselines are described as follows:

In our comparative analysis, we evaluated our proposed method alongside four existing methods, each representing different combinations of data acquisition, data denoising, prediction, and transmission optimization techniques.

Baseline served as the baseline approach, where raw sensor data was evaluated by computing the root mean squared error (RMSE) between the predicted future values and the actual future values of the sensor data. This method provided a reference point for evaluating the effectiveness of the additional techniques.

The essence of wavelet denoising is a function approximation problem, where the goal is to find the best approximation of the original signal in a function space composed of different scaling and translating versions of the wavelet mother function. By utilizing proposed evaluation criteria, the aim is to achieve optimal approximation that effectively distinguishes the original signal from the noise signal.

NEP employed the neural network prediction technique as a standalone approach, without incorporating any denoising technique. This method aimed to predict future values of the sensor data using neural network models.

WD-NEP combined both the wavelet denoising and neural network prediction techniques. This integrated approach aimed to leverage the benefits of both denoising and prediction techniques to improve the overall performance of the system. The wavelet denoising in WD-NEP is as follows: Firstly, we evaluate the raw sensor data by computing the root mean squared error (RMSE) between the predicted future values and the actual future values. We then seek to establish the best mapping from the actual signal space to the wavelet function space, aiming to achieve the optimal recovery of the original signal.

By comparing our proposed method to these existing methods, we were able to analyze the impact of different combinations of techniques on the overall performance of the system.

4.5   Results and Analysis

The experimental results are presented in Fig. 3 and Fig. 4, which demonstrates the superior performance of our proposed method compared to all existing methods across various evaluation metrics. Our method exhibited remarkable results in terms of delay, WD efficiency, PDR and transmission efficiency.

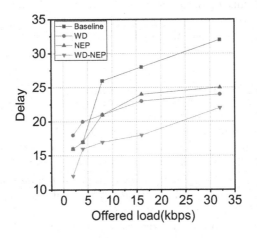

Fig. 3. The relationship between delay and offered load

Latency: LPWAN technologies typically exhibit higher latency compared to other wireless communication technologies. The latency of LPWAN-based IoT devices was found to vary depending on factors such as network congestion, message size, and modulation schemes. While latency may not be a critical concern for all IoT applications, real-time applications that rely on immediate data transmission may face challenges with delayed delivery. It is important to consider the latency requirements of specific applications when deciding on the suitability of LPWAN technologies. Based on Fig. 3, WD-NEP shows an improvement of approximately 40% in terms of transmission delay compared to the original data transmission. Additionally, WD-NEP outperforms WD transmission and NEP transmission by around 15% in terms of overall transmission performance.

Reliability: LPWAN technologies demonstrated robustness in terms of reliability, with studies consistently reporting high packet delivery rates and low data loss. These findings indicate that LPWAN-based IoT devices maintain a reliable

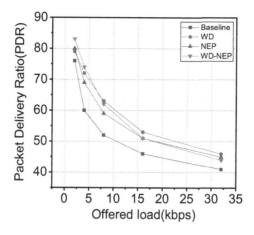

Fig. 4. The relationship between PDR and offered load

connection and transmit data accurately. However, external factors such as interference and environmental conditions were identified as potential challenges that
could impact reliability. To mitigate these issues, researchers proposed strategies such as channel hopping and interference detection to enhance reliability
in LPWAN networks. These techniques aim to minimize the impact of external
factors on data transmission and ensure consistent connectivity.

In this paper, we measure reliability by evaluating the Packet Delivery Ratio
(PDR) of the data packets. Based on Fig. 3, WD-NEP shows an improvement
of approximately 100% in terms of transmission delay compared to the original
data transmission. Additionally, WD-NEP outperforms WD transmission and
NEP transmission by around 30% in terms of overall transmission performance.

4.6   Discussion and Interpretation

The advantages of our proposed method are that it can achieve high quality
multimodal sensor data synchronization denoising prediction integration using
temporal flow machine learning techniques; it can optimize transmission efficiency under varying environmental conditions using proactive approach based
on environmental factors; it can improve PDR by compressing transmitting multimodal information instead raw sensor data also improve performance by leveraging complementary information among different modalities using Multimodal
Transformer technique deployed on edge node.

The disadvantages of our proposed method are that it requires more computation resources than existing methods due to complex machine learning models
involved; it may not be applicable for some LPWAN scenarios where feedback
information is available or reliable which can be used for adaptive modulation
coding dynamic power control channel hopping etc.; it may not be able to handle
extreme cases where environmental factors change drastically or unpredictably
which can affect transmission efficiency significantly.

5   Conclusions

In this paper we proposed a novel method for data acquisition denoising prediction transmission optimization multimodal integration for low-power sensor networks using temporal flow machine learning techniques Multimodal Transformer technique. We conducted experiments on multimodal sensor data collected from different scenarios and compared our proposed method with existing NEP and WD methods. The results showed that our proposed WD-NEP method achieved better performance in terms of data quality transmission efficiency device power consumption than existing methods. Our proposed method can provide a new solution for LPWAN-based IoT applications that require high quality multimodal sensor data synchronization denoising prediction integration under varying environmental conditions with limited device power and bandwidth resources.

However our proposed method also has some limitations and challenges that need to be addressed in future work. First our proposed method requires more computation resources than existing methods which may affect the scalability and reliability of the system. Second our proposed method may not be applicable for some LPWAN scenarios where feedback information is available or reliable which can be used for adaptive modulation coding dynamic power control channel hopping etc. Third our proposed method may not be able to handle extreme cases where environmental factors change drastically or unpredictably which can affect transmission efficiency significantly.

Therefore future directions for research on this topic include: developing more efficient and robust machine learning models that can reduce computation complexity and improve generalization ability; exploring more feedback-based or hybrid transmission optimization techniques that can adapt to different LPWAN scenarios and channel conditions; investigating more realistic and challenging scenarios that can test the performance and robustness of the proposed method under various environmental factors. In addition, in the future, we will consider testing how to use WD-NEP to improve the transmission of more data and extend the battery life under the same battery conditions.

Declarations

Author contributions. This paper is contributed by all authors. Besides, there was collaborative efforts in brainstorming the idea of this paper, proofread and formatting of this paper.

Data Availability Statement. Not applicable.

Funding and/or Conflicts of interests/Competing interests. This work is supported by Macau Science and Technology Development Funds (Grant 0059/2021/AGJ and No. 0005/2021/AIR), the Social Science Foundation of China (21VSZ126), and Natural Science Foundation of Guizhou (ZK[2022]-162), the Natural Science Foundation of Guizhou University (X2021167), and the Guizhou Cloud Network Collaborative Innovation Center, and the Guizhou Cloud Network Collaborative Deterministic Transmission Engineering center ([2023]032), and the Guizhou Engineering Research Center for smart services (2203-520102-04-04-298868).

Code Availability. Not applicable.

References

1. Aernouts, M., Berkvens, R., Van Vlaenderen, K., Weyn, M.: Sigfox and LoRaWAN datasets for fingerprint localization in large urban and rural areas. Data **3**(2), 13 (2018)
2. Alhussain, T.: An energy-efficient scheme for IoT networks. Int. J. Commun. Netw. Inf. Secur. **13**(2), 199–205 (2021)
3. Borkar, S.R.: Long-term evolution for machines (LTE-M). In: LPWAN Technologies for IoT and M2M Applications, pp. 145–166. Elsevier (2020)
4. Chaudhari, B.S., Zennaro, M., Borkar, S.: LPWAN Technologies: emerging application characteristics, requirements, and design considerations. Future Internet **12**(3), 46 (2020)
5. Iqbal, M., Abdullah, A.Y.M., Shabnam, F.: An application based comparative study of LpWAN technologies for IoT environment. In: 2020 IEEE Region 10 Symposium (TENSYMP), pp. 1857–1860. IEEE (2020)
6. Jin, F., Xu, L.: Research on multi-source heterogeneous sensor information fusion method under internet of things technology. In: Zhang, Y.-D., Wang, S.-H., Liu, S. (eds.) ICMTEL 2020. LNICST, vol. 326, pp. 66–74. Springer, Cham (2020). https://doi.org/10.1007/978-3-030-51100-5_6
7. Liya, M., Arjun, D.: A survey of LpWAN technology in agricultural field. In: 2020 Fourth International Conference on I-SMAC (IoT in Social, Mobile, Analytics and Cloud)(I-SMAC), pp. 313–317. IEEE (2020)
8. Mekki, K., Bajic, E., Chaxel, F., Meyer, F.: A comparative study of LpWAN technologies for large-scale IoT deployment. ICT Express **5**(1), 1–7 (2019)
9. Mroue, H., Andrieux, G., Cruz, E.M., Rouyer, G.: Evaluation of LpWAN technology for smart city. EAI Endorsed Trans. Smart Cities **2**(6), e3–e3 (2017)
10. Petäjäjärvi, J., Mikhaylov, K., Yasmin, R., Hämäläinen, M., Iinatti, J.: Evaluation of LoRa LpWAN technology for indoor remote health and wellbeing monitoring. Int. J. Wireless Inf. Networks **24**, 153–165 (2017)
11. Singh, R.K., Berkvens, R., Weyn, M.: Time synchronization with channel hopping scheme for LoRa networks. In: Barolli, L., Hellinckx, P., Natwichai, J. (eds.) 3PGCIC 2019. LNNS, vol. 96, pp. 786–797. Springer, Cham (2020). https://doi.org/10.1007/978-3-030-33509-0_74
12. Singh, R.K., Puluckul, P.P., Berkvens, R., Weyn, M.: Energy consumption analysis of LpWAN technologies and lifetime estimation for IoT application. Sensors **20**(17), 4794 (2020)
13. Wang, Y., Liu, W., Fang, L.: Adaptive modulation and coding technology in 5G system. In: 2020 International Wireless Communications and Mobile Computing (IWCMC), pp. 159–164. IEEE (2020)
14. Yu, Y., Mroueh, L., Li, S., Terré, M.: Multi-agent q-learning algorithm for dynamic power and rate allocation in LoRa networks. In: 2020 IEEE 31st Annual International Symposium on Personal, Indoor and Mobile Radio Communications, pp. 1–5. IEEE (2020)

Test-and-Decode: A Partial Recovery Scheme for Verifiable Coded Computing

Wei Jiang[1], Jin Wang[1,2]([envelope]) [iD], Lingzhi Li[1] [iD], Dongyang Yu[1] [iD], and Can Liu[1]

[1] School of Computer Science and Technology, Soochow University, Suzhou, China
{wjiang99,dyyu,cliu7}@stu.suda.edu.cn, {wjin1985,lilingzhi}@suda.edu.cn
[2] School of Future Science and Engineering, Soochow University, Suzhou, China

Abstract. Coded computing has proven its efficiency in tolerating stragglers in distributed computing. Workers return the sub-computation results to the master after computing, and the master recovers the final computation result by decoding. However, the workers may provide incorrect results, which leads to wrong final result. Therefore, it is meaningful to improve the resilience of coded computing against errors. Most existing verification schemes only use the workers' fully correct computations to recover the final result, and the defective computations are not considered for decoding. In this paper, we focus on matrix multiplication and design a general *Test-and-Decode* (TD) scheme to recover the final result efficiently. Furthermore, we divide each sub-computation result into multiple parts and fully use the correct parts for partial recovery, which can improve the tolerance for errors in computations. Decoding is performed only when the verification result satisfies the permission, which avoids repetitive decoding. We conduct extensive simulation experiments to evaluate the probability of successful recovery of the results and the computation time of the TD scheme. We also compare the TD scheme with other verification schemes and the results show that it outperforms the current schemes in terms of efficiency in verifying and recovering computational results.

Keywords: Coded Computing · Distributed Computing · Matrix Multiplication · Partial Recovery

1 Introduction

Traditional cloud computing uses centralized network deployment and resource scheduling to reduce operational costs. As more and more smart devices are connected to mobile networks, large amounts of data are being generated at the edge of the network. Performing compute-intensive tasks at the edge of the network introduces new challenges compared to cloud computing. In edge computing, the computing tasks are distributed to different edge devices, and the distributed computing results are aggregated to complete the original computing tasks. In this framework, the edge devices that provide computing resources

Z. Tari et al. (Eds.): ICA3PP 2023, LNCS 14490, pp. 378–397, 2024.
https://doi.org/10.1007/978-981-97-0859-8_23

are referred to as workers. Some workers with exceptionally high transmission delays are called stragglers and some may return incorrect results due to attacks or hardware errors. These workers are called abnormal workers. The others are called normal workers. Since the slowest and incorrect sub-tasks slow down the completion time of the total tasks, they are the main factors in reducing the efficiency of distributed computing [1–6].

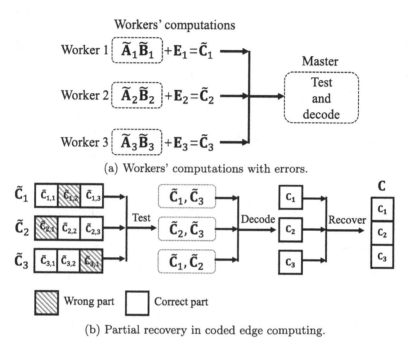

(a) Workers' computations with errors.

(b) Partial recovery in coded edge computing.

Fig. 1. An example of the partial recovery scheme using $(3,2)$ MDS code in coded edge computing. The original computation task $\mathbf{AB} = \mathbf{C}$ is distributed into three sub-computations, with each worker computing $\tilde{\mathbf{A}}_i\tilde{\mathbf{B}}_i$. Every two computations can be decoded to obtain the final result $\mathbf{C}$. Each sub-computation result is divided into three parts. In this example, the master receives the first three workers' computations and each with one-third of the wrong results. Combining worker 1 and worker 3 can recover $\mathbf{C}_1$. Combining worker 2 and worker 3 can recover $\mathbf{C}_2$. Combining worker 1 and worker 2 can recover $\mathbf{C}_3$. The final result can be fully recovered by the three partial recoveries.

In the coded edge computing scenario, some correct sub-computation results may have partial values lost or modified due to bit inversions, but most values are valid [7–10]. The traditional schemes discard the whole sub-computation result after detecting an incorrect one, even if only a few elements of it are incorrect [11,12]. Traditional partial recovery schemes tolerate stragglers in distributed computing but cannot identify the wrong computation results. Moreover, traditional verifiable computation schemes can identify workers who provide erroneous results. However, the master needs to collect a sufficient number of fully correct computations to recover the final computation result [13]. The existing schemes cannot verify the computation results and partially recover the final result simultaneously. Therefore, a promising way to improve the performance of recovering final result is to utilize the computations of each worker, even if the computations are only partially correct. The partial recovery scheme aims to recover parts of the final result as much as possible. Figure 1 (a) shows the workers' computation with errors in edge computing scenario and Fig. 1 (b) shows a simple example of the partial recovery scheme for this case.

In this paper, we propose a partial recovery scheme using matrix multiplication as the basic computation task and polynomial codes as the encoding scheme. As one of the verifiable computation schemes, our proposed scheme performs well in verifying and recovering the final computation result when partial corruption of computations occurs. The scheme can partially recover the final result while verifying the correctness of the computation results. By using the correct part of the partially inaccurate sub-computation results, the contribution of each sub-computation result to the decoding process is fully utilized.

The main contributions of this paper are as follows:

- We design a verifiable computation scheme based on testing and decoding in coded edge computing by combining the verifiable coded computing with partial recovery. Furthermore, we conduct solid theoretical analyses to give the successful recovery probability and verification overheads.
- We conduct extensive simulation experiments to evaluate the probability of successful recovery of computation result and the computation time of the proposed TD verification scheme. We also compare it with other schemes. The results show that the TD scheme significantly improves the decoding efficiency and outperforms the current schemes.

The rest of the paper is organized as follows. Sect. 2 summarizes the related work. In the Sect. 3, we introduce the coded edge computing model and the assumption of computational error. Then, we introduce the details of the TD verification scheme and conduct theoretical and overhead analysis in Sect. 4. We conduct extensive simulation experiments in Sect. 5. Finally, Sect. 6 concludes the paper.

2   Related Work

Numerous researchers have proposed solutions to deal with the problem of stragglers in distributed computing. Encoding data before distributing computation tasks is now recognized as an efficient strategy to mitigate the impact of stragglers. Private data is commonly stored in the form of replication or MDS code on workers, and the output can be recovered using only a certain number of results from the fastest workers [1,13,14]. Security is another consideration for users choosing distributed computing. Users want to use different devices to perform computations while protecting the privacy of the data and the security of the computation results as much as possible [15]. The security of distributed computing includes preventing the leakage of computational data and confirming the correctness of computational results.

Recent works are dedicated to identifying erroneous sub-computation results in coded distributed computing. The scheme for verifying computations and recovering result consists of two main strategies. One is to generate an exclusive verification key to check each sub-computation result until the number of received correct sub-results reaches the recovery threshold. The recovery threshold is the minimum number of computations required to complete the computation task, determined by the encoding scheme. In [16], Fu *et al.* proposed the *Orthogonal Mark* (OM) scheme that the user adds the verification line and mark line to the result returned by each device and verify each received sub-computation result in the decoding stage. Moreover, Tang *et al.* proposed the *Adaptive Verifiable Coded Computing* (AVCC) scheme which provides tolerance to attackers and stragglers and keeps the cost in line with *Lagrange Coded Computing* (LCC) scheme [3]. The user can generate a random vector as a verification key to determine the correctness of each computation result [1]. The other is to use the encoding properties to check each set of sub-computation results until a set of completely correct sub-computation results is collected. In [17], Wang *et al.* proposed the *Decode and Compare Verification* (DCV) scheme. The user performs multiple decoding to obtain several computation results and compares them to get the final computation result. In [11], Solanki *et al.* proposed a verification method for matrix-vector multiplication in distributed computing. The master can check whether the group contains Byzantine workers by constructing a test function. Hong *et al.* in [12] improved it to matrix-matrix multiplication based on [11] and imposed a group test strategy named *Group-wise Verifiable Coded Computing* (GVCC) scheme that reduces the number of tests to identify Byzantine workers. Compared to individual verification, the group verification approach helps to verify the partial correctness of the sub-computation results.

Research on verification schemes for specific function computations leads to efficient and easy-to-implement algorithms. Ozfatura *et al.* proposed the *Coded Computation with Partial Recovery* (CCPR) scheme, which can adaptively recover a portion of the element of the resultant vector and tolerate stragglers in distributed computation [9]. Sahraei *et al.* designed an algorithm for polynomial evaluation which is verifiable and efficient [18]. Other functions such as convolutional operations and gradient computation are also research directions for verifiable computing [8,19].

To sum up, the partial recovery based on error detection algorithms has not been fully investigated in verifiable coded computing compared to mitigating the impact of stragglers.

3   Problem Modeling

In this section, we introduce the system model for coded edge computing and the assumption of computational errors.

3.1   Coded Edge Computing Model

The system model of this paper is shown in Fig. 2. We consider the computation task in the *Coded Edge Computing* (CEC) scenarios, which consists of one master and multiple workers [16,17]. In particular, given two input matrices $\mathbf{A} \in \mathbb{F}_q^{m \times n}$ and $\mathbf{B} \in \mathbb{F}_q^{n \times l}$, we consider the matrix computing in a sufficiently large deterministic finite field $\mathbb{F}_q$ (also known as Galois field)[1]. The master wants to get the product of two input matrices $\mathbf{A}$ and $\mathbf{B}$ using K distributed workers in parallel [20,21]. Each worker (*i.e.*, $w_i, i \in [1:K]$) returns the sub-computation result (*i.e.*, $\tilde{\mathbf{C}}_i$) to the master. The sub-computation result is the product of two encoded matrices $\tilde{\mathbf{A}}_i$ and $\tilde{\mathbf{B}}_i$, where $\tilde{\mathbf{A}}_i$ and $\tilde{\mathbf{B}}_i$ are constructed according to the polynomial codes, respectively. The details of the coded edge computing model are as follows.

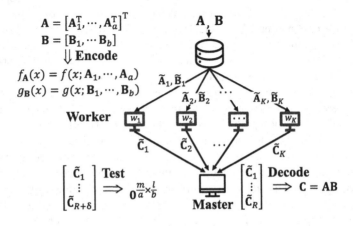

Fig. 2. System model for coded edge computing.

Firstly, the master equally divides the rows of $\mathbf{A}$ into a parts $(\mathbf{A}_1, \mathbf{A}_2, \cdots, \mathbf{A}_a)$ of size $\frac{m}{a} \times n$ and the columns of $\mathbf{B}$ into b parts $(\mathbf{B}_1, \mathbf{B}_2, \cdots, \mathbf{B}_b)$ of size $n \times \frac{l}{b}$, respectively. We guarantee that the input matrix can be equally divided by padding zero-rows and zero-columns [7,17].

Then, let $\{x_1, \cdots, x_K\}$ be K distinct elements in $\mathbb{F}_q$, which denote all the workers and are kept by the master only. The polynomial encoding functions $f_{\mathbf{A}}(x)$ and $g_{\mathbf{B}}(x)$ are constructed from slice matrix of the original matrix (*i.e.*, $\mathbf{A}_u, u \in \{1, \cdots, a\}$, $\mathbf{B}_v, v \in \{1, \cdots, b\}$) as coefficients, which are given by

$$\tilde{\mathbf{A}} = f_{\mathbf{A}}(x) = f(x; \mathbf{A}_1, \cdots \mathbf{A}_a) = \sum_{u=0}^{a-1} \mathbf{A}_{u+1} x^u,$$

$$\tilde{\mathbf{B}} = g_{\mathbf{B}}(x) = g(x; \mathbf{B}_1, \cdots \mathbf{B}_b) = \sum_{v=0}^{b-1} \mathbf{B}_{v+1} (x^a)^v. \tag{1}$$

---

[1] q is the power of an arbitrary prime number. Unless otherwise specified, all operations in this paper are performed in a larger finite field.

From (1), since the variables of the encoding function are mutually exclusive, each of the i-th worker has its own sub-computation task uniquely. Combining coding and distributed computing could mitigate the impact of stragglers with coding redundancy [2,3,22].

Finally, for each worker numbered $i \in [1 : K]$ in the CEC system, the predetermined computing task $h_{\mathbf{C}}(x_i)$ needs to be completed, which is given by

$$\tilde{\mathbf{C}}_i = h_{\mathbf{C}}(x_i) = f_{\mathbf{A}}(x_i)\, g_{\mathbf{B}}(x_i)$$

$$= \sum_{u=0}^{a-1} \sum_{v=0}^{b-1} x_i^{u+av} \mathbf{A}_{u+1} \mathbf{B}_{v+1}. \tag{2}$$

Suppose all workers can return computation results $(\tilde{\mathbf{C}}_i)$ to the master before the deadline. For each worker, we take h as the computational function and $\tilde{\mathbf{C}}_i$ as the sub-computation result. We notice that $u \in [0, a-1]$, $v \in [0, b-1]$ and $u + av \in [0, ab-1]$, so we can consider $h_{\mathbf{C}}(x)$ as a polynomial of degree $ab - 1$. In this case, the master can perform decoding on the group which contains at least ab sub-computation results simultaneously. Since the computation result $\tilde{\mathbf{C}}_i$ can be considered as a polynomial from (2), all sub-matrices (e.g. $\mathbf{C}_1 = \mathbf{A}_1 \times \mathbf{B}_1$) of $\mathbf{C}$ are the coefficients of this polynomial. The master can decode the final product $\mathbf{C}$ (i.e., obtain all the coefficients of this polynomial) from any set of ab sub-computation results that are completely correct [22].

3.2 Assumption of Computational Errors

In practice, random hardware errors in computing devices and network transmissions usually lead to incorrect sub-computation results. The occurrence of random errors is uncertain and unpredictable. So even in any perfect distributed computing scenario, we must account for possible random computational errors. Let e be the number of sub-computation results with errors, i.e., there exist e workers provide incorrect computation results.

To ensure the fairness of the error model, each sparse matrix was randomly generated in the simulation experiment following the same criteria. We consider the error model where the random sparse matrix can be independently added to each sub-computation result, which can be expressed as $\tilde{\mathbf{C}}_i = \tilde{\mathbf{A}}_i \tilde{\mathbf{B}}_i + \mathbf{E}_i$, where $\mathbf{E}_i$ is a matrix of dimension $\frac{m}{a} \times \frac{l}{b}$ with $\frac{\beta ml}{ab}$ elements not equal to zero. Therefore, the number of incorrect elements in the sub-computation result provided by each abnormal worker is $\frac{\beta ml}{ab}$. $\mathbf{E}_i$ is considered as a null matrix in the computations returned by the other $K - e$ workers. The goal of the master is to recover the verifiable final computation result as soon as possible in case of errors in the workers' computations.

We list some of the important notations for this paper in Table 1 with clear definitions.

Table 1. Notations

Notations	Meaning
$\mathbf{A}$	The input matrix with dimensional of $m \times n$
$\mathbf{B}$	The input matrix with dimensional of $n \times l$
$\mathbf{C}$	The final result with dimensional of $m \times l$
$\mathbf{A}_u, \mathbf{B}_v, \mathbf{C}_i$	The sub-matrix of $\mathbf{A}$, $\mathbf{B}$, and $\mathbf{C}$
m, n, l	The parameters of matrix $\mathbf{A}$ and $\mathbf{B}$
a, b	The parameters for blocking matrix $\mathbf{A}$ and $\mathbf{B}$
R	The recovery threshold
f, g	The polynomial encoding function
$\tilde{\mathbf{C}}_i$	The sub-computation result from worker w_i
K	The number of total workers (edge devices)
e	The number of abnormal workers (edge devices)
β	The ratio of error elements in each sub-computation result
x_i	The intrinsic variables of the i-th worker

4   The Test-and-Decode Verification Scheme

In this section, we provide the TD scheme for partial recovery in verifiable coded computing. Hong *et al.* have applied the group testing algorithm to the identification scheme of Byzantine nodes for distributed computing in [12]. In their scheme, the master performs group testing in each group of sub-computation results. Using the intrinsic variables assigned to the coding tasks of each worker in the group, the master can generate the test function. By using the results provided by all workers in the test group as input to the test function, the master can check the correctness of the sub-computation results in a test group.

We apply group test and partial recovery with polynomial codes [22]. Using the results of group tests to guide the decoding process can make full use of the correct part of the sub-computation results, thus reducing the verification rounds of obtaining the final correct result.

Firstly, we introduce the main idea of the TD verification scheme, which includes testing and decoding. Then, we analyze the probability of full recovery from partial recovery and the correctness of the verification. We also give the minimum verification threshold and the number of verification rounds required for full recovery. Finally, we analyze the storage, computation and communication overheads of the proposed verification scheme.

Algorithm 1: Partial Recovery for Coded Edge Computing

Input: $\mathbf{A} \in \mathbb{F}_q^{m \times n}, \mathbf{B} \in \mathbb{F}_q^{n \times l}$

Output: C

1 Encode $\mathbf{A} \to \tilde{\mathbf{A}}_i$, $\mathbf{B} \to \tilde{\mathbf{B}}_i$, $i \in [1, K]$

2 Distributed $\tilde{\mathbf{A}}_i$, $\tilde{\mathbf{B}}_i$ to worker $i \in [1, K]$

3 $\tilde{\mathbf{C}} = \{\}$

4 **for** *worker* $i = 1, 2, \cdots, K$ **do**

5 $\quad$ Compute $\tilde{\mathbf{A}}_i \tilde{\mathbf{B}}_i \in \mathbb{F}_q^{\frac{m}{a} \times \frac{l}{b}}$

6 $\quad$ **return** $\tilde{\mathbf{A}}_i \tilde{\mathbf{B}}_i + \mathbf{E}_i = \tilde{\mathbf{C}}_i$ *to the master*

7 $\quad$ $\tilde{\mathbf{C}} = \tilde{\mathbf{C}} \cup \tilde{\mathbf{C}}_i$

8 **end**

9 Verification initialization $p = 1$, $count = 0$, $\mathbf{C} = \mathbf{0}^{m \times l}$

10 **while** $p \leq \binom{K}{R+\delta}$ **do**

11 $\quad$ Select a test group $\mathbf{G}_p = \{\tilde{\mathbf{C}}_{p,1}, \cdots, \tilde{\mathbf{C}}_{p,R+\delta}\}\} \in \tilde{\mathbf{C}}$

12 $\quad$ $\mathbf{D} = \sum_{k=1}^{R+\delta} \mathbf{S}_{R+\delta,k} \times \tilde{\mathbf{C}}_{p,k}$

13 $\quad$ **for** $i = 1, 2, \cdots, \frac{m}{a}$ **do**

14 $\quad\quad$ **for** $j = 1, 2, \cdots, \frac{l}{b}$ **do**

15 $\quad\quad\quad$ **if** $\mathbf{D}_{i,j} == 0$ *and* $\mathbf{C}_{[i,j]} == 0$ **then**

16 $\quad\quad\quad\quad$ $\mathbf{D}_{i,j} \to \mathbf{C}_{[i,j]}$

17 $\quad\quad\quad\quad$ $count = count + 1$

18 $\quad\quad\quad\quad$ **if** $count == \frac{ml}{ab}$ **then**

19 $\quad\quad\quad\quad\quad$ **return** C

20 $\quad\quad\quad\quad$ **end**

21 $\quad\quad\quad$ **end**

22 $\quad\quad$ **end**

23 $\quad$ **end**

24 $\quad$ $p = p + 1$

25 **end**

26 **return** *False*

4.1 The Main Idea of the TD Verification Scheme

We propose the TD scheme in this subsection and describe the implementation steps.

After each worker completes the given computation task, the master starts retrieving these sub-computation results. We denote $\tilde{\mathbf{C}} = \{\tilde{\mathbf{C}}_1, \cdots, \tilde{\mathbf{C}}_K\}$ as the set of whole sub-computation results. If the number of received computations exceeds the recovery threshold, the master can perform group test. And the master can obtain multiple test results according to the test of different groups. The results obtained from the group test indicate the correctness of the partial value of the final result decoded from this group of sub-computation results with high probability. Choosing whether to decode or not based on the test output avoids invalid decoding. Precisely, each element of the group test output can be mapped to a partial value of the final result.

In the test function of our proposed scheme, if all computation results in a test group are not contaminated or missing, then the test output based on them must be **Negative**. Otherwise, the probability of **Positive** test output is close to 1. Specifically, the master performs the test from the retrieved group of computation results. The master decodes only when the test output is negative. If the number of decoding rounds reaches $\frac{ml}{ab}$, we are sure that we fully recover the final result correctly. Since each round of verification includes testing and decoding steps, we name it as **Test-and-Decode** verification scheme. The detailed algorithm for testing and obtaining verifiable decoding results is described in Algorithm 1.

As the name implies, the two key components of the TD scheme are testing and decoding. These two steps will be described in detail separately.

1) **Test**

All workers are scheduled according to the polynomial codes of tasks in the system model and provide the sub-computation result (*i.e.*, $\tilde{C}_i$) for the master. Each sub-computation result is treated as a polynomial of degree $R-1$. For the master, x is a known value and the coefficients of the polynomial are instead unknowns. So we can construct a system of linear equations to solve for the coefficients [2,3,22]. If the number of received sub-computation results is larger than the recovery threshold (*e.g.*, $\mathbf{G} = \{\tilde{C}_1, \cdots, \tilde{C}_{R+\delta}\}$, $|\mathbf{G}| = R + \delta$, $\delta \in \mathbb{Z}^+$) as follows:

$$
\begin{bmatrix} \tilde{C}_1 \\ \vdots \\ \tilde{C}_R \\ \vdots \\ \tilde{C}_{R+\delta} \end{bmatrix} = \begin{bmatrix} x_1^0 & \cdots & x_1^{R-1} \\ \vdots & \ddots & \vdots \\ x_R^0 & \cdots & x_R^{R-1} \\ \vdots & \ddots & \vdots \\ x_{R+\delta}^0 & \cdots & x_{R+\delta}^{R-1} \end{bmatrix} \cdot \begin{bmatrix} C_1 \\ \vdots \\ C_R \end{bmatrix}.
\tag{3}
$$

Notice that the right-hand side of the equation contains an incomplete Vandermonde matrix. Based on the intrinsic variables of each group of workers, we can generate the Vandermonde matrix $\mathbf{V}$ as

$$
\mathbf{V} = \begin{bmatrix} x_1^0 & \cdots & x_1^{R-1} & \cdots & x_1^{R+\delta-1} \\ \vdots & \ddots & \vdots & \ddots & \vdots \\ x_R^0 & \cdots & x_R^{R-1} & \cdots & x_R^{R+\delta-1} \\ \vdots & \ddots & \vdots & \ddots & \vdots \\ x_{R+\delta}^0 & \cdots & x_{R+\delta}^{R-1} & \cdots & x_{R+\delta}^{R+\delta-1} \end{bmatrix}
\tag{4}
$$

and its inverse matrix $\mathbf{S}$.

We also know that $\mathbf{SV} = \mathbf{I}$. Based on the properties mentioned above, the following equation holds true.

$$
\mathbf{S} \cdot
\begin{bmatrix}
\tilde{\mathbf{C}}_1 \\
\vdots \\
\tilde{\mathbf{C}}_R \\
\vdots \\
\tilde{\mathbf{C}}_{R+\delta}
\end{bmatrix}
= \mathbf{I}_{R+\delta}
\begin{bmatrix}
\mathbf{C}_1 \\
\vdots \\
\mathbf{C}_R \\
\mathbf{0}_1 \\
\vdots \\
\mathbf{0}_\delta
\end{bmatrix},
\tag{5}
$$

where $\mathbf{I}_{R+\delta}$ denotes the identity matrix of rank $R + \delta$ and $\mathbf{0}$ represents the all-zero values matrix with the same dimension as $\mathbf{C}_i, i \in \{1, \cdots, R\}$. If only the last δ rows of $\mathbf{S}$ are taken, the above equation can be simplified to

$$
\begin{bmatrix}
\mathbf{S}_{R+1,1} & \cdots & \mathbf{S}_{R+1,R+\delta} \\
\vdots & \ddots & \vdots \\
\mathbf{S}_{R+\delta,1} & \cdots & \mathbf{S}_{R+\delta,R+\delta}
\end{bmatrix}
\cdot
\begin{bmatrix}
\tilde{\mathbf{C}}_1 \\
\vdots \\
\tilde{\mathbf{C}}_{R+\delta}
\end{bmatrix}
=
\begin{bmatrix}
\mathbf{0}_1 \\
\vdots \\
\mathbf{0}_\delta
\end{bmatrix},
\tag{6}
$$

where $\mathbf{S}_{i,j}$ represents the elements in row i and column j. According to Eqs. (5) and (6), we can express the output of the test function as

$$
\begin{aligned}
\mathbf{D} &= \sum_{k=1}^{R+\delta} \mathbf{S}_{t,k} \times \tilde{\mathbf{C}}_k \\
&= \sum_{k=1}^{R+\delta} \mathbf{S}_{t,k} \left(\sum_{u=0}^{a-1} \sum_{v=0}^{b-1} x_i^{u+av} \mathbf{A}_{u+1} \mathbf{B}_{v+1} \right), \forall t \in [R+1, R+\delta].
\end{aligned}
\tag{7}
$$

In the coded distributed computing task, the minimum group of sub-computation results required for the master to perform decoding is called the decoding group. Similarly, the group of sub-computation results required for the master to perform a test is called test group. The minimum number of sub-computation results required for decoding is also called the recovery threshold of the encoding scheme, and the minimum number of sub-computation results required for testing is called the verification threshold of the verification scheme.

If all the sub-computation results in the test group $\mathbf{G}$ are correct, the output of the test function is $\mathbf{D} = \mathbf{0}^{\frac{m}{a} \times \frac{l}{b}}$. Conversely, if the test group contains incorrect sub-computation results, the output of the test function is $\mathbf{D} \neq \mathbf{0}^{\frac{m}{a} \times \frac{l}{b}}$ with probability $1 - \frac{1}{q}$ [12]. In more detail, the value of each position of the test output (e.g., $\mathbf{D}_{i,j}, i \in [1, \frac{m}{a}], j \in [1, \frac{l}{b}]$) is given by the corresponding position of the tested sub-computation results.

$$
\mathbf{D}_{i,j} = \sum_{k=1}^{R+\delta} \mathbf{S}_{t,k} \times \tilde{\mathbf{C}}_{k_{i,j}}, \forall t \in [R+1, R+\delta].
\tag{8}
$$

If the elements of row i and column j of all sub-computation results matrices in the test group are correct. Then $\mathbf{D}_{i,j} = 0$ holds. Otherwise, $\mathbf{D}_{i,j} \neq 0$. Two cases can make the group testing process erroneous: 1) A group with no erroneous sub-computation results has a non-zero matrix of test result. 2) A group with erroneous sub-computation results has a zero matrix of test result. According to [11], its probability is close to zero as the finite field increases.

The test output matrix consists of $\frac{ml}{ab}$ values, which is the same as the sub-computation result. The key to improving the decoding efficiency lies in fully utilizing the test results of polynomial codes mentioned above.

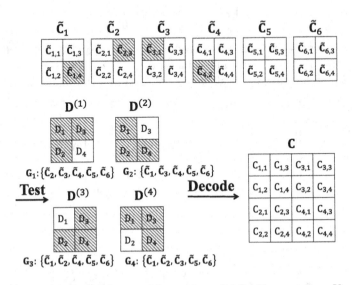

Fig. 3. An Example of TD Scheme with matrix multiplication, suppose $K = 6$, $e = 4$, $\beta = 25\%$, $R = 4$, $\delta = 1$. Choose $\mathbf{G}_1 = \{\tilde{\mathbf{C}}_2, \tilde{\mathbf{C}}_3, \tilde{\mathbf{C}}_4, \tilde{\mathbf{C}}_5, \tilde{\mathbf{C}}_6\}$ for the first test group. According to the test output $\mathbf{D}_4^{(1)}$, the master can recover $\{\mathbf{C}_{1,4}, \mathbf{C}_{2,4}, \mathbf{C}_{3,4}, \mathbf{C}_{4,4}\}$ correctly. Similarly, $\{\mathbf{C}_{1,3}, \mathbf{C}_{2,3}, \mathbf{C}_{3,3}, \mathbf{C}_{4,3}\}$ can be recovered correctly from $\mathbf{G}_2$. The final result $\mathbf{C}$ can be fully recovered by four partial recoveries.

2) *Decode*

Each round of testing requires $R + \delta$ sub-computation results based on the verification scheme and decoding requires R sub-computation results based on the coding scheme [12,22,23]. Thus, we construct a system of linear equations from any R of the $R + \delta$ sub-computation results after testing as follows:

$$\begin{bmatrix} \tilde{\mathbf{C}}_1 \\ \vdots \\ \tilde{\mathbf{C}}_R \end{bmatrix} = \begin{bmatrix} x_1^0 & \cdots & x_1^{R-1} \\ \vdots & \ddots & \vdots \\ x_R^0 & \cdots & x_R^{R-1} \end{bmatrix} \cdot \begin{bmatrix} \mathbf{C}_1 \\ \vdots \\ \mathbf{C}_R \end{bmatrix}. \tag{9}$$

Given an $\frac{m}{a} \times \frac{l}{b}$ dimensional matrix $\mathbf{D}$ as the test output and the parameters of spliting $\mathbf{A}$ and $\mathbf{B}$, (i.e., a and b) after a test round, if an element of output matrix $\mathbf{D}$ is negative (*e.g.*, $\mathbf{D}_{i,j} = 0$), the set of coefficients of the polynomial is one of the solutions of the final result (*e.g.*, $\mathbf{C}_{[i,j]}$), called sub-result. The mapping for decoding $\mathbf{C}$ is defined as

$$\mathbf{D}_{i,j} \to \mathbf{C}_{[i,j]}, \tag{10}$$

where $\mathbf{D}_{i,j}$ represents the element in the ith row and jth column of matrix $\mathbf{D}$ and

$$\begin{aligned}
\mathbf{C}_{[i,j]} = [&\mathbf{C}_{i,j}, \mathbf{C}_{i,j+\frac{l}{b}}, \cdots, \mathbf{C}_{i,j+\frac{l(b-1)}{b}}, \\
&\mathbf{C}_{i+\frac{m}{a},j}, \cdots, \mathbf{C}_{i+\frac{m}{a},j+\frac{l(b-1)}{b}}, \\
&\cdots \\
&\mathbf{C}_{i+\frac{m(a-1)}{a},j}, \cdots, \mathbf{C}_{i+\frac{m(a-1)}{a},j+\frac{l(b-1)}{b}}], \\
&\forall i \in \{1, \cdots, \frac{m}{a}\}, \forall j \in \{1, \cdots, \frac{l}{b}\}.
\end{aligned} \tag{11}$$

The output of the test in the previous step is a matrix of dimension $\frac{m}{a} \times \frac{l}{b}$. According to Eq. (11), the final result can be obtained by decoding at least $\frac{ml}{ab}$ times. Figure 3 is an example of the TD scheme. Since the complexity of the polynomial solution algorithm is linearly related to the size of the input data, our proposed verification scheme does not significantly increase the computational complexity [24]. Vector, matrix, and tensor operations are among the most common computational tasks in parallel computing. The proposed scheme enhances computational efficiency when the user can decompose large-scale complex computations into multiple sub-computational tasks and encode them.

4.2   Theoretical Analysis

In this section, we first prove the successful recovery probability and decoding correctness of the proposed scheme. Then we prove the minimum verification threshold and the number of verification rounds required to recover the final computation result.

Lemma 1. *The final result $\mathbf{C}$ can be fully recovered, iff the all $\frac{ml}{ab}$ sub-results are decoded correctly.*

Proof. The final result $\mathbf{C}$ is a matrix of dimension $m \times l$. Since each sub-computation result is a matrix of dimension $\frac{m}{a} \times \frac{l}{b}$, the decoding process of polynomial codes can be viewed as solving a polynomial of degree $R - 1$ in $\mathbb{F}_q$ for $\frac{ml}{ab}$ times. The sub-result obtained from each decoding contains ab elements according to (11). If all $\frac{ml}{ab}$ sub-results are successfully decoded, the final result $\mathbf{C}$ is considered to be fully recovered.

Remark 1. *If there exist $R + \delta$ entirely correct sub-computation results, since a set of test outputs with negative result is theoretically available, the final result $\mathbf{C}$ can be recovered directly by selecting R out of $R + \delta$ sub-computation results.*

Theorem 1. *The probability that the final result can be fully recovered at each verification round is*

$$P_F = \left(\prod_{\tau=0}^{R+\delta-1} (1 - \frac{\beta e}{K - \tau}) - 1 \right) \beta + 1. \tag{12}$$

Proof. The output of the test function is a matrix with $\frac{ml}{ab}$ elements. To obtain a correct sub-result, all of the test group members should be correct. Since the internal error ratio of the assumed incorrect sub-computation results is β, which means the master needs to select $R + \delta$ elements out of $K - e\beta$ elements. For the correct part of the sub-computation results, a negative test result can be obtained by selecting any $R + \delta$ of the K elements. In the test group, each element of the sub-computation result is independent of the other. Therefore, the probability is

$$\begin{aligned} P_F &= \frac{\binom{K-\beta e}{R+\delta}\beta + \binom{K}{R+\delta}(1 - \beta)}{\binom{K}{R+\delta}} \\ &= \left(\prod_{\tau=0}^{R+\delta-1} (1 - \frac{\beta e}{K - \tau}) - 1 \right) \beta + 1. \end{aligned} \tag{13}$$

Remark 2. *When $\beta = 0$, all sub-computation results are completely correct. The probability of fully recovered is 100%. When $\beta = 1$, all sub-computation results from abnormal workers are completely incorrect. The probability of fully recovered is $\frac{\binom{K-e}{R+\delta}}{\binom{K}{R+\delta}} = \prod_{\tau=0}^{R+\delta-1} (1 - \frac{e}{K-\tau})$.*

Theorem 2. *When the output of the test function is negative, the corresponding sub-result can be decoded correctly.*

Proof. If the sub-computation results in the test group **G** are all correct, then the test output in our proposed scheme is a matrix with all elements zero as follows:

$$\begin{aligned} \mathbf{D} &= \sum_{k=1}^{R+\delta} \mathbf{S}_{t,k} \times \tilde{\mathbf{C}}_k \\ &\triangleq \left[\mathbf{S}_{t,1} \cdots \mathbf{S}_{t,R+\delta} \right] \cdot \begin{bmatrix} x_1^0 & \cdots & x_1^{R-1} \\ \vdots & \ddots & \vdots \\ x_{R+\delta}^0 & \cdots & x_{R+\delta}^{R-1} \end{bmatrix} \cdot \begin{bmatrix} \mathbf{C}_1 \\ \vdots \\ \mathbf{C}_R \end{bmatrix} \\ &= \mathbf{0}^{1 \times R} \cdot \begin{bmatrix} \mathbf{C}_1 \\ \vdots \\ \mathbf{C}_R \end{bmatrix} \\ &= \mathbf{0}^{\frac{m}{a} \times \frac{l}{b}}, \forall t \in [R+1, R+\delta]. \end{aligned} \tag{14}$$

Each test group contains $R + \delta$ sub-computation results. The decoding group consists of any R sub-computation results in the test group. Each decoding group can be constructed as a polynomial with degree $R-1$. Calculating the solution of the polynomial and obtaining the final result are equivalent. If the computations in the test group contain incorrect elements such as $\tilde{\mathbf{C}}_i = \tilde{\mathbf{A}}_i \tilde{\mathbf{B}}_i + \mathbf{E}_i$, the output of the test is $\mathbf{D} = \sum_{k=1}^{R+\delta} \mathbf{S}_{t,k} \times \mathbf{E}_k$, which is positive.

Theorem 3. *The TD scheme has the optimal verification threshold $R+1$.*

Proof. The matrix constructed by all the variables in a decoding group is a full-rank matrix. Therefore, the number of variables in a test group should be more than the recovery threshold, which is a necessary condition for constructing the test function. From Eqs. (3) and (4) we can see that when the size of the test group is $R + \delta$, the matrix with dimensional $(R + \delta) \times R$ is first expanded to a Vandermonde matrix of dimension $(R + \delta) \times (R + \delta)$. When the test group contains more variables, the reliability of the test is higher, and The complexity of verification will become higher. Since the recovery threshold of polynomial codes is R, to reduce the verification complexity, the optimal verification threshold of the TD scheme is $R + 1$.

Remark 3. *The coding methods construct the computational tasks of each worker. Likewise, test functions are constructed from coding methods. It means that in the verification of distributed matrix multiplication, there are constraints on the recovery threshold and the test function input.*

Theorem 4. *The expected number of verification rounds of the TD scheme with probability $1 - \epsilon$ is*

$$E(T) = \log_{1-P_F} \epsilon. \tag{15}$$

Proof. When the size of the test group is $R + \delta$, the final result can be fully recovered at each verification round with probability P_F. Since ϵ is the probability of not getting the final result after T verification rounds. $1 - P_F$ is the probability that the final result is not available. Therefore, we can get $E(T) = \log_{1-P_F} \epsilon$ from $(1 - p_F)^{E(T)} = \epsilon$.

Remark 4. *According to the above theorem, it is known that the number of times required to obtain the final correct result with the same encoding strategy is independent of the parameters of the data matrix and is related to the number of anomalous workers (e), the proportion of incorrect elements (β) and the error probability (ϵ) when the number of workers is constant.*

4.3 Overhead Analysis

In this subsection, we analyze the storage, computation, and communication overheads of the proposed scheme for workers and master, respectively.
1) **Workers:**
Firstly, each worker receives and saves $\tilde{\mathbf{A}}_i$ with dimension $\frac{m}{a} \times n$ and $\tilde{\mathbf{B}}_i$ with dimension $n \times \frac{l}{b}$. Secondly, it calculates the product of the two matrices to get the sub-computation result $\tilde{\mathbf{C}}_i$ with dimension $\frac{m}{a} \times \frac{l}{b}$. Finally, each worker return

the sub-computation result to the master. So the storage, computation, and communication overheads of each worker is $\mathcal{O}(\frac{mn}{a} + \frac{nl}{b})$, $\mathcal{O}(\frac{mnl}{ab})$ and $\mathcal{O}(\frac{mn}{a} + \frac{nl}{b})$.

2) **Master:**

The master has pre-saved the $m \times n$ dimension matrix $\mathbf{A}$ and the $n \times l$ dimension matrix $\mathbf{B}$. The master will generate $\tilde{\mathbf{A}}_i$ and $\tilde{\mathbf{B}}_i$ through slicing and encoding. The complexity of encoding two matrices is $\mathcal{O}(K(\frac{mn}{a} + \frac{nl}{b}))$. Before executing the verification algorithm, the master stores all sub-computation results and the private intrinsic variables of the workers, which is $\mathcal{O}(K(\frac{ml}{ab} + 1))$ in total.

The testing function can be simplified as $\mathbf{D} = \sum_{k=1}^{R+\delta} \mathbf{S}_{R+\delta,k} \times \tilde{\mathbf{C}}_k$ with the complexity of $\mathcal{O}((ab + \delta) \times \frac{ml}{ab})$. Considering the required verification rounds of the TD scheme is T, the complexity of testing is expressed as $\mathcal{O}(Tml(1 + \frac{\delta}{ab}))$.

The decoding process of the master can be regarded as solving the polynomial of degree $R - 1$ in $\mathbb{F}_q$ for $\frac{ml}{ab}$ times. A plain solution is to list the system of equations and then use the Gaussian elimination method, which requires a complexity of $\mathcal{O}(ml(ab)^2)$. Obviously, the complexity can be reduced to $\mathcal{O}(ml \log^2(ab) \log \log(ab))$ by using the faster interpolation method [24]. The storage, computation and communication overheads of the proposed scheme are summarized in Table 2.

Table 2. The Resource Requirements of the TD Scheme

	Storage	Computation	Communication
Worker	$\mathcal{O}(\frac{mn}{a} + \frac{nl}{b})$	$\mathcal{O}(\frac{mnl}{ab})$	$\mathcal{O}(\frac{mn}{a} + \frac{nl}{b})$
Master	$\mathcal{O}(mn + nl + K(\frac{ml}{ab} + 1))$	$\mathcal{O}(K(\frac{mn}{a} + \frac{nl}{b}) + Tml(1 + \frac{\delta}{ab}) + ml \log^2(ab) \log \log(ab))$	$\mathcal{O}(\frac{Kml}{ab})$

5   Simulation

In this section, we conduct simulation experiments to evaluate the performance of the proposed TD verification scheme under different system parameters. Specifically, we evaluate three dimensions: (1) the probability of successful verification, *i.e.*, the probability that the TD scheme obtain the correct final result by verification and decoding; (2) the required time of obtaining the final result by the verification scheme; (3) the number of verification rounds required to recover the correct final result.

We compare the proposed TD scheme with some recently published schemes as follows:

1. *Adaptive Verifiable Coded Computing* (AVCC): A verification scheme based on Freivalds' algorithm, which checks the correctness of the sub-computation results returned individually [1].
2. *Decode and Compare Verification* (DCV): A verification scheme that decodes directly to get several decoded results and compares with each other until λ sets of identical decoded results are obtained [17].

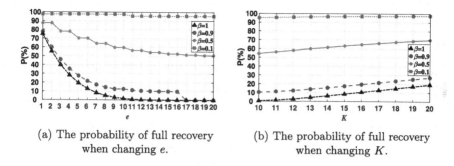

(a) The probability of full recovery when changing e.

(b) The probability of full recovery when changing K.

Fig. 4. The probability of full recovery with different β.

3. *Group-wise Verifiable Coded Computing* (GVCC): A non-colluding attacks identification scheme, which utilizes the properties of Vandermonde encoding matrix to verify each group of sub-computation results [12].

The default values of system parameters in the three sets of experiments are $K = 20$, $e = 5$, $\beta = 20\%$, and the default parameters of the matrix are $m = 600$, $n = 4000$, $l = 40$. For faster verification, the default size of the verification threshold is set to $R + 1$ and $R = 4$. The programs for the coded edge computing with different verification schemes are based on Python and operate on the Intel Core i7-10700 CPU 2.90GHz. The data presented in the plots are the result of averaging 1000 repetitions in the simulation experiments.

5.1 Simulation Result

Firstly, we analyze the probability of full recovery for the proposed TD scheme in Fig. 4. Full recovery of the final result within the deadline is considered a successful verification. When the number of abnormal workers e is small, the scheme maintains a high probability of successful recovery for different error ratios. When β is held constant, the probability of full recovery decreases as the number of abnormal workers increases, which means that the recovery of the final result becomes more difficult. Comparing the different curves, there is a larger fraction of the sub-computation results available due to the decrease in β. Therefore, the probability of successful verification increases with the decrease of β which is shown in Fig. 4 (a) and Fig. 4 (b). When the β is small, the verification scheme consistently hold high probability of full recovery. In particular, $P(\beta = 0.1)$ is about 44.7% larger than $P(\beta = 0.5)$ even $e = K$.

Secondly, we compare the computation time consumption with and without the test when the parameters (m, l, n) of the input matrix are varied. Decoding without test implies the final result with the highest confidence is obtained by comparing the results of different decoding groups. This method is described in detail in [17]. Decoding with test means that the test group is first tested, then the decoding is done based on the test output. The computation time in the experiment includes the time for verification and decoding. In Fig. 5 (a) and Fig. 5

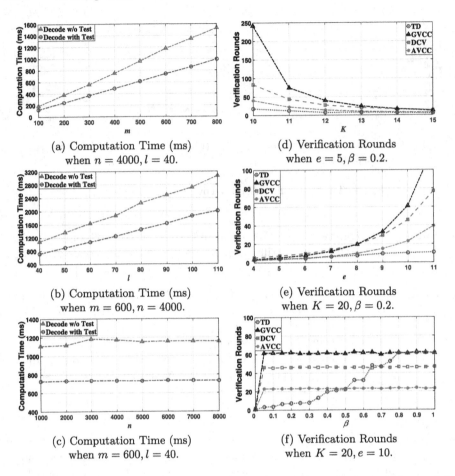

(a) Computation Time (ms)
when $n = 4000, l = 40$.

(b) Computation Time (ms)
when $m = 600, n = 4000$.

(c) Computation Time (ms)
when $m = 600, l = 40$.

(d) Verification Rounds
when $e = 5, \beta = 0.2$.

(e) Verification Rounds
when $K = 20, \beta = 0.2$.

(f) Verification Rounds
when $K = 20, e = 10$.

Fig. 5. The computation time when changing the matrix parameters m, l, n and the verification rounds of TD, GVCC, DCV and AVCC schemes when changing the system parameters K, β, e.

(b), the computation time increases as the size of the input matrix increases (*e.g.*, m, l), because the computational complexity in the verification and decoding phase is related to the size of the input matrix. Although adding additional testing functions before decoding increases the computation time slightly, the testing leads to a much more significant reduction in decoding time than the time consumption of verification. With the increase of m and l, the gap between the consumption time of the two strategies is also getting bigger and bigger. In Fig. 5 (c), decoding after test function checking can save about 35.5% of time consumption when $m = 600$, $l = 40$. The time consumed by two verification strategies does not increase as n increases which implies that the time complexity of the verification scheme is independent of n.

Finally, the required verification rounds comparisons between the proposed TD scheme and the baseline schemes are also shown in Fig. 5. The number of verification rounds for all four schemes decreases as the total number of workers increases or the number of corrupted workers decreases in Fig. 5 (d) and Fig. 5 (e). The TD scheme always has the lowest number of verification rounds. In particular, the TD scheme requires much fewer verification rounds than the other schemes when the value of e/K is larger. In Fig. 5 (f), as the error rate of each abnormal sub-computation result increases, the number of verification rounds of the TD scheme gradually increases, eventually aligning with the GVCC scheme, which has been analyzed in Sect. 4.2. Significantly, the TD scheme has 68.6%, 82.6%, and 88.6% fewer verification rounds than the ACVV, DCV, and GVCC schemes when the error rate of sub-computation results is 20%, respectively.

5.2   Simulation Analysis

Based on the results of the simulations above, we conclude the following:

- The probability of successful recovery is influenced by the proportion of erroneous elements. If the sub-computation result has fewer error elements, this means that more elements can pass the checksum and participate in the decoding of the final result. The TD schemes can also significantly reduce the number of test rounds. Therefore, the performance of TD scheme is better when the abnormal worker has a smaller proportion of error elements.
- The computation time is related to the matrix size. The test-based verification scheme has a significant advantage in computation time compared to the comparison-based verification scheme.
- The TD scheme also alleviates the limitation of the coding scheme on the recovery threshold of the worker. Because the TD scheme is a partial recovery scheme, full recovery of the final result no longer requires fully correct sub-computation results.

6   Conclusion

In this paper, we study the problem of efficient recovery of distributed computation results. We design a general *Test-and-Decode* (TD) scheme combining verifiable computing and partial recovery. It allows the master to verify a group of sub-computation results and selectively recover partial final result multiple times. We demonstrate the correctness of verification and decoding feasibility of the proposed scheme. Extensive simulation experimental results show that the TD scheme is more efficient than the current schemes.

Acknowledgements. This work was supported in part by the National Natural Science Foundation of China (62072321, 61972272), the Six Talent Peak Project of Jiangsu Province (XYDXX-084), the China Postdoctoral Science Foundation (2020M671597), the Jiangsu Postdoctoral Research Foundation (2020Z100), Natural Science Foundation of the Higher Education Institutions of Jiangsu Province (22KJA520007, 20KJB520002), the Collaborative Innovation Center of Novel Software Technology and Industrialization, and Soochow University Interdisciplinary Research Project for Young Scholars in the Humanities.

References

1. Tang, T., Ali, R.E., Hashemi, H., Gangwani, T., Avestimehr, S., Annavaram, M.: Adaptive verifiable coded computing: towards fast, secure and private distributed machine learning. In: Proceedings of IEEE International Parallel Distribution Processing Symposium (IPDPS), pp. 628–638 (2022)
2. Yu, Q., Avestimehr, A.S.: Coded computing for resilient, secure, and privacy-preserving distributed matrix multiplication. IEEE Trans. Commun. **69**(1), 59–72 (2020)
3. Yu, Q., Li, S., Raviv, N., Kalan, S.M.M., Soltanolkotabi, M., Avestimehr, S.A.: Lagrange coded computing: Optimal design for resiliency, security, and privacy. In: Proceedings of the 22nd International Conference on Artificial Intelligence and Statistics (AISTATS), pp. 1215–1225 (2019)
4. Lee, K., Suh, C., Ramchandran, K.: High-dimensional coded matrix multiplication. In: Proceedings of IEEE International Symposium on Information Theory (ISIT), pp. 2418–2422 (2017)
5. Fan, X., Soto, P., Zhong, X., Xi, D., Wang, Y., Li, J.: Leveraging stragglers in coded computing with heterogeneous servers. In: Proceedings of IEEE International Symposium on Quality of Service (IWQoS), pp. 1–10. IEEE (2020)
6. Liu, S., Jiang, C.: A novel prediction approach based on three-way decision for cloud datacenters. Appl. Intell., 1–17 (2023)
7. Dutta, S., Fahim, M., Haddadpour, F., Jeong, H., Cadambe, V., Grover, P.: On the optimal recovery threshold of coded matrix multiplication. IEEE Trans. Inf. Theory **66**(1), 278–301 (2020)
8. Sarmasarkar, S., Lalitha, V., Karamchandani, N.: On gradient coding with partial recovery. In: Proceedings of IEEE International Symposium on Information Theory (ISIT), pp. 2274–2279 (2021)
9. Ozfatura, E., Ulukus, S., Gündüz, D.: Coded distributed computing with partial recovery. IEEE Trans. Inf. Theory **68**(3), 1945–1959 (2021)
10. Qiu, H., et al.: EC-Fusion: an efficient hybrid erasure coding framework to improve both application and recovery performance in cloud storage systems. In: Proceedings of IEEE International Symposium on Parallel and Distributed Processing (IPDPS), pp. 191–201 (2020)
11. Solanki, A., Cardone, M., Mohajer, S.: Non-colluding attacks identification in distributed computing. In: Proceedings of the IEEE Information Theory Workshop (ITW), pp. 1–5 (2019)
12. Hong, S., Yang, H., Lee, J.: Hierarchical group testing for byzantine attack identification in distributed matrix multiplication. IEEE J. Sel. Areas Commun. **40**(3), 1013–1029 (2022)

13. Li, J., Hollanti, C.: Private and secure distributed matrix multiplication schemes for replicated or MDS-coded servers. IEEE Trans. Inf. Forensics Secur. **17**, 659–669 (2022)
14. Yu, Q., Maddah-Ali, M.A., Avestimehr, A.S.: Straggler mitigation in distributed matrix multiplication: fundamental limits and optimal coding. IEEE Trans. Inf. Theory **66**(3), 1920–1933 (2020)
15. Ulukus, S., Avestimehr, S., Gastpar, M., Jafar, S., Tandon, R., Tian, C.: Private retrieval, computing and learning: recent progress and future challenges. IEEE J. Sel. Areas Commun. **40**(3), 729–748 (2022)
16. Fu, M., Wang, J., Zhou, J., Wang, J., Lu, K., Zhou, X.: A null-space-based verification scheme for coded edge computing against pollution attacks. In: Proceedings of IEEE International Conference on Parallel and Distributed Systems (ICPADS), pp. 454–461 (2019)
17. Wang, J., Lu, Z., Fu, M., Wang, J., Lv, K., Jukan, A.: Decode-and-compare: an efficient verification scheme for coded distributed edge computing. IEEE Trans. Cloud Comput., 1–18 (2022)
18. Sahraei, S., Maddah-Ali, M.A., Avestimehr, A.S.: Interactive verifiable polynomial evaluation. IEEE J. Sel. Areas Inf. Theory **2**(1), 317–325 (2021)
19. Dutta, S., Cadambe, V., Grover, P.: Coded convolution for parallel and distributed computing within a deadline. In: Proceedings of IEEE International Symposium on Information Theory (ISIT), pp. 2403–2407 (2017)
20. Bitar, R., Xhemrishi, M., Wachter-Zeh, A.: Fountain codes for private distributed matrix-matrix multiplication. In: Proceedings of International Symposium on Information Theory and Its Applications (ISITA), pp. 480–484 (2020)
21. Hofmeister, C., Bitar, R., Xhemrishi, M., Wachter-Zeh, A.: Secure private and adaptive matrix multiplication beyond the singleton bound. IEEE J. Sel. Areas Inf., Theory (2022)
22. Yu, Q., Maddah-Ali, M., Avestimehr, S.: Polynomial codes: an optimal design for high-dimensional coded matrix multiplication. In: Proceedings of Advance Neural Information Processing Systems(NIPS), pp. 4403–4413 (2017)
23. Chang, W.T., Tandon, R.: On the capacity of secure distributed matrix multiplication. In: Proceedings of IEEE Global Communications Conference (GLOBECOM), pp. 1–6 (2018)
24. Kedlaya, K.S., Umans, C.: Fast polynomial factorization and modular composition. SIAM J. Comput. **40**(6), 1767–1802 (2011)

RecAGT: Shard Testable Codes with Adaptive Group Testing for Malicious Nodes Identification in Sharding Permissioned Blockchain

Dongyang Yu[1], Jin Wang[1,2](✉), Lingzhi Li[1](✉), Wei Jiang[1], and Can Liu[1]

[1] School of Computer Science and Technology, Soochow University, Suzhou, China
{dyyu,wjiang99,cliu7}@stu.suda.edu.cn
[2] School of Future Science and Engineering, Soochow University, Suzhou, China
{wjin1985,lilingzhi}@suda.edu.cn

Abstract. Recently, permissioned blockchain has been extensively explored in various fields, such as asset management, supply chain, healthcare, and many others. Many scholars are dedicated to improving its verifiability, scalability, and performance based on sharding techniques, including grouping nodes and handling cross-shard transactions. However, they ignore the node vulnerability problem, *i.e.*, there is no guarantee that nodes will not be maliciously controlled throughout their life cycle. Facing this challenge, we propose RecAGT, a novel identification scheme aimed at reducing communication overhead and identifying potential malicious nodes. First, shard testable codes are designed to encode the original data in case of a leak of confidential data. Second, a new identity proof protocol is presented as evidence against malicious behavior. Finally, adaptive group testing is chosen to identify malicious nodes. Notably, our work focuses on the internal operation within the committee and can thus be applied to any sharding permissioned blockchains. Simulation results show that our proposed scheme can effectively identify malicious nodes with low communication and computational costs.

Keywords: Permissioned blockchain · Sharding · Coded computation · Group testing

1 Introduction

Permissioned blockchain has emerged as an appropriate architecture concept for business environments, and it is presently arising as a promising solution for distributed cross-enterprise applications. However, it still faces many challenges regarding verifiability [1,2], scalability [3], and performance [4]. To solve these problems, the sharding technique inspired by Spanner [5] is proposed to be integrated with permissioned blockchain, which partitions block data into multiple shards that are maintained by different committees (or "clusters").

Z. Tari et al. (Eds.): ICA3PP 2023, LNCS 14490, pp. 398–418, 2024.
https://doi.org/10.1007/978-981-97-0859-8_24

Existing work [6–12] in sharding permissioned blockchains focuses on how to partition nodes into different committees and efficiently handle cross-shard transactions. But they ignore **the vulnerability of nodes to malicious control**. There is no guarantee that nodes could remain honest[1]. In other words, nodes cannot always behave normally throughout their life cycle. For example, nodes could come under control and turn malicious[2] as a result of cyber attacks such as BGP hijacking [14], DNS attack [15], or Eclipse attack [16]. Many of the previous research [17–20] on malicious node identification has been explored in distributed computing. The common idea of them is to utilize different coding algorithms to check the final computation output and use numerous testing trials to find malicious nodes. However, workers in distributed systems perform intermediate computing tasks and do not maintain any data locally. Moreover, there must be complete trust between the master and workers [21].

In this paper, we consider the node vulnerability problem in sharding permissioned blockchains. We propose the sha**R**d testable **c**ode with **A**daptive **G**roup **T**esting (**RecAGT**), a malicious node identification scheme. Specifically, we first present our shard testable codes by designing polynomial functions to reduce communication overhead. Nodes perform verification based on the properties of testable codes. Then we design an identity proof protocol based on the digital signature as the proof of malicious behaviors. Finally, we use an adaptive group testing algorithm to calculate the required number of test trials. Therefore, the newly-joined node can verify the received messages and recover the original data to improve its ability to identify malicious nodes, which further enhances the security and stability of the sharding permissioned blockchain.

The main contributions of this paper are summarized as follows:

- We propose a new scheme called RecAGT for malicious node identification. It is shown that communication costs and computational complexity will be significantly reduced from $O(n^2 b)$ to $O(\log^2(m) \log\log(m))$ compared to other schemes (Table 1).
- In addition, the administrator could perform adaptive group testing to reduce the number of tests required to identify malicious nodes.
- We conduct theoretical simulations and the results show that our scheme only needs a low number of group tests, which effectively improves the system security and stability.

The rest of the paper is organized as follows. Section 2 discusses related work. Section 3 describes the setup of the permissioned blockchain and introduces the system model. In Sect. 4, we propose our identification scheme and give detailed theoretical analyses. Section 5 analyzes and discusses the experimental results. Finally, Sect. 6 concludes the paper.

---

[1] Honest nodes are those that perform normally following the rules of the system (*e.g.*, read, write or maintain blocks and perform or relay transactions).

[2] The behaviors of malicious (or Byzantine) nodes could censor, reverse, reorder or withhold specific transactions without including them in any block to interfere with the system [13].

Table 1. The communication cost and computational complexity of nodes[a]

	Communication cost	Computational complexity
Uncoded	$O(nb)$	$O(n^2 b)$
CheckSum	$O(b)$	$O(n^2)$
RecAGT	$O(b)$	$O(\log^2(m) \log\log(m))$

[a]b: shard size in bytes, n: size of committee, m: size of coding shard

2 Related Work

Recently, a lot of research has been made on the permissioned blockchain to improve its verifiability, scalability, and performance. Since our scheme is proposed based on the sharding technique, we will discuss the related work in the aspects of sharding and other methods.

Sharding Methods. There have been many studies working on sharding permissioned blockchains. Amiri *et al.* [6] introduce a model to handle intra-shard transactions and their subsequent work [7] uses a directed acyclic graph to resolve cross-shard transaction agreements to improve verifiability. Dang *et al.* [8] design a comprehensive protocol including shard formation and transaction handling to upgrade performance. Huang *et al.* [9] propose an adaptive resource allocation algorithm to efficiently allocate network resources for system stability. Gao *et al.* [10] propose the Pshard protocol, which adopts a two-layer data model and uses a two-phase method to execute cross-shard transactions to ensure safety and liveness. Mao *et al.* [11] propose a locality-based sharding protocol in which they cluster participants based on their geographical properties to optimize inter-shard performance. As we can see, they pay more attention to the operation of blockchain systems. Nonetheless, these approaches overlook the potential actions of individual nodes. In our research, we consider the scenario where nodes might be under malicious control and propose an identification scheme to mitigate these vulnerabilities.

Identification of Malicious Nodes. The problem of malicious node identification has been studied in distributed systems. Yu *et al.* [17] provide resiliency against stragglers and security against Byzantine attacks based on Lagrange codes. Solanki *et al.* [18] design a coding scheme to identify attackers in distributed computing. Hong *et al.* [19] propose locally testable codes to identify Byzantine attackers in distributed matrix multiplication. They also propose a hierarchical group testing [20] in distributed matrix multiplication, making the required number of tests smaller. However, there must be complete trust between the master and workers, which is unsuitable for blockchain. In this paper, we develop an identity proof method that serves as a safeguard against potential malicious behavior.

In view of these unresolved problems, we propose a novel identification scheme named RecAGT, specifically designed to address the issue of identifying malicious nodes effectively.

Remark 1. Our work focuses on reducing communication costs and identifying potential malicious nodes based on their transmitted messages. Therefore, we do not investigate further details about transaction verification and subsequent penalty actions.

3   System Overview

In this section, we introduce the system model based on the permissioned[3] blockchain and explore potential corresponding attacks.

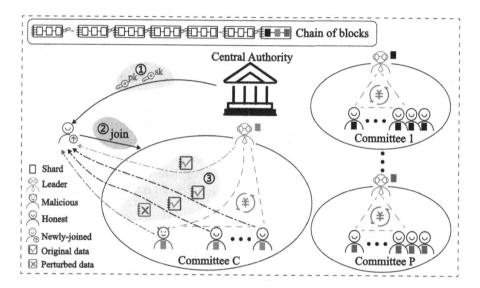

Fig. 1. System model of sharding permissioned blockchain under the existence of malicious nodes

3.1   System Model

The permissioned blockchain is a distributed ledger that cannot be publicly accessed and is only open to users with authorized digital certificates. Nodes perform specific operations granted by the administrator (*Central Authority, CA*). Without losing generality, we assume that there is a *public key infrastructure (PKI)* in the system, that is, *CA* distributes the private (secret) key sk_i and the public key pk_i to each node N_i as identity credentials. Note that each node

---

[3] Generally speaking, permissioned blockchains can be divided into private and consortium blockchains since both of them only allow nodes with identity to join the network. Our study primarily focuses on consortium blockchains due to their alignment with the idea of decentralization.

knows each other's public key via CA. In addition, CA needs to assign a unique scalar x_i to each member for the construction of our identification scheme.

In sharding permissioned blockchains, nodes are partitioned into committees. Essentially, the processing mechanism is the same for each committee since each can be seen as a tiny blockchain system that maintains a subchain. Hence, we will focus on a committee C in the following sections. There are basically two types of nodes: full nodes and light nodes. Full nodes are able to generate and store valid blocks and verify new blocks from other full nodes. Light nodes are able to perform transaction inclusion verification and thus increase blockchain scalability instead of storing the entire ledger. Based on the purpose of reducing communication costs and identifying malicious nodes, the object of our study is the full node.

The overall system model is illustrated in Fig. 1. Each member would maintain the entire shards of its corresponding committee under a permissioned blockchain. But for simplicity of presentation, we only show the shard of the latest block for each node. Each committee handles different transactions in parallel. When a new node enters the network, CA will verify its identity and assign keys (sk and pk) to it (shown in yellow background). Then the new node joins one of the committees by partitioning rules (shown in purple background), and it needs to retrieve all the shards of the committee in order to participate in the transaction processing. However, the node cannot fully trust any members, even the leader, at all. Therefore, it must receive data messages from as many members as possible against interference from potential malicious nodes (shown in grey background).

The data stored by nodes is a chain of sub-blocks, denoted as B, which is actually a set of byte strings of blocks containing a batch of transactions. If we divide them into m subshards, they can be shown as

$$B = \begin{bmatrix} B_1 & B_2 & \cdots & B_m \end{bmatrix}^{\top}.$$

The system has P committees and each committee C includes n members. Other assumptions of our network model are as follows:

1. Credibility of nodes: we assume only CA is honest and has no assumptions about other nodes. It's a weak decentralization. In practice, permissioned blockchains do not defy the principles of decentralization but rather strike a balance between centralized and decentralized requirements.
2. Finite maximum network delay: all requests can be answered in a finite time. In other words, a finite maximum delay δ is assumed. If a node sends a request, it will receive a response message within δ. Otherwise, we assume that the target node is offline or does not work.
3. Network communication:
   - Nodes communicate with other nodes;
   - Point-to-point communications are asynchronous;
   - The communication channel is noiseless.

3.2 Threat Model

When a new node enters the system, its identity undergoes verification by CA. Once the node's identity is authenticated, CA allocates identity credentials to the new node. However, it is possible that committee members behave maliciously to prevent new nodes from joining the system (e.g., they may not send feedback or they may respond to perturbed shard data). In other words, although nodes are authenticated when they join the network, they cannot be fully trusted because there is no guarantee that they will not be maliciously controlled throughout their life cycle. Therefore, it is imperative to devise an efficient identification scheme to identify potential malicious nodes.

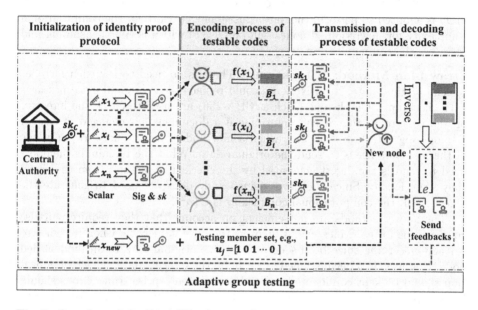

Fig. 2. Overview of the RecAGT scheme. When a new node N_{new} joins the system, CA utilizes PKI to allocate public and private keys, and a scalar along with a testing set to it. CA signs the scalar of N_{new} using its own private key and transmits it to N_{new}. This allows N_{new} to verify and store the received scalar using the CA's public key. Then, N_{new} sends requests to nodes of the testing set. N_{new} generates the historical data of its committee by receiving messages containing encoded shard data with the first and second signatures. If the computing result is incorrect during the verification process, N_{new} will forward feedback to CA based on these messages, aiding in identifying potential malicious nodes.

4   RecAGT Identification Scheme

Building upon the system model outlined in Sect. 3, we propose an identification scheme based on our shard testable codes. In this scheme, newly-joined nodes are provided with encoded shard data to facilitate the retrieval of the original data specific to the committee they are part of. If an inconsistency arises during the verification process, the node will send feedback to CA to help identify potential malicious nodes, as illustrated in Fig. 2. The scheme consists of three key components: shard testable codes, an identity proof protocol, and an adaptive group testing method.

A straightforward data identification scheme, referred to as "**Uncoded**", might involve each committee member sending the original data to the newly-joined node, enabling its participation in blockchain activities. However, malicious nodes could disrupt the normal operation (*e.g.*, sending fraudulent transactions). Therefore, the node has to compare all received data to ensure that there is no perturbed data, which would be equivalent to receiving the entire shard. If MD5 is adopted to perform the data consistency check, then the computational complexity for newly-joined nodes will be $O(n^2 b)$ which is time-consuming and increases the communication overhead.

Straw Man Method: "Checksums". There is no need to ask each member to send original data. Intuitively, we could replace those original data messages with checksums. Checksums (such as SHA-256) are used to check the integrity of web content so that it can determine if a document has changed in case of a tampering attack. Specifically, when a new node joins the network, only one of the members needs to send the original data. What the remaining members need to send is the digest generated by the checksums method from the shard they store locally. Since the newly-joined node does not need to generate the digest itself, the computational complexity is $O(n^2)$.

The above methods are not efficient on account that they require pairwise comparisons for each message. Moreover, they are vulnerable when it comes to privacy-preserving since both send at least one original data to an unidentified node. However, in a permissioned blockchain, the internal data of each organization should be kept confidential meanwhile cross-enterprise transactions should be transparent to all parties. Therefore, in addition to identifying potential malicious nodes, we need to encode data during transmission.

4.1   Shard Testable Codes

To check if the returned data is not perturbed in an efficient way, we first design a method to encode the data that is inspired by polynomial codes [22]. The linear encoding function $\mathbf{f}$ is constructed by using the row-divided sub-matrices B_v and the corresponding scalars x_i as coefficients, which is given by

$$\mathbf{f}(x_i) = \sum_{v=0}^{m-1} B_{v+1}^T x_i^v = \widetilde{B}_i. \tag{1}$$

Accordingly, each member will store the encoded shard (*e.g.*, Node N_i will store $\widetilde{B}_i$ locally).

By combining encoding functions for a committee of size n, we represent the coded shards as a matrix, which is given by

$$\tilde{B} = G \cdot B = \begin{bmatrix} x_1^0 & \cdots & x_1^{m-1} \\ \vdots & \ddots & \vdots \\ x_n^0 & \cdots & x_n^{m-1} \end{bmatrix} \cdot \begin{bmatrix} B_1 \\ \vdots \\ B_m \end{bmatrix} = \begin{bmatrix} \tilde{B}_1 \\ \vdots \\ \tilde{B}_n \end{bmatrix}, \tag{2}$$

where G is the encoding matrix consisting of all scalars assigned to each committee member.

Waiting for $m+1$ encoded shard messages from different nodes to form a test group denoted $\mathbf{u}$ (we will discuss the test group in Sect. 4.3), the newly-joined node collects these messages into a new matrix C. For simplicity of demonstration, we assume that the nodes' indexes that return messages are $[1: m+1]$, which means that their scalars are $[x_i]_{i=1}^{m+1}$. Therefore, the result vector consisting of intermediate messages can be expressed as

$$C = G_u \cdot B = \begin{bmatrix} x_1^0 & \cdots & x_1^{m-1} \\ \vdots & \ddots & \vdots \\ x_{m+1}^0 & \cdots & x_{m+1}^{m-1} \end{bmatrix} \cdot B, \tag{3}$$

where

$$C_i = \sum_{k=0}^{m-1} x_i^k B_{k+1}, \forall i \in [1, m+1], \tag{4}$$

and G_u is the encoding matrix consisting of the $m+1$ encoding vectors.

Based on G_u, we can make Vandermonde matrix V by adding one more column shown in bold, which is given by

$$V = \begin{bmatrix} x_1^0 & \cdots & x_1^{m-1} & \mathbf{x_1^m} \\ \vdots & \ddots & \vdots & \vdots \\ x_{m+1}^0 & \cdots & x_{m+1}^{m-1} & \mathbf{x_{m+1}^m} \end{bmatrix}. \tag{5}$$

Since the Vandermonde matrix is invertible, we denote the inverse of V as S. The inverse of the Vandermonde matrix can be computed as

$$S = \prod_{1 \le j < i \le m+1} \frac{1}{x_i - x_j}. \tag{6}$$

Since the computing process can be viewed as a polynomial interpolation problem [23]. We adopt one of the method [24] whose computational complexity is $O(\log^2(m) \log \log(m))$. Note that the complexity could be further reduced by adopting any variant of interpolation algorithms.

Without loss of generality, let's define matrix U as the first m columns of V, which is another name for G_u. Therefore, $C = U \cdot B = G_u \cdot B$, and by the construction and rules for matrix multiplication, the following equation holds true,

$$S \cdot U = \left[\frac{\mathbf{I}_m}{\mathbf{0}_{1,m}} \right], \tag{7}$$

where $\mathbf{I}_m$ denotes the identity matrix of rank m and $\mathbf{0}_{1,m}$ denotes the null matrix of row size 1 and column size m. Let's denote $S_{j,k}$ as the element of row j and column k in matrix S, we can express Eq. (7) as

$$\sum_{j=1}^{m+1} S_{m+1,j} \times x_j^p = 0, \ \forall p \in [0, m-1]. \tag{8}$$

After the construction of shard testable codes, we will discuss two different scenarios based on the intermediate messages:

1. These messages are all correct, which means the test group set is honest;
2. There is at least one perturbed message, which means some of the group members are malicious.

The Group Members are all Honest. If the test group does not include malicious nodes, with the group and the collected messages serving as $\mathbf{u}_h$ and C_h, respectively, we would get the following test result

$$S \cdot C_h = S \cdot U \cdot B = \left[\frac{\mathbf{I}_m}{\mathbf{0}_{1,m}} \right] \cdot B, \tag{9}$$

which means $S_{m+1} \cdot C = \mathbf{0}_{1,m}$ where S_{m+1} is the row $m+1$ of matrix S. By using Eqs. (7) and (8), we can define the output of test row result as $\mathbf{O} = S_{m+1} \cdot C$.

We choose any m rows of matrix C, denoted as $\overset{m}{C}$, and the corresponding encoding matrix is denoted as $\overset{m}{G}$. Therefore, we can get

$$\overset{m}{C} = \overset{m}{G} \cdot B. \tag{10}$$

Since $\overset{m}{G}$ is a Vandermonde matrix, we represent the inverse of $\overset{m}{G}$ as $(\overset{m}{G})^{-1}$. When we left multiply both sides of Eq. (10) by the inverse of $\overset{m}{G}$, we would get

$$(\overset{m}{G})^{-1} \cdot \overset{m}{C} = (\overset{m}{G})^{-1} \cdot \overset{m}{G} \cdot B = B, \tag{11}$$

which means the newly-joined node could recover the original shard based on our coding scheme.

In the following scenario, we will consider a more complicated case where some malicious nodes will interfere with the shard data to keep the new node from joining the committee, which compromises the scalability of the permissioned blockchain.

The Group Includes at Least One Malicious Node. If there is at least one byzantine node in the group, denoting the group as $\mathbf{u}_b$ and the matrix including perturbed shard data as C_b, we would get the following result

$$S \cdot C_b = \left[\frac{\mathbf{I}_m}{\mathbf{e}_{1,m}} \right] \cdot B, \tag{12}$$

where $\mathbf{e}_{1,m} \neq \mathbf{0}_{1,m}$, which indicates some of the elements in matrix $\mathbf{e}_{1,m}$ are not zero. We can express Eq. (12) as

$$\sum_{j=1}^{m+1} S_{m+1,j} \times x_j^p = 0 + b, \quad \forall p \in [0, m-1], \tag{13}$$

where b is the interfering data.

Lemma 1. *The output of the test result in the last row with malicious group $\mathbf{u}_b$ is $\mathbf{O} \neq 0$, and the output with honest group $\mathbf{u}_h$ is $\mathbf{O} = 0$.*

Proof. Based on our construction, the core of our test computation is the last element, denoted as $\mathbf{O} = S_{m+1} \cdot C$. If the group is a malicious group $\mathbf{u}_b$, then the result is

$$
\begin{aligned}
\mathbf{O} &= S_{m+1} \cdot C_b \\
&= \sum_{j=1}^{m+1} S_{m+1,j} \cdot C_j \\
&= \sum_{j=1}^{m+1} S_{m+1,j} \left(\sum_{k=0}^{m-1} x_j^k B_{k+1} + b_j \right) \\
&= \sum_{j=1}^{m+1} \left(\sum_{k=0}^{m-1} S_{m+1,j} x_j^k B_{k+1} \right) + \sum_{j=1}^{m+1} S_{m+1,j} b_j \tag{14} \\
&= \sum_{j=1}^{m+1} S_{m+1,j} b_j. \quad (\forall k \in [0, m-1]). \tag{15}
\end{aligned}
$$

However, it is possible that the term $\sum_{j=1}^{m+1} S_{m+1,j} b_j$ in (14) could be zero even if the group has malicious nodes. Under this situation and over a finite field $\mathbb{F}_q$, the probability of this exceptional case to occur is at most $1/q$. Thus, the probability approaches zero as the field size q increases.

By contrast, if the group members are all honest, the term $\sum_{j=1}^{m+1} S_{m+1,j} b_j$ in (14) will not exist. Therefore, the result with a honest group $\mathbf{u}_h$ will be $\mathbf{O} = S_{m+1} \cdot C_h = 0$.

4.2   Identity Proof Protocol

We develop an identity proof protocol using the digital signature to ensure the integrity of the node's scalar. What's more, it can be used as evidence if a node perturbs or forges data, in which case the new node cannot compute the inverse of the Vandermonde matrix causing the failure to recover the original shard data.

Digital signature technology is used to encrypt the digest of a message with the sender's private key and transmit it to the receiver along with the plain message. The receiver can only decrypt the encrypted digest with the sender's public key to get d_{new}, and then use the hash function to generate a digest d_{ori} for the received plain text. If d_{new} and d_{ori} are the same, it means that the plain message received is complete and not modified during the transmission, otherwise the message is tampered with. Thus the digital signature can be used to verify the integrity of the information.

Inspired by [25] and based on the characteristic of known public keys in permissioned blockchain, our identity proof protocol using digital signature contains the following components:

- The message $\mathcal{M}$, which is the content to which the signature algorithm may be applied. The content in our protocol is"node's scalar x_i where i is the index of the node;
- A key generation algorithm G, which is used by the central authority to generate credentials for each node;
- A signature algorithm σ, which produces a signature $\sigma(\mathcal{M}, sk_i)$ for a message $\mathcal{M}$ using the secret key sk_i;
- A verification algorithm $\mathcal{V}$, which tests whether $\sigma(\mathcal{M}, sk_i)$ is a valid signature for message $\mathcal{M}$ using the corresponding public key pk_i. In other words, $\mathcal{V}(\sigma, \mathcal{M}, pk_i)$ will be true if and only if it is valid.

Since PKI exists among nodes, each node i could create a digital signature $\sigma(\mathcal{M}, sk_i)$ on message $\mathcal{M}$ with its secret key sk_i. And the signature can be verified by the corresponding public key pk_i which is known to each node in the committee.

Cryptographic Primitives. First, we present some primitives that we use in the rest of the paper.

- hash(msg) - a cryptographically secure hash function that returns the digest of msg (e.g., SHA-256, SHA-512);
- encrypt($hash$, sk) - an encrypted hash function that returns the encrypted hash value (or called signature) for a hash value $hash$ using a secret key sk ;
- decrypt(sig, pk) - a decrypted hash function that returns the hash value of signature result using corresponding public key pk.

Signature Verification. At the initialization phase of each committee, CA will send the plain message $\mathcal{M}$ (which is scalar x_i) and signature $sig_C^i = \sigma(\mathcal{M}, sk_C)$ to each committee member, where i is the index of the target node and sk_C is the secret key of CA. Since the setting of permissioned blockchain where member knows the public key of each other, the member could use CA's public key pk_C to verify if sig_C^i is valid using $\mathcal{V}(sig_C^i, \mathcal{M}, pk_C)$. More details are shown in Algorithm 1. The necessity for a two-step comparison can be attributed to two distinct purposes. First, the first signature guarantees that scalars from other nodes are allocated by CA. Next, the second signature is critical in confirming that the first signature is sent from the intermediary node. An erroneous second signature allows N_{new} to forward the message to CA, offering "evidence" to reveal any suspicious behavior, as the private key necessary for signing the second message is unique to the intermediary node.

4.3   Adaptive Group Testing Method

Group testing originated from World War II for testing blood supplies, Robert Dorfman reduced the number of tests detecting whether the US military draftees had syphilis dramatically by pooling samples [26].

There are two types of group testing methods for identifying members with defects in a group: non-adaptive group testing (NAGT) and adaptive group testing (AGT). NAGT involves pooling samples from multiple individuals and testing them all together as a group, while AGT involves designing the test pools sequentially and adjusting the groups based on previous test results to minimize the number of individual tests needed. We adopt the AGT method to identify potential malicious nodes since NAGT requires a large number of tests, which is time-consuming, and the validation of transactions in the blockchain is very sensitive to time.

In our design, we assume each node is unidentified unless there is an identity proof to ensure its credibility, which means these nodes will be in at least one negative test. The goal is to build a testing set with as few tests T as possible to identify all malicious nodes. In particular, each test $\mathbf{u}$ consists of a group of nodes, and the result is false if data messages transmitted are tampered with. Otherwise, it is correct. Based on our settings, the group testing problem can be formulated as follows

$$\mathbf{y} = \mathbf{M} \circ \mathbf{x}, \tag{16}$$

where (1) $\circ$ denotes the row-wise Boolean operation, and the result is $y_i = 0$ if nodes in the i-th test are all honest or $y_i \neq 0$ instead; (2) $\mathbf{M} \in \mathbb{F}_2^{t \times n}$ is a contact matrix where $\mathbf{M}_{i,j} = 1$ indicates the i-th test contains node N_j.

The goal of group testing is to design a test matrix $\mathbf{M}$ such that the number of tests is as small as possible and it can be expressed as $\mathbf{M} = [\mathbf{u_1} \; \mathbf{u_2} \; \cdots \; \mathbf{u_n}]^\top$. Items included in tests with negative outcomes will be viewed as noninfective and collected into an honest set $\mathcal{H}$. Similarly, those items in positive tests (validation result is wrong) will be collected into the malicious set $\mathcal{S}$. Owing to the property of k-disjunct, each sample includes a different set of tests. By matching the honest set $\mathcal{H}$ and malicious set $\mathcal{S}$, the f defectives can be identified.

The selection of nodes in a test to identify malicious nodes with shard testable codes can be regarded as a group testing problem. Thus the $\circ$ operation in Eq. (16) can be expressed as the matrix multiplication operation of shard messages verification in Eq. (7). And our goal is to minimize the number of tests for identifying malicious nodes. In the general AGT problem, the algorithm designs a set of tests $\{u_1, u_2, \cdots, u_T\}$ to make T as small as possible. Given the outcomes

---

Algorithm 1: Identity proof protocol with digital signature

   ▷ **Phase 1**: Initialization of nodes

1 **As CA of the system**

2 | **foreach** node N_i in committee $\mathcal{C}$ **do**

3 | | $x_i \leftarrow$ generate a random scalar over $\mathbb{F}_q$;

4 | | $sig_C^i \leftarrow \sigma(x_i, sk_C)$;   // encrypt the scalar x_i with sk_C

5 | | send $[x_i, sig_C^i]$ to node N_i;

6 | **end**

7

8 **As other node N_i of the committee**

9 | wait for message $[x_i, sig_C^i]$ from CA;

10 | $pk_C \leftarrow$ query public key of CA;

11 | **if** $\mathcal{V}(sig_C^i, x_i, pk_C) ==$ true **then**

12 | | store the assigned scalar x_i;

13 | **else**False

14 | | request CA to resend the message;

15 | **end**

16

   ▷ **Phase 2**: New node N_{new} joins the committee $\mathcal{C}$

17 **As CA of the system**

18 | $x_{new} \leftarrow$ generate a new random scalar over $\mathbb{F}_q$;

19 | $sig_C^{new} \leftarrow \sigma(x_{new}, sk_C)$;

20 | send $[x_{new}, sig_C^{new}]$ to node N_{new};

21

22 **As other node N_i of the committee**

23 | $sig_i^{new} \leftarrow \sigma(sig_C^i, sk_i)$;   // new signature for sig_C^i with sk_i

24 | send $[x_i, sig_C^i, sig_i^{new}]$ to node N_{new};

25

   ▷ **Phase 3**: Signature verification by newly-joined node N_{new}

26 wait for messages from nodes of the committee $\mathcal{C}$;

27 **foreach** message $[x_i, sig_C^i, sig_i^{new}]$ from node N_i **do**

28 | $pk_i \leftarrow$ query public key of N_i from CA;

29 | $pk_C \leftarrow$ query public key of N_{leader} from CA;

30 | **if** $\mathcal{V}(sig_i^{new}, sig_C^i, pk_i) \wedge \mathcal{V}(sig_C^i, x_i, pk_C)$ **then**

31 | | store the scalar x_i locally for decoding operation;

32 | **else if** $\neg(\mathcal{V}(sig_i^{new}, sig_C^i, pk_i))$ **then**   // the second signature is invalid, there may be some error during transmission

33 | | require N_i to resend the message;

34 | **else** // the initial signature is inconsistent with x_i

35 | | send $[x_i, sig_C^i, sig_i^{new}]$ to CA; // make it as fraud-proof to punish the sender N_i

36 | **end**

37 **end**

---

of these tests, the honest set and malicious set will be generated, achieving the goal of malicious node identification.

In a simple method, CA can fix m honest nodes and check one unidentified node whether honest or not if CA knows $m+1$ honest nodes with trials before. Thus the remaining $N-m-1$ nodes can be identified in the same way. Dorfman [26] proposes a simple procedure which partitions the whole E items containing f defectives into $\sqrt{Ef}$ subsets, each of size $\sqrt{E/f}$. Hence, the number of tests that Dorfman's procedure requires is at most

$$T = \sqrt{Ef} + f\sqrt{\frac{E}{f}} = 2\sqrt{Ef}. \tag{17}$$

Following Dorfman's procedure, the key is the number of trials $\widehat{T}$ that finds the first honest group whose result of the last row in Eq. (7) is $\mathbf{O} = 0$.

Theorem 1. *Given a committee of n nodes, if there are f malicious nodes where $n \geq m + f + 1$, the probability of having no malicious nodes in a test group of size $m+1$ is*

$$P(H = 0) = \prod_{i=0}^{m}(1 - \frac{f}{n-i}), \tag{18}$$

where H is the number of malicious nodes.

Proof. We first compute the number of ways to pick $m+1$ chunks among the set of non-malicious nodes, *i.e.*, $\binom{n-f}{m+1}$. Similarly, we can produce the total number of ways to pick any $m+1$ samples out of the total number of samples, *i.e.*, $\binom{n}{m+1}$. Therefore, the probability can be computed as

$$P(H = 0) = \frac{\binom{n-f}{m+1}}{\binom{n}{m+1}}$$

$$= \prod_{i=0}^{m}(1 - \frac{f}{n-i}). \tag{19}$$

If we define *Malice Ratio* (R_f) to represent the proportion of malicious nodes to the total number of nodes (f/n), then we can rewrite Equation (19) as:

$$P(H = 0) = \prod_{i=0}^{m}\left(1 - \frac{R_f \times n}{n-i}\right). \tag{20}$$

After thorough analysis and computation, we have observed that regardless of the total number of nodes within a committee, when the *Malice Ratio* does not exceed $1/5$ and the number of encoding shards is limited to 2 or fewer, there is at least a 50% probability that all members of the testing set are honest nodes. Under such circumstances, system stability and security can be guaranteed.

Accordingly, the probability of failing to find the first honest group with $\widehat{T}$ trials is $(1 - P(H = 0))^{\widehat{T}}$. Assume there is a error probability ρ, and we want to make $(1 - P(H = 0))^{\widehat{T}} \leq \rho$. Since $0 < (1 - P(H = 0)) \leq 1$, $0 <$

$T \leq \log_{1-P(H=0)} \rho$. Therefore, the maximum number of trials to find the first non-malicious group with ρ is

$$\widehat{T} = \log_{1-P(H=0)} \rho. \tag{21}$$

Theorem 2. *For malicious node identification in committees of sharding permissioned blockchain, adaptive group testing with shard testable codes can identify all malicious nodes by T testing trials, where T can be written as*

$$T \leq \log_{1-P(H=0)} \rho + 2\sqrt{(n-m-1)f}, \tag{22}$$

where $P(H = 0)$ is given in Eq. (18) of Lemma 1.

Proof. If CA finds the first honest group of size $m+1$ in a committee by $\widehat{T}$ trials. Based on Dorfman's procedure, CA divides the remaining $n-m-1$ unidentified nodes into $\sqrt{(n-m-1)f}$ subgroups, each of size $\sqrt{(n-m-1)/f}$. For a newly-joined node, CA requires one of the subgroups to send shard messages to it so that the new node can perform verification and send feedback to CA. With the results of group testing, CA could test those suspicious nodes separately and identify them as malicious if the outcome is wrong.

Hence, by using Dorfman's procedure, the total number of group testing trials is at most

$$T \leq \widehat{T} + \sqrt{(n-m-1)f} + f\sqrt{\frac{n-m-1}{f}}$$
$$= \log_{1-P(H=0)} \rho + 2\sqrt{(n-m-1)f}. \tag{23}$$

Remark 2. The second term in Eq. (23) is from the original adaptive testing method [26]. It can be improved by other well-design schemes, *e.g.*, HGBSA [27]. But the core idea is similar. For simplicity of demonstration, we will not discuss the variant of adaptive group testing algorithms in this paper.

4.4 Cost and Complexity

Communication Cost. Following our RecAGT scheme, the newly-joined node has three parts of communication costs: (1) the initialization identity proof from CA; (2) $m + 1$ encoded shard messages from the assigned group testing members, and (3) $m + 1$ identity proof messages consisting of the scalar x_i, the first signature from CA and the second signature from the member. Thus the communication cost for a newly-joined node can be computed as

$$(w + z + s) + ((m+1) \times \frac{b}{m}) + ((m+1) \times (w + 2z)) = O(b), \tag{24}$$

where w is the size of scalar, z is the size of digital signature, s is the size of secret key, m is the size of testable codes and b is the size of original shard.

Computational Complexity. At the core of computational complexity is the decoding complexity, *i.e.*, computing the inverse of the Vandermonde matrix. Therefore the computational complexity is $O(\log^2(m) \log \log(m))$.

5   Experiments

In this section, we conduct extensive experiments to evaluate the performance of our proposed identification scheme under different parameters. We also compare our identification scheme with others.

5.1   Setup for Parameters

Table 2. Different settings with respect to P, n, f/n and m

	# Committees (P)	# each committee(n)	malicious ratio (f/n)	# shard-coding size (m)
Setting 1	300	6	0.2 (1)	2
Setting 2	70	24	0.125 (3)	3
Setting 3	25	72	0.05 (4)	8
Setting 4	4	450	0.01 (5)	10

To simulate a more practical and realistic permissioned blockchain, we adopt a similar experimental configuration referring to PShard [10] and Omniledger [28]. There are four settings under a network of 1800 nodes, as shown in Table 2. Specifically, the configuration of setting 3 means that there are 25 committees of each 72 members, and the adversary ratio is 0.05 ($0.05 \times 72 \approx 4$).

5.2   Simulation Results and Analyses

The key parameter is the shard-coding size. A larger m means more divided sub-shards and thus larger members in each group. But it will also increase the computational complexity of encoded matrix construction and other computation. Therefore, we run simulations of the theoretical results of Theorem 1, which is shown in Fig. 3. It is apparent from Fig. 3(a) that the successful probability decreases with the increasing of the shard-coding size. The reason why the blue line approaches 0 is that $n \geq (m + f + 1)$ which can be derived easily from Eq. (19). For a reasonable trade-off between group size and computational complexity, the choice of m for settings is shown in the last column of Table 2.

Next, we investigate the impact of varying ratios (f/n and m/n) on $P(H = 0)$. The results are depicted in Fig. 3(b). There is a gradual decline in the probability as m/n increases. Moreover, the probability experiences a significant drop as the value of f/n increases. The result of Eq. (19) agrees with our guess where m represents the number of terms, f is the numerator, and n is the denominator. Since the value of each term is in the range of [0: 1], the result of cumulative multiplication will get smaller if each of them increases.

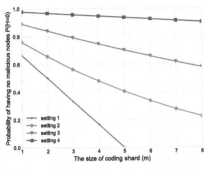 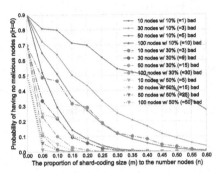

(a) Probability $P(H = 0)$ with changing m

(b) Probability $P(H = 0)$ with different ratio m/n

Fig. 3. The influence of different settings about m on the probability of having no malicious nodes in group testing $P(H = 0)$.

Table 3. Experiment parameter settings

Parameter	Value	Notes
checksum value d (Bytes, B)	16	MD5 algorithm is used here
secret key s (B)	128	1024 bits in general
scalar w (B)	1	
digital signature z (B)	256	
committee size n	1, 50, 100	
shard-coding size m	$0.1n$	For simplicity, the ratio of m/n is chosen by 0.1

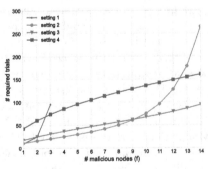 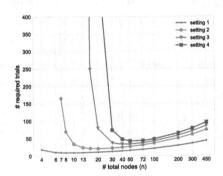

(a) Number of trials with changing f

(b) Number of trials with changing n

Fig. 4. The influence of different parameters on the number of required group testing trials T.

Additionally, Fig. 4 shows the number of required group testing trials with different system parameters specified in Table 2 and we set $\rho = 0.01$ empirically.

In Fig. 4(a), we show the number of trials when f varies from 1 to 14 in different settings. It is obvious that the number of trials to identify all malicious nodes increases along with f. The reason why the blue line stops at $f = 3$ is because $f \leq n - m - 1$. Lines of setting 2, 3, and 4 in Fig. 4(b) do not begin at $n = 6$ because of the restriction of $n \geq (m + f + 1)$. All lines begin at a high point and then show a trend from decline to rise because the ratio of $(m + f)/n$ approaches 1 at the beginning, then T reaches the minimum when the ratio approaches $1/4$.

In Fig. 5, we numerically evaluate communication costs and computational decoding speed of RecAGT, and compare them with the other two methods. Referring to some deployments of permissioned blockchain, the specific parameters in our experiments are summarized in Table 3.

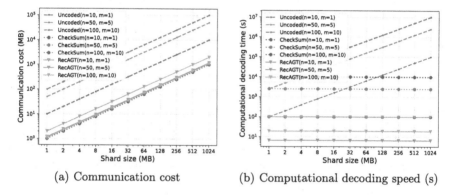

(a) Communication cost   (b) Computational decoding speed (s)

Fig. 5. Comparison of Uncoded, CheckSum, and RecAGT for communication cost and computational decoding speed.

Communication Cost. From Fig. 5(a), it becomes evident that the Uncoded scheme incurs a substantially higher communication cost than the others. This discrepancy arises from requiring all members to transmit the original shard data to the new node. It is worth mentioning that when n is small, the communication cost of the RecAGT exceeds that of the CheckSum. But with the increasing n, the difference between them is notably small. The reason is that the CheckSum scheme needs to send the original data and checksum values of the rest, which is $(b + d \times n)$. And the decisive term in RecAGT is $\lceil (m + 1) \times b/m \rceil$ that is explained in Eq. (24). As n and m increase, other factors could be ignored, so these two schemes achieve almost the same cost. However, compared with the CheckSum scheme, our RecAGT scheme encodes the original shard data and keeps the internal data of each organization confidential.

Computational Decoding Speed. In Fig. 5(b), we observe a significant quadratic increase in computing time for the Uncoded scheme. To illustrate, when the shard size reaches 256 MB, the computation time exceeds an unacceptable 10, 000 s. The lines of CheckSum and RecAGT appear constant because they do not directly manipulate shard data. CheckSum uses check codes to validate the consistency of data received from other members. The RecAGT scheme uses the inverse of the Vandermonde matrix to identify perturbed data. Our adoption of the polynomial interpolation method demonstrates superior speed and efficiency compared to the other two schemes, as evidenced by our experimental results.

6 Conclusion

In this paper, we propose RecAGT scheme for the identification of potential malicious nodes, which focuses on reducing communication overhead and identifying potential malicious nodes. Specifically, we design the shard testable codes to encode original data. And we come up with an identity proof using the digital signature and choose an adaptive group testing method to make the required number of trials as small as possible. The simulation results demonstrate that our proposed RecAGT scheme can efficiently identify malicious nodes and reduce communication and computational costs.

Acknowledgements. This work was supported in part by the National Natural Science Foundation of China (62072321, 61972272), the Six Talent Peak Project of Jiangsu Province (XYDXX-084), the China Postdoctoral Science Foundation (2020M671597), the Jiangsu Postdoctoral Research Foundation (2020Z100), Suzhou Planning Project of Science and Technology (SS202023), the Future Network Scientific Research Fund Project (FNSRFP-2021-YB-38), Natural Science Foundation of the Higher Education Institutions of Jiangsu Province (22KJA520007, 20KJB520002), the Collaborative Innovation Center of Novel Software Technology and Industrialization, and Soochow University Interdisciplinary Research Project for Young Scholars in the Humanities.

References

1. Rebello, G.A.F., Camilo, G.F., Guimarães, L.C., de Souza, L.A.C., Duarte, O.C.M.: Security and performance analysis of quorum-based blockchain consensus protocols. In: 2022 6th Cyber Security in Networking Conference (CSNet), pp. 1–7. IEEE (2022)
2. Amiri, M.J., Duguépéroux, J., Allard, T., Agrawal, D., El Abbadi, A.: Separ: Towards regulating future of work multi-platform crowdworking environments with privacy guarantees. In: Proceedings of the Web Conference 2021, pp. 1891–1903 (2021)
3. Amiri, M.J., Agrawal, D., El Abbadi, A.: Permissioned blockchains: properties, techniques and applications. In: Proceedings of the 2021 International Conference on Management of Data, pp. 2813–2820 (2021)
4. Gorenflo, C., Lee, S., Golab, L., Keshav, S.: Fastfabric: scaling hyperledger fabric to 20 000 transactions per second. Int. J. Network Manage **30**(5), e2099 (2020)
5. Corbett, J.C., et al.: Spanner: Google's globally distributed database. ACM Trans. Comput. Syst. (TOCS) **31**(3), 1–22 (2013)

6. Amiri, M.J., Agrawal, D., El Abbadi, A.: On sharding permissioned blockchains. In: 2019 IEEE International Conference on Blockchain (Blockchain), pp. 282–285. IEEE (2019)
7. Amiri, M.J., Agrawal, D., El Abbadi, A.: Sharper: sharding permissioned blockchains over network clusters. In: Proceedings of the 2021 International Conference on Management of Data, pp. 76–88 (2021)
8. Dang, H., Dinh, T.T.A., Loghin, D., Chang, E.C., Lin, Q., Ooi, B.C.: Towards scaling blockchain systems via sharding. In: Proceedings of the 2019 International Conference on Management of Data, pp. 123–140 (2019)
9. Huang, H., et al.: Elastic resource allocation against imbalanced transaction assignments in sharding-based permissioned blockchains. IEEE Trans. Parallel Distrib. Syst. **33**(10), 2372–2385 (2022)
10. Gao, J., et al.: Pshard: a practical sharding protocol for enterprise blockchain. In: Proceedings of the 2022 5th International Conference on Blockchain Technology and Applications, pp. 110–116 (2022)
11. Mao, C., Golab, W.: Geochain: a locality-based sharding protocol for permissioned blockchains. In: 24th International Conference on Distributed Computing and Networking, pp. 70–79 (2023)
12. Zheng, P., Xu, Q., Zheng, Z., Zhou, Z., Yan, Y., Zhang, H.: Meepo: multiple execution environments per organization in sharded consortium blockchain. IEEE J. Sel. Areas Commun. **40**(12), 3562–3574 (2022)
13. Falazi, G., Khinchi, V., Breitenbücher, U., Leymann, F.: Transactional properties of permissioned blockchains. SICS Softw.-Intensive Cyber-Physical Syst. **35**(1–2), 49–61 (2020)
14. Ekparinya, P., Gramoli, V., Jourjon, G.: The attack of the clones against proof-of-authority. arXiv preprint arXiv:1902.10244 (2019)
15. Saas, M., et al.: Exploring the attack surface of blockchain: a comprehensive survey. IEEE Commun. Surv. Tutorials **22**(3), 1977–2008 (2020)
16. Davenport, A., Shetty, S., Liang, X.: Attack surface analysis of permissioned blockchain platforms for smart cities. In: 2018 IEEE International Smart Cities Conference (ISC2), pp. 1–6. IEEE (2018)
17. Yu, Q., Li, S., Raviv, N., Kalan, S.M.M., Soltanolkotabi, M., Avestimehr, S.A.: Lagrange coded computing: Optimal design for resiliency, security, and privacy. In: The 22nd International Conference on Artificial Intelligence and Statistics, pp. 1215–1225. PMLR (2019)
18. Solanki, A., Cardone, M., Mohajer, S.: Non-colluding attacks identification in distributed computing. In: 2019 IEEE Information Theory Workshop (ITW), pp. 1–5. IEEE (2019)
19. Hong, S., Yang, H., Lee, J.: Byzantine attack identification in distributed matrix multiplication via locally testable codes. In: 2022 IEEE International Symposium on Information Theory (ISIT), pp. 560–565. IEEE (2022)
20. Hong, S., Yang, H., Lee, J.: Hierarchical group testing for byzantine attack identification in distributed matrix multiplication. IEEE J. Sel. Areas Commun. **40**(3), 1013–1029 (2022)
21. Zhao, X., Lei, Z., Zhang, G., Zhang, Y., Xing, C.: Blockchain and distributed system. In: Web Information Systems and Applications: 17th International Conference, WISA 2020, Guangzhou, China, September 23–25, 2020, Proceedings 17, pp. 629–641. Springer (2020)
22. Yu, Q., Maddah-Ali, M., Avestimehr, S.: Polynomial codes: an optimal design for high-dimensional coded matrix multiplication. In: Advances in Neural Information Processing Systems, 30 (2017)

23. Verde-Star, L.: Inverses of generalized vandermonde matrices. J. Math. Anal. Appl. **131**(2), 341–353 (1988)
24. Kedlaya, K.S., Umans, C.: Fast polynomial factorization and modular composition. SIAM J. Comput. **40**(6), 1767–1802 (2011)
25. Kaur, R., Kaur, A.: Digital signature. In: 2012 International Conference on Computing Sciences, pp. 295–301. IEEE (2012)
26. Dorfman, R.: The detection of defective members of large populations. Ann. Math. Stat. **14**(4), 436–440 (1943)
27. Hwang, F.K.: A method for detecting all defective members in a population by group testing. J. Am. Stat. Assoc. **67**(339), 605–608 (1972)
28. Kokoris-Kogias, E., Jovanovic, P., Gasser, L., Gailly, N., Syta, E., Ford, B.: Omniledger: a secure, scale-out, decentralized ledger via sharding. In: 2018 IEEE Symposium on Security and Privacy (SP), pp. 583–598. IEEE (2018)

GLAM-SERP: Building a Graph Learning-Assisted Model for Soft Error Resilience Prediction in GPGPUs

Xiaohui Wei, Jianpeng Zhao, Nan Jiang, and Hengshan Yue

College of Computer Science and Technology, Jilin University, Changchun, China
{weixh,yuehs}@jlu.edu.cn, {jpzhao21,jiangnan22}@mails.jlu.edu.cn

Abstract. Due to their efficient data-parallel computing capabilities, General-Purpose Graphics Processing Units (GPGPUs) have become increasingly prevalent in deep learning and scientific computing domains. Because of the growing chip integration density, GPGPUs are becoming more susceptible to soft errors, which can cause catastrophic results in safety-critical systems. Consequently, conducting a GPGPU program error resilience analysis is essential to provide guidance for enhancing reliability. Unfortunately, traditional analysis methods such as Statistical Fault Injection (SFI) suffer from colossal time overhead, while machine learning-based resilience prediction methods are constrained in characterizing program error propagation behavior. To address the above challenges, we propose GLAM-SERP, a *G*raph *L*earning-*A*ssisted *M*odel for *S*oft *E*rror *R*esilience *P*rediction in GPGPUs. Our critical insight is that the error resilience of GPGPU instructions is related to their inherent properties and error propagation characteristics. Thus, we construct a Dependency Graph (DG) for the GPGPU program, where nodes represent individual GPGPU instructions, node features capture the resilience characteristics of each instruction, and graph edges depict the error propagation pathway between GPGPU instructions. Based on the established DG, we then drive a Graph Attention Network (GAT) to predict Silent Data Corruption (SDC) proneness of GPGPU instructions under the soft error influences. The experimental results demonstrate that our approach achieves an average prediction performance of 94.14% on individual programs and 89.50% on unseen programs, showcasing its accurate and general error resilience prediction capability in GPGPUs.

Keywords: GPGPU · Soft Error · Error Resilience · Graph Neural Network

This work is supported by the National Natural Science Foundation of China (NSFC) (Grants No.62302190, No.62272190 and No.U19A2061).

Z. Tari et al. (Eds.): ICA3PP 2023, LNCS 14490, pp. 419–435, 2024.
https://doi.org/10.1007/978-981-97-0859-8_25

1   Introduction

Evolving GPU architecture designs have devoted dense hardware structures and energy-efficient data-parallel computing models, driving GPUs omnipresent in scientific computing domains, also called General-Purpose computing on GPUs (GPGPUs). However, with the scaling of transistor counts, the highly integrated GPGPU exacerbates high system vulnerability incurred by high-energy particle strikes or electrical noise (i.e., soft errors) [1,2]. In particular, the soft errors arising in GPGPUs applied in stringently safety-critical systems (e.g., autonomous vehicles) may incur fateful consequences, which makes GPGPU reliability a critical concern [3,4]. Consequently, it is imperative to characterize the resilience of GPGPU programs under the perturbation of soft errors, which helps programmers or architects identify the vulnerable parts of GPGPU programs and provides helpful guidance for various fault tolerance mechanisms design (e.g., selective instruction duplication [5,6]).

Conventionally, Statistical Fault Injection (SFI) is employed to understand the error resilience of programs [7,8]. People typically randomly inject faults into a subset of the program's fault sites and monitor the corrupted program's symptoms to check the severity of soft errors. However, unlike CPU programs, GPGPU programs with thousands of threads spawn billions of possible fault sites. Traditional SFI typically involves massive fault injection trials (e.g., 60K) [7] to obtain a statistically significant resilience profile, which introduces ponderous resources- and time- overhead as each injection campaign needs a complete program execution [9].

Alternatively, recent studies [10–13] have utilized Machine Learning (ML) techniques to expedite the SFI process, which focus on identifying reliability-related heuristic features to train classifiers for program error resilience estimation. While these works provide attractive alternatives to expensive SFI, they possess the following limitations: 1) These methods are constrained in their ability to characterize program error propagation behavior by only characterizing error propagation information as heuristic features. 2) These approaches typically build application-specific error analytical models, which may result in significant accuracy degradation when applied to unseen GPGPU programs.

To tackle the above limitations, in this study, we introduce GLAM-SERP, an efficient and general *Graph Learning-Assisted Model for Soft Error Resilience Prediction* in GPGPUs. Our critical insight is that the error resilience of GPGPU instructions is determined by both their intrinsic properties and the dependency relationships between them. Therefore, we construct a Dependency Graph (DG) where nodes represent instructions, node features capture their error resilience characteristics, and edges convey their dependency relationships for error propagation. Subsequently, we acquire a limited set of DG node labels through a restricted number of fault injections and drive a Graph Attention Network (GAT) model on the built DG to predict instruction-wise error resilience. As the experimental results exhibit, our approach achieves an average prediction performance of 94.14% on individual programs and 89.50% on previously unseen programs, highlighting the accurate and general capabilities of our approach.

We summarize our main contributions as follows:

- Firstly, we explore a set of heuristic features associated with the error resilience for GPGPU instructions.
- Secondly, we construct a Dependency Graph (DG) to capture error propagation based on the data, control, and memory dependencies among GPGPU instructions.
- Finally, we drive a GAT model based on the established DG to predict the error resilience of GPGPU instructions.

2 Background

2.1 GPGPU Programming Model

This study concentrates on GPGPU applications based on Compute Unified Device Architecture (CUDA) of NVIDIA. GPGPUs organize their threads in three hierarchies: kernel, Cooperative Thread Array (CTA), and warp. Multiple computation functions called kernels are launched by the host CPU and executed on GPGPU. Each kernel contains several CTAs as the granularity for assigning threads to a streaming multiprocessor (SM). The threads of CTAs are uniformly managed and scheduled in groups of 32 threads called warps. All threads in a warp execute the same instruction parallelly, following a SIMT paradigm. We can gain insight into the execution of these instructions by profiling the intermediate code known as Parallel Thread Execution (PTX) in the compiling toolchain. The NVCC compiler of NVIDIA generates PTX codes that are weakly coupled to hardware architectures, guaranteeing compatibility for architecture iteration.

2.2 Fault Model

In this paper, we focus on transient hardware faults occurring in computational components (e.g., registers) on GPGPUs, specifically excluding memories and caches that are protected by Error Correction Codes (ECCs) [14]. Aligned with the other researches [4,12,15], we adopt a single-bit fault model due to the low soft error occurring probability and the similar effect between single- and multi-bit fault models on resilience analysis [16]. We flip a single bit at one register to simulate the effect of particle strikes and acquire one of the three outcomes: (1) Masked, where the program executes correctly and produces golden (i.e., error-free) results. (2) Silent Data Corruption (SDC), where the program executes successfully but produces erroneous results silently. (3) Detected Unrecoverable Error (DUE), where the program runs into a hang or crash. Emphasizing its importance, SDC in GPGPU programs poses a significant risk to safety-critical applications as it corrupts the program without manifesting apparent system symptoms. Therefore, it is crucial to identify the SDC-prone parts of GPGPU programs, which is the primary focus of our work.

2.3   Graph Neural Networks

Recently, Graph Neural Networks (GNNs) have almost emerged as the predominant approach in node classification tasks [17,18]. The proposed GNN models can be broadly classified into two categories: transductive GNNs and inductive GNNs. Transductive GNNs require training on a fixed graph and are incapable of predicting unseen data. On the contrary, inductive GNNs concentrate on partial graph structure rather than the whole, owning a transfer ability instinctively. In this work, we exploit an inductive GNN called GAT [18] to assist resilience prediction for GPGPUs.

3   Graph Data Generating Strategy

Due to the inefficiency of SFI, in this study, we aim to construct an efficient GAT-based model for instruction-wise error resilience estimation. To achieve this objective, we generate trainable graph data based on the Intermediate Representation (IR) code of GPGPU programs. Firstly, we excavate the instruction heuristic features (i.e., node features in DG) relevant to SDC proneness in Sect. 3.1. Secondly, to capture the error propagation behaviors, we construct the graph structure based on the data, control, and memory dependencies (i.e., edges in DG) among instructions in Sect. 3.2. Eventually, taking into consideration the execution behaviors of different threads in GPGPUs, we calculate weights for DG edges based on their respective execution counts in Sect. 3.3, thereby obtaining weighted edges in DG.

3.1   Extract Heuristic Features for Graph Nodes

To facilitate graph construction, each instruction with a destination register is abstracted as a graph node. As the example shown in Fig. 2a and Fig. 2b, we denote the instruction in the nth line as I_n, the node representing it as N_n, and the edge from N_i to N_j as $E_{i,j}$. This section introduces resilience-related heuristic features for the graph nodes from the perspective of *instruction inherent property*, *computational pattern*, and *graph node centrality* [19].

Instruction Inherent Property Features. The first feature dimension we consider is the inherent properties of the instructions. To explore these features, we perform fault injections on GPGPU instructions and display the benchmark-averaged results in Fig. 1, which demonstrates that different types of instructions exhibit variations in their susceptibility to soft errors. Therefore, leveraging the properties of instructions as features, as listed in Table 1, can effectively estimate the susceptibility to SDC of GPGPU instructions.

Firstly, we observe that floating-point instructions (e.g., FFMA, FMUL) are more prone to SDC compared to integer instructions (e.g., IADD, IMUL). As depicted in Fig. 1, the SDC rate of floating-point instructions is measured at 71.30%, whereas for integer instructions, it is 34.60%. The observed discrepancy can be explained

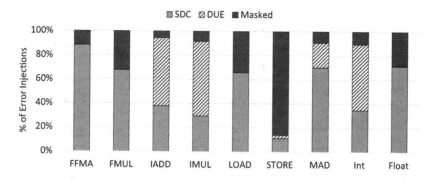

Fig. 1. Soft error distribution across different types of instructions.

by the frequent utilization of specific integer instructions for address calculations and particular floating-point instructions for numerical computations. Errors occurring in address calculations may incur memory access violations or out-of-bounds problems, exhibiting a high DUE rate (56.53% for IADD), while errors happening in numerical computations tend to cause deviations in the calculation results, demonstrating high SDC rates (88.39% for FFMA).

Afterward, we discover that LOAD instructions have a higher likelihood (65.79%) of triggering SDC, whereas the STORE instructions tend to mask errors (86.20%). This is because errors in LOAD instructions may lead to deviations in the initial values, subsequently resulting in inaccurate final results. On the other hand, many STORE instructions are considered dead-store [11,12] because their successors do not actually use their stored values.

Lastly, we observed that shifting and logical operations have a tendency to mask errors in their source registers. For instance, in the instruction "shl.b32 %r4, %r3, 8", errors occurring in the high 8 bits of %r3 are masked and do not impact the outcome. Similarly, logical operations that involve patterns such as "AND with 0" and "OR with 1" on a bit-by-bit basis can also mask errors in the source registers of the instructions.

Computational Pattern Features. Beyond the inherent properties of instructions, we have identified specific critical code snippets (e.g., thread ID computations) that exhibit fixed computation patterns. These code snippets possess resilience-related characteristics and constitute subgraphs within DG. Based on this observation, we explore the features associated with computational patterns and present them in Table 1.

We emphasize that errors in a branch control instruction may result in a deviation from the intended branch (i.e., SDCs), while in loop control instructions may disrupt the proper termination of the loop, resulting in program hangs. As illustrated in Fig. 3, the blue box contains an AND instruction, which functions as a branch control instruction with a significant SDC rate of 98.04%. Conversely, in Fig. 2a, line 6 represents a loop control instruction exhibiting a notably lower

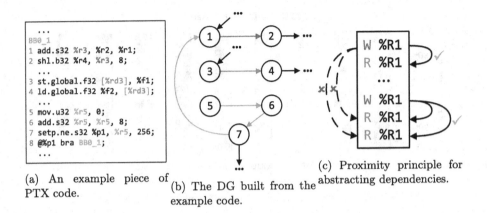

(a) An example piece of PTX code.

(b) The DG built from the example code.

(c) Proximity principle for abstracting dependencies.

Fig. 2. The example of constructing a DG.

SDC rate of 3.51%. Moreover, whether an instruction is inside a loop also affects the distribution of soft errors.

Besides, we observe that the instructions for calculating thread IDs exhibit a high SDC proneness. As depicted in Fig. 3, all launched threads commence by computing their thread IDs and subsequently directing the flow of execution through branching. Errors in the thread ID calculation (marked in yellow) may propagate to the registers marked in blue responsible for branch controlling, resulting in deviations from the intended execution flow, and leading to inaccurate outputs. For example, the instruction MAD in thread ID calculation exhibits a high SDC rate of 69.99% as illustrated in Fig. 1.

Graph Node Centrality Features. Beyond the two feature dimensions mentioned above, the significance of graph node centrality is also demonstrated in previous studies [19], which is often overlooked in GATs' attention calculation. To better capture the potential characteristics of SDC proneness, we augment the instruction features by incorporating node centrality measures, such as indegree, out-degree, and total degree. These features capture the semantic relationships and importance of nodes within the attention mechanism and are listed in Table 1.

3.2 Abstract Dependencies as Graph Edges

In this section, we introduce a methodology for constructing directed edges in DG to represent the data, control, and memory dependencies among GPGPU instructions, which enables us to sufficiently capture the error propagation behaviors in GPGPU.

Data Dependency. We create a directed edge if the destination register of an instruction is used as a source register by its successor. For example, the

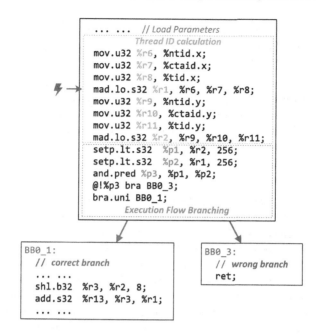

Fig. 3. Errors in thread ID calculation may induce threads to the wrong branches.

Table 1. Heuristic Feature Set

Instruction Inherent Property			Computational Pattern		
No	Type	Description	No	Type	Description
1	Int	Number of operands	1	Bool	Is address-computing pattern
2	Int	Number of operand bits	2	Bool	Is arithmetic-computing pattern
3	Bool	Is an integer operation	3	Bool	Is thread ID calculation pattern
4	Bool	Is a floating-point operation	4	Bool	Is a data movement operation
5	Bool	Is a mul/div operation	5	Bool	Is a branch control operation
6	Bool	Is an add/sub operation	6	Bool	Is a loop control operation
7	Bool	Is a logical operation	7	Bool	Is in a loop pattern
8	Bool	Is a shifting operation	**Graph Node Centrality**		
9	Bool	Is a load operation	No	Type	Description
10	Bool	Is a store operation	1	Int	Graph node's in-degree
11	Bool	Is a shared mem-load operation	2	Int	Graph node's out-degree
12	Bool	Is a shared mem-store operation	3	Int	Graph node's total degree

destination register of I_1 in Fig. 2a is used as a source register in I_2, which indicates the data dependency between them. As a result, we establish the edge $E_{1,2}$ between N_1 and N_2 to represent this data dependency.

Control Dependency. The control dependency is determined by the branches and loops in the program. For example, Fig. 2a depicts a loop example. If the value of the condition register %p1 is true, meaning the loop is not terminated, the program will proceed to execute instruction I_1, which corresponds to the edge $E_{7,1}$ in Fig. 2b.

Memory Dependency. We construct a directed edge between store and load instructions that access the same memory address. For example, in the case of I_3 and I_4 in Fig. 2a, where data is stored and loaded to/from the same memory space, we create the edge $E_{3,4}$ to represent this memory dependency.

It is important to note that the abstraction of the three dependencies should adhere to the proximity principle illustrated in Fig. 2c. For instructions that read from or write to the same register or memory space, each write only affects the read executing before the next write. In the example shown in Fig. 2a, edges $E_{5,6}$ and $E_{6,7}$ would be generated, while the edge $E_{5,7}$ would be excluded. This accurate modeling of the data and control flow in GPGPU programs enables a precise representation of error propagation, thereby promoting the learning of GPGPU instruction resilience knowledge by the subsequent GAT model.

3.3   Assign Weights to Graph Edges

In GPGPU programs, the execution behavior of threads (e.g., the sequence of executed instructions) can vary due to different branch conditions, leading to different execution counts of GPGPU instructions and varying importance of dependency edges in DG. As a result, we propose two approaches in this section, namely *Edge Counting* and *Thread Counting*, to evaluate the relative importance of dependency edges in the DG.

Edge Counting Approach. An individual instruction in GPGPU programs may spawn multiple instances due to the loop segment. This repetition pattern leads to increased execution counts along the associated dependency paths, thereby exacerbating the propagation of errors throughout the entire path. Considering this issue, we obtain the Dynamic Instruction Sequence (DIS) based on thread execution traces and calculate weights for all edges in DG accordingly. We represent the execution count of an edge $E_{i,j}$ in a DIS as $EC(i, j, DIS)$. As depicted in Fig. 4, the instructions in the benchmark *2mm* from lines 57 to 97 are executed sequentially 32 times by each thread, whose edge counts in the graph are accordingly set to 32.

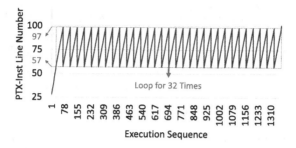

Fig. 4. Thread execution sequence with 32 loops in *2mm*.

Thread Counting Approach. The threads following the same execution flow share a standard DIS. As a result, we select a Representative Thread (RepThread) for each thread group that shares the same DIS. For a brief, we denote the thread ID of a RepThread as RepThread-ID and its DIS as RepDIS. For an individual GPGPU program, we can obtain a set of distinct RepDISs and their respective counts of executing threads. Assuming the GPGPU program has n RepDISs, for a specific $RepDIS_k$, we denote the count of threads executing it as $TC(RepDIS_k)$. Table 2 illustrates the existence of 7 distinct RepDISs in the benchmark *hotspot*, each with varying executing instruction counts and thread counts. Based on this information, it can be concluded that most threads (44.44%) execute $RepDIS_7$, while a small fraction of threads (3.39%) execute $RepDIS_1$.

Table 2. Thread execution information of *hotspot*.

RepDIS	RepThread-ID	Instruction Count	Thread Count	Thread Ratio
1	1	109	312	3.39%
2	2	128	1152	12.50%
3	3	147	1636	17.75%
4	243	166	640	6.94%
5	47	197	740	8.03%
6	227	216	640	6.94%
7	35	271	4096	44.44%

Based on these two approaches above, we can assign weight to every edge in DG by computing its total execution count. In detail, we denote $W_{i,j}$ as the weight of an edge $E_{i,j}$ in DG, which can be calculated by Eq. 1.

$$W_{i,j} = \sum_{k=1}^{n} ((EC(i,j,RepDIS_k) \times TC(RepDIS_k)) \tag{1}$$

It is worth mentioning that the impact between all launched threads, like error propagation via shared memory, is also considered as node features listed in Table 1 of Sect. 3.1. Our GAT model captures this error propagation characteristic of GPGPU during the aggregating and updating processes, thereby improving prediction accuracy.

4   Proposed Approach: GLAM-SERP

In this chapter, we introduce GLAM-SERP, which stands for a Graph Learning-Assisted Model for Soft Error Resilience Prediction in GPGPUs. Our model exhibits high prediction performance and generalization capability in estimating the soft error resilience of GPGPU instructions, which allows us to take further steps in improving reliability and stability by implementing additional protective measures, such as selective instruction duplication [6]. As demonstrated in Fig. 5, GLAM-SERP comprises three stages: *GPGPU Program Profiling*, *Graph Data Generation*, and *Prediction Model Training*.

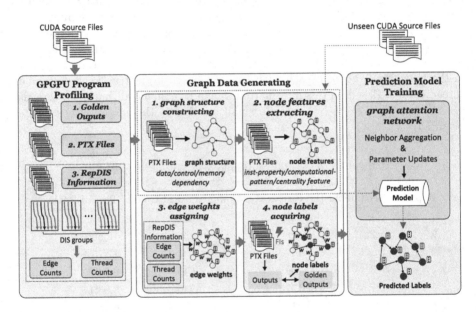

Fig. 5. An overview of GLAM-SERP.

4.1   GPGPU Program Profiling Stage

In the first stage, we first run the fault-free program to obtain the golden outputs and PTX files, which serve as references for the graph data construction and random fault injections in the subsequent stage. By comparing the corrupted program outputs with the golden output, we can assess the severity of the impact caused by the injected faults. Then we employ profiling techniques to collect statistical trace information from the execution of GPGPU threads, involving the edge count and thread count for each RepDIS, which will be used to calculate DG edge weights in the next stage.

4.2   Graph Data Generating Stage

In the second stage, we generate graph data for training a graph learning model, including the necessary topology structure, feature vectors, directed edge weights, and the labels of partial DG nodes. Precisely, we follow the methodology outlined in Sect. 3 to construct nodes for instructions and establish dependency edges based on the mentioned three dependencies. After that, we profile individual instructions to extract node feature vectors. Then we assign weight values to the edges based on the statistical information of RepDIS collected in the first stage. Finally, we perform fault injections on a subset of instructions to obtain the ground truth. This involves introducing a bit flip in the destination registers of a dynamic instance of each instruction, simulating the effect of high-energy particle attacks on the program. By executing the corrupted program and monitoring its symptoms, we can observe the resulting error distribution and assign labels to the selected instructions.

4.3   Prediction Model Training Stage

In the final stage, we drive a GAT model on the generated graph data to estimate the soft error resilience in GPGPUs. GAT is chosen for its ability to capture relationships between nodes in a graph, generalize across different graph structures and tasks, and learn transferable node representations. During the neighbor information aggregating and parameter updating process, the GAT leverages self-attention mechanisms to learn node representations by computing attention coefficients that capture the significance of neighboring nodes with different weights. Multiple attention heads are employed to capture diverse node relationships and learn attention weights independently. The information from these heads is then aggregated and fused, enabling the model to incorporate different perspectives and capture complex graph patterns effectively.

In summary, the GAT model encodes features of GPGPU program instructions related to soft error resilience into high-dimensional embedding vectors, enabling accurate classification into SDC-prone (class 1) and non-SDC-prone (class 2) instructions. For an unseen program, we construct the graph structure and extract the node features following the same steps as before. By leveraging our trained model, we can make predictions without the need for additional fault injections, thereby avoiding the substantial time overhead.

5   Evaluation

In this chapter, we first provide comprehensive explanations of the experimental setup in Sect. 5.1. Subsequently, we present the evaluation results and analysis of both our model and a previous work [6], focusing on the aspects of prediction performance and generalization ability in Sect. 5.2.

5.1 Experimental Setup

We conduct our experiments on a host running Ubuntu 18.04 with an Intel(R) Xeon(R) E5-2620 v4 16-core CPU (2.10GHz), 64GB of RAM, and an NVIDIA Tesla P40 GPU. The host system is equipped with graphics driver version 530.30.02 and CUDA version 12.1.

To get instruction execution traces of all launched GPGPU threads, we modify the GPGPU simulating tool GPGPU-Sim [20] to record all thread DISs, several RepDISs, and their executing thread counts. Moreover, to obtain partial DG node labels, we should perform a limited number of fault injection campaigns. Due to the PTX code being architecture-independent and offering higher-level abstractions, making it easier to develop and ensuring compatibility with future extensions, we choose to interrupt the compilation to expose the PTX codes, introduce random bit-flipping, and resume compilation to generate a faulty executable. By comparing its output and the golden output, we can determine if an SDC error has occurred. In addition, to ensure the coverage of diverse instruction dependency patterns in GPGPU programs, we select representative benchmarks from the Polybench [21] and Rodinia [22] suites. These benchmarks cover various domains, and their corresponding information is provided in Table 3.

Besides, we carefully tune and optimize several parameters in our graph learning experiments. Our model configuration consists of a two-layer GAT with 8 attention heads in the multi-headed attention mechanism. The hidden layer features are set to the dimension of 256. The final layer of the model is responsible for two-category classification. During the training process, we utilize the Cross Entropy Loss as our loss function to calculate the training loss. To optimize the model parameters, we employ the Adam optimizer with a learning rate of 0.001, a weight decay coefficient of 5e-4, and betas values of (0.9, 0.99).

5.2 Evaluation Results

In this section, we compare our GLAM-SERP's performance and generalization capability of identifying SDC-prone instructions with the approach presented in the previous work [6], which extracts error propagation-related features and

Table 3. Information of selected benchmarks.

Benchmark Name	Kernel Count	RepDIS Count	Input Size	Launched Thread Number	Domain	Description
2mm	2	2	256×256	65536	Linear Algebra	Two matrix multiplications.
conv2D	1	2	128×128	16384	Convolution Operation	2D Convolution computation.
correlation	4	259	256×256	66304	Data Mining	Correlation computation.
covariance	3	130	128×128	4352	Data Mining	Covariance computation.
hotspot	1	7	64	9216	Physics Simulation	Processor temperature estimation.
mvt	2	2	2048	4096	Linear Algebra	Matrix vector product and transpose.
pathfinder	1	19	10000,10,2	51200	Graph Algorithm	Shortest path finding

drives a Support Vector Machine (SVM) model. We select the metrics *Precision*, *Recall*, and *F1-score* to evaluate the performance. A higher Precision implies fewer false positives (i.e., predicting the non-SDC-prone as SDC-prone), while a higher Recall implies fewer false negatives (i.e., predicting the SDC-prone as non-SDC-prone). The F1-Score combines Precision and Recall into a single metric, offering a balanced evaluation of the model's performance. Therefore, a higher F1-Score indicates a more reliable and effective model.

Prediction Performance. To verify the prediction performance of GLAM-SERP on individual benchmarks, we randomly divide the graph data into a training set and a test set for each benchmark. The training set is utilized to train the GAT node classification model, while the test set is employed to evaluate its prediction performance. Based on the performance results presented in Table 4, we note that specific benchmarks, such as *2mm* and *conv2D*, demonstrate relatively straightforward execution patterns. This attribute facilitates GLAM-SERP in capturing the characteristics associated with potential SDC proneness, resulting in a high average performance of 97.13% on these benchmarks, surpassing the previous work's achievement by 3.01%. Conversely, in benchmarks such as *covariance*, *hotspot* and *pathfinder*, which involve intricate branches and loops, the model faces greater difficulty in accurately acquiring the knowledge about instruction SDC proneness. Consequently, the performance attained on these benchmarks tends to be lower. Nonetheless, GLAM-SERP consistently maintains a high average accuracy of 92.53% on them, which is approximately on par with the average accuracy of 92.79% accomplished by the previous study.

In summary, GLAM-SERP achieved an average F1-score of 94.14% on seven benchmarks, which closely aligns with the 93.72% achieved by the previous work [6]. The remarkable prediction performance of our model, as evidenced by its strong agreement with the fault injection results in predicting the SDC proneness of instructions, firmly demonstrates the capability to accurately identify SDC-prone instructions.

Table 4. Prediction performance on individual benchmarks.

Benchmark	GLAM-SERP			Previous Work [6]		
	Precision	Recall	F1-Score	Precision	Recall	F1-Score
2mm	95.60%	98.62%	97.09%	92.86%	93.57%	93.21%
conv2D	96.33%	98.00%	97.16%	92.57%	97.61%	95.02%
correlation	91.30%	95.45%	93.33%	94.44%	95.24%	94.84%
covariance	90.42%	94.57%	92.45%	92.50%	91.67%	92.08%
hotspot	92.38%	93.27%	92.82%	93.75%	97.73%	95.70%
mvt	96.64%	91.10%	93.79%	97.62%	91.67%	94.55%
pathfinder	91.53%	93.10%	92.31%	89.83%	91.38%	90.60%
VAR	0.00070	0.00074	**0.00044**	0.00056	0.00077	**0.00034**
AVG	93.46%	94.87%	**94.14%**	93.37%	94.12%	**93.72%**

Generalization Performance. In addition to evaluating the prediction performance on individual benchmarks, it is essential to examine the model's generalization ability to handle unseen data. Therefore, we assess the model's performance across benchmarks by designating one benchmark at a time as the test set and training the prediction model using the remaining benchmarks.

According to Table 5, GLAM-SERP achieved an average generalization performance of 95.87% on benchmarks *2mm* and *conv2D* when they were used as the test set, with the remaining benchmarks serving as the training set. This represents a notable improvement of 6.76% compared to the previous work's performance of 89.11%. Conversely, when employing more complex benchmarks such as *covariance*, *hotspot* or *pathfinder* as the test set, GLAM-SERP achieves an average generalization performance of 86.26%, demonstrating a significant improvement of 9.58% compared to the previous work's performance of 76.68%. This improvement highlights the limitations of the previous work when handling certain complex applications, which our model lacks. For example, the previous work's Precision of only 73.47% for *covariance* leads to the unnecessary protection of numerous instructions when applying the selective instruction duplication policy, resulting in increased overhead. Additionally, its Recall for *hotspot* is as low as 75.29%, causing the failure to identify and protect a significant number of SDC-prone instructions, which leaves the program highly vulnerable to soft errors. Nevertheless, our GLAM-SERP exhibits remarkable generalization performance on these complex applications, achieving Precision of over 80% and Recall of over 85% for each of them.

On the other hand, GLAM-SERP achieved an average F1-Score of 89.50% across all seven benchmarks, representing a slight decrease of 4.62% compared to its application-specific model on individual benchmarks. In contrast, the previous approach achieved 82.70%, experiencing a significant reduction of 11.02%. This is because the previous approach inadequately characterizes the error propagation

Table 5. Generalization performance on unseen benchmarks.

	GLAM-SERP			Previous Work [6]		
Benchmark	Precision	Recall	F1-Score	Precision	Recall	F1-Score
2mm	95.83%	97.18%	96.50%	90.14%	88.89%	89.51%
conv2D	93.75%	96.77%	95.24%	94.83%	83.33%	88.71%
correlation	80.13%	90.98%	85.21%	81.05%	82.80%	81.91%
covariance	82.41%	95.70%	88.56%	73.47%	83.72%	78.26%
hotspot	82.57%	86.54%	84.51%	74.42%	75.29%	74.85%
mvt	89.39%	92.19%	90.77%	94.83%	83.33%	88.71%
pathfinder	83.61%	87.93%	85.71%	76.27%	77.59%	76.92%
VAR	0.00381	0.00182	**0.00237**	0.00903	0.00198	**0.00390**
AVG	86.81%	92.47%	**89.50%**	83.57%	82.14%	**82.70%**

behaviors solely based on heuristic features, whereas our approach effectively captures the error propagation behaviors using graph structures.

In summary, GLAM-SERP outperforms the previous work [6] with an average improvement of 6.80%, suggesting that the traditional machine learning-based approach is limited to specific applications, whereas our approach significantly improves generalization performance in identifying SDC-prone instructions.

6 Related Work

In recent years, several studies have utilized machine learning methods to predict the resilience of soft errors [6,10,12,13]. PRISM [10] introduced linear regression and K-NN models to predict fault rates based on instruction types in GPU programs. G-SEPM [12] drove four ML models to estimate the resilience of individual bits, providing an alternative to massive fault injections. Another work [6] incorporated instruction resilience characteristics and error propagation behaviors to develop an SVM model for estimating SDC proneness in GPGPUs. Additionally, the previous research [13] considered both program characteristics and hardware performance metrics and leveraged both regression and classification models for application-wise vulnerability prediction.

The mentioned techniques are application-specific and lack the capability of characterizing natural error propagation behaviors. They commonly represent error propagation information as scalar or vector features, which may not adequately capture the intricate data or control flow involved in practical error propagation scenarios. However, errors in GPGPUs propagate in a manner that follows a topological-structured dependency flow, which can be intuitively and effectively represented by graph structures.

7 Conclusion

In this study, we propose GLAM-SERP, a Graph Learning-Assisted Model for Soft Error Resilience Prediction in GPGPUs. We first investigate the heuristic features of GPGPU instructions related to error resilience and construct DG to represent instruction dependencies for error propagation. We further enhance the DG by assigning weights to edges based on thread execution information. Building upon this, we obtain partial DG node labels through a limited number of fault injections and drive a GAT to effectively learn a prediction model for GPGPU instruction-wise error resilience estimation. The comparative analysis with previous work [6] reveals that GLAM-SERP achieves a remarkable level of accuracy, with an average F1-Score of 94.14%, and demonstrates strong generality, with an average F1-Score of 89.50%.

References

1. Tan, J., Goswami, N., Li, T., Fu, X.: Analyzing soft-error vulnerability on GPGPU microarchitecture. In: 2011 IEEE International Symposium on Workload Characterization (IISWC). IEEE (2011). https://doi.org/10.1109/iiswc.2011.6114182

2. Tan, J., Fu, X.: RISE: improving the streaming processors reliability against soft errors in GPGPUs. In: Proceedings of the 21st International Conference on Parallel Architectures and Compilation Techniques. ACM (2012). https://doi.org/10.1145/2370816.2370846

3. Li, G., et al.: Understanding error propagation in deep learning neural network (DNN) accelerators and applications. In: Proceedings of the International Conference for High Performance Computing, Networking, Storage and Analysis. ACM (2017). https://doi.org/10.1145/3126908.3126964

4. Yang, L., Nie, B., Jog, A., Smirni, E.: Enabling software resilience in GPGPU applications via partial thread protection. In: 2021 IEEE/ACM 43rd International Conference on Software Engineering (ICSE). IEEE (2021). https://doi.org/10.1109/icse43902.2021.00114

5. Kalra, C., Previlon, F., Rubin, N., Kaeli, D.: ArmorAll: compiler-based resilience targeting GPU applications. ACM Trans. Archit. Code Optim. **17**(2), 1–24 (2020). https://doi.org/10.1145/3382132

6. Wei, X., Jiang, N., Wang, X., Yue, H.: Detecting SDCs in GPGPUs through an efficient instruction duplication mechanism. In: Qiu, H., Zhang, C., Fei, Z., Qiu, M., Kung, S.-Y. (eds.) KSEM 2021. LNCS (LNAI), vol. 12817, pp. 571–584. Springer, Cham (2021). https://doi.org/10.1007/978-3-030-82153-1_47

7. Nie, B., Yang, L., Jog, A., Smirni, E.: Fault site pruning for practical reliability analysis of GPGPU applications. In: 2018 51st Annual IEEE/ACM International Symposium on Microarchitecture (MICRO). IEEE (2018). https://doi.org/10.1109/micro.2018.00066

8. Previlon, F.G., Kalra, C., Tiwari, D., Kaeli, D.R.: PCFI: program counter guided fault injection for accelerating GPU reliability assessment. In: 2019 Design, Automation & Test in Europe Conference & Exhibition (DATE). IEEE (2019). https://doi.org/10.23919/date.2019.8714781

9. Lu, Q., Pattabiraman, K., Gupta, M.S., Rivers, J.A.: SDCTune: a model for predicting the SDC proneness of an application for configurable protection. In: Proceedings of the 2014 International Conference on Compilers, Architecture and Synthesis for Embedded Systems. ACM (2014). https://doi.org/10.1145/2656106.2656127

10. Kalra, C., Previlon, F., Li, X., Rubin, N., Kaeli, D.: PRISM: predicting resilience of GPU applications using statistical methods. In: SC18: International Conference for High Performance Computing, Networking, Storage and Analysis. IEEE (2018). https://doi.org/10.1109/sc.2018.00072

11. Wei, X., Yue, H., Gao, S., Li, L., Zhang, R., Tan, J.: G-SEAP: analyzing and characterizing soft-error aware approximation in GPGPUs. Future Gener. Comput. Syst. **109**, 262–274 (2020). https://doi.org/10.1016/j.future.2020.03.040

12. Yue, H., Wei, X., Li, G., Zhao, J., Jiang, N., Tan, J.: G-SEPM: building an accurate and efficient soft error prediction model for GPGPUs. In: Proceedings of the International Conference for High Performance Computing, Networking, Storage and Analysis. ACM (2021). https://doi.org/10.1145/3458817.3476170

13. Topcu, B., Oz, I.: Predicting the soft error vulnerability of GPGPU applications. In: 2022 30th Euromicro International Conference on Parallel, Distributed and Network-based Processing (PDP). IEEE (2022). https://doi.org/10.1109/pdp55904.2022.00025

14. Wei, X., Yue, H., Tan, J.: LAD-ECC: energy-efficient ECC mechanism for GPGPUs register file. In: 2020 Design, Automation & Test in Europe Conference & Exhibition (DATE). IEEE (2020). https://doi.org/10.23919/date48585.2020.9116503

15. Hari, S.K.S., Tsai, T., Stephenson, M., Keckler, S.W., Emer, J.: SASSIFI: an architecture-level fault injection tool for GPU application resilience evaluation. In: 2017 IEEE International Symposium on Performance Analysis of Systems and Software (ISPASS). IEEE (2017). https://doi.org/10.1109/ispass.2017.7975296

16. Sangchoolie, B., Pattabiraman, K., Karlsson, J.: One bit is (Not) enough: an empirical study of the impact of single and multiple bit-flip errors. In: 2017 47th Annual IEEE/IFIP International Conference on Dependable Systems and Networks (DSN). IEEE (2017). https://doi.org/10.1109/dsn.2017.30

17. Kipf, T.N., Welling, M.: Semi-supervised classification with graph convolutional networks. CoRR abs/1609.02907 (2016). http://arxiv.org/abs/1609.02907

18. Velickovic, P., Cucurull, G., Casanova, A., Romero, A., Liò, P., Bengio, Y.: Graph attention networks. CoRR abs/1710.10903 (2017). http://arxiv.org/abs/1710.10903

19. Ying, C., et al.: Do transformers really perform badly for graph representation? Adv. Neural. Inf. Process. Syst. **34**, 28877–28888 (2021)

20. Bakhoda, A., Yuan, G.L., Fung, W.W.L., Wong, H., Aamodt, T.M.: Analyzing CUDA workloads using a detailed GPU simulator. In: 2009 IEEE International Symposium on Performance Analysis of Systems and Software. IEEE (2009). https://doi.org/10.1109/ispass.2009.4919648

21. Pouchet, L.N.: PolyBench: the polyhedral benchmark suite (2012). http://www.cs.ucla.edu/pouchet/software/polybench

22. Che, S., et al.: Rodinia: a benchmark suite for heterogeneous computing. In: 2009 IEEE International Symposium on Workload Characterization (IISWC). IEEE (2009). https://doi.org/10.1109/iiswc.2009.5306797

THRCache: DRAM-NVM Multi-level Cache with Thresholded Heterogeneous Random Choices

Tao Tao, Zhiwen Xiao$^{(\boxtimes)}$ (iD), Jibin Wang, Jing Shang, and Zhihui Wu

China Mobile Information Technology Center, Beijing 100033, China
{taotao,xiaozhiwen,wangjibin,shangjing,wuzhihui}@chinamobile.com
http://it.10086.cn

Abstract. Caching is essential for accelerating data access, balancing storage cluster load and improving quality of service. However, single-node cache may become a bottleneck as the storage system scales up. Distributed caching is proposed to provide caching services for large-scale storage cluster. However, existed cache mechanisms based on cache partitioning or cache replication may lead to load imbalance and high coherency overhead. We propose THRCache, a multi-level heterogeneous distributed cache mechanism which combines the speed of DRAM with the high-capacity of NVM to cache more easily accessible data. THRCache implements cache allocation for different cache layers by independent hash functions, routes query with a threshold random heterogeneous selection, and introduces a prefetching mechanism based on data access correlation. In this paper, we implement a prototype of THRCache and demonstrate through experiments that THRCache has higher cache hit rate and throughput than existed distributed cache architectures under different workloads.

Keywords: Distributed Caching · DRAM-NVM Hybrid Storage · Load Balance · Cache Allocation · Data Prefetching

1 Introduction

With the development of cloud computing, big data and distributed computing technologies, large-scale storage systems have become vital tools for storing and processing big data. However, request loads are often highly skewed in the real world, with a few objects receiving most of the requests. For example, previous studies have shown that 10% of items in Facebook's Memcached deployment account for 60–90% of queries [1]. Such skewed workloads may overload some nodes, leading to data loss, service unavailability and other problems that threaten the stability and scalability of the storage system.

Caching is a technique that accelerates data access and balances load in large-scale Internet storage systems. These systems often use memory-based caching systems such as Memcached [2] and Redis [3], or local caching libraries such

© The Author(s), under exclusive license to Springer Nature Singapore Pte Ltd. 2024
Z. Tari et al. (Eds.): ICA3PP 2023, LNCS 14490, pp. 436–455, 2024.
https://doi.org/10.1007/978-981-97-0859-8_26

as Caffeine [4] and Ehcache [5], to meet their service-level objectives (SLOs). Research has shown that a solitary high-speed cache node, storing only the most frequently accessed data of $O(n \log n)$, effectively balances the load of n storage nodes [6]. Nevertheless, in the context of large-scale storage systems, the constrained throughput and capacity of a single cache node can emerge as a system bottleneck, underscoring the significance of cache service scalability in enhancing the performance of such systems. Consequently, the deployment of multiple cache nodes for distributed caching has emerged as a critical strategy for augmenting cache service scalability. The majority of current distributed caching architectures employ homogeneous cache nodes (either DRAM-based or NVM-based) to construct distributed caching services like DistCache [7] and NVMCache [8]. These services utilize a single storage medium for caching objects, resulting in data with varying access frequencies coexisting within the same storage medium. When caching an equivalent data volume, the utilization of a storage medium with faster read and write capabilities squanders resources for data with lower access frequencies. Conversely, deploying a storage medium with greater capacity impedes efficient caching of data with higher access frequencies due to slower read and write speeds. Consequently, a caching system reliant on a sole storage medium falls short in satisfying the distinct demands of data with diverse access frequencies. In contrast, heterogeneous caching systems excel in resolving this concern. Within a heterogeneous caching system, dissimilar data types are allocated to distinct storage media, facilitating the optimization of storage resource allocation. With the rapid development of NVM technology, some NVM devices such as flash-based solid-state drives (SSDs) have become mainstream data storage devices for computers due to their increasing speed, density, and decreasing cost [9]. Moreover, there has been a lot of research on introducing NVM into KV caching/storage [10–17], some NVM devices such as PCM and ReRAM have read/write latency comparable to DRAM [18], but lower cost-per-byte and non-volatility. DRAM-NVM hybrid hierarchical distributed caches can provide high-speed and reliable caching services by leveraging the fast access of DRAM and the high capacity, non-volatile, and low read/write latency of NVM. Several studies [19–21] have proposed DRAM-NVM hybrid caching mechanism which NVM acts as a secondary cache to handle cache-miss data accesses in DRAM and the design of cache allocation strategies considers more about DRAM.

Therefore, providing a distributed caching for large-scale storage systems poses several challenges. One of these challenges is how to handle skewed workloads and changes in data access frequency. Another challenge is how to take full advantage of the low read/write latency and large capacity of NVM to design a more efficient cache allocation strategy. In this paper, we propose a DRAM-NVM multi-level heterogeneous distributed caching architecture called THRCache (Thresholded Heterogeneous Random Choices Cache), which introduces a THRC mechanism and a prefetching mechanism. The THRC mechanism uses a thresholded power-of-d-choices algorithm [22,23] for cache allocation and query routing. Meanwhile, the prefetching mechanism enables prefetching of related data. This architecture makes full use of the fast access of DRAM and the large capacity of NVM, which can store a number of copies of data that varies with the hotness. In THRCache, We use a high-speed cache to cache extremely hot data and store warming data in a slower but larger cache, just like the multi-level cache in a CPU. So NVM serves not only as a secondary cache to store warm data and handle unhit queries in DRAM, but also as a primary cache to share the load for DRAM cache. Furthermore, THRCache introduces a prefetching mechanism to prefetch related data into NVM based on the correlation of data access patterns to improve the system cache hit rate. Experiments show that THRCache has higher stability compared to existing multi-layer caching mechanisms while improving throughput by up to 30%.

The main contributions of this paper are as follows:

- Presenting THRCache, a DRAM-NVM multi-level heterogeneous distributed caching mechanism for large-scale storage system with highly skewed load.
- Proposing a THRC mechanism for multi-level DRAM-NVM caching architectures. Considering the fast access of DRAM and the large capacity of NVM, this mechanism reduces the load on the DRAM cache nodes and increase the throughput of the system.
- Proposing a THRC-based data prefetching mechanism. NVM prefetches data decided by DRAM load, data heat, and correlation of data accesses to improve cache hit ratio and system throughput.
- Building a prototype of THRCache. Experimental results show that THRCache improves throughput by up to 30% compared to other multi-layer caching mechanisms under different workloads.

2   Related Work

In this section, we present existing research on caching and analyze the characteristics of these mechanisms.

2.1   Distributed Caching

Large-scale storage systems with many clusters often use a distributed caching architecture to provide caching services. Distributed caching relies on multiple cache nodes to enhance the scalability of cache service. A good cache allocation strategy can balance the load across different cache nodes and avoid issues such as node failure due to overload. Moreover, an effective strategy can also optimize the efficiency of cache usage and improve the performance of the storage service system. Therefore, the cache allocation strategy determines the performance, stability and scalability of large-scale storage systems. Traditional strategies include cache partitioning and cache replication. Cache partitioning directly divides the object space among cache nodes, and each cache node is responsible for caching the hot objects in its own partition. However, multiple hot data may be assigned to the same cache node in the situation of load skewing, leading to an imbalance in the load across nodes. Cache replication ensures load balancing by replicating the hot objects to all cache nodes. Yet, since data objects have multiple replicas, cache updates must be made on each copy, requiring a significant cost to ensure cache consistency. As for large object-intensive workloads, some research like InfiniCache [24], EC-Cache [25], POCache [26] employs erasure coding to provide large object caching for data-intensive cluster computing workloads. However, the encoding and decoding operations of erasure coding add non-negligible latency to access the in-memory cache.

2.2 Multi-level Caching

To achieve load balancing of cache nodes and reduce the overhead of cache coherency, DistCache [7] proposed a multi-layer distributed caching mechanism which uses independent hash functions to assign hot objects to two layers of cache nodes according to the cache topology and query routing. This method uses adaptive query routing with power-of-two-choices [22], which ensures intra-cluster and inter-cluster load balancing while scaling cache services to a large scale. NVMCache [8] presents an NVM-based multi-layer caching mechanism that leverages the read/write asymmetry and prioritize NVM access to enhance overall access performance. It also avoids local intensive write operations by a wear-balancing scheme that balances both load and wear to extend the overall service lifetime of the NVM-based cache. However, DistCache and NVMCache have identical structure and capacity for the two cache layer nodes, and both use the same storage media. The correlation of data access and the difference in data heat are not taken into account, and data with low heat in the cache will be replaced. Moreover, in real-world application scenarios, data access patterns often exhibit correlation. That is, when one data is accessed, another data is more likely to be accessed. If the relevant data can be preloaded into the cache, the system's cache hit rate will increase, enhancing the system performance.

3 THRCache Distributed Cache

THRCache is a multi-level heterogeneous distributed caching architecture that leverages the large capacity of NVM to reduce the load of DRAM and improve the space utilization of the system by prefetching data based on data popularity and DRAM and NVM load. Figure 1 shows an overview of the THRCache architecture, which consists of five parts: Controller, DRAM Cache, NVM Cache, Prefetcher and Storage Servers. The Controller receives data queries and routes them to cache nodes or storage nodes. It also maintains recent data access records and information on frequently accessed data to provide contextual data for prefetching. DRAM Cache and NVM Cache are responsible for caching hot data. Storage Servers provide persistent storage for all data in the system. Prefetcher performs data prefetching based on the contextual information of system operation. We describe each part in more detail in the following sections.

Cache Controller. The Controller receives queries from clients and distributes them to the appropriate node based on the query type and cache routing policy. Read queries are sent to the cache node, while write queries are sent to Storage Servers. The Controller logs the information of recently accessed data, such as key, access type, and timestamp. Moreover, the Controller maintains the keys of frequently accessed data in each DRAM cache node based on the log. This information enables the system to comprehend the trend of data access and provide contextual information for data prefetching. The Controller only stores the logs of recent data access and keys of a small amount of hot data, which can be placed in DRAM, which has low storage and I/O costs, thus avoiding becoming a system bottleneck.

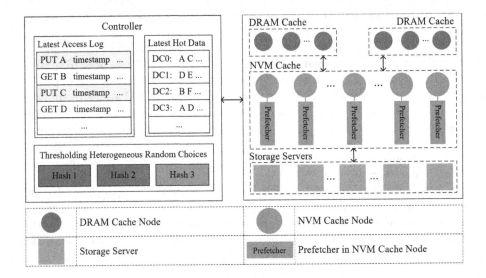

Fig. 1. THRCache distributed caching architecture

DRAM Cache. DRAM Cache comprises multiple DRAM cache nodes that handle queries for the hottest data in the storage system and balance the load for the storage nodes. THRCache has two DRAM Cache layers, and each cache layer uses independent hash functions to partition hot data. The independent hash functions enable us to locate the cache node position where data is stored in each cache layer. Moreover, if hot data is concentrated in one cache node in one layer, it will likely be distributed to multiple cache nodes in the other layer due to the independence of the two hash functions. When the Controller decides to forward a query to DRAM Cache, it always chooses the node with lower load.

NVM Cache. The NVM Cache consists of multiple high-capacity NVM cache nodes, and data allocation and query routing also rely on an independent hash function. Data queries are forwarded to the NVM Cache in two cases: the first case is when there is a cache miss in the DRAM cache node, the DRAM cache node forwards the query to the NVM Cache and in this perspective the NVM Cache acts as the secondary cache of the system; the second case is when the Controller decides to send the query directly to the NVM Cache without going through the DRAM Cache. This happens when the DRAM cache node is under high load: the Controller uses three independent hash functions and get two DRAM cache nodes and one NVM cache node where the data is located, and then determines that the DRAM cache node is overloaded based on the current load of the DRAM cache node. So, the query is sent to the NVM cache node, and at this time the NVM Cache and DRAM Cache together act as the primary cache together.

Prefetcher in NVM. Each NVM cache node has a built-in Prefetcher to perform prefetching operations. The Controller directly selects the NVM cache node. The main principle is that the direct selection of the NVM cache node by the Controller indicates that the load on both DRAM cache nodes where the requested data is located is high, and there is a high probability that queries will be sent to the NVM cache node in the next period of time, so the Prefetcher can prefetch the hot data that may be accessed later. As a result, Prefetcher takes full advantage of the high capacity of the NVM cache nodes.

Storage Servers and Clients. Storage Servers persistently store key-value data in the entire system. Clients provide a simple data access interface, which encapsulates data access as data request packets sent to the Controller and generates return values to the interface caller based on the Controller's data response packets.

4   THRCache Caching Mechanism

In this section, we first introduce the THRC algorithm used for cache allocation and query routing and the prefetching mechanism based on THRC, and then elaborate on handling read and write queries.

4.1   THRC Mechanism

DistCache and NVMCache employ the standard d-left algorithm [22] for cache allocation and query routing, aimed at achieving load balance within a homogeneous cache node configuration. The standard d-left algorithm distributes the load evenly among cache nodes of comparable performance by initially hashing requests into a predefined set of cache node groups and subsequently mapping them to specific cache nodes. In contrast, THRCache comprises a mix of heterogeneous DRAM cache nodes and NVM cache nodes, with NVM cache nodes typically exhibiting lower theoretical performance than DRAM cache nodes.

Thus, to attain an optimal load balance, it is expected that DRAM cache nodes will handle a larger portion of requests in comparison to NVM nodes. Consequently, we introduce THRC, a threshold-enhanced version of the d-left algorithm designed to mitigate performance disparities between NVM and DRAM cache nodes. In THRC, the core concept of query routing revolves around augmenting the likelihood of selecting DRAM cache nodes while diminishing the probability of selecting NVM cache nodes, accomplished via a threshold mechanism. The fundamental concept is that a query is directed to an NVM cache node solely when the load disparity between the DRAM cache node and the NVM cache node exceeds a predetermined threshold.

Specifically, the target cache node of one query will be selected from one of the three cache nodes which are determined by the three independent hash functions used by the two DRAM Cache and the NVM Cache, respectively. Assume that the hash functions used in the two DRAM Cache are h_1 and h_2, and the hash function used in the NVM Cache is h_3. Data d is mapped to DRAM cache node DC_1, DRAM cache node DC_2 and NVM cache node NC_3, i.e., $h_1(d) = DC_1$, $h_2(d) = DC_2$ and $h_3(d) = NC_3$. When the Controller receives a query for data d, it looks up the load table to obtain the load L_1 of DC_1, the load L_2 of DC_2 and the load L_3 of NC_3. If $min(L_1, L_2)/L_3 > T$ is satisfied where T is the chosen threshold value, Controller forwards the query to NC_3, otherwise it forwards the query to the less loaded cache node in NC_1 and NC_2.

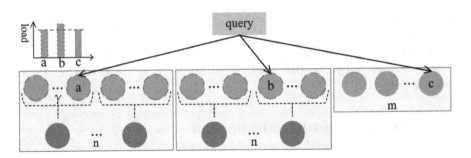

Fig. 2. Thresholded Heterogeneous Random Choices

Figure 2 illustrates how THRC works. There are two DRAM Caches with n DRAM cache nodes each and an NVM Cache with m NVM cache nodes. The theoretical throughput of a DRAM cache node is denoted as T_{DRAM} and the theoretical throughput of an NVM cache node is denoted as T_{NVM}. A DRAM cache node is simulated as $\gamma = T_{DRAM}/T_{NVM}$ NVM cache nodes. The query is routed to two simulated NVM cache nodes a with load L_a, b with load L_b, and one real NVM cache node c with load L_c. This query is eventually absorbed by the node with minimum load $L_{min} = min(L_a, L_b, L_c)$. If there are multiple nodes with minimum load L_{min}, the leftmost node will absorb the query [27]. In Fig. 2, the query is processed by node a because the load of nodes a and c are equal to L_{min} and a is the leftmost node. Setting the number of nodes in the NVM Cache (m) to γn results in a total of $s = 3\gamma n$ NVM cache nodes (comprising both simulated and actual nodes) within the system. This configuration can be likened to a 3-left scheme, where each query is allocated to the cache node with the lowest load among three nodes drawn from distinct cache layers through a hashing process. In cases of multiple cache nodes sharing the lowest load, the query is directed to the leftmost cache node. P. Berenbrink et al. [28] have provided a proof (in Eq. 1) demonstrating that the load of the most heavily burdened node, denoted as L_{max}, in a d-left configuration (where d equals 3 in our context) closely approximates the average load q/s with a high level of confidence. Here, q represents the number of queries, and ϕ_d serves as a constant proportional to d, satisfying the condition $1.61 \leq \phi_d < 2$ when $d \geq 2$.

$$L_{max} \leq \frac{q}{s} + \frac{lnlns}{d \cdot ln\phi_d} \pm \Theta(1) = \frac{q}{3\gamma n} + \frac{lnln3\gamma n}{3 \cdot ln\phi_3} \pm \Theta(1), \ w.h.p \qquad (1)$$

where, ϕ_d is a constant proportional to d and $1.61 \leq \phi_d < 2$ when $d \leq 2$.

According to previous analysis of the d-left based on the fluid limit theorem [22], we can obtain the differential equations:

$$\begin{cases} w_{i=0}^k(t) = \frac{1}{d}, & i = 0, k = 0, 1, \ldots, d-1 \\ w_i^k(0) = 0, & i \geq 1, k = 0, 1, \ldots, d-1 \\ \frac{dw_i^k(t)}{dt} = d^d(w_{i-1}^k(t) - w_i^k(t)) \prod_{j=k+1}^{d-1} w_{i-1}^j \prod_{l=0}^{k-1} w_i^l, & i \geq 1, k = 0, 1, \ldots, d-1 \end{cases} \qquad (2)$$

where $w_i^k(t)$ is the fraction of the s cache nodes that have load at least i and are in the kth group when ts queries have been routed. $W_i(t)$ is the sum of these d fraction $w_i^{k=0,1,\ldots,d-1}(t)$ which denotes the overall fraction of the s cache nodes that have load at least i:

$$W_i(t) = \sum_{k=0}^{d-1} w_i^k \qquad (3)$$

Table 1 shows the distribution of $W_i(t)$ when $d = 3$. Taking the column where $t = 6$ as an example, the fraction of cache nodes with load at least 6 is 8.3387e-1 and at least 7 is 1.7824e-1, but at least 8 is 2.3648e-5 which is negligible. It can be seen that most nodes have the load which is extremely close to the average load. Owing to the excellent performance of the d-left algorithm, the DRAM cache nodes in THRC are not actually simulated as multiple slow NVM cache nodes. THRCache can simply use the average load of the DRAM cache node $L_{avg} = \frac{L_{DRAM}}{\gamma}$ to approximate the load of a simulated NVM cache node. The performance of the system is related to the chosen threshold T: (a) when $T = \gamma$ and the number of NVM cache nodes $m \geq \gamma n$, THRC is an approximation of 3-left; (b) when $T > \gamma$, as T increases, the number of queries routed to the NVM Cache will gradually decrease and THRC gradually degrades to 2-left problem; (c) when $T < \gamma$, more queries will be routed to the NVM Cache, and it will become the system bottleneck.

Table 1. The Distribution of $W_j(t)$

t \ j	1	2	3	4	5	6
1	8.3791e-1	9.9088e-1	9.9954e-1	9.9998e-1	≈ 1	≈ 1
2	1.6208e-1	8.3431e-1	9.8915e-1	9.9939e-1	9.9997e-1	≈ 1
3	1.0852e-5	1.7479e-1	8.3393e-1	9.8837e-1	9.9934e-1	9.9996e-1
4	≈ 0	2.0482e-5	1.7736e-1	8.3388e-1	9.8816e-1	9.9932e-1
5	≈ 0	≈ 0	2.1915e-5	1.7819e-1	8.3388e-1	9.8858e-1
6	≈ 0	≈ 0	≈ 0	2.2734e-5	1.7824e-1	8.3387e-1
7	≈ 0	≈ 0	≈ 0	≈ 0	2.3648e-5	1.7824e-1
8	≈ 0	≈ 0	≈ 0	≈ 0	≈ 0	2.3648e-5
9	≈ 0	≈ 0	≈ 0	≈ 0	≈ 0	≈ 0

4.2 Prefetching Mechanism

Each NVM cache node houses a Prefetcher tasked with identifying the data to prefetch and executing prefetch operations to retrieve data from storage servers into the NVM cache nodes.

Prefetched Object. The prefetched object can be either location-related or access-pattern-related to the queried data. Suppose the queried data is d and the Controller routes this query to a NVM cache node which implies there are two unselected DRAM cache nodes $h_1(d) = DC_1$, $h_2(d) = DC_2$ with high load. The following queries for data which are mapped to DC_1 or DC_2 will be routed to NVM cache nodes with high probability due to the high load of DC_1 and DC_2. These data mentioned above are location-related to d. The access-related data to d are determined by the contextual information which includes the latest access log and the distribution of queries. The Prefetcher does not immediately fetch all the data to be prefetched from the storage servers but sends the prefetch command asynchronously to the Prefetcher built in other NVM cache nodes where the data is mapped to. Multiple Prefetcher forms a distributed prefetcher layer.

Prefetching Timing. When the Controller directs a query to an NVM cache node using the THRC algorithm, it concurrently transmits contextual information, comprising the most recent access log and the keys of the most frequently accessed data in the two respective, yet unselected, DRAM cache nodes. The NVM cache node then transfers this contextual information to its integrated prefetcher to initiate the prefetching process. The Controller's direct selection of the NVM cache node indicates elevated or rising load in the associated, as-yet-unselected DRAM cache nodes, prompting subsequent queries to be directly routed to the NVM cache node. This rationale underlines our timing choice for prefetching.

4.3   Cache Access Handling

Cache Queries Handling. The Controller calculates the cache node containing the target data based on the hash functions h_1 and h_2 used in the DRAM cache and h_3 used in the NVM cache, and selects the cache node to forward it according to the THRC method.

If the Controller selects a DRAM cache node, the query is forwarded to the corresponding DRAM cache node, and if it hits the cache, the value of the requested data is returned directly; If it does not hit the cache, the DRAM cache node forwards the request to the NVM cache node, which will handle the query. If the Controller selects an NVM cache node, the query is forwarded to the corresponding NVM cache node, and if the cache hits, the value of the data is returned directly; if it does not hit the cache, the NVM cache node forwards the query to the storage server. Figure 3(a) shows an example. The client sends a read query to the Controller to read object C. Assuming that C is cached in DRAM cache node DC_1, DRAM cache node DC_3 and NVM cache node NC_3. The Controller uses THRC to decide whether to select DC_1, DC_3 or NC_3. If there is a cache hit, the cache node (DC_1, DC_3 or NC_3) is returned directly to the client. When it does not hit the cache, different actions are taken depending on the cache node selected by the Controller. Figure 3(b) shows a case where the Controller selects the DRAM cache node. The Controller forwards the query to DC_1, and if it does not hit the DC_1 cache, DC_1 forwards the query to the NVM cache node NC_2, which returns the data if the NC_2 cache hits, or forwards the query to the back-end storage server if the NC_2 cache does not hit. Figure 3(c) shows the case where the Controller selects the NVM cache node, and the Controller forwards the query to NC_3, and if it does not hit the NC_3 cache the query is forwarded to the back-end storage server.

Cache Update Handling. Cache update processing is done through a two-phase update protocol to ensure cache consistency. The first phase invalidates all replicas and the second phase updates all replicas. Figure 4 illustrates an example where object D's copies in the DRAM cache and NVM cache are invalidated by a query from the Phase 1 storage server. After the first phase the server can update its primary replica and return a successful update message to the client. In the second phase, the data in the cache node is updated.

5   Experiment and Evaluation

5.1   THRCache Implementation

We implemented a THRCache prototype using the Go language, including a Controller, a Cache (DRAM Cache and NVM Cache), Storage Servers and a Client. DRAM Cache and NVM Cache use three independent hash functions to partition cache objects. Such skewed workload is prevalent in benchmarking key-value stores [29,30] and it is supported by production system measurements [1].

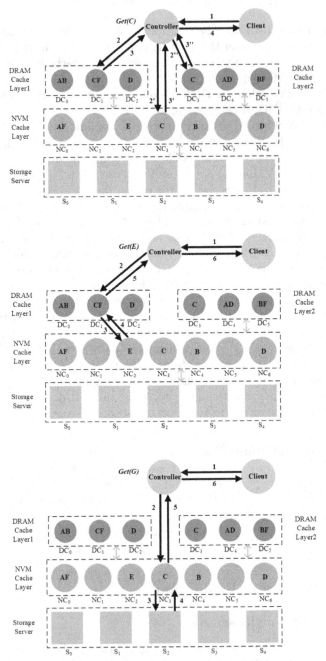

(c) Cache miss in NVM node working as primary cache

Fig. 3. Cache queries handling

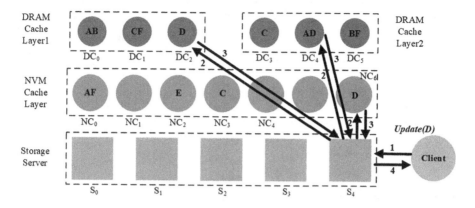

Fig. 4. Cache update handling

Controller. Controller maintains a load table that records the real-time load for each cache node and the three independent hash functions used by THRC. Controller also keeps track of the latest access log and the key of hottest data in each DRAM cache node for prefetching. The response packet from a cache node contains its real-time load. The key of the hottest data will also be contained in the response packet if it is from a DRAM cache node. Controller extracts the information from the response packets to keep the local records up to date.

Cache. DRAM cache nodes and NVM cache nodes have different cache sizes and throughputs. We simulate DRAM cache nodes and NVM cache nodes by allocating space of different sizes to Redis and using a rate-limiter to control the throughput size.

Storage Server and Client. Storage servers store all the data in the system. The client generates queries with different distributions and write ratios.

5.2 Experiment Setup

System Workloads. We use real-world highly-skewed workloads in the evaluation. The skewed workload is subject to the Zipf distribution with skew parameters of 0.9, 0.95 and 0.99. Such skewed workload is prevalent in benchmarking key-value stores [29,30] and it is supported by production system measurements [1].

As shown in Fig. 5, some data are accessed more frequently under the skewed workload. We store a total of one million data objects in Storage Servers. The default skew parameter is set to 0.99 to show that THRCache also performs well under extreme scenarios. The experiments evaluate the performance of the system under different workloads by varying the skewness parameter and the write rate.

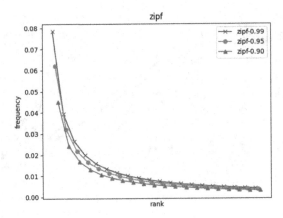

Fig. 5. Cache update handling

Baseline and Metric. The baselines are DistCache [7] and a traditional multi-level caching mechanism where NVM only serves as a secondary cache. For the convenience of discussion, we denote the traditional multi-level cache as TML-Cache. DistCache is a distributed caching mechanism that provides load balancing for large-scale storage systems. It provides better performance under different workloads compared to cache replication and cache partitioning. We implemented DistCache by mapping hot data to cache nodes with independent hash functions in each of the two DRAM cache layers and routing queries with the power-of-two-choices algorithm. TMLCache is implemented by adding a secondary NVM cache layer to DistCache.

We normalize the system throughput to the maximum throughput as a performance evaluation metric. In addition, we use the imbalance index to measure the imbalance of the caching system, where a larger imbalance index indicates a more unbalanced cache. The imbalance index is defined as:

$$\lambda = \frac{(L_{max} - \overline{L})}{\overline{L}} \tag{4}$$

where L_{max} denotes the maximum load of the cache nodes in the system and $\overline{L}$ denotes the average load of the cache nodes.

Parameter Setting. Here are 2 DRAM Cache nodes with 4 DRAM cache nodes each and a NVM Cache with 26 NVM cache nodes in the THRCache. Each DRAM cache node can hold 50 data objects, and a NVM cache node can hold 200 data objects; thus, THRCache provides a cache size of 5600 data objects. We populate 40 Storage Servers with a total number of one hundred thousand key-value data. The throughput of a DRAM cache node to a NVM cache node is set to 2:1 by rate limiters. The skewness parameter of the default workload is 0.99 and the write ratio is 0, which means that the default workload is read-only. The default number of prefetched objects is 20, which includes 10 Access-Related data and 10 Location-Related data.

According to the analysis in the Sect. 4, we only need to deploy 8 NVM cache nodes while 26 nodes are used in the experiment. This is because NVM cache nodes need to handle queries that are missed in DRAM cache nodes. The theoretical throughput of a DRAM cache node to a NVM cache node is more accurately denoted by:

$$\gamma = \frac{T_{DRAM}}{T_{NVM} - L_{secondary}} \tag{5}$$

where $L_{secondary}$ is the load of a NVM cache node caused by handling queries which are missed in DRAM cache nodes. According to the conclusion proved in [5], $L_{secondary}$ can be bounded by:

$$L_{secondary} \leq \frac{1 + \sqrt{1 + 2\alpha^2 \frac{m \log m}{c-1}}}{2} \cdot \frac{R_{DRAM}}{m} \tag{6}$$

where c is the cache size of DRAM Cache, m is the number of NVM cache nodes, α is a small constant and R_{DRAM} is the query rate less than $2n \cdot T_{DRAM}$ which is the sum of $2n$ DRAM cache nodes' theoretical throughput. We set the lower bound on m as shown in Eq. 7 so that even if $L_{secondary}$ achieves its upper bound, THRCache also can be modeled by 3-left.

$$\begin{cases} m \geq \gamma \bullet n \\ \gamma = \dfrac{T_{DRAM}}{T_{NVM} - L_{secondary}} \\ L_{secondary} = \dfrac{1 + \sqrt{1 + 2\alpha^2 \frac{m \log m}{c-1}}}{2} \cdot \dfrac{R_{DRAM}}{m} \\ R_{DRAM} = 2n \bullet T_{DRAM} \end{cases} \tag{7}$$

Table 2 summarizes the default parameters in the experiment.

Table 2. Experiment Parameters and Default Values

Parameter	Default value	Explanation
Threshold	6.5	The threshold of THRC is set to 6.5 which is equal to 26/4.
Skewness	0.99	The default workload is subject to Zipf-0.99 which is close to the real-world workload.
Access-Related	10	The number of prefetched Access-Related data.
Location-Related	10	The number of prefetched Location-Related data.
Write ratio	0	The default workload is read-only. We also investigated the performance of THRCache under workloads with different write ratios

5.3   Experiment Results and Analysis

Impact of Workload Skew. Our performance evaluation of the THRC mechanism is based on throughput. In Fig. 6(a), the system throughput of THRCache, TMLCache, and DistCache is depicted under various load skews. In scenarios with skewed workloads, THRCache exhibits substantially superior system throughput compared to TMLCache and DistCache. Moreover, as data access becomes more skewed, THRCache attains even higher system throughput. Figure 6(b) illustrates the maximum load borne by DRAM cache nodes in THRCache, TMLCache, and DistCache across varying workload distributions. Notably, the load on DRAM cache nodes in THRCache is significantly lower than that in TMLCache and DistCache.

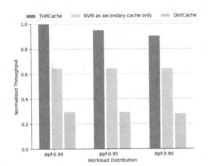
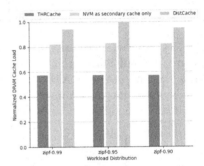

(a) System workload under different query distributions

(b) DRAM load under different query distributions

Fig. 6. Impact of workload skew

In contrast to TMLCache, THRCache introduces both the THRC mechanism and prefetching mechanism. These enhancements enable NVM Cache to alleviate the load on DRAM Cache and elevate the NVM Cache's hit rate, consequently enhancing system throughput. In comparison to DistCache, THRCache incorporates NVM as a secondary cache, leveraging NVM's ample capacity to store additional data for future access. This integration, combined with the THRC mechanism, facilitates the sharing of the load between NVM Cache and DRAM Cache, thereby mitigating the risk of DRAM cache nodes becoming a system bottleneck. As the distribution of data access becomes more skewed, the likelihood of accessing location-related and access-pattern-related data through prefetching increases, resulting in higher system throughput for THRCache.

Impact of Threshold. Figure 7 shows the variation of system throughput and DRAM load with T. When T is small, a large number of queries are forwarded

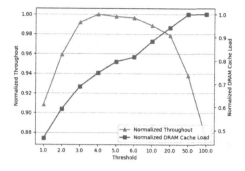

Fig. 7. System throughput and DRAM load at different thresholds

to NVM. The load on DRAM cache nodes is low, and thus the NVM tends to become the system bottleneck, resulting in low system throughput. As T increases, more queries are forwarded to DRAM cache, the utilization of DRAM cache gradually increases, and the load gradually rises. When T is large, fewer queries are sent to NVM, which cannot share the load for DRAM effectively, and DRAM tends to become the bottleneck of the system, resulting in lower system throughput. In this experimental environment, the maximum system throughput is achieved when T is around 6 where DRAM utilization is high but it has not become the system bottleneck.

Figure 8 shows the variation of the system load balance with T. As the value of T increases, the imbalance index becomes smaller, indicating that the load of DRAM cache nodes becomes more balanced. When $T = 6$, the imbalance index reaches a smaller value, and after $T > 10$, the imbalance index is basically stable with the growth of T having little impact on the system load balancing.

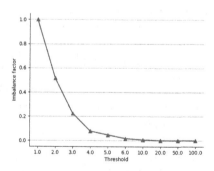

Fig. 8. Read balance at different thresholds

When $L_{NVM} \times T > min\,(L_{DRAM_1}, L_{DRAM_2})$, the query is sent to the less loaded node in $DRAM_1$ and $DRAM_2$, which serves to regulate the load balance of the two nodes.

When $L_{NVM} \times T \leq min(L_{DRAM_1}, L_{DRAM_2})$, the query is sent to NVM cache node and L_{nvm} gradually rises. When T is small, more queries need to be sent to NVM cache node to make $L_{NVM} \times T > min(L_{DRAM_1}, L_{DRAM_2})$ hold; when T is large, $L_{NVM} \times T > min(L_{DRAM_1}, L_{DRAM_2})$ will be satisfied faster, with more queries sent to the less loaded nodes in DRAM1 and DRAM2, regulating the load balance of both. When $T = 10$, THRC is an approximation of the 3-left problem and the imbalance index reaches a smaller value. As T increases further, the THRC problem is gradually approximated as a 2-left problem, and the imbalance index remains basically unchanged.

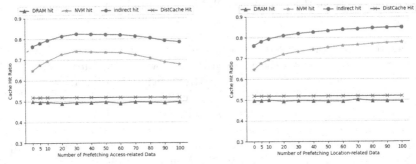

(a) Impact of the number of Prefetching Access-related Data

(b) Impact of the number of Prefetching Location-related Data

Fig. 9. Cache hit rate for different amounts of related data

Impact of Prefetching. Figure 9 illustrates how the cache hit rate in the prefetching mechanism is influenced by the quantity of prefetched data. Figure 9(a) and Fig. 9(b) depict the impact of prefetched access-pattern-related data and location-related data quantities on the cache hit rate, respectively. "DRAM hit" and "NVM hit" represent the proportions of queries sent to DRAM and NVM, respectively, that result in cache hits. "THRCache hit" denotes the proportion of all queries directed to THRCache that achieve cache hits, irrespective of whether they reside in DRAM or NVM. "DistCache hit" represents the cache hit rate specific to DistCache. In our experimental investigation of the influence of the quantity of prefetched location-related data on the cache hit rate, we intentionally refrain from prefetching access-related data to eliminate its impact, and vice versa. With an increasing quantity of prefetched related data, the DRAM cache hit rate remains consistent, while the NVM cache hit rate rises, consequently elevating the overall THRCache cache hit rate. Although the DRAM cache hit rate in THRCache closely approximates that of DistCache, the overall system cache hit rate experiences a significant increase due to the incorporation of the NVM cache and prefetching mechanism.

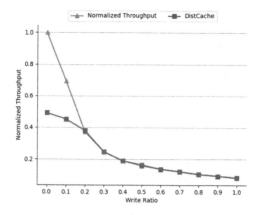

Fig. 10. System throughput at different write ratios

Impact of Write Ratio. Figure 10 shows the variation of THRCache and DistCache throughput for different write ratios. As the write ratio increases, the throughput of both gradually decreases, and eventually Storage Server becomes the bottleneck of the system and the throughput of both converges. In the case of lower write ratio, THRCache only handles one more copy consistency than DistCache, but thanks to THRC mechanism and prefetching strategy based on location and semantic correlation, the throughput of THRCache is still higher than DistCache.

6   Conclusion

In this paper, we present THRCache, a DRAM-NVM hybrid distributed caching mechanism that provides load balancing for heterogeneous cache nodes and improve the space utilization of NVM. THRCache utilizes the THRC mechanism to solve the cache allocation and cache routing problems of heterogeneous nodes and uses the prefetching mechanism based on the DRAM load and data popularity in NVM cache to improve the space utilization making full use of the high speed of DRAM and the large capacity of NVM. We implemented a prototype of THRCache on the Redis KV database and evaluated the throughput performance and cache hit ratio improvement of THRCache.

Acknowledgements. Please place your acknowledgments at the end of the paper, preceded by an unnumbered run-in heading (i.e. 3rd-level heading).

References

1. Atikoglu, B., Xu, Y., Frachtenberg, E., Jiang, S., Paleczny, M.: Workload analysis of a large-scale key-value store. In: Proceedings of the 12th ACM SIGMETRICS/PERFORMANCE Joint International Conference on Measurement and Modeling of Computer Systems, pp. 53–64 (2012)
2. Memcached (2023). http://memcached.org/. Accessed 17 May 2023
3. Redis (2023). http://redis.io/. Accessed 17 May 2023
4. Manes, B.: A high performance caching library for Java 8 (2016). https://github.com/ben-manes/caffeine
5. Luck, G., Suravarapu, S., King, G., Talevi, M.: Ehcache distributed cache system. http://www.ehcache.org/. Accessed 17 May 2023
6. Fan, B., Lim, H., Andersen, D.G., Kaminsky, M.: Small cache, big effect: provable load balancing for randomly partitioned cluster services. In: Proceedings of the 2nd ACM Symposium on Cloud Computing, pp. 1–12 (2011)
7. Liu, Z., et al.: DistCache: provable load balancing for large-scale storage systems with distributed caching. In: FAST. vol. 19, pp. 143–157 (2019)
8. Cai, Z., Lin, J., Liu, F., Chen, Z., Li, H.: NVMCache: wear-aware load balancing nvm-based caching for large-scale storage systems. In: 2020 IEEE Intl Conf on Parallel & Distributed Processing with Applications, Big Data & Cloud Computing, Sustainable Computing & Communications, Social Computing & Networking (ISPA/BDCloud/SocialCom/SustainCom), pp. 657–665. IEEE (2020)
9. Liu, H.K., et al.: A survey of non-volatile main memory technologies: state-of-the-arts, practices, and future directions. J. Comput. Sci. Technol. 36, 4–32 (2021)
10. Chen, H., Ruan, C., Li, C., Ma, X., Xu, Y.: SpanDB: a fast, cost-effective LSM-tree based KV store on hybrid storage. In: FAST. vol. 21, pp. 17–32 (2021)
11. Li, C., Chen, H., Ruan, C., Ma, X., Xu, Y.: Leveraging NVMe SSDs for building a fast, cost-effective, LSM-tree-based KV store. ACM Trans. Storage (TOS) 17(4), 1–29 (2021)
12. Dong, S., Kryczka, A., Jin, Y., Stumm, M.: RocksDB: evolution of development priorities in a key-value store serving large-scale applications. ACM Trans. Storage (TOS) 17(4), 1–32 (2021)
13. Eisenman, A., et al.: Flashield: a hybrid key-value cache that controls flash write amplification. In: NSDI, pp. 65–78 (2019)
14. Liu, J., Chai, Y., Qin, X., Xiao, Y.: PLC-cache: endurable SSD cache for deduplication-based primary storage. In: 2014 30th Symposium on Mass Storage Systems and Technologies (MSST), pp. 1–12. IEEE (2014)
15. Jiang, D., Che, Y., Xiong, J., Ma, X.: uCache: a utility-aware multilevel SSD cache management policy. In: 2013 IEEE 10th International Conference on High Performance Computing and Communications & 2013 IEEE International Conference on Embedded and Ubiquitous Computing, pp. 391–398. IEEE (2013)
16. Yoon, S.K., Youn, Y.S., Kim, J.G., Kim, S.D.: Design of DRAM-NAND flash hybrid main memory and Q-learning-based prefetching method. J. Supercomput. 74, 5293–5313 (2018)
17. Ozawa, K., Hirofuchi, T., Takano, R., Sugaya, M.: Fogcached: a DRAM/NVMM hybrid KVS server for edge computing. IEICE Trans. Inf. Syst. 104(12), 2089–2096 (2021)
18. Mittal, S., Vetter, J.S.: A survey of software techniques for using non-volatile memories for storage and main memory systems. IEEE Trans. Parallel Distrib. Syst. 27(5), 1537–1550 (2015)

19. Xia, F., Jiang, D., Xiong, J., Sun, N.: HiKV: a hybrid index Key-Value store for DRAM-NVM memory systems. In: 2017 USENIX Annual Technical Conference (USENIX ATC 17), pp. 349–362. USENIX Association, Santa Clara, CA (2017)
20. Kim, M., Kim, B.S., Lee, E., Lee, S.: A case study of a dram-nvm hybrid memory allocator for key-value stores. IEEE Comput. Archit. Lett. **21**(2), 81–84 (2022)
21. Li, Y., et al.: A multi-hashing index for hybrid dram-nvm memory systems. J. Syst. Architect. **128**, 102547 (2022)
22. Mitzenmacher, M.: The power of two choices in randomized load balancing. IEEE Trans. Parallel Distrib. Syst. **12**(10), 1094–1104 (2001)
23. Wang, S., Luo, J., Wong, W.S.: Improved power of two choices for fat-tree routing. IEEE Trans. Netw. Serv. Manage. **15**(4), 1706–1719 (2018)
24. Wang, A., et al.: InfiniCache: exploiting ephemeral serverless functions to build a cost-effective memory cache. In: Proceedings of the 18th USENIX Conference on File and Storage Technologies, pp. 267–282 (2020)
25. Rashmi, K., Chowdhury, M., Kosaian, J., Stoica, I., Ramchandran, K.: EC-Cache: load-balanced, low-latency cluster caching with online erasure coding. In: Osdi. vol. 16, pp. 401–417 (2016)
26. Zhang, M., Wang, Q., Shen, Z., Lee, P.P.: POCache: toward robust and configurable straggler tolerance with parity-only caching. J. Parallel Distrib. Comput. **167**, 157–172 (2022)
27. Vöcking, B.: How asymmetry helps load balancing. J. ACM (JACM) **50**(4), 568–589 (2003)
28. Berenbrink, P., Czumaj, A., Steger, A., Vöcking, B.: Balanced allocations: the heavily loaded case. In: Proceedings of the Thirty-Second Annual ACM Symposium on Theory of Computing, pp. 745–754 (2000)
29. Li, J., Nelson, J., Michael, E., Jin, X., Ports, D.R.: Pegasus: tolerating skewed workloads in distributed storage with {In-Network} coherence directories. In: 14th USENIX Symposium on Operating Systems Design and Implementation (OSDI 20), pp. 387–406 (2020)
30. Takruri, H., Kettaneh, I., Alquraan, A., Al-Kiswany, S.: {FLAIR}: Accelerating reads with {Consistency-Aware} network routing. In: 17th USENIX Symposium on Networked Systems Design and Implementation (NSDI 20), pp. 723–737 (2020)

Efficient Multi-tunnel Flow Scheduling for Traffic Engineering

Renhai Xu, Wenxin Li(✉), and Keqiu Li

Tianjin Key Laboratory of Advanced Networking, College of Intelligence and Computing, Tianjin University, Tianjin, China
toliwenxin@tju.edu.cn

Abstract. Network link failures can cause heavy congestion and skewed bandwidth usage. It is crucial for traffic engineering to switch multi-tunnel flow from faulty to non-faulty tunnels. Most existing studies on network link failures focus on achieving optimal network performance in terms of bandwidth usage, network congestion, and reliability, while neglecting to adequately consider flow completion time for applications. We propose *AptTE*, an efficient scheduler for multi-tunnel flows in traffic engineering aimed at reducing flow completion time and improving bandwidth utilization, even in the event of network link failure. The essence of *AptTE* lies in employing minimum impact scheduling to minimize the overall impact of scheduling one flow on subsequent flow scheduling. Simulation results demonstrate that *AptTE* enhances bandwidth utilization, reduces average completion time for multi-tunnel flows, and exhibits superior adaptability to network link failures compared to alternative approaches.

Keywords: Multi-tunnel flow · Minimum impact · Traffic engineering

1 Introduction

Network is expected to provide good performance even if there is a network failure. Thus, the dynamic adjustment of traffic distribution across network paths is crucial to traffic engineering (TE) and has been extensively researched in numerous studies [1–14]. Many service providers (e.g. Amazon, Google, Microsoft) select transport paths and allocate bandwidth for their application traffic via a centralized controller [2,7,8,13,15]. These application flows frequently present multiple tunnels that can be chosen for transportation. Therefore, the scheduling of multi-tunnel flows has become a pressing issue for TE [1,2,16].

Most prior research has focused on network congestion and reliability [1,2, 7,8,17–19], overlooking the performance of the application. Such as FFC [1], guarantees bandwidth reservation for application flows with less than K failures in network links, while ensuring a congestion-free network. However, the likelihood of a network link failure in practice is minimal, with a probability of less than 1%. As a result, FFC conservatively reserves bandwidth, hindering the full utilization of the network's bandwidth resources.

© The Author(s), under exclusive license to Springer Nature Singapore Pte Ltd. 2024
Z. Tari et al. (Eds.): ICA3PP 2023, LNCS 14490, pp. 456–473, 2024.
https://doi.org/10.1007/978-981-97-0859-8_27

Additionally, the existing heuristics for scheduling flows, namely SFF (smallest flow first) and SEF (shortest expected time first) [20–22], are inadequate for multi-tunnel flow scheduling. This is because multi-tunnel flow has multiple optional transport paths. Multiple tunnels may intersect within the flow, while tunnels in different flows can also intersect with each other. For complex applications in which multiple tunnel flows share and compete for the network's bandwidth resources, utilizing the shortest- or smallest-first policy to minimize the average flow completion time (FCT) is inappropriate.

Therefore, this paper proposes *AptTE* to schedule multi-tunnel flow with the goal of reducing flow completion time and improving network bandwidth utilization. The key of *AptTE* is based on a minimum impact scheduling policy, i.e., scheduling one multi-tunnel flow to minimize the impact on the total completion time of subsequent flows. In large-scale simulations, *AptTE* speeds up the average FCT by up to 3.97× and improves bandwidth utilization by 2.02×, compared to the FFC(k=1) [1].

In summary, the main highlights of this paper include:

- First, we proposed a minimal cut mathematical model to allocate bandwidth for multi-tunnel flow. Furthermore, we argue and analyze that scheduling multi-tunnel flows to reduce FCT and improve bandwidth utilization is difficult when utilizing traditional heuristic approaches such as shortest- or smallest-first, i.e., SFF and SEF.
- We formalize the impact of multi-tunnel flow scheduling and present a minimum impact scheduling algorithm, which we refer to as *AptTE*. It minimizes the total impact of scheduling one multi-tunnel flow on subsequent multi-tunnel flows to reduce the average FCT and improve network bandwidth utilisation.
- We have conducted large-scale simulations to evaluate the performance of *AptTE*, in terms of reducing the average FCT and improving network utilization. The results demonstrate that *AptTE* achieves significantly performance than other methods. Additionally, the results of the experiment indicate that *AptTE* better accommodates potential network failures than FFC(k=1), SFF, and SEF.

2   Motivation

FFC [1] switches flow f from the failed tunnel to the non-failed tunnel by reserving bandwidth. For example, if a flow f has three tunnels, each with a bandwidth capacity of $10Gb/s$. And if the FFC wants to guarantee a congestion-free network with at most one damaged tunnel of f, then the FFC will reserve $20Gb/s$ of bandwidth for flow f, so that no matter which tunnel fails, flow f will also get $20Gb/s$. However, it has been observed over time that the link is normal 99% of the time and fails 1% of the time. The bandwidth reserved by FFC for flow f is too conservative, resulting in a maximum bandwidth of $20Gb/s$ for f at 99% of the time, when in fact the maximum bandwidth available is $30Gb/s$. This is illustrated by an example below.

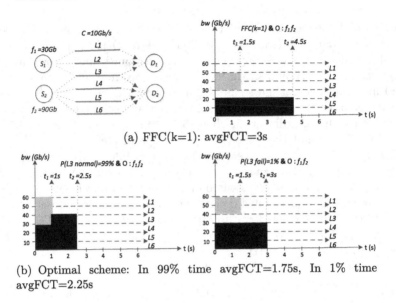

(a) FFC(k=1): avgFCT=3s

(b) Optimal scheme: In 99% time avgFCT=1.75s, In 1% time avgFCT=2.25s

Fig. 1. A motivating example, where (a) shows the FFC scheme and (b) shows the optimal scheme.

In the example, flow f_1 is a flow that transmits data from S_1 to D_1 with a data size of $30Gb$ and f_1 has three non-intersecting tunnels that pass through links $L1$, $L2$ and $L3$ respectively. Similarly, f_2 is a flow from S_2 to D_2 with a data size of $90Gb$ and has four disjoint tunnels passing through links $L3$, $L4$, $L5$ and $L6$ respectively. Flow f_1 and f_2 both start at the $0s$ and they compete for bandwidth on link $L3$, which has a bandwidth capacity of $10Gb/s$. As shown in Fig. 1(a), FFC(k=1) scheduling flows in the case of at most one failure link. Considering that the data size of f_1 is less than f_2, so that the optimal scheduling order is $O : f_1 f_2$. Flow f_1 gets priority usage of link $L3$ and the maximum reserved bandwidth of $20Gb/s$. It ends at $t_1 = 1.5s$. Then flow f_2 can reserve a maximum bandwidth of $20Gb/s$, completes at $t_2 = 4.5s$. Finally, the average completion time of these two flows is $avgFCT_{FFC(k=1)} = 3s$.

However, for a long period of time, $L3$ is normal state in 99% of the time and failed in the remaining time. As a result, FFC(k=1) is too conservative to reserve bandwidth for flows f_1 and f_2 when at most one link fails. As shown in Fig. 1(b), with $L3$ at 99% of normal time, the optimal scheduling for f_1 and f_2 is shown on the left. Flow f_1 uses $L3$ first and gets maximum bandwidth of $30Gb/s$. It ends at $t_1 = 1s$. Then f_2 can use the bandwidth on $L3$. So that, flow f_2 gets maximum bandwidth of $30Gb/s$ until $t_1 = 1s$, then changes to $40Gb/s$. f_2 ends at $t_2 = 2.5s$. In this case, their average completion time is $avgFCT_{L3\ normal} = 1.75s$. Similarly, if $L3$ is failed, under the optimal scheduling, the completion times of f_1 and f_2 are $t_1 = 1.5s$ and $t_2 = 3s$. As a result, their average completion time is $avgFCT_{L3\ fail} = 2.25s$.

From the above example, it is easy to see that regardless of whether the link $L3$ is in a normal or faulty state, the average completion time of the optimal scheduling method is smaller than that of the FFC(k=1) method. The main reason is that FFC(k=1) reserves bandwidth for flows conservatively, and does not make full use of the network bandwidth. Based on these problems, we focus on how to schedule multi-tunnel flows in a WAN where network links are likely to fail. In order to achieve the goal of fully utilising the network bandwidth and reducing the average completion time of flows, we propose a algorithm *AptTE* based on minimum impact scheduling.

3   *AptTE* overview

For a brief introduction to *AptTE*, we divide it into two parts, as shown in Fig. 2. One part is the *Agent* on the traffic sender server and the other part is the traffic scheduler *Master* on the network central controller.

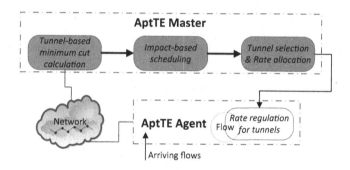

Fig. 2. An overview of *AptTE* architecture.

The process of running *AptTE*. First, flows arrive at the *Agent*, and *Agent* relays flow information (the size and tunnels) over the network to the *Master*. Then, *Master* calculates the minimum cut (i.e., the maximum available bandwidth) for each flow based on the network's link state and the flow information. Next, *Master* calculates the impacts of all flows, and prioritizes scheduling the least impact flow. Next, *Master* selects the corresponding tunnels for the minimum-impact flow and calculates the bandwidth of each tunnel. Finally, *Master* informs *Agent* of the tunnels and their bandwidth rates needed for minimum-impact flow, and *Agent* performs the rate adjustment of the tunnels.

3.1   Minimum Cut Calculation Based on Tunnels

In this paper, a flow $f(s, d)$ is a traffic that transmits data from a start point s to an end point d, with one or more tunnels. All of the tunnels in a flow can form a local network. How to achieve the maximum available bandwidth from the small local network of a flow, i.e., the minimum cut, is the first task of the *AptTE Master*. If using traditional min-cut max-flow algorithms such as the *Dinitz* and *Edmonds-Karp* algorithms, however, the augmented paths are not necessarily the tunnels of the flow. Thus, the min-cut is not necessarily the maximum available bandwidth of the flow. In order to obtain the minimum cut based on tunnels, *AptTE* proposes the corresponding mathematical model for the case of possible failure of the network link.

3.2   Minimum Impact Scheduling

The scheduling order of flows affects the allocation of network bandwidth and the average completion time of flows. In order to make decisions about the scheduling order of flows, we first build a general mathematical model for scheduling multi-tunnel flows. And then we analyze the difficulty of solving the mathematical model, which is an NPC problem but is difficult to solve due to the complex and large scale constraints. Finally, We propose a heuristic scheduling algorithm based on the minimum impact scheduling. We observe that regardless of the traffic scheduling method, scheduling one flow has an impact on the subsequent flows' scheduling. And therefore, *AptTE* minimizes the impact in order to improve bandwidth utilization and reduce the average completion time of flows.

4   Minimum Cut Based on Tunnel

In this section, we present the minimum cut calculation for multi-tunnel flow. The minimum cut of a flow refers to the maximum bandwidth that the flow can currently obtain. At each scheduling cycle, *AptTE* controller calculates the bandwidth demand for each flow by minimum cut.

4.1   Problems with Traditional Minimum Cut Algorithms

The traditional algorithm for solving the minimum cut problem is equivalent to solving the maximum flow problem. Such as *Dinitz* and *Edmonds-Karp* algorithms, both search for augmented paths until no more augmented paths can be searched. And the sum of the available bandwidth on all augmented paths is taken as the minimum cut of the network graph. However, the traditional algorithms are not applicable to multi-tunnel flow. First, the augmented path is not necessarily the tunnel of the flow. Second, even if the traditional algorithm searches for augmented paths along the tunnels of the flow, it does not necessarily obtain the minimum cut due to tunnels in a flow may intersect each other.

4.2   Minimum Cut Mathematical Model

Through the analysis of traditional minimum cut algorithms, it is concluded that even when finding augmented paths along the tunnels of a flow, the exact minimum cut of the multi-tunnel flow is not always obtained. With this challenge,

Table 1. Symbols and definitions.

Symbol	Definition
V	the set of network equipments (e.g., switch, router, machine) in WAN
E	the set of network links
T	the set of time slots
C_e	the bandwidth capacity on the outgoing/incoming link of each $e \in E$
C_e^t	the available bandwidth on the outgoing/incoming link of each $e \in E$ at time slot $t \in T$
q_e	$q_e = 1$ denotes the state of link $e \in E$ is fail, otherwise, the state of link e is normal
p_e	the probability of state $q_e = 1$
$f_{i,j}$	the flow f from the site i to j $i,j \in V$
$S_{i,j}, F_{i,j}$	the start time and the end time of flow $f_{i,j}$
$FCT_{i,j}$	the completion time of flow $f_{i,j}$, $FCT_{i,j} = F_{i,j} - S_{i,j}$
$D_{i,j}$	the flow volume of $f_{i,j}$
$\mathcal{K}_{i,j}$	the tunnels set of flow $f_{i,j}$
$x_{i,j}^{k,t}$	the number of bandwidth units allocated to the tunnel $k \in \mathcal{K}_{i,j}$ at time slot $t \in T$
$Q_{i,j}^{k,t}$	the state of tunnel k at time slot t $k \in \mathcal{K}_{i,j}$, $t \in T$, $0 \le Q_{i,j}^{k,t} \le 1$
$\delta(e \in k)$	$\delta(e \in k) = 1$, $e \in E$ is a link of tunnel $k \in \mathcal{K}_{i,j}$, otherwise, link e is not belong to tunnel k

we mathematically formalise the minimum cut problem. As shown in Table 1, the relevant notations are defined.

Then, the minimum cut problem for the flow $f_{i,j}$ at time t can be mathematically formalized as the following mathematical model:

$$\text{Maximize}_{\boldsymbol{x}} \quad \sum_{k \in \mathcal{K}_{i,j}} x_{i,j}^{k,t} \tag{1}$$

Subject to:

$$\sum_{k \in \mathcal{K}_{i,j}} \delta(e \in k) x_{i,j}^{k,t} \le C_e^t, \ \forall e \in E \tag{2}$$

$$0 \le x_{i,j}^{k,t}, \ \forall k \in \mathcal{K}_{i,j} \tag{3}$$

The $\sum_{k \in \mathcal{K}_{i,j}} x_{i,j}^{k,t}$ in the objective function (1) is the sum of the bandwidths of all tunnels of $f_{i,j}$, so that the objective function solves for the maximum bandwidth of $f_{i,j}$. The conditional inequality (2) states that the sum of the bandwidths of all tunnels of $f_{i,j}$ on link e cannot exceed the available bandwidth of link e. The conditional inequality (3) indicates that the bandwidth of any tunnel for traffic $f_{i,j}$ cannot be negative. If we consider that some links in the WAN may fail at time t, then the conditional inequality (2) would be modified as follows:

$$\sum_{k \in \mathcal{K}_{i,j}} \delta(e \in k) \cdot Q_{i,j}^{k,t} \cdot x_{i,j}^{k,t} \leq C_e^t, \ \forall e \in E \tag{4}$$

$$Q_{i,j}^{k,t} = \prod_{e \in k} (q_e^t p_e^t + (1 - q_e^t)(1 - p_e^t)), \ \forall k \in \mathcal{K}_{i,j} \tag{5}$$

where the conditional inequality (4) added to $Q_{i,j}^{k,t}$ denotes the state of tunnel k of the flow $f_{i,j}$ at time t, which is determined by the state of all network links that constitute tunnel k. Equation (5) reflects the state of tunnel k at time t by the state of all the links of the tunnel k in $f_{i,j}$.

Computational complexity of the minimal cut mathematical model. Typically, the number of tunnels ($|\mathcal{K}_{i,j}|$) of traffic $f_{i,j}$ is not very large in a wide area network, and hence the number of variables $x_{i,j}^{k,t}$ in the mathematical model is not very large. Moreover, the length of any tunnel of $f_{i,j}$ is finite and not very large, not exceeding ($|V| - 1$), so that the number of conditional inequalities (2) or (4) does not exceed $|\mathcal{K}_{i,j}| \times (|V| - 1)$. Thereby, the minimum cut mathematical model is small in size and is a linear programming problem that can be solved in a very short time.

5   *AptTE* Scheduling

This section first mathematically formalises the scheduling problem for multi-tunnel flows into a mathematical model. The mathematical model is an NPC polynomial-time solvable problem, however, it is difficult to solve in time-sensitive networks due to the large size of the conditional inequalities. For this reason, we focus on heuristic scheduling algorithms, and propose the minimum impact scheduling algorithm.

5.1   Mathematical Model for Multi-tunnel Flow Scheduling

Multi-tunnel flows compete with each other for network bandwidth resources, making flow scheduling critical for bandwidth utilization and average time to completion of flows. For this reason, we first build a mathematical model for the multi-tunnel flow scheduling problem. Before mathematically formalising the flow scheduling problem, we move on to introduce a few key notational definitions. As shown in Table 1, the start time of flow $f_{i,j}$ is defined as $S_{i,j}$ and the end time is defined as $F_{i,j}$, then the transmission completion time of flow $f_{i,j}$ is $FCT_{i,j} = F_{i,j} - S_{i,j}$. The data size of the flow $f_{i,j}$ is defined as $D_{i,j}$. The capacity bandwidth of the network link is defined as C_e. Generally, the multi-tunnel flow scheduling problem can then be formalised in a mathematical model as:

$$\text{Minimize}_{x} \quad \sum_{i \in V} \sum_{j \in V} FCT_{i,j} \tag{6}$$

Subject to:

$$\sum_{t=S_{i,j}}^{F_{i,j}} \sum_{k \in \mathcal{K}_{i,j}} Q_{i,j}^{k,t} \cdot x_{i,j}^{k,t} \geq D_{i,j}, \ \forall i,j \in V \tag{7}$$

$$\sum_{i,j \in V} \sum_{k \in \mathcal{K}_{i,j}} \delta(e \in k) x_{i,j}^{k,t} \leq C_e, \ \forall e \in E, \ \forall t \in \mathcal{T} \tag{8}$$

$$0 \leq x_{i,j}^{k,t}, \ \forall i,j \in V, \ \forall k \in \mathcal{K}_{i,j}, \ \forall t \in \mathcal{T} \tag{9}$$

$$Q_{i,j}^{k,t} = \prod_{e \in k} (q_e^t p_e^t + (1 - q_e^t)(1 - p_e^t)), \ \forall i,j \in V, \ \forall k \in \mathcal{K}_{i,j}, \ \forall t \in \mathcal{T} \tag{10}$$

The objective function (6) solves for the minimum total completion time of all flows. And the variable is the bandwidth per tunnel for each flow at time t. Conditional inequality (7) is a data volume constraint stating that the total bandwidth of the flow at all times needs to be greater than or equal to the data volume of the flow. Conditional inequality (8) is the bandwidth capacity constraint, stating that the total bandwidth of flows on link e at any time must not be greater than the capacity bandwidth of link e. Conditional inequality (9) ensures that the bandwidth allocated to each tunnel for each flow is a non-negative number. Equation (10) is the state condition that indicates the state of tunnel k by uniting the states of all links on tunnel k. It is noting that the state of link e can be expressed as $q_e^t p_e^t + (1 - q_e^t)(1 - p_e^t)$;

This mathematical model is difficult to solve directly. First, the total number of variables, $|V|^2 \times |\mathcal{K}|_{max} \times |\mathcal{T}|$, maybe a large number. This is because, while the maximum number of tunnels $|\mathcal{K}|_{max}$ and nodes number $|V|$ are small but $|\mathcal{T}|$ maybe large. Second, the conditional inequalities are more complex, such as inequality (7), which contains the completion time of the flow in the objective function. In fact, it has been shown that the mathematical model is a polynomial time computable NPC problem. However, the size of the constraints and their complexity make the problem difficult to solve directly.

Due to the difficulty of solving the model directly, we consider some heuristics to approximate the solution. There are two common heuristics for flow scheduling: one is the smallest flow first scheduling SFF, and the other is the shortest expected completion time first scheduling SEF. The expected completion time is equal to the amount of data remaining in the flow divided by the maximum bandwidth currently available. However, SFF and SEF are not optimal heuristics. This is because they schedule multi-tunnel flow f_1 in such a way that multi-tunnel flow f_2 can transmit data through a tunnel that does not intersect f_1. This indicates that scheduling f_1 has an impact on the scheduling of f_2, but not the maximum impact, because f_2 is still able to transmit data during scheduling f_1. In order to better approach optimal scheduling, this paper proposes *AptTE* minimum impact scheduling, which minimizes the total impact of scheduling a previous flow on the completion time of subsequent flows, and aims to improve network bandwidth utilization and reduce the average completion time of flows.

5.2   Minimum Impact Scheduling Algorithm

Before presenting *AptTE MIF*, we first mathematically formalize the impact of flow scheduling. In particular, the impact of scheduling flow f_i on subsequent flow f_j can be calculated as follows. Let the network be G before scheduling f_i. And the maximum bandwidth that can be offered to flow f_i, i.e. the minimum cut, is Cut_i. Then the expected completion time of f_i is $E_i = f_i/Cut_i$. Similarly, the maximum bandwidth that can be offered to flow f_j by network G is Cut_j. And the expected completion time for f_j is $E_j = f_j/Cut_j$. When scheduling traffic f_i, the network G allocates Cut_i bandwidth to traffic f_i such that f_i ends at the expected completion time E_i. At this point, defining the residual network as G'. Let the minimum cut of f_j in network G' be Cut_j'. Then the expected completion time E_j' of f_j after scheduling traffic f_i can be obtained by calculating as following:

$$E_j' = \begin{cases} E_i + \dfrac{f_j - E_i \times Cut_j'}{Cut_j}, & f_j > E_i \times Cut_j' \\[2ex] \dfrac{f_j}{Cut_j'}, & f_j \leq E_i \times Cut_j' \end{cases} \tag{11}$$

Thus, the impact of scheduling flow f_i on flow f_j is:

$$E_j' - E_j = \begin{cases} \dfrac{E_i \times (Cut_j - Cut_j')}{Cut_j}, & f_j > E_i \times Cut_j' \\[2ex] \dfrac{f_j \times (Cut_j - Cut_j')}{Cut_j \times Cut_j'}, & f_j \leq E_i \times Cut_j' \end{cases} \tag{12}$$

In the above equations, the minimum cuts of the flows f_i and f_j can both be solved by a linear programming mathematical model of the minimum cut.

There are three algorithms for minimum impact scheduling. Algorithm 1 computes the minimum cut of flow f and the transmission rate (bandwidth) of each tunnel at time t network G_t. Algorithm 2 picks a particular flow f such that scheduling f minimizes the impact on other flows at time t network G_t. Algorithm 3 continuously schedules the flows picked by algorithm 2 at time t network G_t. It allocates bandwidth for the selected flow f and updates network G_t. Then it informs *AptTE Agent* to perform rate adjustment for each tunnel of flow f. The details of the three algorithms are as follows:

Algorithm 1, calculates the minimum cut and bandwidth of each tunnel for the flow f at time t network G_t. In algorithm 1, $f.tunnel_\mu$ denotes the μ-th tunnel of flow f and $f.tunnel_\mu.bw$ denotes the rate of tunnel μ. The specific steps of algorithm 1 are as follows: Step 2, gets the minimum cut of f and the rate of each tunnel by the minimum cut model Eq. (1). Steps $3-8$, consider the relationship between the amount of data and the minimum cut of flow f.

Algorithm 1. RateOfTunnel

Input: Flow f, Network G_t
Output: The minimum cut Cut_f and the rate of each tunnel of the flow f at time slot t

1: **function** TUNNEL-RATE(Flow f, Network G_t)
2:     Get the rate of each $f.tunnel_\mu \in f$ and the minimum cut Cut_f by using the maths mode (1) on the network G_t
3:     $f.FCT = \frac{f.size}{Cut_f}$
4:     **if** $f.FCT > 1$ **then**
5:         $f.FCT = 1$
6:     **end if**
7:     **for** each $f.tunnel_\mu \in f$ **do**
8:         $f.tunnel_\mu.bw = f.tunnel_\mu.bw \times f.FCT$
9:     **end for**
10:     **return** f
11: **end function**

If $f.size \geq Cut_f$, then the network G_t is assigned to each tunnel of f at the rate calculated by the minimum cut model. Otherwise, the network G_t allocates $f.tunnel_\mu.bw \times f.FCT$ bandwidth to each tunnel μ.

Algorithm 2, picks a particular flow f such that scheduling f minimizes the total impact on other subsequent flows at time t and network G_t. Inside the algorithm, steps 2–7, solve for the minimum cut of each unselected flow and the rate of all their tunnels, and compute the expected completion time E_i of flow i. Steps 8–11, tentatively select a flow i as an alternative, calculate the minimum cut and the rate of each tunnel for flow i under G_t, and let the network G_t pre-allocate bandwidth to flow i, after which the network G_t will change to G'_t. Steps 12–16, calculate the total impact of scheduling flow i on subsequent unscheduled flows. Steps 17–23, find the special flow f that minimizes the total impact of scheduling f on all other flows.

Algorithm 3 At time t network initial G_t, step 2 uses algorithm 2 each time to select the flow f that has the least total impact on the rest of the subsequent flows. Steps 3–7, allocate bandwidth to f and update the network G_t to G'_t. Step 8, notifies *AptTE Agent* to perform rate adjustment for all tunnels of flow f. Steps 9–12, determine whether f is completed based on whether flow f has data left. Step 13, marks flow f as having been picked and scheduled. When all scheduling is finished at time t, step 15 saves the state of all flows.

With the above algorithms, at time t and initial network G_t, *AptTE* picks particular flow f at a time such that scheduling f has a minimum total impact on the completion time of other subsequent flows, thus achieving the goal of improving network utilization and reducing the average completion time of the flows. We perform a computational complexity analysis of *AptTE*. Assume that the total number of flows is n. The minimum cut cost of computing flow f is small and not considered. Then, the *AptTE* algorithm takes a total of n rounds to determine a scheduling sequence. The first round selects a particular flow f_1 from

---

Algorithm 2. MinImpFirst

Input: The set of flows F, Network G_t
Output: The flow f to be scheduled
1: **function** MIF(F, G_t)
2:     **for** $i \in F$ and $i.color == false$ and $i.end == false$ **do**
3:         TUNNEL-RATE(i, $G(t)$)
4:         $E_i = \frac{i.size}{Cut_i}$
5:     **end for**
6:     $f = NULL, f \in F$
7:     $MinTotalInf = INF$
8:     **for** $i \in F$ and $i.color == false$ and $i.end == false$ **do**
9:         TUNNEL-RATE(i, G_t)
10:         Suppose the network G_t allocates bandwidth to flow i, and then G_t changes to G'_t, but it does not notify the AptTE agent to re-adjust the rate of each $i.tunnel_\mu \in i$
11:         $TotalInf = 0$
12:         **for** $j \in F$ and $j \neq i$ and $j.color == false$ and $j.end == false$ **do**
13:             TUNNEL-RATE(j, G'_t)
14:             Use the equation (12) to calculate E'_j
15:             $TotalInf = TotalInf + (E'_j - E_j)$
16:         **end for**
17:         **if** $MinTotalInf > TotalInf$ **then**
18:             $MinTotalInf = TotalInf$
19:             $f = i$
20:         **end if**
21:         G'_t trace back to G_t
22:     **end for**
23:     **return** f
24: **end function**

---

n flows such that the total impact on the other subsequent flows is minimized, so that a total of n scheduling cases need to be viewed and $(n-1)$ impacts need to be computed for each scheduling case. As a result, the first round has computational complexity $n(n-1)$. The second round selects a particular f_2 from all remaining flows that are not f_1 such that the total impact on the other subsequent flows is minimized. There are $(n-1)$ scheduling cases need to be viewed, and $(n-2)$ impacts need to be computed for each scheduling case. As a result, the second round computational complexity is $(n-1)(n-2)$ ... and so on, then the total number of computations needed is $n(n-1) + (n-1)(n-2) + \cdots + 2 \times 1 = \{n^2 + (n-1)^2 + \cdots + 2^2\} - \{n + (n-1) + \cdots + 2\} = \frac{n(n+1)(2n+1)}{6} - \frac{n(n+1)}{2} = \frac{n^3 - n}{3}$, so that the computational complexity of the *AptTE* algorithm is $O(\frac{n^3-n}{3})$. In fact, there are not many concurrently scheduled multi-tunnel flows in a computation cycle of 5 minutes (n is not large), enables *AptTE* to meet most situations.

Algorithm 3. AptTE

Input: The set of flows F, The network G_t
Output: The rates of all tunnels of all flows at time slot t
1: **function** AP2TE($Timeslot\ t$)
2:     **while** f=MIF(F, G_t) is not NULL **do**
3:         TUNNEL-RATE(f, G_t)
4:         **for** $f.tunnel_\mu \in f$ **do**
5:             $f.size = f.size - f.tunnel_\mu.bw$
6:         **end for**
7:         Allocate bandwidth to flow f, Update network G_t
8:         Notify the AptTE agent to re-adjust the rate of each $f.tunnel_\mu \in f$
9:         **if** $f.size \leq 0$ **then**
10:             $f.end = true$
11:             $f.FCT = f.FCT + (t - f.arriveTime)$
12:         **end if**
13:         $f.color = true$
14:     **end while**
15:     **return** F
16: **end function**

---

6 Performance Evaluation

In this section, we conduct comprehensive simulations to evaluate the performance of *AptTE*. Our evaluation centers around the following questions:

- **How effective is *AptTE* in reducing FCT?** Through evaluation in various network topologies, *AptTE* achieves an average completion time reduction of 3.94×, 1.56× and 1.70× compared to FFC(k=1), SFF and SEF respectively.
- **How does *AptTE* perform in improving bandwidth utilization?** Our results demonstrate that *AptTE* improves bandwidth utilization by a factor of 2.02×, 1.83× and 1.93× compared to FFC(k=1), SFF and SEF respectively.
- **Is *AptTE* better adapted to situation where the network link is likely to fault?** By measuring tthe deviation in completion time for flows under normal and failed link conditions, *AptTE* shows smaller deviation than SFF and SEF, indicating better suitability in situations where network link failure is probable.

6.1 Simulation Setup

Network Topology And Workloads: We make reference to four different network topologies for TeaVaR [2], namely B4, IBM, ATT and MWAN. Specifically, the B4 network has 12 nodes and 38 links, the IBM network has 18 nodes and 48 links, the ATT network has 25 nodes and 112 links, and the MWAN network has 30 nodes and 75 links. In addition, the bandwidth capacity of the network link is set to $10Gb/s$. For the probability of network link failures, the same as in the TeaVaR [2], is obtained based on the Weibull distribution as a statistical over a long period of time. We generate workloads based on the Pareto distribution consistent with the work [16].

Comparing Schemes: We compare *AptTE* with the following three schemes:

- **FFC(k=1):** FFC(k=1) [1] conservatively reserves bandwidth to avoid congestion in the case of at most one link, leaving network bandwidth underutilized. The reason is that the reserved bandwidth is less than the available bandwidth of the network.
- **SFF:** Smallest Flow First Scheduling, a common traffic scheduling method, determines the order of flow based on the flow size. SFF fails to take into account the impact of scheduling a smaller flow on the subsequent larger flows, resulting in the possibility that scheduling larger flows first may be better for reducing average completion time.
- **SEF:** Shortest Expected First Scheduling, is similar to SEBF of Varys [20], which prioritizes the flow with the shortest expected completion time. SEF can reduce the average completion time of the flow to some extent. However, it also does not consider the impact of scheduling the flow with the shortest expected time on the subsequent flows, and there is a case that the flow with the longer expected time is better to be scheduled first.

Performance Metrics: For comparison, we define three performance metrics. First, the *Factor of Improvement* $= \frac{The\ avgFCT\ of\ the\ other\ scheme}{The\ avgFCT\ of\ AptTE}$. Second, the amount of data transferred per time on the network, called it *Throughput*. Third, the *Deviation value* $= \frac{avgFCT_{fail} - avgFCT_{normal}}{avgFCT_{fail}} \times 100\%$. The smaller the deviation, the more resilient the method is to network failures.

6.2   Simulation Results

We evaluate the performance of different algorithm schemes under different workloads and network topologies, with respect to FCT and bandwidth utilization. Furthermore, We use the deviation of the average completion time of flows in both normal and failed network environments to check whether *AptTE* can be more applicable to network failures.

Average FCT: Figure 3 first shows that the ratios of the other methods to *AptTE* exceeded 1 for both the average FCT and the 95th percentile FCT, indicating that *AptTE* effectively reduces the flow completion time. Where the ratio of FFC(k=1) to *AptTE* is the largest. The reason is the fact that FFC(k=1) conservatively reserves bandwidth for traffic, resulting in not fully utilizing the network bandwidth. Compared to FFC(k=1), *AptTE* is able to reduce the average FCT by a factor of 3.97 and the 95th percentile FCT by a factor of 2.49. By comparison, SEF has been found to have a greater improvement factor than SFF. The reason for this is that multi-tunnel flow scheduling has the potential for the tunnels of the flows to intricately intersect, making SEF scheduling of a

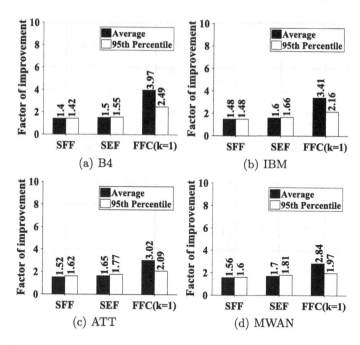

Fig. 3. The factor of improvements in average FCT and 95th percentile using SFF, SEF and FFC(k=1) over *AptTE*, across various workloads and network topologies.

flow with a shorter expected completion time have a greater direct impact on subsequent flows with longer expected completion times. However, both SFF and SEF fail to focus on the impact of scheduling one flow on other flows in the case of intersecting multi-tunnel flows. As a result, the average and 95th percentile flow FCTs are both longer than those of *AptTE*. In addition, the simulation results show that *AptTE* is applicable to different networks and traffic matrices.

CDF of per Flow FCT: To comprehend the microscopic FCT performance level, we plotted CDFs of per-flow FCT for various schemes across distinct workloads and network topologies, depicted in Fig. 4. The horizontal axis of each CDF curve is the FCT in timeslot, and each timeslot is for 5 min. The vertical axis is the ratio of the number of flows less than or equal to a certain completion time to the total number of flows. Due to the small number of high-data flows and longer completion times, the CDF curves in Fig. 4 display progressively decreasing slopes over time. By observation, the CDF curves of *AptTE* are all higher than SFF, SEF and FFC(k=1). This indicates that the number of flow completions in *AptTE* is greater than other schemes for a specified timeslot t with the same workload, network topology and link state. This validates that *AptTE* is effective in reducing the average FCT. In addition, the FFC(k=1) curve is the lowest, indicating that FFC(k=1) is at a disadvantage relative to the other schemes in reducing the FCT. The reason is still that FFC(k=1) adopts an overly

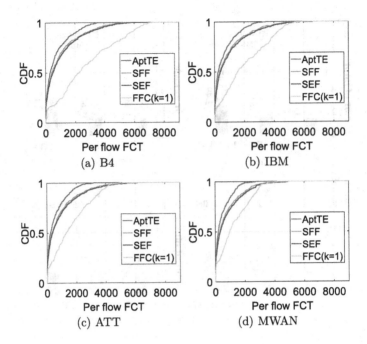

Fig. 4. CDFs of per flow FCT achieved by *AptTE*, FFC(k=1), SFF and SEF across various workloads and network topologies.

conservative bandwidth reservation mechanism and cannot fully utilize the bandwidth. For example, in Fig. 4(b), the percentage of completed flows transmitted within timeslots 0 to 2000 is 93.67%, 86%, 84% and 57.67% for *AptTE*, SFF, SEF, and FFC(k=1), respectively. This indicates that *AptTE* is able to transmit 93.67% of the flows to completion within the timeslot 2000, while FFC(k=1) still leaves more than 42% of the flows untransmitted. This once again validates *AptTE*'s effectiveness in reducing FCT. Furthermore, *AptTE* exhibits universal applicability due to its lower average flow completion time in comparison with the other three schemes across various network states and workloads.

Bandwidth utilization-*Throughput*: To test the potential impact of *AptTE* on network bandwidth utilization, we utilized data transfer rate as a measurement metric. Figure 5 illustrates our findings. First, the curves for all four different schemes show a decreasing trend. The reason for this is that the traffic in the network ends gradually over time and the total bandwidth used by the network decreases. Even so, the curve for *AptTE* is higher than the other schemes and ends earlier. This demonstrates that *AptTE* has the capability to enhance the network's bandwidth. Second, FFC(k=1) has the lowest curve and transmits the least amount of data per time slot. The conservative reservation of bandwidth for flow by FFC(k=1) contributes to its longest average FCT. Finally, the results presented in the four subfigures of Fig. 5 demonstrate that *AptTE* consistently outperforms other schemes in terms of the amount of data transmitted in each

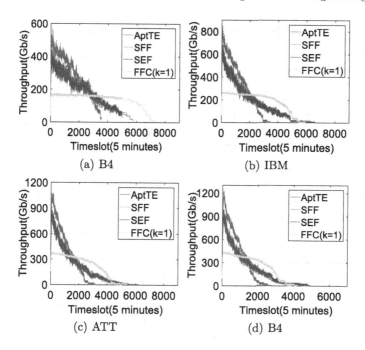

Fig. 5. The amount of data transferred in each timeslot by *AptTE*, FFC(k=1), SFF and SEF across various workloads and network topologies.

time slot, across various workloads and network topologies. Therefore, *AptTE* is an effective method for enhancing network throughput.

Effectiveness of *AptTE*'s error-tolerant design: We will now analyze the fault-tolerance performance of *AptTE* by measuring the difference in average FCT under normal conditions versus network link failure. To achieve this objective, we utilize *AptTE*, SFF, SEF, and FFC(k=1) for scheduling flows in both link failure and non-failure scenarios. The resulting deviation values of the average FCTs are displayed in Fig. 6. It is evident that FFC(k=1) has the smallest deviation value. The reason for this is that, in order to guarantee that the network is not congested even if at most one link fails, FFC(k=1) reserves a relatively stable bandwidth for each flow. In other words, the FFC(k=1) mechanism already takes into account the possibility of link failures in advance. Thus, the average completion time of the flows scheduled by FFC(k=1) remains largely unchanged, even in the event of link failures. Thereby, the value of deviation in FFC(k=1) is smallest. The deviation value of *AptTE* is the smallest except for FFC(k=1). This is because *AptTE* aims to reduce the overall scheduling impact on subsequent flows, regardless of link status, which is not a consideration in SFF or SEF. Furthermore, it can be inferred that *AptTE* is proficient in handling network failures as evidenced by its second smallest deviation value of the average FCT across all four network topologies.

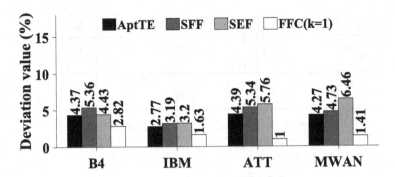

Fig. 6. Deviation values across various workloads and network topologies for *AptTE*, SFF, SEF and FFC(k=1).

7 Conclusions

In this paper, we present *AptTE*, a novel TE solution that schedules multi-tunnel flows to reduce the average flow completion time and improve network bandwidth utilization. In *AptTE* design, we apply the heuristic of minimizing the impact of scheduling one flow on subsequent flows. This approach facilitates increased bandwidth utilization of the network and reduces the average flow completion time for multi-tunnel flows. We conducted thorough simulations over four realistic network topologies to exhibit that *AptTE* is more efficient at reducing the average FCT and enhancing bandwidth utilization than the current TE schemes. It is argued that *AptTE* is better suited to handle network link failures.

Acknowledgments. We thank the reviewers of ICA3PP 2023, whose feedback helped improve this paper. This work was supported partly by the NSFC under Grant 62202325.

References

1. Liu, H.H., Kandula, S., Mahajan, R., et al.: Traffic engineering with forward fault correction. In: Proceedings of the 2014 ACM Conference on SIGCOMM, pp. 527–538 (2014)
2. Bogle, J., Bhatia, N., Ghobadi, M., et al.: TEAVAR: striking the right utilization-availability balance in WAN traffic engineering. In: Proceedings of the ACM Special Interest Group on Data Communication, pp. 29–43 (2019)
3. Akyildiz, I.F., Lee, A., Wang, P., Luo, M., Chou, W.: A roadmap for traffic engineering in SDN-OpenFlow networks. Comput. Netw. **71**, 1–30 (2014)
4. Alizadeh, M., et al.: CONGA: distributed congestion-aware load balancing for datacenters. In: ACM SIGCOMM, pp. 503–514 (2014)
5. Benson, T., Anand, A., Akella, A., Zhang, M.: MicroTE: fine grained traffic engineering for data centers. In: ACM CoNEXT (2011)
6. Fortz, B., Thorup, M.: Internet traffic engineering by optimizing OSPF weights. In: IEEE INFOCOM (2000)

7. Hong, C.-Y., et al.: achieving high utilization with software driven WAN. In: ACM SIGCOMM (2013)
8. Jain, S., et al.: B4: experience with a globally-deployed software defined WAN. In: ACM SIGCOMM (2013)
9. Jiang, W., Zhang-Shen, R., Rexford, J., Chiang, M.: Cooperative content distribution and traffic engineering in an ISP network. In: ACM SIGMETRICS (2009)
10. Kandula, S., Katabi, D., Davie, B., Charny, A.: Walking the tightrope: responsive yet stable traffic engineering. In: ACM SIGCOMM (2005)
11. Kandula, S., Menache, I., Schwartz, R., Babbula, S.R.: Calendaring for wide area networks. In: Bustamante, F.E., Hu, Y.C., Krishnamurthy, A., Ratnasamy, S. (eds.). ACM SIGCOMM (2014)
12. Kumar, P., et al.: Semi-oblivious traffic engineering: the road not taken. In: USENIX NSDI (2018)
13. Liu, H.H., Kandula, S., Mahajan, R., Zhang, M., Gelernter, D.: Traffic engineering with forward fault correction. In: ACM SIGCOMM (2014)
14. Zhang, H., et al.: Guaranteeing deadlines for inter-data center transfers. In: IEEE/ACM TON (2017)
15. Alok Kumar, et al.: BwE: flexible, Hierarchical Bandwidth Allocation forWAN Distributed Computing. In: ACM SIGCOMM (2015)
16. Kumar, P., et al.: Semi-oblivious traffic engineering: the road not taken. In: 15th USENIX Symposium on Networked Systems Design and Implementation (NSDI 18) (2018)
17. Al-Fares, M., Radhakrishnan, S., Raghavan, B., Huang, N., Vahdat, A.: Hedera: dynamic flow scheduling for data center networks. In: NSDI10
18. Curtis, A.R., Mogul, J.C., Tourrilhes, J., Yalag, P., Sharma, P., Banerjee, S.: DevoFlow: scaling flow management for high-performance networks. In: SIGCOMM11
19. Halperin, D., Kandula, S., Padhye, J., Bahl, P., Wetherall, D.: Augmenting data center networks with multi-gigabit wireless links. In: SIGCOMM11
20. Chowdhury, M., Zhong, Y., Stoica, I.: Efficient coflow scheduling with Varys. In: Proceedings of the 2014 ACM Conference on SIGCOMM (2014)
21. Alizadeh, M., et al.: pFabric: minimal near-optimal datacenter transport. In: SIGCOMM (2013)
22. Hong, C.-Y., et al.: Finishing flows quickly with preemptive scheduling. In: SIGCOMM (2012)

An Updatable Key Management Scheme for Underwater Wireless Sensor Networks

Zhiyun Guan⬥, Junhua Wu(✉)⬥, Guangshun Li, and Tielin Wang⬥

School of Computer Science, Qufu Normal University, Rizhao 276826, China
shdwjh@163.com

Abstract. In recent years, underwater wireless sensor networks (UWSNs) have emerged as promising network model for various marine exploration, detection, and protection applications. The problems of frequent node movement and high battery energy consumption have led to the design of terrestrial wireless sensor networks (TWSNs) that are not applicable to underwater. There is an urgent need to develop new and efficient security mechanisms. On the other hand, nodes drift with the current. It is extremely challenging to balance the energy constraints while ensuring secure and private data transfer between moving nodes. In this paper, we propose an updatable key management scheme for UWSNs. A group mobility model suitable for underwater is used in a network model based on a hierarchical structure. We process the communication through a symmetric polynomial-based key pre-distribution scheme. In addition, we design a dynamic key update scheme based on reverse hash chains to extend the network security survival time. The obtained results show that our approach saves storage and communication overhead while providing highly secure connectivity and good resistance to node capture.

Keywords: Underwater wireless sensor networks (UWSNs) · Key management · Security · Symmetric polynomials

1 Introduction

In recent years, the application of underwater wireless sensor networks (UWSNs) has become increasingly widespread with the development of hydroacoustic sensor network technology. UWSNs are networks of sensors deployed to perform collaborative monitoring tasks on a given body of water. These nodes are distributed underwater to sense water quality, pressure, temperature, and other relevant factors [1]. The sensed data can be used for oceanographic data collection

Supported by the National Natural Science Foundation of China (61832012, 12271295, 62072273), the Natural Science Foundation of Shandong Province with Grants (ZR2022MF304, ZR2021MF075, ZR2021QF050, ZR2019ZD10), and the Key Research and Development Program Project of Shandong Province (2022CXPT055, 2019GGX1050).

[2,3], water pollution monitoring [4], offshore exploration [5], disaster prevention [6,7], guidance navigatio [8], and tactical surveillance applications [9,10], etc., which can be used directly or indirectly for the benefit of mankind. However, the hardware limitations in UWSNs with their own restricted computing, communication, and storage capabilities and characteristics such as unreliable hydroacoustic channels, dynamic network structures, and weak link connectivity pose unique challenges to research. UWSNs are usually deployed in unattended or even extremely harsh environments, making the deployed sensor nodes highly vulnerable to various sabotages. Therefore, node security is one of the important research topics in UWSNs to ensure node collaboration and network reliability and has been a hot topic of interest for researchers in recent years [11].

Reliable key management is the cornerstone of security services in various wireless sensor networks (WSNs), and efficient key management is the top priority for securing other security techniques. The traditional public key algorithm, although computationally feasible, requires significant computing power and storage space, and it consumes about three orders of magnitude more energy than symmetric key encryption [12]. Therefore, using symmetric keys is one of the most suitable methods for implementing secure exchanges in UWSNs because of its low energy consumption and simple hardware requirements. Besides the security of the nodes that must be taken into account, another issue in UWSNs is mobility. Apart from some stationary nodes mounted on surface buoys, most underwater sensor nodes drift with them under the action of external factors such as currents, ocean currents, and winds [13]. It is empirically observed that underwater objects move at 2–3 knots (or 36 km/hr) under typical underwater conditions [5]. To thoroughly model the protocol used for UWSNs, it is necessary to use a mobility model that accurately represents the mobile nodes in the network.

The main aim of this paper is to improve the security and performance of UWSNs while taking into account the mobility of underwater nodes. We develop a suitable key management scheme for UWSNs and apply the group mobility model in a network model based on the hierarchical structure. In UWSNs that lack a priori knowledge, we use symmetric polynomial key pre-distribution to establish pairwise keys for the communicating parties, overcoming the drawback of using multivariate polynomials for key generation. In addition, we design a dynamic key update scheme based on time-slice mechanism and reverse hash chain to ensure that the network can operate securely for a long time. The performance analysis indicates that our approach is more suitable for application in UWSNs as it reduces communication overhead and storage overhead while ensuring sufficient security and network connectivity, and extends network lifetime compared to existing key management schemes.

The rest of this paper is organized as follows: Sect. 2 presents some related work, Sect. 3 presents our proposed scheme, Sect. 4 gives security and performance analysis, and finally, some conclusions are drawn in Sect. 5.

2 Related Work

With the application and development of WSNs, many researchers have developed various key management schemes. However, the underwater environment is more complex and dynamic, resulting in most of the techniques are not applicable to UWSNs. Due to the immaturity of hardware and networking technology, there is no complete system for key management research in UWSNs. It is necessary to propose a key management scheme suitable for UWSNs based on the one in WSNs, and key distribution is one of the core steps of key management.

Eschenauer and Gligor [14] proposed the first random key pre-distribution scheme, called the E-G scheme. In this scheme, each node is pre-loaded with k randomly selected keys from a large key pool P to form a key ring and the key's corresponding identifier. After deployment, each node constructs a secret session key by exchanging the identifier list of pre-distributed keys with each other and with neighboring nodes. This scheme offers improved energy efficiency compared to public key-based security schemes. However, a limitation of the E-G scheme is that different sensor nodes may store overlapping identical keys.

Chan et al. [15] improved the E-G scheme and proposed the q-composite random key pre-distribution scheme. This scheme requires at least q common pre-distribution keys between two nodes to establish a shared key, and the E-G scheme is a special case of the q-composite scheme ($q = 1$). And, instead of simply picking an identical public key as the shared key like the E-G scheme, this scheme uses some hash of all identical public keys as the shared key: $K_{i,j} = Hash\,(K_1 \parallel K_2 \parallel ... \parallel K_{q'})$, where $K_1, K_2, ...K_{q'}$ are the same $q'\,(q' \geq q)$ public key between nodes i and j. Although this scheme improves the network resilience, it reduces the network security connection coverage because two adjacent nodes share at least q keys to establish a secure link.

All the above schemes establish a secure connection between any two nodes in distributed WSNs. Recently, several scholars have conducted research on hierarchical WSNs. WSNs with a hierarchical structure only need to ensure secure connections between cluster heads and their members, which not only greatly reduces the communication overhead, but also has better network performance.

Shen et al. [16] proposed a scalable key pre-distribution (SKPD) scheme based on a three-layer hierarchical network model to design two symmetric polynomial functions for computing paired keys between cluster heads and between cluster heads and sensor nodes, respectively. The scheme can fully resist node capture attacks and provide network connectivity with reduced storage and communication overhead.

Gholizadeh et al. [12] proposed an efficient key distribution (EKDM) mechanism based on symmetric polynomials that implements node addition and key freshness and reduces energy consumption and computational overhead while guaranteeing high flexibility with appropriate security levels, but the scheme requires more storage overhead.

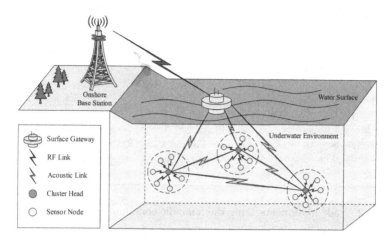

Fig. 1. Network architecture of UWSNs.

3   Proposed Key Management Scheme in UWSNs

In this section, we formalize the network model, nomadic mobility model, and use it as a basis to propose an updatable key management scheme for UWSNs.

3.1   Network Model

From practical considerations, a hierarchical network architecture with low communication cost is more suitable for UWSNs than a distributed network architecture [9,17]. In this paper, we propose a network model for UWSNs based on a hierarchical architecture, as shown in Fig. 1. To reduce overload and delay, we divide the network into multiple clusters and classify the nodes within the network into the base station, surface gateway, cluster heads, and sensor nodes according to their different functions.

Base Station (BS). Base station deployed on land, the main function is to manage the nodes within the network, along with being the final object of the data sensed by the sensor nodes. As a node manager, the base station is trusted by the nodes within the network and other nodes need to be initialized by the base station before joining the network.

Surface Gateway (SG). Surface gateway acts as a collector, aggregator, and forwarder of data from the underwater network in the proposed UWSNs and is the intermediary between the underwater and terrestrial networks, responsible for transmitting the data received from the cluster head to the base station located on land.

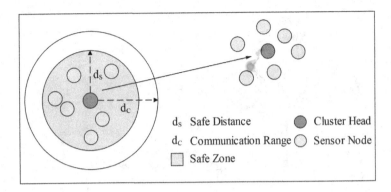

Fig. 2. Node movements using the nomadic community mobility model.

Cluster Head (CH). Cluster heads mainly used to simply process the data sensed by the sensor nodes in the cluster and pass the information to the surface gateway, with more powerful computing and storage capabilities.

Sensor Node (S). Sensor Nodes are deployed underwater to sense data from the nearby environment, store the collected data, and send it to the cluster head at regular intervals. Usually, sensor nodes can only sense and transmit simple data, their computational and storage capabilities are weak and their energy is limited to perform complex operations and data processing.

3.2 Nomadic Mobility Model

In our proposed scheme sensor nodes are deployed in water at different depths, so the nodes must be subject to external influences that cause changes in position. The most suitable community mobility model that fits these properties is the nomadic community mobility model. In this model each community moves to a place with its own reference point, while each node has its own personal space and each node is free to move within its range.

The nomadic community mobility model used in our scheme is shown in Fig. 2. The cluster head moves to reach a new location over a period of time with the effect of the wave shown by the solid arrow, and the sensor nodes within the cluster follow the movement of the cluster head. Once a sensor node drifts away from the safe zone of the cluster head, a connection to the new cluster head can be established at the next time interval using pre-loaded information.

3.3 Proposed Key Management Scheme

Based on the proposed network model, we will introduce the key management scheme for UWSNs in detail, which mainly includes five phases: system initialization, key pre-distribution, inter-cluster key establishment, intra-cluster key establishment and key update.

Initialization Phase. Before deployment, we divide the deployed UWSNs into n clusters, each of which contains a cluster head $\{CH_i\}_{i=1}^n$ and m sensor nodes $\{S_j\}_{j=1}^m$. First, BS chooses two distinct cryptographic hash functions: H_1/H_2 : $\{0,1\}^* \longrightarrow \{0,1\}^{len}$, where len is the bit length of the output. Then, a unique identity is assigned to each node, including itself, denoted as ID_*.

After that, two different t-degree symmetric polynomials are randomly generated over a finite field F_p, one is $f_{CH}(x,y,z) \in F_p[x,y,z]$ for establishing pairwise keys between CHs and one is $f_{CH_i}(x,y) \in F_p[x,y]$ for computing secret shares between CH_i and sensor nodes.

Finally, BS divides the time into a series of identical time intervals T using the time slicing mechanism and generates a time sequence $\{T_u\}_{u=0}^{l_{time}}$ where every two consecutive time points T_{u-1} and T_u form the u-th time interval T_u.

Key Pre-distribution Phase. In this phase, BS pre-loads different secret messages into different levels of nodes. For each cluster head $\{CH_i\}_{i=1}^n$ pre-loads before the deployment its own parameters: (i) the symmetric key K_{CH_i-SG} for communication exchange between CH_i and SG, (ii) the polynomial share $f_{CH}(ID_i,y,r)$ generated by the trivariate symmetric polynomial $f_{CH}(x,y,z)$, the cluster head identity ID_i, and the random number r generated by BS and made $z=r$, and (iii) the polynomial share $f_{CH_i}(ID_i,y)$ generated by the bivariate symmetric polynomial $f_{CH_i}(x,y)$ and the cluster head identity ID_i.

To minimize the key storage overhead for sensor nodes with limited hardware resources in UWSNs, in our proposed scheme, each sensor node S_j is pre-loaded with only one temporary key k_{S_j-CH} for communication with the physical cluster head (cluster head to which the sensor node belongs after network deployment), which is generated as follows:

BS randomly selects $c(c \geq 1)$ polynomials from n polynomials $\{f_{CH_i}(x,y)\}_{i=1}^n$ and needs a larger c in order to obtain sufficient security. For convenience, the proposed scheme sets $c=2$ and randomly selects polynomials $f_{CH_u}(x,y)$ and $f_{CH_v}(x,y)$. Then, evaluates the polynomials $f_{CH_u}(x,y)$ where $(x=ID_u, y=ID_j)$ and $f_{CH_v}(x,y)$ where $(x=ID_v, y=ID_j)$ to obtain the partial secret values s_1 and s_2 of the generated k_{S_j-CH} and pre-loads it into S_j. k_{S_j-CH} is calculated by performing the following equation:

$$s_1 = f_{CH_u}(ID_u, ID_j) \tag{1}$$

$$s_2 = f_{CH_v}(ID_v, ID_j) \tag{2}$$

$$k_{S_j-CH} = H_1(s_1 \oplus s_2) \tag{3}$$

After the key pre-distribution phase, different nodes in the UWSNs store different secret shares and keys. Considering that the sensor networks are arranged underwater in a complex environment and lack of a-priori knowledge of post-deployment configuration, i.e., it is not possible to know in advance which nodes will be within communication range of each other after deployment, we do not predict which cluster head's communication range the sensor nodes will belong to after deployment.

Inter-cluster Key Establishment Phase. In the inter-cluster key establishment phase, the cluster heads (CH_a and CH_b) interact to establish pairwise keys for authentication and message reliability between them. The process involves the following steps:

CH_a and CH_b exchange identity with each other. Then, CH_a calculate from ID_b and the pre-loaded polynomial share $f_{CH}(ID_a, y, r)$ to obtain $K_{CH_a - CH_b}$ by performing the following equation:

$$K_{CH_a - CH_b} = H_1 \left(f_{CH} \left(ID_a, ID_b, r \right) \right) \tag{4}$$

CH_b is calculated in the same way to obtain $K_{CH_b - CH_a}$ by performing the following equation:

$$K_{CH_b - CH_a} = H_1 \left(f_{CH} \left(ID_b, ID_a, r \right) \right) \tag{5}$$

Due to the symmetry of the polynomial known, thus CH_a and CH_b establish a unique pair of keys $K_{CH_a - CH_b}$ that are used to authenticate both parties and protect the communication data.

$$K_{CH_a - CH_b} = K_{CH_b - CH_a} \tag{6}$$

Intra-cluster Key Establishment Phase. This phase is performed by the cluster head and the sensor nodes and aims at establishing pairwise keys between each cluster head and the sensor nodes within the cluster to ensure intra-cluster communication. We assume that the cluster head CH_i and the sensor node S_j within the cluster need to establish symmetric keys as follows:

First, S_j sends its identity ID_j and the identity of its stored cluster head ID_u and ID_v to the physical cluster head CH_i. Upon receiving message, CH_i send ID_j to CH_u and CH_v, respectively, requesting the corresponding secret shares;

Upon receiving the request message, CH_u calculate from ID_j and pre-stored polynomials $f_{CH_u}(ID_u, y)$ to obtain the secret share s_1 and send encrypted s_1 to CH_i : $E_{K_{CH_i - CH_u}}(s_1)$. CH_v generates s_2 in the same way, encrypted and sent to the CH_i : $E_{K_{CH_i - CH_v}}(s_2)$.

CH_i decrypt with the pairwise key to obtain the secret share s_1 and s_2 respectively and compute the temporary key $k_{S_j - CH_i}$ for communication with S_j by Eq. (3).

Then, CH_i selects the corresponding element $e_{i,0}$ as the seed element of the hash function according to the time interval T_0, and computes a reverse hash chain consisting of a sequential set of hash values $\{R_{i,j}\}_{j=1}^{l_{hash}}$:

$$R_{i,j} = H_2 \left(R_{i,j+1} \right) \tag{7}$$

$$R_{i,l_{hash}} = H_2 \left(e_{i,0} \right) \tag{8}$$

where $1 \leq j \leq l_{hash} - 1$. The elements $R_{i,j}$ in the reverse hash chain are encrypted using the temporary key sent to S_j: $E_{k_{S_j - CH_i}}(R_{i,j})$, and the symmetric key with the sensor node S_j is computed:

$$K_{S_j - CH_i} = H_1 \left(k_{S_j - CH_i} \| R_{i,j} \right) \tag{9}$$

Finally, S_j decrypts the message using the pre-loaded temporary key and calculates the symmetric key between them by Eq. (9), and subsequently removes $R_{i,j}$.

Key Update Phase

Inter-cluster Key Update. To enhance the security of UWSNs, BS periodically updates the parameters r of the trivariate symmetric polynomial $f_{CH}(ID_i, y, r)$. The new random number updated by BS is noted as r' and is sent to the cluster heads through SG using symmetric key encryption r' with the cluster heads: $E_{K_{CH_i-SG}}(r')$. Cluster heads within the network update r' upon receipt, ensuring that they all update their respective polynomial shares in a timely manner.

Intar-cluster Key Update. To prevent the attacker from analyzing the ciphertext and maintain the security of communication links within each cluster, the session keys of the sensor nodes and the cluster head in each cluster need to be updated within the elapsed time interval T. After the time interval T has elapsed, CH_i selects the element $e_{i,u}$ corresponding to the time interval T_u as the input to the hash function and uses the hash value in the generated reverse hash chain for the communication link between the cluster and each sensor.

4 Security and Performance Analysis

We simulated using NS-2 V2.35, an open source extensible tool for network simulation with libraries that create the feel of a real-time environment for UWSNs. The algorithms were coded in C++ and added to the NS-2 platform. There are 20 clusters in this simulation and each cluster contains 50 sensor nodes, 1000 nodes are deployed in an area of 500 m * 500 m and the MAC layer protocol is IEEE 802.11.

4.1 Storage Overhead

Table 1 compares the storage overhead of different schemes. We assume that the number of cluster heads and sensor nodes in UWSNs are n and m, respectively, and usually we have $n \ll m$. Assume that each key occupies k space, and for storing any polynomial share the space is s. It can be seen that the storage overhead of EKPD and SKPD schemes is more than our scheme.

Figure 3 compares the storage overheads of different schemes in different network sizes. The overhead caused by key management does not increase significantly as the network size increases. In the E-G and q-composite schemes, each sensor node has to pre-store m keys, occupying $m*k$ space. The number of m depends on the key pool size P and the probability of sharing at least one key between two sensor nodes, and their key storage overhead becomes larger as the network size increases. In our scheme, each sensor node stores only one key after the network is deployed, regardless of the network size. The above analysis shows that our scheme has better scalability in terms of storage overhead.

Table 1. Storage overhead of different schemes.

	EKDM	SKPD	Proposed scheme
CH	$k + (2 * Q + 1) * s$	$k + 2 * s$	$k + 2 * s$
SN	$2 * k$	$k + s$	k

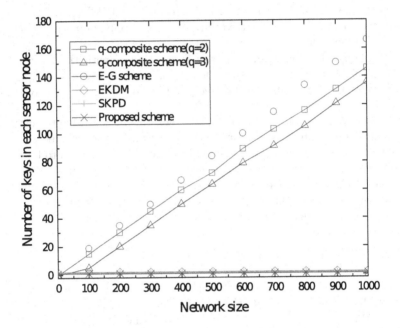

Fig. 3. Comparison of storage overhead.

4.2 Communication Overhead

Figure 4 compares the communication overhead of the different schemes. The SKPD scheme has the lowest communication overhead since all shares are assigned to each sensor node and cluster head. In our proposed scheme, the sensor node sends its ID and the two pre-loaded cluster head IDs to the physical cluster head, and the physical cluster head gets the key with a much shorter handshake message than most schemes. It can be seen that we incur less communication overhead compared to the EKPD scheme.

4.3 Security Analysis

Figure 5 compares the ability of different schemes to resist node capture attacks. It can be seen that in a general key pool-based scheme, the information held by the adversary about the global key pool increases as the number of captured sensor nodes increases. However, in schemes based on symmetric polynomials of suitable order, the adversary cannot obtain the symmetric key even if the secret

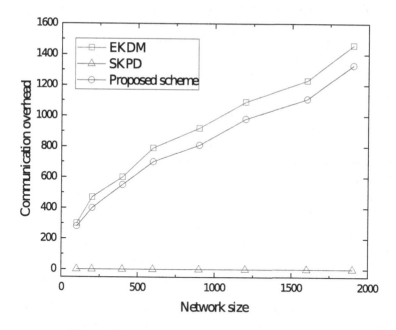

Fig. 4. Comparison of communication overhead.

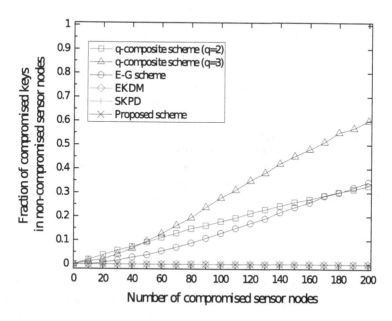

Fig. 5. Comparison of network resistance to capture.

stored by all sensor nodes is compromised. In our proposed scheme, each pair of communicating nodes have a unique pairwise key, and we design a key update scheme, where normal nodes periodically update their symmetric polynomials and keys. In conclusion, our scheme achieves perfect resilience no matter how many sensor nodes are compromised.

Table 2. Feature Comparison.

	EKDM	SKPD	Proposed scheme
Storage overhead	High	Low	Low
Communication overhead	High	Low	Medium
Mobility	Yes	No	Yes
Node addition	Yes	No	Yes
Key update	Yes	No	Yes
Robust resistance to capture attacks	High	High	High

Table 2 compares the different schemes from several aspects. It can be seen that our proposed scheme tradeoffs between memory overhead and communication overhead while taking into account the node mobility compared to other schemes, and supports the addition of new nodes and key updates. Therefore, our proposed scheme is more suitable for UWSNs.

5   Conclusion

In this paper, we propose an updatable key management scheme suitable for use in UWSNs. A nomadic community mobility model is applied in a network model based on a hierarchical structure. A symmetric key is established for both communicating parties through symmetric polynomial key pre-distribution based on a symmetric polynomial key. The defects of multivariate polynomial key generation are successfully overcome, while guaranteeing the network to work securely for a long time. Performance evaluation and security analysis show that our scheme has better scalability, resistance to node attacks, less overhead, and more suitable for application in UWSNs than existing key management schemes.

References

1. Cicioğlu, M., Çalhan, A.: Performance analysis of cross-layer design for internet of underwater things. IEEE Sens. J. **22**(15), 15429–15434 (2022)
2. Wei, X., Guo, H., Wang, X., Wang, X., Qiu, M.: Reliable data collection techniques in underwater wireless sensor networks: a survey. IEEE Commun. Surv. Tutorials **24**(1), 404–431 (2021)

3. Zhuo, X., Liu, M., Wei, Y., Yu, G., Qu, F., Sun, R.: Auv-aided energy-efficient data collection in underwater acoustic sensor networks. IEEE Internet Things J. **7**(10), 10010–10022 (2020)
4. Luo, J., Chen, Y., Wu, M., Yang, Y.: A survey of routing protocols for underwater wireless sensor networks. IEEE Commun. Surv. Tutorials **23**(1), 137–160 (2021)
5. Jahanbakht, M., Xiang, W., Hanzo, L., Azghadi, M.R.: Internet of underwater things and big marine data analytics-a comprehensive survey. IEEE Commun. Surv. Tutorials **23**(2), 904–956 (2021)
6. Li, S., Qu, W., Liu, C., Qiu, T., Zhao, Z.: Survey on high reliability wireless communication for underwater sensor networks. J. Netw. Comput. Appl. **148**, 102446 (2019)
7. Hu, C., Pu, Y., Yang, F., Zhao, R., Alrawais, A., Xiang, T.: Secure and efficient data collection and storage of IoT in smart ocean. IEEE Internet Things J. **7**(10), 9980–9994 (2020)
8. Cheng, M., Guan, Q., Ji, F., Cheng, J., Chen, Y.: Dynamic-detection-based trajectory planning for autonomous underwater vehicle to collect data from underwater sensors. IEEE Internet Things J. **9**(15), 13168–13178 (2022)
9. Li, Y., Zhang, Y., Li, W., Jiang, T.: Marine wireless big data: efficient transmission, related applications, and challenges. IEEE Wirel. Commun. **25**(1), 19–25 (2018)
10. Han, G., Long, X., Zhu, C., Guizani, M., Zhang, W.: A high-availability data collection scheme based on multi-auvs for underwater sensor networks. IEEE Trans. Mob. Comput. **19**(5), 1010–1022 (2019)
11. Diamant, R., Casari, P., Tomasin, S.: Cooperative authentication in underwater acoustic sensor networks. IEEE Trans. Wireless Commun. **18**(2), 954–968 (2018)
12. Gholizdeh, I., Amiri, E., Javidan, R.: An efficient key distribution mechanism for large scale hierarchical wireless sensor networks. In: 2019 27th Iranian Conference on Electrical Engineering (ICEE), pp. 1553–1559. IEEE
13. Coutinho, R.W., Boukerche, A., Vieira, L.F., Loureiro, A.A.: Underwater wireless sensor networks: a new challenge for topology control-based systems. ACM Comput. Surv. (CSUR) **51**(1), 1–36 (2018)
14. Eschenauer, L., Gligor, V.D.: A key-management scheme for distributed sensor networks. In: Proceedings of the 9th ACM Conference on Computer and Communications Security, pp. 41–47. CCS '02, Association for Computing Machinery, New York, NY, USA (2002). https://doi.org/10.1145/586110.586117
15. Chan, H., Perrig, A., Song, D.: Random key predistribution schemes for sensor networks. In: 2003 Symposium on Security and Privacy, 2003, pp. 197–213. IEEE (2003)
16. Shen, A., Guo, S., Chien, H.Y., Guo, M.: A scalable key pre-distribution mechanism for large-scale wireless sensor networks. Concurr. Comput.: Pract. Exper. **21**(10), 1373–1387 (2009)
17. Duan, R., Du, J., Jiang, C., Ren, Y.: Value-based hierarchical information collection for auv-enabled internet of underwater things. IEEE Internet Things J. **7**(10), 9870–9883 (2020)

Author Index

Printed in the United States
by Baker & Taylor Publisher Services